Kamakhya Prasad Ghatak, Madhuchhanda Mitra

Nanomaterials

Also of interest

Superconductors at the Nanoscale
From Basic Research to Applications
Wördenweber, Moshchalkov, Bending, Tafuri (Eds.), 2017
ISBN 978-3-11-045620-2, e-ISBN 978-3-11-045680-6

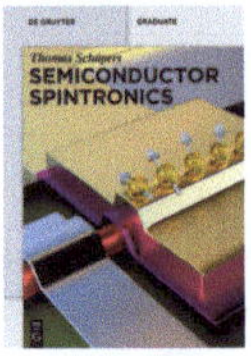

Semiconductor Spintronics
Schäpers, 2016
ISBN 978-3-11-036167-4, e-ISBN 978-3-11-042544-4

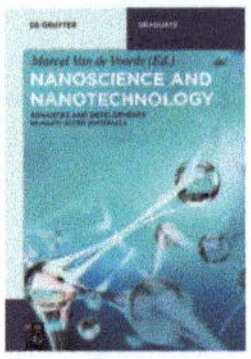

Nanoscience and Nanotechnology
Advances and Developments in Nano-sized Materials
Van de Voorde (Ed.), 2018
ISBN 978-3-11-054720-7, e-ISBN 978-3-11-054722-1

Electrons in Solids
Mesoscopics, Photonics, Quantum Computing, Correlations, Topology
Bluhm, Brückel, Morgenstern, von Plessen, Stampfer, 2018
ISBN 978-3-11-043831-4, e-ISBN 978-3-11-043832-1

Kamakhya Prasad Ghatak,
Madhuchhanda Mitra

Nanomaterials

Volume 1: Electronic Properties

DE GRUYTER

Authors
Prof. Kamakhya Prasad Ghatak
University of Engin. & Management
Institute of Engin. & Management
Kolkata and Jaipur
India
kamakhyaghatakcu@gmail.com

Prof. Madhuchhanda Mitra
University of Calcutta
Department of Applied Physics
92 A.P.C. Road
Kolkata-700009
India
madhuchhanda94@rediffmail.com

ISBN 978-3-11-060922-6
e-ISBN (PDF) 978-3-11-061081-9
e-ISBN (EPUB) 978-3-11-060935-6

Library of Congress Control Number: 2018950592

Bibliographic information published by the Deutsche Nationalbibliothek
The Deutsche Nationalbibliothek lists this publication in the Deutsche Nationalbibliografie; detailed bibliographic data are available on the Internet at http://dnb.dnb.de.

Typesetting: Integra Software Services Pvt. Ltd.
Printing and binding: CPI books GmbH, Leck
Cover image: Science Photo Library/University of Sydney/Ammrf

www.degruyter.com

To know is to know that to know is not to know and that not to know is to know.

From the desk of the first author:
This book is dedicated to Professor Dr. S. Chakrabarti, the Founding Director of the Institute of Engineering & Management and Chancellor of the University of Engineering &Management, an entrepreneur, innovative Mechanical Engineer, born academician, admirable administrator, a great thinker and above all a great soul to whom the first author remains ever grateful as a Member of the IEM &UEM family.

Sir, your inspiration such creativity bring
I must not change my state with the King.

From the desk of the second author:
The second author dedicates this book to her late father Professor Moni Mitra to whom she is ever grateful in the real sense of the term.

Preface

Knowledge is proud, he knows too much but the wise is humble, he knows no more.

The creation of *Nanomaterials*, the subset of the generalized set Nanoscience and Nanotechnology, is based on following two important concepts:

1. The symmetry of the wave-vector space of the charge carriers in electronic materials having various band structures is being reduced from a 3D closed surface to a quantized 2D closed surface, quantized non-parabolas and fully quantized wave vector space leading to the formation of 0D systems such as ultra thin films (UFs), doping superlattices, inversion and accumulation layers, quantum wells (QWs), quantum well super-lattices, carbon nano-tubes, nano wires (NWs), quantum wire super-lattices, magnetic quantization, magneto size quantization, quantum dots (QDs), magneto inversion and accumulation layers, magneto quantum well super-lattices, magneto NIPIs, quantum dot super-lattices and other field aided nanostructures.
2. The advent of modern experimental methods namely Fine Line Lithography (FLL), Metallo-Organic Chemical Vapor Deposition (MOCVD), Molecular Beam Epitaxy (MBE), etc for fabricating the low-dimensional nanostructured systems.

Nanomaterials have gained much interest in Nanoscience and Nanotechnology because of their importance to unlock both new scientific revelations and multi-dimensional all together unheard technological applications. In UFs, the quantization of the motion of the carriers in the direction perpendicular to the surface exhibits the two-dimensional motion of the charge carriers and the third direction is being quantized. Another one-dimensional structure known as NW has been proposed to investigate the physical properties in these materials where the carrier gas is quantized in two transverse directions, and they can move only in the longitudinal direction. As the concept of quantization increases from 1D to 3D, the degree of freedom of the free carriers decreases drastically and the total density-of-states (DOS) function changes from Heaviside step function to the Dirac's delta function forming QDs which, in turn, depend on the carrier energy spectra in different materials. An enormous range of important applications of such low-dimensional structures for modern physics in the quantum regime, along with a rapid increase in computing power, has generated considerable interest in the study of the optical properties of quantum effect devices based on various new materials of reduced dimensionality. Examples of such new applications include quantum switches, quantum registers, quantum sensors, hetero-junction field-effect, quantum logic gates, quantum well and quantum wire transistors, quantum cascade lasers, high-frequency microwave circuits, high-speed digital networks, high-resolution terahertz spectroscopy, advanced integrated circuits, super-lattice photo-oscillator, super-

https://doi.org/10.1515/9783110610819-201

lattice photo-cathodes, resonant tunneling diodes and transistors, super-lattice coolers, thermoelectric devices, thin film transistors, micro-optical systems, intermediate-band solar cells, high performance infrared imaging systems, optical modulators, optical switching systems, single electron/molecule electronics, nanotube based diodes, and other nanoelectronic devices [1–14].

In volume one of this book, we shall study few electronic properties of optoelectronic nanomaterials having various band structures under different physical conditions in the presence of intense photon field with the use of the Heisenberg's Uncertainty Principle (HUP).

With the advent of nanophotonics, there has been considerable interest in studying the optical processes in semiconductors and their nanostructures in the presence of intense light waves [15]. **It appears from the literature that the investigations in the presence of external intense photo-excitation have been carried out on the assumption that the carrier energy spectra are invariant quantities under strong external light waves, which is not fundamentally true.** The physical properties of semiconductors in the presence of strong light waves which alter the basic dispersion relations have relatively been much less investigated in [16–17] as compared with the cases of other external fields and in optoelectronics the influence of strong light waves is needed for the characterization of the low-dimensional optoelectronic devices.

In Chapter 1, we study the carrier contribution to the elastic constants (CEC) of the nanomaterials, which has been studied in the last 30 years mainly by Ghatak et al. [18–45] and few others [46–47]. In this regard, we wish to note that the theory for determining the carrier contributions to the elastic constants in p-type Si already exists [46]. It has been shown that the carrier contribution of the second- and third-order elastic constants depends on the density of states function (DOS) [46]. Sreedhar and Gupta [47] formulated the same for small gap materials whose energy band structures are defined by the two-band model of Kane. It has therefore different values in various materials and varies with the electron concentration, with doping, with the thickness of ultra-thin films and with temperature for semiconductors and their heterostructures having various carrier energy spectra. The nature of these variations has been investigated in the literature [18–48]. Some of the significant features that have emerged from these studies are as follows:

a) The CEC changes monotonically with electron concentration in bulk materials.
b) The nature of the variations is significantly affected by the band non – parabolicity.

The said contribution has significantly different values in ultra-thin films.

It is well known that heavy doping and carrier degeneracy are the keys to unlock the important properties of semiconducting materials, and they are especially instrumental in dictating the characteristics of Ohomic contacts and Schottky contacts, respectively [49]. It is an amazing fact that although the heavily doped semiconductors (HDS) have been investigated in the literature, the study of the corresponding dispersion relations (DRs) of HDS ***is still one of the open research problems***. Our method is not at all related with the DOS technique as used in the aforementioned works. From the electron energy spectrum, one can obtain the DOS but the DOS technique, as used in the literature cannot generate the dispersion laws. ***Therefore, our study is more fundamental than those in the existing literature, because the Boltzmann transport equation, which controls the study of the charge transport properties of the semiconductor devices, can be solved if and only if the carrier energy spectra is known.***

In Chapter 1, we study the CECs in heavily doped optoelectronic materials under different physical conditions not by using the difficult density-of-state function approach but by applying the simplified HUP to formulate the electron statistics (ES) which, in turn, determines the CECs. In addition, we present the suggestion for the experimental determination of CECs for materials having arbitrary dispersion laws. The CECs have different forms for different materials and changes under one-, two- and three-dimensional quantum confinement of the charge carriers. In this context, it may be written that the available reports on the said areas cannot afford to cover even an entire chapter containing the detailed investigations regarding the CECs in semiconductors and their quantized structures. It is important to note that the effects of quantizing magnetic field (B) on the band structures of compound.

Semiconductors are most striking than that of the parabolic one and are easily observed in experiments. A number of interesting physical features originate from the significant changes in the basic energy wave vector relation of the carriers caused by the magnetic field. The valuable information could also be obtained from experiments under magnetic quantization regarding the important physical properties such as Fermi energy and effective masses of the carriers, which affect almost all the transport properties of the electron devices [15] of various materials having different carrier dispersion relations [16–17].

It is worth remarking that the effects of crossed electric and quantizing magnetic fields on the transport properties of semiconductors having various band structures have relatively been less investigated as compared with the corresponding magnetic quantization, although the study of the cross fields are of fundamental importance with respect to the addition of new physics and the related experimental findings in modern quantum effect devices. It is well known that in the presence of electric field (E_0) along x-axis and the quantizing magnetic field (B) along z-axis, the dispersion laws of the carriers in semiconductors become modified and for which the carrier moves in both the z and y directions, respectively. The motion along y-direction is purely due to the presence of E_0 along x-axis and in the absence of electric field, the effective electron mass along y-axis tends to infinity indicating the fact that the electron motion along y-axis is forbidden. The effective electron mass of the isotropic, bulk semiconductors having parabolic energy bands exhibits mass anisotropy in the presence of cross fields and this anisotropy depends on the electron energy, the magnetic quantum number, the electric and the magnetic fields, respectively, although the effective electron mass along z- axis is a constant quantity. In 1966, Zawadzki and Lax [48] derived the expression of the dispersion relation of the conduction electrons for III-V semiconductors in accordance with the two-band model of Kane under cross-fields configuration, which generates the interest to study this particular topic of solid state science in general [50].

With the advent of modern experimental techniques of fabricating nanomaterials as already noted, it is also possible to grow semiconductor super-lattices (SLs) composed of alternative layers of two different degenerate layers with controlled thickness [51]. These structures have found wide applications in many new devices: photodiodes [52], photoresistors [52], transistors [53], light emitters [54], tunneling devices [55], etc. [56–68]. The investigations of the physical properties of narrow gap SLs have increased extensively; since they are important for optoelectronic devices and because of the quality of heterostructures, there is considerable improvement when involving narrow gap materials. It may be written in this context that the doping SLs are crystals with a periodic sequence of ultrathin film layers [69, 70] of the same semiconductor with the intrinsic layer in between together with the opposite sign of doping. All the donors will be positively charged and all the acceptors are negatively charged. This periodic space charge causes a periodic space charge potential that quantizes the motions of the carriers in the z-direction together with the formation of the sub-band energies.

It is well known that the electrons in bulk semiconductors in general have three-dimensional freedom of motion. When these electrons are confined to a one-dimensional potential well whose width is of the order of the carrier wavelength, the motion in that particular direction gets quantized while that along the other two directions remains as free. Thus, the energy spectrum appears in the shape of discrete levels for the one-dimensional quantization, each of which has a continuum for the two-dimensional free motion. The transport phenomena of such one-dimensional confined carriers have recently studied [71] with great interest. For the metal-oxide-semiconductor (MOS) structures, the work functions of the metal and the semiconductor substrate are different and the application of an external voltage at the metal-gate causes the change in the charge density at the oxide semiconductor interface leading to a bending of the energy bands of the semiconductor near the surface. As a result, a one-dimensional potential well is formed at the semiconductor interface. The spatial variation of the potential profile is so sharp that for considerable large values of the electric field, the width of the potential well becomes of the order of the de Broglie wavelength of the carriers. The Fermi energy, which is near the edge of the conduction band in the bulk, becomes nearer to the edge of the valance band at the surface creating inversion layers. The energy levels of the carriers bound within the potential well get quantized and form electric sub bands. Each of the sub-bands corresponds to a quantized level in a plane perpendicular to the surface leading to a quasi-two-dimensional electron gas. Thus, the extreme band bending at low temperature allows us to observe the quantum effects at the surface [71].

It is well known that Keldysh [72] first suggested the fundamental concept of a SL, although it was successfully experimental realized by Esaki and Tsu [73]. The importance of SLs in the field of nanoelectronics has already been described in [74–76]. The most extensively studied III–V SL is the one consisting of alternate layers of GaAs and $Ga_{1-x}Al_xAs$ *owing to the relative ease of fabrication. The GaAs layer forms quantum wells and* $Ga_{1-x}Al_xAs$ *form potential barriers. The III–V SL's are attractive for the realization of high speed electronic and optoelectronic devices [77]. In addition to SLs with usual structure, SLs with more complex structures such as II–VI [78], IV–VI [79] and HgTe/CdTe [80] SL's have also been proposed. The IV–VI SLs exhibit quite different properties as compared to the III–V SL due to the peculiar band structure of the constituent materials [81]. The epitaxial growth of II–VI SL is a relatively recent development and the primary motivation for studying the mentioned SLs made of materials with the large band gap is in their potential for optoelectronic operation in the blue [81]. HgTe/CdTe SLs have raised a great deal of attention since 1979, when as a promising new material for long wavelength infrared detectors and other electro-optical applications [82]. Interest in Hg-based SLs has been further increased as new properties with potential device applications were revealed [83]. These features arise from the unique zero band gap material HgTe [84] and the direct band gap semiconductor CdTe which can be described by the three band mode of Kane [85]. The combination of the aforementioned materials with specified dispersion relation makes HgTe/CdTe SL very attractive, especially because of the possibility to tailor the material properties for various applications by varying the energy band constants of the SLs. In addition to it, for effective mass SLs, the electronic sub-bands appear continually in real space [86].*

We note that all the aforementioned SLs have been proposed with the assumption that the interfaces between the layers are sharply defined, of zero thickness, that is, devoid of any interface effects. The SL potential distribution may be then considered as a one-dimensional array of rectangular potential wells. The aforementioned advanced experimental techniques may produce SLs with physical interfaces between the two materials crystallographically abrupt; adjoining their interface will change at least on an atomic scale. As the potential form changes from a well (barrier) to a barrier (well), an intermediate potential region exists for the electrons. The influence of finite thickness of the interfaces on the electron dispersion law is very important, since the electron energy spectrum governs the electron transport in SLs.

In Chapter 1, we further investigate the CEC in the presence of magnetic quantization, cross-fields configuration, QWs, NWs, QDs, magneto size quantization, inversion and accumulation layers, magneto inversion and magneto

accumulation layers, doping SLs, magneto doping SLs, QWHD, NWHD and QDHD effective mass SLs, magneto QWHD effective mass SLs, magneto HD effective mass SLs, QWHD, NWHD and QDHD SLs with graded interfaces, magneto QWHD SLs with graded interfaces and magneto HD SLs with graded interfaces of opto electronic materials, respectively.

In Chapter 2, we shall study the photoemission from optoelectronic nanomaterials. It is well known that the Einstein's photo emission (EP) is a physical phenomenon and occupies a singular position in the whole arena of Nanoscience and Nanotechnology and related disciplines in general and whose importance has already been established since the inception of Einstein's photoelectric effect (for which Einstein won Nobel Prize in 1921), which in recent years finds extensive applications in modern optoelectronics, characterization and investigation of condensed matter systems, photoemission spectroscopy and related aspects in connection with the investigations of the optical properties of nanostructures [87–91]. Interest in low-dimensional silicon nanostructures also grew up and gained momentum, after the discovery of room temperature photoluminescence and electroluminescence of silicon nanowires in porous silicon [87]. Work on ultrathin layers of $SiSiO_2$ superlattices resulting into visible light emission at room temperature clearly exhibited low-dimensional quantum confinement effect [88] and one of the most popular techniques for analyzing the low-dimensional structures is to employ photoemission techniques. Recent observation of room temperature photoluminescence and electroluminescence in porous silicon has stimulated vigorous research activities in silicon nanostructures [89].

It is worth remarking that in the methods as given in the literature, the physics of photoemission has been incorporated in the lower limit of the photoemission integral and assuming that the band structure of the bulk materials becomes an invariant quantity in the presence of photoexcitation necessary for Einstein's photoelectric effect. The basic band structure of semiconductors changes in the presence of intense external light waves in a fundamental way, which has been incorporated mathematically in Chapter 2, in addition to the appropriate fixation of the lower limit of the photoemission integral for the purpose of investigating the EP.

In Chapter 2, we study the EP from the HD optoelectronic materials, under magnetic quantization and also from quantum well, Nanowire and quantum dots of the said compounds. In addition, we study the EP from HD effective mass quantum well SLs under magnetic quantization, HD effective mass nanowire SLs, HD effective mass quantum box SLs and the magneto EP from HD effective mass SLs of optoelectronic materials, respectively. Under the conditions of extreme degeneracy, the invariant band structure concept in the presence of light waves and certain other limiting constraints all the results of this chapter for the EP assumes the well-known form [93] $J = (2\pi\alpha_0 e m_c g_v / h)(\nu - \nu_0)^2$, ($\alpha_0$ is the probability of photo emission, e is the magnitude of the electron charge, m_c is the effective electron mass at the edge of the conduction band, g_v is the valley degeneracy, h is the Planck constant, v is the frequency of the incident photon and v_0 is the threshold frequency), which indicates the fact current density is independent of temperature and when the energy of light quantum is much greater than the work function the material, the condition of extreme degeneracy is reached.

It is well known that the diffusivity to mobility ratio (DMR) occupies a central position in the whole field of solid-state device electronics and the related sciences since the diffusion constant (a quantity very useful for device analysis where exact experimental determination is rather difficult) can be obtained from this ratio by knowing the experimental values of the mobility. The classical value of the DMR is equal to $(k_BT/|e|)$, $((k_B,T$, and $|e|$ are Boltzmann's constant, temperature and the magnitude of the carrier charge, respectively). This relation in this form was first introduced by Einstein to study of the diffusion of gas particles and is known as the Einstein relation [95, 96]. It appears that the DMR increases linearly with increasing$T$and is independent of electron concentration. This relation is applicable for both types of charge carriers only under nondegenerate carrier concentration, although its validity has been suggested erroneously for degenerate materials [97]. Landsberg first pointed out that the DMR for degenerate semiconductors is essentially determined by their energy band structures [98, 99]. This relation is useful for semiconductor homostructures [100, 101], semiconductor-semiconductor heterostructures [102, 103], metalssemiconductor heterostructures [104, 112] and insulator-semiconductor heterostructures [113–116]. The nature of the variations of the DMR under different physical conditions has been studied in the literature [92, 94–96, 98, 99, 105, 117–142, 143].

In Chapter 3, we study the DMR in Optoelectronic nanomaterials in the presence of intense photon fields under magnetic quantization, cross-fields configurations and also in quantum wells and nanowires together with bulk specimens respectively. **Chapter 3 presents the suggestion for the experimental determination of DMR for materials having arbitrary dispersion laws.**

It is well known that the screening length (SL) of the carriers in semiconductors is a very important quantity characterizing the screening of the Coulomb field of the ionized impurity centers by the free carriers [144]. It affects many of the special features of modern nanodevices, the carrier mobilities under different mechanisms of scattering, and the carrier plasmas in semiconductors [145]. The SL is a very good approximation to the accurate self-consistent screening in presence of band tails and is also used to illustrate the interaction between the colliding carriers in Auger effect in solids [144]. The classical value of the SL is equal to $[\varepsilon_{sc}k_BT/(e^2n_0)]^{1/2}$ (ε_{sc}, k_B, T, e, and n_0 are the semiconductor permittivity, the Boltzmann's constant, the temperature, the magnitude of the carrier charge, and the electron concentration, respectively) which is valid for both the carriers. In this conventional form, the DSL decreases with increasing carrier concentration at a constant temperature and this relation holds only under the condition of carrier non-degeneracy. It is interesting to note that the under the condition of extreme degeneracy, the expression of SL for materials having parabolic energy bands can be written as $L_D = (\pi^{2/3}\hbar\sqrt{\varepsilon_{sc}})(eg_v^{1/3}3^{1/6}n_0^{1/6}\sqrt{m_c})^{-1}$($\hbar$, m_c and g_v are Dirac constant, effective electron mass at the edge of the conduction band and valley degeneracy respectively). Thus we observed that in this case the result is independent of temperature, but depends on n_0, g_v and m_c. Besides, the indices of inverse electron variation changes from half in the former case to one-sixth in the latter case. Since the performance of the electron devices at the device terminals and the speed of operation of modern switching transistors are significantly influence by the degree of carrier degeneracy present in these devices, the simplest way of analyzing such devices taking into account of the degeneracy of the band is to use the appropriate SL to express the performance at the device terminal and switching speed in terms of the carrier concentration [146].

In this chapter we shall study the SL in quantum wells and also under magnetic quantization and cross fields configuration together with bulk specimens of HD optoelectronic materials under intense photon fields. **Chapter 4 presents the suggestion for the experimental determination of SL for materials having arbitrary dispersion laws.**

At field strengths of the order of $10^8 V/m$ (below the electrical breakdown), the potential barriers at the surfaces of different materials usually become very thin resulting in field emission of the electrons due to the tunnel effect. With the advent of field emission (FE) in 1928 [147], the same has been extensively studied under various physical conditions with the availability of a wide range of materials and with the facility for controlling the different energy band constants under different physical conditions and also finds wide applications in materials and related sciences [148–156]. In Chapter 5, we study the field emission from HD III-V, ternary and quaternary materials under magnetic quantization, the HD NWs of the same materials, HD effective mass SLs under magnetic quantization, NWs of the said HD SLs, the HD SLs with graded interfaces both under magnetic quantization and the NWs of the said HD SLs, respectively. Chapter 6 presents the conclusion and the future research as pertinent to this book.

It is needless to say that this monograph is based on the “iceberg principle” [157] and the rest of which will be explored by the researchers of different appropriate fields. Since there is no existing report devoted solely to the study of electronic properties by applying the HUP for HD quantized structures to the best of our knowledge, and we earnestly hope that this book will be a useful reference source for the present and the next generation of the readers and the researchers of materials and allied sciences in general. We have discussed enough regarding electronic properties in different quantized HD materials, although lots of new computer-oriented numerical analysis are being left for the purpose of being computed by the readers, to generate the new graphs and the inferences from them which all together is a sea in itself. The production of error-free first edition of any book from every point of view is a permanent member of impossibility theorems; therefore, despite our joint concentrated efforts for couple of years together with the seasoned team of De Gruyter, the same stands very true for this monograph also. **Various expressions and few chapters of this book have been appearing for the first time in printed form.** The suggestions from the readers for the development of the book will be highly appreciated for the purpose of inclusion in the future edition, if any. **We have presented 200 open research problems for the graduate students, PhD aspirants, researchers and engineers in this pinpointed research topic.** We strongly hope that alert readers of this monograph will not only solve the said problems by removing all the mathematical approximations and establishing the appropriate uniqueness conditions, but also will generate new research problems both theoretical and experimental and, thereby, transforming this monograph into a solid book. Incidentally, our readers after reading this book will easily understand that how little is presented and how much more is yet to be investigated in this exciting topic, which is the signature of coexistence of new physics, advanced mathematics combined with the inner fire for performing creative researches in this context.

In this monograph, the readers will get much information regarding the influence of quantization on the electronic properties in HD low-dimensional materials having different band structures. Although the name of the book is an example of extremely high Q-factor, from the content, one can easily infer that it should be useful in graduate courses on materials science, condensed matter physics, solid states electronics, nanoscience and technology and solid-state sciences and devices in many universities and institutions, in addition to both Ph.D. students and researchers in the aforementioned fields. Last but not the least, the authors hope that their combined humble effort will kindle the desire to delve deeper into this fascinating and deep topic by any one engaged in materials research and quantum effect device development either in academics or in industries.

References

[1] KP Ghatak, Journal of Advanced Physics, 1, 84 (2012); N Paitya, S Bhattacharya, D De, KP Ghatak, *Quantum Matter*, 1, 63 (2012); D De, S Bhattacharaya, S Ghosh, KP Ghatak, *Advanced Science, Engineering and Medicine*, 4, 211 (2012)

[2] S Chakrabarti, M Chakraborty, KP Ghatak, Reviews in Theoretical Science, 4, 10 (2016); N Paitya, KP Ghatak, *Quantum Matter*, 5, 191 (2016); B Chatterjee, S Chakrabarti, SK Sen, M Mitra, KP Ghatak, *Quantum Matter*, 5, 85 (2016); M Mitra, M Chakraborty, S Debbarma, S Chakraborty, SK Sen, B Chatterjee, KP Ghatak, *Quantum Matter*, 5, 58 (2016); TN Sen, KP Ghatak, *Journal of Nanoscience and Nanotechnology*, 16, 1229 (2016)

[3] S Debbarma, KP Ghatak, *Journal of Nanoscience and Nanotechnology*, 16, 1095 (2016); B Chatterjee, S Chakrabarti, M Chakraborty, KP Ghatak, Reviews in *Theoretical Science*, 3, 428 (2015); SM Adhikari, KP Ghatak, *Quantum Matter*, 4, 599 (2015); S Chakrabarti, B

Chatterjee, S Debbarma, KP Ghatak, *Journal of Nanoscience and Nanotechnology*, 15, 6460 (2015)

[4] SM Adhikari, A Karmakar, KP Ghatak, Reviews in Theoretical Science, 3, 273 (2015); M Mitra, B Chatterjee, KP Ghatak, *Journal of Computational and Theoretical Nanoscience*, 12, 1898 (2015); KP Ghatak, LS Singh, K Sarkar, N Debbarma, M Debbarma, *Materials Focus*, 4, 85 (2015)

[5] M Chakraborty, KP Ghatak, Quantum Matter, 4, 104 (2015); K Sarkar, M Chakraborty, S Chakravarti, B Chatterjee, KP Ghatak, *Journal of Nanoengineering and Nanomanufacturing*, 5, 43 (2015); S Debbarma, KP Ghatak, *Reviews in Theoretical Science*, 3, 16 (2015); SM Adhikari, A Karmakar, KP Ghatak, *Journal of Nanoengineering and Nanomanufacturing*, 11, 2499 (2014)

[6] B Chatterjee, N Debbarma, S Debbarma, S Chakrabarti, KP, *Advanced Science, Engineering and Medicine*, 6, 1177 (2015); S Debbarma, S Chakravarti, N Debbarma, M Mitra, KP Ghatak, *Journal of Advanced Physics*, 3, 213 (2014); S Debbarma, N Debbarma, B Chatterjee, SM Adhikari, KP Ghatak, Advanced Science, Engineering and Medicine, 6, 1024, (2014); S Chakrabarti, SK Sen, S Chakraborty, LS Singh, KP Ghatak, *Advanced Science, Engineering and Medicine*, 6, 1042 (2014)

[7] SM Adhikari, A Sakar, KP Ghatak, Quantum Matter, 2, 455 (2013); SBhattachrya, N Paitya, KP Ghatak, *Journal of Computational and Theoretical Nanoscience*, 10, 1999 (2013); N Paitya, KP Ghatak, *Reviews in Theoretical Science*, 1, 165 (2013); SM Adhikari, KP Ghatak, *Quantum Matter*, 2, 296 (2013); SM Adhikari, KP Ghatak, *Journal of Advanced Physics*, 2, 130 (2013); KP Ghatak, PK Bose, S Bhattacharya, A Bhattacharjee, D De, S Ghosh, S Debbarma, N Paitya, *Quantum Matter*, 2, 83 (2013)

[8] S Choudhury, SM Adhikari, D De, S Bhattacharya, KM Chatterjee, KP Ghatak, S Saha, (2013); SM Adhikari, D De, JK Baruah, S Chowdhury, KP Ghatak, *Advanced Science Focus*, 1, 57 (2013); SM Adhikari, KP Ghatak, *Journal of Nanoengineering and Nanomanufacturing*, 3, 48 (2013); S Bhattacharya, D De, S Ghosh, KP Ghatak, *Journal of Computational and Theoretical Nanoscience*, 10, 664 (2013); PK Bose, S Bhattacharya, D De, N Paitya, KP Ghatak, *Advanced Science, Engineering and Medicine*, 5, 245 (2013)

[9] KP Ghatak, S Bhattacharya, A Mondal, S Debbarma, P Ghorai, A Bhattacharjee, Quantum Matter, 2, 25 (2013); N Paitya, KP Ghatak, Journal of Advanced Physics, 1, 161 (2012); N Paitya, KP Ghatak, *Journal of Nanoscience and Nanotechnology*, 12, 8985 (2012); N Paitya, KP Ghatak, *Journal of Nanoengineering and Nanomanufacturing*, 2, 347 (2012); PK Bose, N Paitya, S Bhattacharya, D De, S Saha, KM Chatterjee, S Pahari, KP Ghatak, *Quantum Matter*, 1, 89 (2012); S Bhattacharya, D De, N Paitya, SM Adhikari, KP Ghatak, *Advanced Science Letters*, 16, 348 (2012)

[10] N Paitya, S Bhattacharya, D De, S Ghosh, KP Ghatak, *Journal of Nanoengineering and Nanomanufacturing*, 2, 211 (2012); S Debbarma, A Bhattacharjee, S Bhattacharyya, A Mondal, N Paitya, S Bhattacharya, D De, KP Ghatak, *Advanced Science, Engineering and Medicine*, 4, 96 (2012); S Bhattacharya, D De, SM Adhikari, KP Ghatak, Superlattices and Microstructures, 51, 203 (2012); S Bhattacharya, D De, SM Adhikari, S Saha, KM Chatterjee, S Choudhury, KP Ghatak, Superlattices and Microstructures, 50, 609 (2011); Debashis De, Sitangshu Bhattacharya, SM Adhikari, A Kumar, PK Bose, KP Ghatak, *Beilstein Journal of Nanotechnology*, 2, 339 (2011); S Bhattacharya, S Choudhury, KP Ghatak, *Superlattices and Microstructures*, 48, 257 (2010); S Pahari, S Bhattacharya, D De, SM Adhikari, A Niyogi, A Dey, N Paitya, SC Saha, KP Ghatak, PK Bose, Physica B: *Condensed Matter*, 405, 4064 (2010); KP Ghatak, S Bhattacharya, S Singha Roy, LJ Singh, *Nonlinear Optics*, 32, 307 (2010); KP Ghatak, S Bhattacharya, SK Biswas, A Dey, AK Dasgupta, Physica Scripta, 75, 820 (2007); S Mukherjee, SN Mitra, PK Bose, AR Ghatak, A Neogi, JP Banerjee, A Sinha, M Pal, S Bhattacharya, KP Ghatak, *Journal of Computational and Theoretical Nanoscience*, 4, 550 (2007); S Mukherjee, D De, DJ Mukherjee, S Bhattacharya, A Sinha, KP Ghatak, Physica B: *Condensed Matter*, 393, 347 (2007)

[11] S Bhattarchya, D De, S Ghosh, P Banerjee, S Saha, M Mitra, B Nag, M Pal, SK Biswas, KP Ghatak, *Journal of Computational and Theoretical Nanoscience* 7, 1066 (2010); D De, A Kumar, SM Adhikari, S Pahari, N Islam, P Banerjee, SK Biswas, S Bhattacharya, KP Ghatak *Superlattices and Microstructures*, 47, 377 (2010); A Kumar, S Chowdhury, SM Adhikari, S Ghosh, M Mitra, D De, A Sharma, S Bhattacharya, A Dey, KP Ghatak, *Journal of Computational and Theoretical Nanoscience*, 7, 115 (2010); A Kumar, S Choudhury, S Saha, S Pahari, D De, Santanu Bhattacharya, KP Ghatak, *Physica B: Condensed Matter*, 405, 472 (2010)

[12] S Pahari, S Bhattacharya, S Roy, A Saha, D De, KP Ghatak, Superlattices and Microstructures, 46, 760 (2009); S Pahari, S Bhattacharya, KP Ghatak, *Journal of Computational and Theoretical Nanoscience*, 6, 2088 (2009); KP Ghatak, S Bhattacharya, S Pahari, D De, R Benedictus, *Superlattices and Microstructures*, 46, 387 (2009); KP Ghatak, S Bhattacharya, S Pahari, SN Mitra, PK Bose, D De, *Journal of Physics and Chemistry of Solids*, 70, 122 (2009)

[13] D De, KP Ghatak, *Journal of Materials Science: Materials in Electronics*, 20, 185 (2009); S Bhattacharya, R Sarkar, D De, S Mukherjee, S Pahari, A Saha, S Roy, NC Paul, S Ghosh, KP Ghatak, *Journal of Computational and Theoretical Nanoscience*, 6,112 (2009); S Bhattacharya, NC Paul, D De, KP Ghatak, Physica B: Condensed Matter, 403, 4139 (2008); S Bhattacharya, S Pahari, R Sarkar, S Ghosh, KP Ghatak, Physica B: Condensed Matter, 403, 3635 (2008); KP Ghatak, S Bhattacharya, D De, PK Bose, SN Mitra, S Pahari, Physica B: *Condensed Matter*, 403, 2930 (2008)

[14] KP Ghatak, S Bhattacharya, D De, R Sarkar, S Pahari, A Dey, AK Dasgupta, SN Biswas, *Journal of Computational and Theoretical Nanoscience*, 5, 1345 (2008); KP Ghatak, S Bhattacharya, KM Singh, S Choudhury, S Pahari, *Physica B: Condensed Matter*, 403, 2116 (2008); KP Ghatak, S Bhattacharya, S Bhowmik, R Benedictus, S Choudhury, *Journal of Applied Physics* 103, 94314 (2008); S Choudhury, D De, S Mukherjee, A Neogi, A Sinha, M Pal, SK Biswas, S Pahari, S Bhattacharya, KP Ghatak, *Journal of Computational and Theoretical Nanoscience*, 5, 375 [15] (2008); KP Ghatak, S Bhattacharya, S Bhowmik, R Benedictus, S Choudhury, *Journal of Applied Physics*, 103, 34303 (2008); KP Ghatak, S Bhattacharya, *Journal of Applied Physics*, 102, 73704 (2007)

[15] N. Miura, *Physics of Semiconductors in High Magnetic Fields*,Series on Semiconductor Science and Technology (Oxford University Press, USA, 2007); KHJ Buschow, FR. de Boer, *Physics of Magnetism and Magnetic Materials* (Springer, New York, 2003); D Sellmyer, R Skomski (Eds.), *Advanced Magnetic Nanostructures* (Springer, New York, 2005); JAC Bland, B Heinrich (Eds.), *Ultrathin Magnetic Structures III: Fundamentals of Nanomagnetism (Pt. 3)* (Springer-Verlag, Germany, 2005); B. K. Ridley, *Quantum Processes in Semiconductors*, 4th edn (Oxford publications, Oxford, 1999); JH. Davies, *Physics of Low Dimensional Semiconductors* (Cambridge University Press, UK, 1998); S. Blundell, *Magnetism in Condensed Matter*, Oxford Master Series in Condensed Matter Physics (Oxford University Press, USA, 2001); C Weisbuch, B Vinter, *Quantum Semiconductor Structures: Fundamentals and Applications* (Academic Publishers, USA, 1991); D Ferry, *Semiconductor Transport* (CRC, USA, 2000); M Reed (Ed.), *Semiconductors and Semimetals: Nanostructured Systems* (Academic Press, USA, 1992); T Dittrich, *Quantum Transport and Dissipation* (Wiley-VCH Verlag GmbH, Germany, 1998); AY Shik, *Quantum Wells: Physics & Electronics of Two Dimensional Systems* (World Scientific, USA, 1997).

[16] KP Ghatak, M Mondal, *Zietschrift fur Naturforschung A*41a, 881 (1986); KP Ghatak, M Mondal, Journal of Applied Physics, 62, 922 (1987); KP Ghatak, SN Biswas, *Physica Status Solidi (b)* 140, K107 (1987); KP Ghatak, M Mondal, *Physica Status Solidi (b)* 139, 195 (1987); KP. Ghatak, M Mondal, *Physica Status Solidi (b)*148, 645 (1988); KP Ghatak, SN Biswas, *Journal of Low Temperature Physics* 78, 219 (1990); KP. Ghatak, M Mondal, *Physics Status Solidi (b)*160, 673 (1990); KP. Ghatak, A Ghoshal, B Mitra, *Nouvo Cimento D* 13D, 867 (1991); KP Ghatak,

M Mondal, *Physics Status Solidi (b)*148, 645 (1989); KP Ghatak, *Nouvo Cimento D* 13D, 1321 (1992); KP Ghatak, SN Biswas, *Nonlinear Optics* 4, 347 (1993)
[17] AN Chakravarti, KP Ghatak, A Dhar, S Ghosh, *Phys. Stat. Sol. (b)*105, K55 (1981); AN Chakravarti, AK Choudhury, KP Ghatak, *Physica Status Solidi (a)*63, K97 (1981); AN Chakravarti, AK Choudhury, KP Ghatak, S Ghosh, A Dhar, *Applied Physics* 25, 105 (1981); AN Chakravarti, KP Ghatak, GB Rao, KK Ghosh, *Phys. Stat. Sol. (b)* 112, 75 (1982); AN. Chakravarti, KP Ghatak, KK Ghosh, HM Mukherjee, *Physica Status Solidi. (b)* 116, 17 (1983); M Mondal, KP Ghatak, *Physica Status Solidi. (b)* 133, K143 (1984); M Mondal, KP Ghatak, *Physica Status Solidi. (b)*126, K47 (1984); M Mondal, KP. Ghatak, *Physica Status Solidi. (b)* 129, K745 (1985); M Mondal, KP Ghatak, *Physics Screen* 31, 615 (1985); M Mondal, KP Ghatak, *Physica Status Solidi (b)* 135, 239 (1986); M Mondal, KP Ghatak, *Physica Status Solidi. (b)*135, K21 (1986); M Mondal, S Bhattacharyya, KP Ghatak, *Applied Physics A* 42A, 331 (1987); B Mitra, A Ghoshal, KP Ghatak, *Physica Status Solidi(b)* 150, K67 (1988); M Mondal, KP Ghatak, *Physica Status Solidi. (b)* 146, K97 (1988); B Mitra, KP Ghatak, *Solid State Electronics* 32, 515 (1989); B Mitra, KP Ghatak, *Physics Letters* 135A, 397 (1989)
[18] KP Ghatak, B Mitra, *Physica Scripta*, 46, 182 (1992)
[19] B Nag, KP Ghatak, *Journal of Physics and Chemistry of Solids*, 58, 427 (1997)
[20] KP Ghatak, B Nag, *Physica Status Solidi (b)* 205, 519 (1998)
[21] KP Ghatak, JY Siddiqui, B Nag, *Physics Letters A* 282, 428 (2001)
[22] KP Ghatak, SN Biswas, *MRS Proceedings*, 308, 445, (1993)
[23] KP Ghatak, JP Banerjee, PK Chakrabarty, B Nag, *Journal of Wave Material Interaction* 11, 166 (1996)
[24] LJ Singh, S Choudhary, A Mallik, KP Ghatak, *Journal of Computational and Theoretical Nanoscience*, 2, 287 (2005)
[25] A Mallik, KP Ghatak, S Choudhary, LJ Singh, *Journal of Computational and Theoretical Nanoscience*, 2, 287 (2005)
[26] KP Ghatak, *International Journal of Electronics*, 71, 239 (1991)
[27] KP Ghatak, *Acta Physica Hungarica*, 68, 253 (1990)
[28] B Mitra, KP Ghatak, *Physica Status Solidi (b)*, 154, K35 (1989)
[29] KP Ghatak, PK Bose, *Journal of Wave Material Interaction*, 12, 53 (1997)
[30] SM Adhikari, KP Ghatak, *Journal of Nanoengineering and Nanomanufacturing*, 3, 48 (2013)
[31] B Chatterjee, S Chakrabarti, SK Sen, M Mitra, KP Ghatak, *Quantum Matter*, 5, 85 (2016)
[32] KP Ghatak, S Bhattacharya, SS Roy, LJ Singh, *Nonlinear Optics*, 32, 307 (2010)
[33] D Baruah, S Choudhury, KM Singh, KP Ghatak, *Journal of Physics: Conference Series*, 61, 80 (2007)
[34] KP Ghatak, SN Biswas, *Acta physica Slovaca*, 43, 425 (1993)
[35] KP Ghatak, M Mondal, *Physica Status Solidi (b)*, 170, 57 (1992)
[36] KP Ghatak, DK Basu, B Nag, *Journal of Physics and Chemistry of Solids*, 58, 133 (1997)
[37] KP Ghatak, *Physica Status Solidi (b)*, 154, K29 (1989)
[38] KP Ghatak, JP Banerjee, B Goswami, B Nag, *Nonlinear Optics-Reading*, 16, 241 (1996)
[39] B Nag, KP Ghatak, *Nonlinear Optics-Reading*, 19, 1 (1998)
[40] M Mitra, M Chakraborty, S Debbarma, S Chakraborty, SK Sen, *Quantum Matter*, 5, 58 (2016)
[41] KP Ghatak, B Nag, *Nanostructured Materials*, 10, 923 (1998)
[42] M Mitra, M Chakraborty, S Debbarma, S Chakraborty, SK Sen, *Quantum Matter*, 5, 58 (2016)
[43] B Nag, KP Ghatak, *Physica Scripta*, 54, 657 (1996)
[44] KP Ghatak, S Dutta, DK Basu, B Nag, *Il Nuovo Cimento D*, 20, 227 (1998)
[45] S Bhattacharya, S Chowdhury, S Ghoshal, SK Biswas, D De, KP Ghatak, *Journal of Computational and Theoretical Nanoscience*, 3, 423 (2006); KP Ghatak, S Bhattacharya, S Pahari, SN Mitra, PK Bose, D De, *Journal of Physics and Chemistry of Solids*, 70, 122 (2009);

KP Ghatak, S Karmakar, D De, S Pahari, SK Charaborty, SK Biswas, *Journal of Computational and Theoretical Nanoscience*, 3, 153 (2006); KP Ghatak, JP Banerjee, B Nag, *Journal of Applied Physics*, 83, 1420 (1998); KP Ghatak, SK Biswas, D De, S Ghosal, S Chatterjee, *Physica B*, 353, 127 (2004); KP Ghatak, JP Banerjee, D Bhattacharyya, B Nag, *Nanotechnology*, 7, 110 (1996); KP Ghatak, B Nag, G Mazumder, *MRS Proceedings*, 379, 109,(1995)

[46] RW Keyes, *IBM Journal of Research and Development* 5, 266 (1961)

[47] AK Sreedhar, SC Gupta, *Physics Review*, 5B, 1360 (1972).

[48] W Zawadzki, B Lax XE "Lax", *Physics Review Letter*, 16, 1001 (1966).

[49] SN Mohammad, *Journal of AppliedPhysics*, 97, 063703 (2005); K Suzue, SN Mohammad, ZF Fan, W Kim, O Aktas, AE Botchkarev, H Morkoç, *Journal of Applied Physics*, 80, 4467 (1996); SN Mohammad, ZF Fan, W Kim, O Aktas, AE Botchkarev, A Salvador, H Morkoç, *Electronics Letters*, 32, (598) (1996); Z Fan, SN Mohammad,W Kim, O Aktas, AE Botchkarev, K Suzue, H Morkoç, *Journal of Electronics Materials*, 25, 1703 (1996); C Lu, H Chen, X Lv, X Xia, SN Mohammad, *Journal of Applied Physics* 91, 9216 (2002)

[50] B Mitra, A Ghoshal, KP Ghatak, *Physics Status Solidi (b)*, 154, K147 (1989); KP Ghatak, B Goswami, M Mitra, B Nag, *Non. Opt. and Quan. Opt.*, 16, 9 (1996); M Mondal, KP. Ghatak, *Annalen der Physik*, 46, 502 (1989); MJ Harrison, *Physics Review A*, 29, 2272 (1984); J Zak, W Zawadzki, *Physics Review*, 145, 536 (1966); W Zawadzki, QH Vrehen, B Lax XE "Lax", *Physics Review*, 148, 849 (1966); QH Vrehen, W Zawadzki, M Reine, *Physics Review*, 158, 702 (1967); MH Weiler, W Zawadzki B Lax XE "Lax", *Physics Review*, 163, 733 (1967); W Zawadzki, J Kowalski, *Physics Review Letters*, 27, 1713 (1971); C Chu, M-S Chu, T Ohkawa, *Physics Review Letters*, 41, 653 (1978); P Hu, CS Ting, *Physics Review*, B36, 9671 (1987); EI Butikov, AS Kondratev, AE Kuchma, *Soviet Physics Solid State*, 13, 2594 (1972); M Mondal, KP Ghatak, *Physica Status Solidi (b)*, 133, K67 (1986); M Mondal, NChattopadhyay, KP Ghatak, *Journal of Low temperature Physics*, 66, 131 (1987); KP Ghatak, M Mondal, *Zeitschrift fur Physik B*, 69, 471 (1988); M Mondal, KP Ghatak, *Physics Letters A*, 131A, 529 (1988); M Mondal, KP Ghatak, *Physica Status Solidi (b) Germany*, 147, K179 (1988); B Mitra, KP Ghatak, *Physics Letters* 137A, 413 (1989); B Mitra, KP Ghatak, *Physica Status Solidi (b)*, 164, K13 (1991); KP Ghatak, B Mitra, *International Journal of Electronics*, 70, 345 (1991); KP Ghatak, N Chattopadhyay, SN Biswas, *Proceedings of Society of Photo-optical and Instrumentation Engineers (SPIE)*, 836, Optoelectronic materials, Devices, Packaging and Interconnects, 203 (1988); KP Ghatak, M Mondal, S Bhattacharyya, *SPIE*, 1284, 113 (1990); KP Ghatak, *SPIE*, 1280, *Photonic Materials and Optical Bistability*, 53 (1990); KP Ghatak, SN Biswas, *SPIE*, Growth and Characterization of Materials for Infrared Detectors and Nonlinear Optical Switches, 1484, 149 (1991); KP Ghatak, *SPIE, Fiber Optic and Laser Sensors IX*, 1584, 435 (1992)

[51] NG Anderson, WD Laidig, RM Kolbas, YC Lo, *Journal of Applied Physics*, 60, (2361) (1986); D De, S Bhattacharya, SM Adhikari, A Kumar, PK Bose, KP Ghatak, *Beilstein Journal of Nanotechnology*, 2, 339 (2012); D. De, A Kumar, SM Adhikari, S Pahari, N Islam, P Banerjee, SK Biswas, S Bhattacharya, KP Ghatak, *Superlattices and Microstructures*, 47, 377 (2010); S Pahari, S Bhattacharya, KP Ghatak, *Journal of Computer and Theory Nanoscience*, 6, 2088 (2009); KP Ghatak, J Mukhopadhyay, JP Banerjee, *SPIE Proceedings Series*, 4746, 1292 (2002); KP Ghatak, S Dutta, DK Basu, B Nag, Il *Nuovo Cimento* D 20, 227 (1998); KP Ghatak, B. De, *Materials Research Society Proceedings*, 300, 513 (1993); KP Ghatak, B Mitra, *Il Nuovo Cimento*, D 15, 97 (1993); KP Ghatak, *International Society of Optics and Photonics Proceedings of Society Photo Optics Instrumental Engineers*, 1626, 115 (1992); SN Biswas, KP Ghatak, *International Journal of Electronics Theory Experiment*, 70, 125 (1991); B Mitra, KP Ghatak, *Physics Letter A.*, 142, 401 (1989); KP Ghatak, B Mitra, A Ghoshal, *Phyics Status Solidi (b)*, 154, K121 (1989); B Mitra, KP Ghatak, *Physics Status Solidi (b)*, 149, K117 (1988)

[52] F Capasso, *Semiconductors, and Semimetals*, 22, 2 (1985)

[53] F Capasso, K Mohammed, AY Cho, R Hull, AL Hutchinson, *Applied Physics Letters*, 47, 420 (1985)
[54] F Capasso, RA Kiehl, *Journal of Applied Physics*, 58, 1366 (1985)
[55] K Ploog, GH Doheler, *Advance Physics*, 32, 285 (1983)
[56] F Capasso, K Mohammed, AY Cho, *Applied Physics Letter*, 48, 478 (1986)
[57] R Grill, C Metzner, GH Döhler, *Physics Review B*, 63, 235316 (2001); *Physics Review B*, 61, 614 (2000)
[58] AR Kost, MH Jupina, TC Hasenberg, EM Garmire, *Journal ofApplied Physics*, 99, 023501 (2006)
[59] AG Smirnov, DV Ushakov, VK Kononenko, *Proceedings of SPIE*, 4706, 70 (2002)
[60] DV Ushakov, VK Kononenko, IS Manak, *Proceedings of SPIE*, 4358, 171 (2001)
[61] JZ Wang, ZG Wang, ZM Wang, SL Feng, Z Yang, *Physics Review B*, 62, 6956 (2000)
[62] AR Kost, L West, TC Hasenberg, JO White, M Matloubian, GC Valley, *Applied Physics Letters*, 63, 3494 (1993)
[63] S Bastola, SJ Chua, SJ Xu, *Journal of Applied Physics*, 83, 1476 (1998)
[64] ZJ Yang, EM Garmire, D Doctor, *Journla of Applied Physics*, 82, 3874 (1997)
[65] GH Avetisyan, VB Kulikov, ID Zalevsky, PV Bulaev, *Proceedings of SPIE*, 2694, 216 (1996)
[66] U Pfeiffer, M Kneissl, B Knüpfer, N Müller, P Kiesel, GH Döhler, JS Smith, *Applied Physics Letters*, 68, 1838 (1996)
[67] HL. Vaghjiani, EA Johnson, MJ Kane, R Grey, CC Phillips, *Journal of Applied Physics*, 76, 4407 (1994)
[68] P Kiesel, KH Gulden, A Hoefler, M Kneissl, B Knuepfer, SU Dankowski, P Riel, XX Wu, JS Smith, GH Doehler, *Proceedings of SPIE*, 85, 278 (1993)
[69] GH Doheler, *Physics Scripts*, 24, 430 (1981)
[70] S Mukherjee, SN Mitra, PK Bose, AR Ghatak, A Neoigi, JP Banerjee, A Sinha, M Pal, S Bhattacharya, KP Ghatak, *Journal of Computer Theory Nanoscience*, 4, 550 (2007)
[71] JJ Quinn, PJ Styles (ed.) *Electronic Properties of Quasi Two Dimensional Systems*, North Holland, Amsterdam, (1976); GA Antcliffe, RT Bate, RA Reynolds, Proceedings of the International Conference, *Physics of Semi-metals and Narrow-Gap semiconductors* (ed.) DL Carter, RT Bate, Pergamon Press, Oxford, 499 (1971); G Paasch, T Fiedler, M Kolar, I Bartos, *Physics Status Solidi (b)*, 118, 641 (1983); Th Lindner, G Paasch, *Journal of Applied Physics*, 102, 054514 (2007); S Lamari, *J Applied Physics*, 91, 1698 (2002); KP Ghatak, M Mondal, *Journal of Applied Physics*, 70, 299 (1991); BMitra, KP Ghatak, *Solidi State Electronics*, 32, 177 (1989); M Mondal, KP Ghatak, *Physics Scripts*, 31, 613 (1985); M Mondal, KP Ghatak, *Acta Physics Polon*, A 67, 983 (1985); M Mondal, KP Ghatak, *Physics Status Solidi (b)*, 128, K21 (1985); KP Ghatak, M Mondal, *Physics Status Solidi (b)*, 135, 819 (1986); M Mondal, KP Ghatak, *Physics Status Solidi (b)*, 139, 185 (1987); KP Ghatak, N Chatterjee, M Mondal, *Physics Status Solidi (b)*, 139, K25 (1987); KP Ghatak, A Ghosal, *Physics Status Solidi (b)*, 151, K135 (1989); KP Ghatak, N Chattopadhyay, M Mondal, *Applied Physics* A 48, (365) (1989)
[72] LV Keldysh, Soviet Phys. Solid State, 4, 1658 (1962)
[73] L Esaki, R Tsu, *IBM J. Research and Develop*, 14, 61 (1970)
[74] R Tsu, *Superlattices to Nanoelectronics*, Elsevier, (2005); EL Ivchenko, G Pikus, *Superlattices and other Heterostructures*, Springer, Berlin (1995)
[75] G Bastard, *Wave Mechanics Applied to Heterostructures*, Editions de Physique, Les Ulis, France, (1990)
[76] KP Ghatak, B De, MRS Proceedings, Cambridge University Press 299, 65 (1994); KP Ghatak, B De, MRS Proceedings, Cambridge University Press, 242, 377 (1992); KP Ghatak, San Diego,'91, San Diego, CA, International Society of Optics Photonics, 282 (1991); KP Ghatak, Properties of II-VI Semiconductors: Bulk Crystals, Epitaxial Films, Quantum Well Structures, and Dilute Magnetic Systems, *Materials Research Society*, 161, 371 (1990); KP Ghatak, SN Biswas, MRS Proceedings,

Cambridge University Press (1989) 161; S Bhattacharyya, KP Ghatak, S Biswas, OE/Fibers' 87, 73 (1987)

[77] KV Vaidyanathan, RA Jullens, CL Anderson, HL Dunlap, *Solid State Electron*, 26, 717 (1983)

[78] BA Wilson, *IEEE, J. Quantum Electron*, 24, 1763 (1988)

[79] M Krichbaum, P Kocevar, H Pascher, G Bauer, *IEEE, Journal of Quantum Electronics*, 24, 717 (1988)

[80] JN Schulman, TC McGill, *Applied Physics Letters*, 34, 663 (1979)

[81] H Kinoshita, T Sakashita, H Fajiyasu, *Journal of Applied Physics*, 52, 2869 (1981)

[82] L Ghenin, RG Mani, JR Anderson, JT Cheung, *Physics Review*. B, 39, 1419 (1989)

[83] CA Hoffman, JR Mayer, FJ Bartoli, JW Han, JW Cook, JF Schetzina, JM Schubman, *Physics Review B.*, 39, 5208 (1989)

[84] VA Yakovlev, Soviet Physics Semiconductor, 13, 692 (1979)

[85] EO Kane, *Journal of Physics Chemistry Solids*, 1, 249 (1957)

[86] H Sasaki, *Physics. Review B*, 30, 7016 (1984)

[87] LT Canham, *Applied Physics Letters* 57, 1046 (1990)

[88] ZH. Lu, DJ. Lockwood, JM Baribeam, *Nature*, 378, 258 (1995)

[89] AG Cullis, LT Canham, PDO Calocott, *Journal of Applied Physics*, 82, 909 (1997)

[90] M Cardona, L Ley, *Photoemission in Solids 1 and 2, Topics in Applied Physics*, vols. 26, 27, (Springer-Verlag, Germany, 1978); S Hüfner, *Photoelectron Spectroscopy* (Springer, Germany, 2003); S Hüfner, (Ed.), *Very High Resolution Photoelectron Spectroscopy*, Lecture Notes in Physics, Vol. 715 (Springer-Verlag, Germany, 2007); DW Lynch, CG Olson, *Photoemission Studies of High-Temperature Superconductors* (Cambridge University Press, UK, 1999); *Photoemission and the Electronic Properties of Surfaces* ed. B Feuerbacher, B Fitton, RF Willis (Wiley, New York, 1978); W Schattke, MAV Hove, *Solid-State Photoemission and Related Methods: Theoryand Experiment*, (Wiley, USA, 2003); VV Afanas'ev, *Internal Photoemission Spectroscopy: Principles and Applications* (Elsevier, North Holland, 2010); DJ Lockwood, *Light Emission in Silicon in Silicon Based Materials and Devices*, vol 2, ed HS Nalwa (Academic Press, San Diego, USA, 2001)

[91] KP Ghatak, D. De, S Bhattacharya, *Photoemission from Optoelectronic Materials and their Nanostructures*, Springer Series in Nanostructure Science and Technology (Springer, 2009)

[92] B Mitra, KP Ghatak, *Physics Letters A*, 141, 81 (1989); B Mitra, KP Ghatak, *Solid-State Electronics*, 32, 810 (1989); KP Ghatak, B Mitra, *Physics Letters*, A 156, 233 (1991); S Choudhury, LJ Singh, KP Ghatak, *Nanotechnology*, 15, 180 (2004); KP Ghatak, SN Biswas *Journal of Applied Physics*, 70, 299 (1991); KP Ghatak, AK Chowdhury, S Ghosh, AN Chakravarti, *Applied Physics*. 23, 241 (1980); KP Ghatak, DK Basu, B Nag, *Journal of Physics and Chemical of Solids*, 58, 133 (1997); PK Chakraborty, GC Datta, KP Ghatak, *Physica B: Condensed Matter*, 339, 198 (2003); KP Ghatak, A Ghoshal, B Mitra, *Il Nuovo Cimento D*, 14, 903 (1992); B Mitra, A Ghoshal, KP Ghatak, *Physics Status Solidi (b)*, 154, K147 (1989); KP Ghatak, M Mondal, *Journal of Magetism and Magnetic Material* 74, 203 (1988); KP Ghatak, S Bhattacharya, S Pahari, D De, S Ghosh, M Mitra, *Annalen der Physik*, 17, 195 (2008); KP Ghatak, B Mitra, *Physica Scripta*, 42, 103 (1990); B Mitra, KP Ghatak, *Solid-State Electronics*, 32, 515 (1989); KP Ghatak, S Bhattacharya, S Bhowmik, R Benedictus, S Choudhury, *Journal of Applied Physics*, 103, 094314 (2008); B Nag, KP Ghatak, *Phys Scrip*, 54, 657 (1996); M Mondal, S Bhattacharya, KP Ghatak, *Appl Phys. A*, 42, 331 (1987); KP Ghatak, M Monda, *l Zeitschrift für Physik B Condensed Matter*, 64, 223 (1986); AN Chakravarti, KP Ghatak, A Dhar, KK Ghosh, S Ghosh, *Applied Physics A*, 26, 165 (1981)

[93] RK Pathria, *Statistical Mechanics*, 2nd edn. (Butterworth-Heinemann, Oxford, 1996)

[94] KP Ghatak, S Bhattacharya, *Debye Screening Length: Effects of Nanostructured Materials;* Springer Tracts in Modern Physics (Springer, Heidelberg, 2014) 255; S Bhattacharya, KP Ghatak, *Effective Electron Mass in Low Dimensional Semiconductors*, Springer Series in

Materials Sciences, (Springer, Heidelberg, 2013) 167; S Bhattacharya, KP Ghatak, *Fowler-Nordheim Field Emission: Effects in Semiconductor Nanostructures*, Springer Series in Solid state Sciences (Springer, Heidelberg, 2012) 170; KP Ghatak,S Bhattacharya, *Thermo Electric Power In Nano structured Materials Strong Magnetic Fields*, Springer Series in Materials Science (Springer, Heidelberg, 2010) 137; KP Ghatak, D De, S Bhattacharya, *Photoemission from Optoelectronic Materials and their* Nanostructures, Springer Series in Nanostructure Science and Technology (Springer, Heidelberg, 2009)

[95] KP Ghatak, S Bhattacharya, D De, *Einstein Relation in Compound* Semiconductors *and Their Nanostructures; Springer Series in Materials Science* (Springer, Heidelberg, 2009) 116

[96] A Einstein, *Annalen der Physik*, 17, 549 (1905); H Kroemer, *IEEE Transactions on Electron Devices*, 25, 850 (1978); W Nernst, *Z Physics Chemistry*, 2, 613 (1888); JS Townsend, *Transacations Royol .Society*, 193A, 129 (1900); C Wagner, *Z Physik Chemistry*, B21, 24 (1933); C Herring, MH Nichols, *Reviews of Modern Physics*, 21, 185 (1949); PT Landsberg, *Thermodynamics and Statistical Mechanics* (Oxford University Press, Oxford 1978); *In Recombination in Semiconductors* (Cambridge University Press, U.K. 1991)

[97] RW Lade, *Proceedings of IEEE*, 52, 743 (1965)

[98] PT Landsberg, *Proceedings of Royal Society* A 213, 226(1952); *Proceedings of Physics Society* A 62, 806 (1949)

[99] PT Landsberg, *European Journal of Physics*, 2, 213 (1981)

[100] CH Wang, A Neugroschel, *IEEE Electron Development Letters*, ED-11, 576 (1990)

[101] IY Leu, A Neugroschel, *IEEE Transcations of Electron Development* ED-40, 1872 (1993)

[102] F Stengel, SN Mohammad, H Morkoç, *Journal of Applied Physics*, 80, 3031 (1996)

[103] HJ. Pan, WC Wang, KB. Thai, CC. Cheng, KH. Yu, KW Lin, CZ Wu, WC Liu, *Semiconductor Science Technology*, 15, 1101 (2000)

[104] SN Mohammad, *Journal of Applied Physics*, 95, 4856 (2004)

[105] VK Arora, *Applied Physics Letters*, 80, 3763 (2002)

[106] SN Mohammad, *Journal of Applied Physics*, 95, 7940 (2004)

[107] SN Mohammad, *Philosophical Magzine*, 84, 2559 (2004)

[108] SN Mohammad, *Journal Applied Physics*, 97, 063703 (2005)

[109] K. Suzue, SN Mohammad, ZF Fan, W. Kim, O. Aktas, AE Botchkarev, H. Morkoç, *Journal of Applied Physics* 80, 4467 (1996)

[110] SN Mohammad, ZF Fan, W Kim, O Aktas, AE Botchkarev, A Salvador, H Morkoç, *Electronics Letters*, 32, 598(1996)

[111] Z Fan, SN Mohammad,W Kim, O Aktas, AE Botchkarev, K Suzue, H Morkoç, *Journal of Electronics Materials*, 25, 1703 (1996)

[112] C Lu, H Chen, X Lv, X Xia, SN Mohammad, *Journal of Applied Physics*, 91, 9216 (2002)

[113] SG Dmitriev, Yu V Markin, *Semiconductors*, 34, 931 (2000)

[114] M Tao, D Park, SN Mohammad, D Li, AE Botchkerav, H Morkoç, *Philosophical Magazine B*, 73, 723 (1996)

[115] DG Park, M Tao, D Li, AE Botchkarev, Z Fan, SN Mohammad, H Morkoç, *Journal of Vaccine Science and Technology B*, 14, 2674 (1996)

[116] Z Chen, DG Park, SN Mohammad, H Morkoç, *Applied Physics Letters*, 69, 230 (1996)

[117] AN Chakravarti, DP Parui, *Physics Letters*, 40A, 113 (1972)

[118] AN Chakravarti, DP Parui, *Physics Letters*, 43A, 60 (1973); BR Nag, AN Chakravarti, *Solid State Electronics*, 18, 109 (1975), *Physics Status Solidi (a)* 22, K153 (1974)

[119] BR Nag, AN Chakravarti, PK Basu, *Physics Status Solidi (a)*, 68, K75 (1981)

[120] BR Nag, AN Chakravarti, *Physics Status Solidi (a)*, 67, K113 (1981)

[121] AN Chakravarti, BR Nag, *Physics Status Solidi (a)*, 14, K55 (1972); *International Journal of Electronics*. 37, 281 (1974); *Physics Status Solidi (a)*, 14, K23 (1972); AN Chakravarti, DP Parui,

Canad., *Journal Physics*, 51, 451 (1973); D Mukherjee, AN Chakravarti, BR Nag, *Physics Status Solidi (a)*, 26, K27 (1974); AN Chakravarti, KP Ghatak, A Dhar, KK Ghosh, S Ghosh, *Applied Physics* A26, 165 (1981); S Ghosh, AN Chakravarti, *Physics Status Solidi (b)*, 147, 355 (1988); KP Ghatak, AK Choudhury, S Ghosh, AN Chakravarti, *Applied Physics*, 23, 241 (1980); AN Chakravarti, KK Ghosh, KP Ghatak, HM Mukherjee, *Physics Status Solidi (b)*, 118, 843 (1983); AN Chakravarti, KP Ghatak, KK Ghosh, GB Rao, *Physics Status Solidi (b)*, 111, K61 (1982); AN Chakravarti, KP Ghatak, KK Ghosh, HM Mukherjee, S Ghosh, *Phys Stat Sol (b)*, 108, 609 (1981); AN Chakravarti, KP Ghatak, S Ghosh, A Dhar, *Physics Status Solidi (b)*, 105, K55 (1981); AN. Chakravarti, AK Choudhury, KP Ghatak, D. Roy Choudhury, *Physics Status Solidi ((b)*, 59, K211 (1980); AN Chakravarti, AK Chowdhury, KP Ghatak, DR Choudhury, *Acta Physics Polonica*, A 58, (251) (1980); KP Ghatak, AK Chowdhury, S Ghosh, AN Chakravarti, *Physics Status Solidi (b)*, 99, K55 (1980); AN Chakravarti, AK Chowdhury, KP Ghatak, DR Choudhury, *Czech. Journal of Physics B* 30, 1161 (1980); AN Chakravarti, KP Ghatak, A Dhar, KK Ghosh, S Ghosh, *Acta Physics Polonica A*, 60, 151 (1981); AN Chakravarti, AK Chowdhury, KP Ghatak, A Dhar, DR Choudhury, Czech. *Journal of Physics B*, 31, 905 (1981); AN Chakravarti, KP Ghatak, KK Ghosh, S Ghosh, HM Mukherjee, Czech. *Journal of Physics B*, 31, 1138 (1981)

[122] PT Landsberg, *Journal ofApplied Physics* 56, 1696 (1984); PT Landsberg, AG Guy, *Physics Review B*, 28, 1187 (1983); PT Landsberg, *Physics Review.* B 33, 8321 (1986); Y Roichman, N Tessler, *Applied Physiscs Letters* 80, 1948 (2002); JMH Peters, *European Journal of Physics* 3, 19 (1982); H Van Cong, S Brunet, S. Charar, *Physics Status Solidi B*, 109, (K1) (1982); H. Van Cong, *Physics Status Solidi A*, 56, 395 (1979); H. Van Cong, *Solid State Electronics*, 24,495 (1981)

[123] SN Mohammad, STH Abidi, *Journal Applied Physics* 61, 4909 (1987); *Solid State Electron* 27, 1153 (1985); *Journal of Applied Physics* 56, 3341 (1984); MA Sobhan, SN Mohammad, *Journal of Applied Physics* 58, 2634 (1985); SN Mohammad, AV Bemis, *IEEE Transacations of Electronics Development* ED-39, 282 (1992); SN Mohammad, RL Carter, *Philosophical Magazine* 72, 13 (1995); SN Mohammad, *Solid State Electronics* 46,203 (2002); SN Mohammad, J Chen, JI Chyi, H Morkoç, *Applied Physics Letters* 56, 937 (1990)

[124] PT Landsberg, SA *Hope Solid State Electronics* 20, 421 (1977); SA. Hope, G Feat, PT Landsberg, *J Physics A Mathematical and General* 14, 2377 (1981)

[125] W Elsäber, EO. Göbel, *Electronics Letters*, 19, 335 (1983); R Hilfer, A Blumen, *Physics Review.* A 37, 578 (1988); TG Castner, *Physics Review.* B 55, 4003 (1997); E Barkai, VN Fleurov, *Physics Review E.*, 58, 1296 (1998); TH Nguyen, SK O'Leary, *Applied Physics Letters* 83, 1998 (2003); TH Nguyen, SK O'Leary, *Journal of Applied Physics* 98, 076102 (2005); CG Rodrigues, ÁR Vasconcellos, R Luzzi, *Journal of Applied Physics* 99, 073701 (2006)

[126] RK Jain, *Physics Status Solidi, (a)* 42, K221 (1977); BA Aronzon, EZ Meilikhov, *Physics Status Solidi (a)*, 19, 313 (1973)

[127] S Choudhury, D De, S Mukherjee, A Neogi, A Sinha, M Pal, SK Biswas, S Pahari, S Bhattacharya, KP Ghatak, *Journal of Computer Theory Nanoscience*, 5, 375 (2008); SM Adhikari, KP Ghatak, *Quantum Matter*, 2, 296 (2013); S Pahari, S Bhattacharya, D De, SM Adhikari, A Niyogi, A Dey, N Paitya, KP Ghatak, *Physica B: Condensed Matter*, 405, 4064 (2010); KP Ghatak, S Bhattacharya, S Pahari, D De, R Benedictus, *Superlattices Microsoft*, 46, 387 (2009)

[128] JP Bouchaud, A Georges, *Physics Reports*, 195, 127 (1996); V Blickle, T Speck, C Lutz, U Seifert, C Bechinger, *Physics Review Letter*, 98, 210601 (2007); Y Kang, E Jean, CM Fortmonn, *Applied Physics Letters*, 88, 112110 (2006); F Neumann, YA Genenko, HV Seggern, *Journal of Applied Physics*, 99, 013704 (2006); J. van de Lagemaat, *Physics Review B.*, 73, 235319 (2005); Q Gu, EA Schiff, S Grneber, F Wang, R Schwarz, *Physics Review Letters*, 76, 3196 (1996); MY Azbel, *Physics Review B.*, 46, 15004 (1992)

[129] AH Marshak, *Solid State Electron*, 30, 1089 (1987); AH Marshak, CMV *Vliet, Proceedings of IEEE*, 72, 148 (1984); CMV Vliet, A van der Zeil, *Solid State Electronics*, 20, 931 (1977)
[130] A Khan, A Das, *Applied Physiscs A*, 89, 695 (2007)
[131] O Madelung, *Semiconductors: Data Handbook*, 3rd Edn. Springer (2004); M Krieehbaum, P Kocevar, H Pascher, G Bauer, *IEEE QE*, 24 1727 (1988)
[132] KP Ghatak, SN Biswas, *Journal of Applied Physics*, 70, 4309 (1991); KP Ghatak, B Mitra, M Mondal, *Annalen der Physik*, 48, (283) (1991); B Mitra, KP Ghatak, *Physica Scripta*, 42, (103) (1990); B Mitra, KP Ghatak, *Physics Letters.*, 135A, (397) (1989); KP Ghatak N Chattopadhyay, M Mondal, *Journal of Applied Physics.*, 63, (4536) (1988); *Journal of Low Temperature Physics.*, 73, (321) (1988)
[133] KP Ghatak, D Bhattacharyya, *Physics Letters* A184, (366) (1994)
[134] KP. Ghatak, D Bhattacharyya, *Physica Scripta* 52, (343) (1995); M Mondal, KP Ghatak, The *Journal of Magnetism and Magnetic Materials*, 62, (115) (1986)
[135] KP Ghatak, B Nag, D Bhattacharyya, *Journal of Low Temperature Physics* 14, (1) (1995)
[136] KP Ghatak, M Mondal, *Thin Solid Films*, 148, (219) (1987)
[137] KP Ghatak, SN Biswas, *Nanostructured Materials*, 2, (91) (1993)
[138] KP Ghatak, *Influence of Band Structure on Some Quantum Processes in Nonlinear optical Semiconductors*, D. Eng. Thesis, (1991), Jadavpur University, Kolkata, India.
[139] KP Ghatak, N Chattropadhyay, M Mondal, *Applied Physics* A, 44, (305) (1987)
[140] SN Biswas, KP Ghatak, *Proceedings of the Society of Photo-Optical and Instrumentation Engineers (SPIE), Quantum Well and Superlattice Physics*, 792, 239 (1987); KP Ghatak, M Mondal, S Bhattacharyya, *SPIE*, 1284, 113 (1990); KP Ghatak, S Bhattacharyya, M Mondal, *SPIE*, 1307, 588 (1990).
[141] KP Ghatak, B De, *Defect Engineering in Semiconductor Growth, Processing and Device technology Materials Research Society Proceedings*, (MRS) Spring meeting, 262, 911 (1992); S. Bhattacharya, KP Ghatak, SN Biswas, *Optoelectronic Materials, Devices, Packaging Interconnects, SPIE*, 836, 72 (1988)
[142] M Mondal, KP Ghatak, *Journal of Physics C (Solid State.)*, 20, 1671 (1987); M Mondal, SN Banik, KP Ghatak, *Canadian Journal of Physics*. 67, 72 (1989); KP Ghatak, M Mondal, *Journal of Applied Physics*. 70, 1277 (1992)
[143] S Bhattacharya, KP Ghatak, *Fowler-Nordheim Field Emission: Effects in Semiconductor Nanostructures*, Springer Series in Solid state Sciences 170 (Springer 2012); S Bhattacharya, KP Ghatak, *Effective Electron Mass in Low Dimensional Semiconductors*, Springer Series in Materials Sciences 167 (Springer 2013); KP Ghatak,S Bhattacharya, *Thermo Electric Power In Nano Structured Materials Strong Magnetic Fields*, Springer Series in Materials Science 137 (Springer 2010); KP Ghatak, S Bhattacharya, D De, *Photoemission from Optoelectronic Materials and their Nanostructures*, Springer Series in Nanostructure Science and Technology (Springer 2009); KP Ghatak, S Bhattacharya, D De, *Einstein Relation in Compound Semiconductors and Their Nanostructures*, Springer Series in Materials Science 116 (Springer 2009).
[144] PT Landsberg, *European Journal of Physics*, 2, 213 (1981)
[145] RB Dingle, *Philosophical Magzine*, 46, 813 (1955); D Redfield, MA Afromowitz, *ibid.* 19, 831 (1969); HC Casey, F Stern, *Journal of Applied Physics* 47, 631 (1976).
[146] SN Mohammad, *Journal Physics C* 13, 2685 (1980).
[147] RH Fowler, L Nordheim, *Proceedings ofRoyal Society of London*,Series-A, 119, 173 (1928); AVan Der Ziel, Solid State Physical Electronics (Prentice-Hall, Inc., Englewood Cliffs (N.J.) 1957 (p. 176)
[148] B Mitra,KP Ghatak, *Physics Letters A*, 357, 146 (1990); 142A, 401 (1989); KP Ghatak, M Mondal, *Journal of Magnetism and Magnetic Materials.*, 74, 203 (1988); KP Ghatak, B Mitra, *Physics*

Letters, 156A, 233, (1991); KP Ghatak, A Ghosal, SN Biswas, M Mondal, Proceedings of SPIE, USA, 1308, 356 (1990)

[149] VT Binh, Ch Adessi, *Physics Review Letters*, 85, 864 (2000); RG Forbes,Ultramicroscopy, 79, 11 (1999); JW Gadzuk, EW Plummer, *Review Modern Physics*, 45, 487 (1973); JM Beebe, B Kim, JW Gadzuk,CD Frisbie, JG Kushmerick, *Physics Review Letters*, 97, 026801 (1999); Y Feng, JP Verboncoeur, *Physics Plasmas*, 12, 103301 (2005); WS Koh, LK Ang, *Nanotechnology*, 19,235402 (2008); MRazavy, *Quantum Theory of Tunneling* (World Scientific Publishing Co. Pte. Ltd, Singapore 2003)

[150] SI Baranchuk, NV Mileshkina,*Soviet Physics Solid State*, 23, 1715 (1981); PG Borzyak, AA dadykin, Soviet Physics Docklady, 27,335 (1982); S Bono, RH Good, Jr, *Surface Sci*, 134, 272 (1983); SM Lyth, SRP Silva, *Applied Physics Letters*, 90, 173124 (2007); C Xu,X Sun, *International Journal of Nanotechnology*, 1,452 (2004); SD Liang, L Chen, *Physics Review Letters*, 101, 027602(2008)

[151] Erwin C. Heeres, Erik PAM Bakkers, Aarnoud L Roest, Monja Kaiser, Tjerk H, Oosterkamp, Niels de Jonge, *Nano Letters*, 7, 536 (2007); L Dong, J Jiao, DW Tuggle, JM Petty, SA Elliff, M Coulter, *Appl. Physics Letters*, 82, 1096 (2003); SY Li, P Lin, CY Lee, TY Tseng, *Journal of Applied Physics*., 95, 3711 (2004); NN Kulkarni, J Bae, CK Shih, SK Stanley, SS Coffee, JG Ekerdt, *Applied Physics Letters*, 87, 213115 (2005)

[152] K Senthil, K Yong, *Material Chemistry and Physics*, 112, 88, (2008); R Zhou, HC Chang, V Protasenko, M Kuno, AK Singh, D Jena, H Xing, *Journal of Applied Physics*, 101, 073704 (2007); KS Yeong, JTL Thong, *Journal of Applied Physics*., 100, 114325 (2006); CH Oon, SH Khong, CB Boothroyd, JTL Thong, *Journal of Applied Physics*, 99, 064309 (2006)

[153] BH Kim, MS Kim, KT Park, JK Lee, DH Park, J Joo, SG Yu, SH Lee, *Applied Physics Letters*, 83, 539 (2003); ZS Wu, SZ Deng, NS Xu, J Chen, J Zhou, *Journal of Chen Applied Physics Letters*, 80, 3829 (2002); YW Zhu, T Yu, FC Cheong, XJ Xu, CT Lim, VBC Tan, JTL Thong, CH Sow, *Nanotechnology*, 16, 88 (2005); YW Zhu, HZ Zhang, XC Sun, SQ Feng, J Xu, Q Zhao, B Xiang, RM Wang, DP Yu, *Applied Physics. Letters*, 83, 144 (2003); S Bhattacharjee, T Chowdhury, *Applied Physics Letters*, 95, 061501 (2009); S Kher, A Dixit, DN Rawat, MS Sodha, *Applied Physics Letters* 96, 044101 (2010)

[154] I Shigeo, W Teruo, O Kazuyoshi, T Masateru, U Satoshi, N Norio, *Journal of Vaccine Science and Technology* B:, 13, 487 (2009); CA Spindt, I Brodie, L Humphrey, ER Westerberg, *Journal of Applied Physics*., 47, 5248 (2009)

[155] Q Fu, AV Nurmikko, LA Kolodziejski, RL Gunshor, JW Wu, *Applied Physics Letters*, 51, 578 (2009)

[156] C Majumdar, MK Bose, AB Maity, AN Chakravarti, *Physics Status Solidi* (b), 141, 435 (1987); MK Bose, C Majumdar, AB Maity, AN Chakravarti, *Physics Status Solidi*. (b) 143, 113 (1987)

[157] A Pais, J Robert, *Oppenheimer* (Oxford University Press, U.K 2006) (p. xviii)

Acknowledgments

> Success is the result of continuous effort in the right direction. This stone is broken by the last stroke. This does not mean that the first stroke was useless

Acknowledgment by Kamakhya Prasad Ghatak

I am grateful to A.N. Chakravarti, formerly Head of the Department of the Institute of Radio Physics and Electronics of the University of Calcutta, my mentor and an ardent advocator of the concept of theoretical minimum of Landau, who convinced a twenty 1 years old Circuit theorist that Condensed Matter Physics, in general, is the hidden dual dance of quantum mechanics, statistical mechanics together with advanced mathematical techniques, and even to appreciate the incredible beauty, he not only placed a stiff note for me to derive all the equations in the Monumental Course of Theoretical Physics, the Classics of Landau–Lifshitz together with the two-volume classics of Morse-Feshbach 44 years ago but also forced me to stay in the creative critical zone of research till date. I express my gratitude to Late B.R. Nag, formerly Head of the Departments of Radio Physics and Electronics and Electronic Science of the University of Calcutta, to whom I am ever grateful as a student and research worker, the Semiconductor Grandmaster of India for his three books on semiconductor science in general and more than 200 research papers (many of them are absolutely in honor's class) which still fire my imagination. I consider myself to be rather fortunate to study Mathematics under the direct influence of Late S.C. Dasgupta, formerly Head of the Department of Mathematics of the then Bengal Engineering College, Shibpur (presently Indian Institute of Engineering Science and Technology) and M. Mitra (both of them were formidable Applied Mathematicians with deep physical insight and could solve any problem from the two-volume classics of Morse-Feshbach instantly on the blackboard) of the said Department of Mathematics for teaching me deeply the various methods of different branches of Applied Mathematics with special emphasis to **analytic number theory** when I was pursuing the bachelor degree in the branch of Electronics and Telecommunication Engineering 45 years ago. I offer my thanks to Late S.S. Boral, formerly Founding Head of the Department of Electronics and Telecommunication Engineering of the then Bengal Engineering College, Shibpur, for teaching me the Theoretical Acoustics from the Classic of Morse and Ingard by urging me to solve the problems. I am grateful to S. K. Sen, formerly Head of the Department of Electrical Engineering of Bengal Engineering College, Shibpur, Ex-Vice-Chancellor of Jadavpur University and Ex-Power Minister of West Bengal State for teaching me deeply Non-linear network analysis and synthesis and Non-linear control systems in general. I am indebted to Late C.K. Majumdar, Head of the Department of Physics of the University of Calcutta to light the fire for Theoretical Physics.

https://doi.org/10.1515/9783110610819-202

I am grateful to all my friends and colleagues for the last 40 years from my school days till date for forming my inner fire to solve independently the problems from five volumes Berkley Physics Course, three volumes Sakurai, ten volumes of Addison-Wesley Modular Series on Solid State Devices, two volumes Cohen-Tannoudji et al., three volumes Edwards, three volumes Carslaw, six volumes Guillemin, three volumes Tuttle, two volumes Scott, two volumes Budak, five volumes Chen ………… It is curious to note that they are insisting me in the real sense of the term to verify all the entries of Gradshteyn-Ryzhik and three volumes Bateman manuscript project for the last 44 years. It may be noted that academic output of a nice scholar is the response of a good quality RC coupled amplifier in the mid-band zone whereas the same for a fine research worker is the Dirac's delta function. Incidentally, I can safely say that I belong to neither as a consequence of the academic surroundings which this life presents to me. I express my warm gratitude to H. L. Hartnagel, D. Bimberg, W. L. Freeman and W. Schommers for various academic interactions spanning over the last three decades. I remember the sweet memories of P.N. Robson, I.M. Stephenson, V.S. Letokhov, J. Bodnar and K.F. Brennan with true reverence for inspiring me with the ebullient idea that the publications of the research monographs from internationally reputed Publishers containing innovative concepts is of prime importance to excel in creative research activity.

In 30th December, 2006, I wrote a letter to Late P.T. Landsberg (popularly known as P.T.L to his scientific friends) for writing a book about the applications of the Heisenberg's Uncertainty Principle in investigating the electronic properties of Nano materials which can at least transform me in to a better scientist but P.T.L in turn, requested me repeatedly to complete the proposed book at the earliest and often expressed his desire to write a foreword for this book. Incidentally due to previous other heavy academic and related commitments I was unable to finish this present monograph and his wife Mrs. Sophia Landsberg in her letter of desperate sadness dated January 12, 2009 informed me that from 2008 onward due to Alzheimer's disease her husband had to give up scientific works. The disappearance of P.T.L from my research scenario is a big shock to me which I have to bear till my own physical disappearance. I still remember his power packed words that "The definition of Kamakhya Prasad Ghatak (K.P.G) = M.Tech. + Ph.D. + D. Engg. + more than 300 research papers in nano-science and technology in SCI Journals + more than 60 research papers and International Conferences of SPIE and MRS of USA + Ph.D. guide of more than two dozens of Ph.D. students + many research monographs form Springer-Verlag, Germany + HOD + Dean+ Senior Professor ………. is a **big zero to P.T.L if K.P.G. cannot live in a world full with new creative concepts and not merely repeating the past successes, since past success is a dead concept at least to P.T.L.**

In 21st December, 1974, A. N. Chakravarti (an internationally recognized expert on Einstein Relation in general), my mentor in my first interaction with him emphatically energized me by making myself acquainted with the **famous Fermi-Dirac**

Integrals and telling me that **I must master "SEMICONDUCTOR STATISTICS"** (Pergamon Press, London, 1962) by **J. S. Blakemore** for my initiation in semiconductor physics. Later on I came in deep touch with K. Seeger, the well-known author of the book "Semiconductor Physics" (Springer Series in Solid State Sciences, vol. 40, Springer-Verlag, Germany, 1982). The solid Mathematical Physicist Late P. N. Butcher has been a steady hidden force since 1983 before his sad demise with respect to our scripting the series in band structure dependent properties of nano-structured materials. Both P.T.L and P. N. Butcher visited the Institute of Radio Physics and Electronics of the University of Calcutta, my ALMA MATER where I started my research as a M. Tech. student and later on as a faculty member. I formed permanent covalent bonds with J. S. Blakemore, K. Seeger, P. N. Butcher and P.T.L through letters (Pre email era) and these four first class semiconductor physicist, in turn, shared with pleasure their vast creative knowledge of Semiconductors and related sciences with a novice like me.

I offer special thanks to Late N. Guhachoudhury of Jadavpur University for instilling in me the thought that the academic output = ((desire X determination X dedication) – (false enhanced self ego pretending like a true friend although a real unrecognizable foe)) although a thank you falls in the band gap regime for my beloved better half See who really formed the backbone of my long unperturbed research career, since in accordance with "Sanatan" Hindu Dharma, the fusion of marriage has transformed us to form a single entity, where the individuality is being lost. I am grateful to all the members of my research group (from 1983 till date) for not only quantum confining me in the infinitely deep quantum wells of **Ramanujan and Rabindranath** but also inspiring me to teach quantum mechanics and related topics from the eight volumes classics of Greiner et al.

In this context, from the flash back of my memory I wish to offer my indebtedness to M. Mondal, the first member of my research team who in 1983 joined with me to complete his Ph. D work under my guidance when R. K. Poddar, the then Vice-chancellor of the University of Calcutta selected me as a Lecturer (presently Assistant Professor) of the famous Department of Radio Physics and Electronics. In 1987, S. K. Sen, the then Vice-chancellor of Jadavpur University accepted me as the Reader (presently Associate Professor) in the Department of Electronics and Telecommunication Engineering and since then a flood of young researchers (more than 12 in numbers consisting of B. Mitra, A. Ghoshal, D. Bhattacharya, A. R. Ghatak,.......) around me created my research team and insisted me to work with them at the @ of 16 h per day including holidays in different directions of research for the purpose of my creative conversion from an ordinary engineer to a360^0research scientist and consequently I enjoyed the high rate of research publications in different reputed International journals in various aspects of band structure dependent properties of quantized structures. It is nice to note that the said young talented researchers obtained their respective Ph. D degree under my direct supervision. Incidentally in 1994, R. K. Basu, the then Vice-chancellor of the University of Calcutta selected me as a Professor in the Department of Electronic Science and another flood of research over helmed me in a new direction of Nano Science and Nano Technology. The persons responsible for this change include S. Datta, S. Sengupta, A. Ali.......The 11th and 12th names of this golden series are S. Bhattacharya and D. De respectively who, in turn formed permanent covalent bonds with me not only with respect to research (S. Bhattacharya and D. De are respectively the co-authors of seven and two Monographs in different series of Springer) but also in all aspects of life in general.

It is curious to note that after serving 18 years as a Professor in the Department of Electronic Science, in 2012, P. K. Bose, and the then Director of the National Institute of Technology, Agartala requested me to join as a Professor and Departmental Head in Electronics and Communication Engineering. Being my life-long friend, I accepted his offer (and later on as a DEAN) and more than ten young scholars around me again directed my research in an altogether new direction. In 2015 the respected Director Professor S. Chakrabarti of Institute of Engineering & Management in Saltlake City Kolkata, has kindly offered me the position of Research Director and Senior Professor in his famous Institute in the academic fag end of my life to complete my last run towards the creative knowledge temple with my new young research workers and I am grateful to him for his creative gesture. In my 40+ years of teaching life (I have the wide experience of teaching Engineering Physics, Applied Mechanics(from engineering statics up to nonlinear mechanics including indeterminate structures) and 70% of the courses of Electronics and Telecommunication and Electrical Engineering respectively) and 40+ years of research life (mostly in Materials Science, Nano-science and Number Theory), I have finally realized that the teaching without research is body without brain and research without teaching is body without blood **although my all time hero, creatively prolific number theorist Godfrey Harold Hardy in his classic book entitled "A Mathematician's Apology" (Cambridge University Press, 1990) tells us "I hate teaching".**

Incidentally, one young theoretician friend of mine often tells me that many works in theoretical semiconductor science are based on the following seven principles:

- Principles of placing the necessary and sufficient conditions of a proof in the band gap regime.
- Principles of interchange of the summation and the integration processes and unconditioned convergences of the series and the integrals.
- Principles of random applications of one electron theory and super-position theorem in studying the properties of semiconductors, although the many body effects are very important together with the fact that the nature is fundamentally nonlinear in nature.
- Principles of using the invariant band structure concept of semiconductors even in the presence of strong external fields (light, electric, heavy doping etc.) and the

random applications of perturbation theory, which is in a sense quantum mechanical Taylor series without considering the related consequences.
- Principle of random applications of the binomial theorem without considering the important concept of branch cut.
- Principle of little discussion regarding the whole set of materials science comprising of different compounds having various band structures under different physical conditions as compared with the simplified two band model of Kane for III-V semiconductors.
- Principle of using the Fermi's golden rule, the band structure, and the related features, which are valid for non-degenerate semiconductors to materials having degenerate carrier concentrations directly.
- Although my young friend is a purist in his conjecture, there are no doubt certain elements of truth inside his beautiful comments. I hope that our readers will present their intricate and advanced theories after paying due weight age of his aforementioned seven principles.

I offer special thanks to the members of my research team for placing their combined effort towards the development of this book in the DO-LOOP of a computer and critically reading the manuscript in its last phase before sending it to C. Ascheron, Ex-Executive Editor Physics, Springer-Verlag. Last but not the least; I am grateful for ever to our life long time tested friend S. Sanyal, Ex-Principal, Lake School for Girls, Kolkata for not only motivating me at rather turbulent moments of my academic carrier but also instilling in me the concept that **the ratio of total accumulated knowledge in my chosen field of research to my own contribution in my topic of research tends to infinity at any time and is the definition of non-removable pole in the transfer function of my life.**

As always, myself with the members of my research team are grateful for ever to Dr. C. Ascheron, Ex-Executive Editor Physics, Springer Verlag, Germany, in the real sense of the term for his inspiration and priceless technical assistance from the very start of our first monograph from Springer. C. Ascheron is the hidden force behind the publications of nine Monographs, the collective output of myself and my research group for the last 40 years, from Springer. Dr. Ascheron has proposed my name for being a future author in the STEM program of De Gruyter and forwarded the manuscript of this book and later on Dr. K. Berber-Nerlinger, Acquisitions Editor, Physics of De Gruyter for accepting our book proposal. I am grateful in the real sense of the term to Dr. Vivien Schubert, the Project Editor, Physical Sciences, Anett Rehner, Production Editor and the seasoned team of De Gruyter for the overall detailed and minute supervision without which the publication of this book would be a mere dream. Last but not the least, the senior Author expresses his heart felt gratitude to Triparna Datta , a very young member of his research group for inspiration, support and overall condensation of this book. Naturally, the authors are responsible for non-imaginative shortcomings. We firmly believe that our Mother Nature has propelled this Project in her own unseen way in spite of several insurmountable obstacles.

Acknowledgment by Madhuchanda Mitra

None of us is as intelligent as all of us

It is a great pleasure to express my gratitude to Professor K.P. Ghatak for instigating me to carry out extensive researches in the fields of Nano Science and Nano Technology. I am thankful to the faculty members and staffs of Department of Applied Physics, University of Calcutta for support and cooperation. My family members especially my husband and son deserve a very special mention for forming the backbone of my long research carrier. Lastly I wish to offer special thanks and respect to B. Nag of Department of Applied Physics for his constant support, motivation and guidance.

Kolkata, India
27th February, 2018

K.P. Ghatak,
M. Mitra

Contents

About the Authors

Vision without execution, is just hallucination

Professor Dr. Engg. Kamakhya Prasad Ghatak is the First Recipient of the Degree of Doctor of Engineering of Jadavpur University in 1991 since the University inception in 1955 and in the same year he received the Indian National Science Academy visiting fellowship to IIT-Kharagpur. He is the principal co-author of more than 300 research papers on Semiconductor Nano-science and Technology in eminent peer-reviewed SCI Journals and more than 60 research papers in the Proceedings of SPIE, MRS and many of his papers are being cited many times. Professor Ghatak is the invited Speaker of SPIE, MRS, etc., the referee and Editor of different eminent Journals. At present, the h-index, i10-index, total citations and the maximum citation of a research paper within 6 years of its publication of Professor Ghatak are 33, 165, 4969 and 354 respectively. He has produced more than two dozens of PhD candidates in various aspects of materials and nano-sciences and is the principal co-author of the NINE research monographs from Springer -Verlag between 2009 and 2016; among them one from Springer Series in Solid State sciences (Vol.-170), three from Springer Series in Materials Science, (Vols-116,137 and 167), one from Nanostructure Science and Technology and four from Springer Tracts in Modern Physics (Vols-255, 260, 262 and 265). He is the solo author of two research monographs (Vols-7 and 8) from World Scientific in series on the Foundations of Natural Science and Technology. He is the Principal Editor of the two edited monographs entitled "Bismuth: Characteristics, Production and Applications: Series in Materials Science and Technologies" and "Quantum Dots and Quantum Cellular Automata: Recent Trends and Applications: Series in Nanotechnology Science" from NOVA, USA respectively. The All Indian Council for Technical Education has selected the first Research and Development project in his life for the best project award in Electronics and in 2012, the University Grant Commission recommended a research project to him and placed him at the top in the list of awardees. His 40 years teaching experience include Collision Theory, Engineering Mathematics, Applied Mechanics, Non-Linear circuits and Non-Linear Transport respectively. His present research interests are semiconductor nano science and number theory respectively. His brief CV has been enlisted in many biographical references of USA and UK and for more details please visit http://www.amazon.com/Kamakhya-Prasad-Ghatak/e/B003B09OEY

Dr. Madhuchhanda Mitra received her B.Tech, M.Tech and Ph.D (Tech) degrees in 1987, 1989 and 1998, respectively, from the University of Calcutta, Calcutta, India. She is a recipient of "Griffith Memorial Award" of the University of Calcutta. She is the principal co-author of 150 scientific research papers in International peer reviewed journals and is the supervisor of sixteen PhD candidates. Her present research

https://doi.org/10.1515/9783110610819-204

interests are nano science and technology, identification of different biomarkers and biomedical signal processing which includes feature extraction, compression, encryption and classification of electrocardiography (ECG) and photoplethysmography (PPG) signals. At present she is Professor in the Department of Applied Physics, University of Calcutta, India, where she has been actively engaged in both teaching and research in Instrumentation.

Symbols

α	Band non-parabolicity parameter
a	The lattice constant
a_0, b_0	The widths of the barrier and the well for SLs structures
A_0	The amplitude of the light wave
$\overrightarrow{A}$	The vector potential
$A(E, n_z)$	The area of the constant energy 2D wave vector space for ultrathin films
B	Quantizing magnetic field
B_2	The momentum matrix element
b	Bandwidth
c	Velocity of light
C_1	Conduction band deformation potential
C_2	A constant which describes the strain interaction between the conduction and valance bands
ΔC_{44}	Second order elastic constant
ΔC_{456}	Third order elastic constant
δ	Crystal field splitting constant
Δ_0	Interface width
$\Delta(1B)$	Period of SdH oscillation
d_0	Superlattice period
$D_0(E)$	Density-of-states (DOS) function
$D_B(E)$	DOS function in magnetic quantization
$D_B(E, \lambda)$	DOS function under the presence of light waves
d_x, d_y, d_z	Nano thickness along the x, y and z-directions
$\Delta_{\|\|}$	Spin-orbit splitting constants parallel
$\Delta_{\perp}$	Spin-orbit splitting constants perpendicular to the C-axis
Δ	Isotropic spin-orbit splitting constant
d^3k	Differential volume of the$k$space
$\in$	Energy as measured from the center of the band gap
ε	Trace of the strain tensor
ε_0	Permittivity of free space
ε_∞	Semiconductor permittivity in the high frequency limit
ε_{sc}	Semiconductor permittivity
ΔE_g	Increased band gap
$\|e\|$	Magnitude of electron charge
E	Total energy of the carrier
E_0, ζ_0	Electric field
E_g	Band gap
E_i	Energy of the carrier in the ith band
E_{ki}	Kinetic energy of the carrier in the ith band
E_F	Fermi energy
E_{Fs}	Fermi energy in the presence of magnetic quantization
E_n	Landau sub band energy
E_{FB}	Fermi energy in the presence of size quantization
$\bar{E}_{Fn}$	Fermi energy for nipis
E_{FSL}	Fermi energy in superlattices
$\overrightarrow{\varepsilon}_s$	Polarization vector

https://doi.org/10.1515/9783110610819-205

E_{FQWSL}	Fermi energy in quantum wire super lattices with graded interfaces
E_{F_L}	Fermi energy in the presence of light waves
$E_{F_{BL}}$	Fermi energy under quantizing magnetic field in the presence of light waves
E_{F2DL}	2D Fermi energy in the presence of light waves
E_{F1DL}	1D Fermi energy in the presence of light waves
E_{g_0}	Un-perturbed band-gap
$Erfc$	Complementary error function
Erf	Error function
E_{F_h}	Fermi energy of HD materials
$\bar{E}_{hd}$	Electron energy within the band gap
F_s	Surface electric field
$F(V)$	Gaussian distribution of the impurity potential
$F(\eta)$	One parameter Fermi-Dirac integral of order j
f_0	Equilibrium Fermi-Dirac distribution function of the total carriers
f_{0i}	Equilibrium Fermi-Dirac distribution function of the carriers in the ith band
g_v	Valley degeneracy
G	Thermoelectric power under classically large magnetic field
G_0	Deformation potential constant
g^*	Magnitude of the band edge g- factor
h	Planck's constant
$\hat{H}$	Hamiltonian
$\hat{H}'$	Perturbed Hamiltonian
$H(E - E_n)$	Heaviside step function
$\hat{i}, \hat{j}$ and $\hat{k}$	Orthogonal triads
i	Imaginary unit
I	Light intensity
j_{ci}	Conduction current contributed by the carriers of the ith band
k	Magnitude of the wave vector of the carrier
k_B	Boltzmann's constant
λ	Wavelength of the light
$\bar{\lambda}_0$	Splitting of the two spin-states by the spin-orbit coupling and the crystalline field
$\bar{l}, \bar{m}, \bar{n}$	Matrix elements of the strain perturbation operator
L_x, L_z	Sample length along x and z directions
L_0	Super lattices period length
L_D	Debyescreening length
m_1	Effective carrier masses at the band-edge along x direction
m_2	Effective carrier masses at the band-edge along y direction
m_3	The effective carrier masses at the band-edge along z direction
m_2'	Effective- mass tensor component at the top of the valence band (for electrons) or at the bottom of the conduction band (for holes)
m_i^*	Effective mass of the i^t charge carrier in the ith band
$m_{\|}^*$	Longitudinal effective electron masses at the edge of the conduction band
$m_{\perp}^*$	Transverse effective electron masses at the edge of the conduction band
m_c	Isotropic effective electron masses at the edge of the conduction band
$m_{\perp,1}^*$, $m_{\|,1}^*$	Transverse and longitudinal effective electron masses at the edge of the conduction band for the first material in superlattice
m_r	Reduced mass

m_v	Effective mass of the heavy hole at the top of the valance band in the absence of any field
n	Landauquantum number
n_x, n_y, n_z	Size quantum numbers along the x, y and *z*-directions
n_{1D}, n_{2D}	1D and 2D carrier concentration
n_{2Ds}, n_{2Dw}	2D surface electron concentration under strong and weak electric field
$\bar{n}_{2Ds}, \bar{n}_{2Dw}$	Surface electron concentration under the strong and weak electric field Quantum limit
n_i	Mini-band index for nipi structures
$N_{nipi}(E)$	DOS function for nipi structures
$N_{2DT}(E)$	2D DOS function
$N_{2D}(E, \lambda)$	2D DOS function in the presence of light waves
$N_{1D}(E, \lambda)$	1D DOS function in the presence of light waves
n_0	Total electron concentration
$\bar{n}_0$	Electron concentration in the electric quantum limit
n_i	Carrier concentration in the ith band
P	Isotropic momentum matrix element
P_n	Available noise power
$P_{\parallel}$	Momentum matrix elements parallel to the direction of crystal axis
$P_{\perp}$	Momentum matrix elements perpendicular to the direction of crystal axis
$\overrightarrow{r}$	Position vector
S_i	Zeros of the Airy function
$\vec{s}_0$	Momentum vector of the incident photon
t	Time scale
t_c	Tight binding parameter
T	Absolute temperature
$\tau_i(E)$	Relaxation time of the carriers in the ith band
$u_1(\vec{k}, \vec{r}), u_2(\vec{k}, \vec{r})$	Doubly degenerate wave functions
$V(E)$	Volume of *k* space
V_0	Potential barrier encountered by the electron
$V(\vec{r})$	Crystal potential
x, y	Alloy compositions
z_t	Classical turning point
μ_i	Mobility of the carriers in the ith band
μ	Average mobility of the carriers
$\zeta(2r)$	Zeta function of order $2r$
$\Gamma(j+1)$	Complete Gamma function
η	Normalized Fermi energy
η_g	Impurity scattering potential
ω_0	Cyclotron resonance frequency
θ	Angle
μ_0	Bohr magnetron
ω	Angular frequency of light wave
$\uparrow', \downarrow'$	Spin up and down function

1 Heisenberg's uncertainty principle (HUP) and the carrier contribution to the elastic constants in heavily doped (HD) optoelectronic nanomaterials in the presence of intense light waves

Time is more valuable than money. You can get more money, but you cannot get more time at all.

1.1 Introduction

In recent years, there has been considerable interest in studying the optical processes in semiconductors and their nanostructures in the presence of strong external photo-excitation [1]. It may be noted that the works have been done on the basis of the fundamental assumption that the carrier energy wave vector dispersion relations are invariant quantities in the presence of intense light waves, *which is not basically true.* The physical properties of nanomaterials in the presence of light waves, which change the basic $E-\vec{k}$ E and $\vec{k}$ are the carrier energy and carrier wave vector, respectively, relation, have relatively less investigated in the literature [2–5]. In this chapter, we shall study the CECs in HD III–V, ternary, and quaternary semiconductors on the basis of newly formulated electron dispersion law under external photoexcitation.

In Section 1.2, we have formulated the CECs of the conduction electrons of HD III–V, ternary, and quaternary materials in the presence of light waves whose unperturbed electron energy spectrum is described by the three-band Kane model in the absence of band tailing. The III–V compounds find applications in infrared detectors [6], quantum dot light-emitting diodes [7], quantum cascade lasers [8], QWs wires [9], optoelectronic sensors [10], high electron mobility transistors [11], and so on. The electron energy spectrum of III–V semiconductors can be described by the three- and two-band Kane models [12, 13], together with the models of Stillman et al. [14], Newson and Karobe [15], and Palik et al. [16], respectively. In this context, it may be noted that the ternary and quaternary compounds enjoy the singular position in the entire spectrum of optoelectronic materials. The ternary alloy $Hg_{1-x}Cd_xTe$ is a classic narrow gap compound. The band gap of this ternary alloy can be varied to cover the spectral range from 0.8 to over 30 μm [17] by adjusting the alloy composition. $Hg_{1-x}Cd_xTe$ finds extensive applications in infrared detector materials and photovoltaic detector arrays in the 8–12 μm wave bands [18]. The aforementioned uses have generated the $Hg_{1-x}Cd_xTe$ technology for the experimental realization of high-mobility single crystal with specially prepared surfaces. The same compound has emerged as an optimum choice for illuminating the narrow subband physics because the relevant material constants can easily be experimentally measured [19].

https://doi.org/10.1515/9783110610819-001

Besides, the quaternary alloy $In_{1-x}Ga_xAs_yP_{1-y}$ lattice matched to InP also finds wide use in the fabrication of avalanche photodetectors [20], heterojunction lasers [21], light-emitting diodes [22] and avalanche photodiodes [23], field effect transistors, detectors, switches, modulators, solar cells, filters, and new types of integrated optical devices made from the quaternary systems [24].

In the same section, we have studied the CEC for the said HD materials in the presence of external photoexcitation when the unperturbed energy spectra are defined by the two-band Kane model and that of parabolic energy bands in the absence of band tails respectively for the purpose of relative comparison. Section 1.2.2 discusses about the opto-CECs in the said HD materials under magnetic quantization. Section 1.2.3 describes about the opto-CECs in the presence of crossed electric and quantizing magnetic fields. In Section 1.2.4, the opto-CECs in QWs in HD Kane-type semiconductors are discussed. In Section 1.2.5, the CECs in doping superlattices of HD Kane-type semiconductors in the presence of light waves are investigated. In Section 1.2.6, the CECs in QDs of HD Kane-type semiconductors in the presence of light waves are studied. Section 1.2.7 discusses about the magneto-CECs in QWs of HD Kane-type semiconductors in the presence of light waves. In Section 1.2.8, the CECs in accumulation and inversion layers of Kane-type semiconductors in the presence of light waves are studied. In Section 1.2.9, the CECs in NWs of HD of Kane-type semiconductors in the presence of light waves are studied. Section 1.2.10 deals with the magneto-CECs in accumulation and inversion layers of Kane-type semiconductors in the presence of light waves. Section 1.2.11 describes about the magneto-CECs in doping superlattices of Kane-type semiconductors in the presence of light waves. Section 1.2.12 discusses about the CECs in QWHD effective mass superlattices (EMSLs) of Kane-type semiconductors. Section 1.2.13 deals with the CECs in NWHD EMSLs of Kane-type semiconductors. Section 1.2.14 provides details regarding the magneto-CECs in HD EMSLs of Kane-type semiconductors in the presence of light waves. In Section 1.2.15, the magneto-CECs in QWHD EMSLs of Kane-type semiconductors in the presence of light waves are studied. In Section 1.2.16, the CECs in QWHD superlattices of Kane-type semiconductors with graded interfaces in the presence of light waves are studied. Section 1.2.17 deals with CECs in NWHD superlattices of Kane-type semiconductors with graded interfaces in the presence of light waves. In Section 1.2.18, the CECs in quantum dot HD superlattices of Kane-type semiconductors with graded interfaces in the presence of light waves are studied. In Section 1.2.19, the magneto-CECs in HD superlattices of Kane-type semiconductors with graded interfaces in the presence of light waves are studied. Section 1.2.20 discusses about the magneto-CEC in QWHD superlattices of Kane-type semiconductors with graded interfaces in the presence of light waves. In Section 1.3 the suggestions of the experimental determination of the CECs are provided. Section 1.4 contains the summary and conclusion of this chapter. Section 1.5 presents open research problems.

1.2 Theoretical background

1.2.1 The CEC in the presence of light waves in HD III–V, ternary, and quaternary semiconductors

The Hamiltonian $(\hat{H})$ of an electron in the presence of light wave characterized by the vector potential $\vec{A}$ can be written as follows [3]:185

$$\hat{H} = \left[\left|(\hat{p} + |e|\vec{A})\right|^2 / 2m\right] + V(\bar{r}) \tag{1.1}$$

where $\hat{p}$ is the momentum operator, $V(\bar{r})$ is the crystal potential, and m is the free electron mass. Eq. (1.1) can be expressed as

$$\hat{H} = \hat{H}_0 + \hat{H}' \tag{1.2}$$

where $\hat{H}_0 = \frac{\hat{p}^2}{2m} + V(\bar{r})'$
and

$$\hat{H}' = \frac{|e|}{2m}\bar{A}.\hat{p} \tag{1.3}$$

The perturbed Hamiltonian $\hat{H}'$ can be written as

$$\hat{H}' = \left(\frac{-i\hbar|e|}{2m}\right)(\bar{A}.\nabla) \tag{1.4}$$

where $i = \sqrt{-1}$ and $\hat{p} = -i\hbar\nabla$

The vector potential $(\vec{A})$ of the monochromatic light of plane wave can be expressed as

$$\vec{A} = A_0\vec{\varepsilon}_s \cos(\vec{s}_0.\vec{r} - \omega t) \tag{1.5}$$

where A_0 is the amplitude of the light wave, $\vec{\varepsilon}_s$ is the polarization vector, $\vec{s}_0$ is the momentum vector of the incident photon, $\vec{r}$ is the position vector, ω is the angular frequency of light wave, and t is the time scale. The matrix element of $\hat{H}'_{nl}$ between initial state $\psi_1(\vec{q}, \vec{r})$ and final state $\psi_n(\vec{k}, \vec{r})$ in different bands can be written as

$$\hat{H}'_{nl} = \frac{|e|}{2m}\langle n\vec{k}|\vec{A}\cdot\hat{p}|l\vec{q}\rangle \tag{1.6}$$

Using eqs. (1.4) and (1.5), we can rewrite eq. (1.6). The first matrix element of eq. (1.7) can be written as

$$\left\langle n\vec{k}|e^{(i\vec{s}_0.\vec{r})}\nabla|l\vec{q}\right\rangle = \int e^{\left(i[\vec{q}+\vec{s}_0-\vec{k}].\vec{r}\right)} i\vec{q}u_n^*(\vec{k}.\vec{r})u_l(\vec{q},\vec{r})d^3r + \int e^{\left(i[\vec{q}+\vec{s}_0-\vec{k}].\vec{r}\right)} u_n^*(\vec{k},\vec{r})\nabla u_l(\vec{q},\vec{r})d^3r \tag{1.7}$$

The functions $u_n^* u_l$ and $u_n^* u_l$ are periodic. The integral over all space can be divided into the sum over unit cells times an integral over a single unit cell. It is assumed that the wavelength of the electromagnetic wave is sufficiently large so that if $\vec{k}$ and $\vec{q}$ are within the Brillouin zone, $(\vec{q}+\vec{s}_0-\vec{k})$ is not a reciprocal lattice vector.

Therefore, we can write eq. (1.8) as

$$< n\vec{k}\left|e^{\left(i\vec{s}_0-\vec{r}\right)}\nabla\right|l\vec{q}> = \left[\frac{(2\pi)^3}{\Omega}\right]\left\{i\vec{q}\delta(\vec{q}+\vec{s}_0-\vec{k})\delta_{nl}+\delta(\vec{q}+\vec{s}_0-\vec{k})u_n^*\int_{cell}(\vec{k},\vec{r})\nabla u_1(\vec{q},\vec{r})d^3r\right\}$$
$$= \left[\frac{(2\pi)^3}{\Omega}\right]\left\{\delta(\vec{q}+\vec{s}_0-\vec{k})\int_{cell}u_n^*(\vec{k},\vec{r})\nabla u_1(\vec{q},\vec{r})d^3r\right\} \tag{1.8}$$

where Ω is the volume of the unit cell and $\int u_n^*(\vec{k},\vec{r})u_l(\vec{q},\vec{r})d^3r = \delta(\vec{q}-\vec{k})\delta_{nl} = 0$, since $n \neq l$.

The delta function expresses the conservation of wave vector in the absorption of light wave and $\vec{s}_0$ is small compared to the dimension of a typical Brillouin zone and we set $\vec{q} = \vec{k}$.

From eqs. (1.7) and (1.8), we can write

$$\hat{H}'_{nl} = \frac{|e|A_0}{2m}\vec{\varepsilon}_s \cdot \hat{p}_{nl}(\vec{k})\delta(\vec{q}-\vec{k})\cos(\omega t) \tag{1.9}$$

where $\hat{p}_{nl}(\vec{k}) = -i\hbar\int u_n^*\nabla u_1 d^3t = \int u_n^*(\vec{k},\vec{r})\hat{p}u_l(\vec{k},\vec{r})d^3r$

Therefore, we can write

$$\hat{H}'_{nl} = \frac{|e|A_0}{2m}\vec{\varepsilon}_s \cdot \hat{p}_{nl}(\vec{k}) \tag{1.10}$$

where $\vec{\varepsilon} = \vec{\varepsilon}_s \cos\omega t$.

When a photon interacts with a semiconductor, the carriers (i.e., electrons) are generated in the bands that are followed by the interband transitions. For example, when the carriers are generated in the valence band, the carriers then make an interband transition to the conduction band (CB). The transition of the electrons within the same band, that is, $\hat{H}'_{nn} = \langle n\vec{k}|\hat{H}'|n\vec{k}\rangle$ is neglected. Because, in such a case, that is, when the carriers are generated within the same bands, photons are lost by recombination within the aforementioned band, resulting in zero carriers.

Therefore,

$$\langle n\vec{k}|\widehat{H}'|n\vec{k}\rangle = 0 \tag{1.11}$$

With $n = c$ stands for CB and $l = v$ stands for valance band (VB), the energy equation for the conduction electron can approximately be written as

$$I_{11}(E) = \left(\frac{\hbar^2 k^2}{2m_c}\right) + \frac{\left(\frac{|e|A_0}{2m}\right)^2 \langle|\vec{\varepsilon}\cdot\hat{p}_{cv}(\vec{k})|^2\rangle_{av}}{E_c(\vec{k}) - E_v(\vec{k})} \tag{1.12}$$

where $I_{11}(E) \equiv E(aE+1)(bE+1)/(cE+1)$, $a \equiv 1/E_{g0}$, $a \equiv 1/E_{g0}$, E_{g0} is the unperturbed band gap, $b \equiv 1/(E_{g0}+\Delta)$, $c \equiv 1/(E_{g0}+2\Delta/3)$, and $\langle|\vec{\varepsilon}\cdot\hat{p}_{cv}(\vec{k})|^2\rangle_{av}$ represents the average of the square of the optical matrix element (OME).

For the three-band Kane model, we can write

$$\xi_{1k} = E_c(\vec{k}) - E_v(\vec{k}) = \left(E_{g0}^2 + E_{g0}\hbar^2k^2/m_r\right)^{1/2} \tag{1.13}$$

where m_r is the reduced mass and is given by $m_r^{-1} = (m_c)^{-1} + m_v^{-1}$, and m_v is the effective mass of the heavy hole at the top of the VB in the absence of any field.

The doubly degenerate wave functions $u_1(\vec{k},\vec{r})$ and $u_2(\vec{k},\vec{r})$ can be expressed as [16]

$$u_1(\vec{k},\vec{r}) = a_{k+}\left[(is)\downarrow'\right] + b_{k+}\left[\frac{X' - iY'}{\sqrt{2}}\uparrow'\right] + c_{k+}\left[Z'\downarrow'\right] \tag{1.14}$$

and

$$u_2(\vec{k},\vec{r}) = a_{k-}\left[(is)\uparrow'\right] - b_{k-}\left[\frac{X' + iY'}{\sqrt{2}}\downarrow'\right] + c_{k-}\left[Z'\uparrow'\right] \tag{1.15}$$

s is the s-type atomic orbital in both unprimed and primed coordinates, and $\downarrow'$ indicates the spin down function in the primed coordinates,

$a_{k\pm} \equiv \beta[E_{g0} - (\gamma_{0k\pm})^2(E_{g0} - \delta')]^{1/2}(E_{g0} + \delta')^{-1/2}$, $\beta \equiv \left[(6(E_{g0} + 2\Delta/3)(E_{g0} + \Delta))/\chi\right]^{1/2}$, $\chi \equiv (6E_{g0}^2 + E_{g0}\Delta + 4\Delta^2)$, $\xi_{1k} = E_c(\vec{k}) - E_v(\vec{k}) = E_{g0}\left[1 + 2(1 + m_c m_v)I_{11}(E)E_{g0}\right]^{1/2}$, $\delta' \equiv (E_{g0}^2\Delta)(\chi)^{-1}$, X', Y' and Z' are the p-type atomic orbitals in the primed coordinates, $\uparrow'$ indicates the spin-up function in the primed coordinates, $b_{k\pm} \equiv \rho\gamma_{0k\pm}$, $\rho \equiv (4\Delta^2/3\chi)^{1/2}$, $c_{k\pm} \equiv t\gamma_{0k\pm}$, and $t \equiv [6(E_{g0} + 2\Delta/3)^2/\chi]^{1/2}$.

We can, therefore, write the expression for the OME as

$$OME = \hat{p}_{cv}(\vec{k}) = \langle u_1(\vec{k}, \vec{r})|\hat{\mathrm{p}}|\mathrm{u}_2(\vec{k}, r')\rangle \tag{1.16}$$

Since the photon vector has no interaction in the same band for the study of inter-band optical transition, we can therefore write

$$\langle S|\hat{p}|S\rangle = \langle X|\hat{p}|X\rangle = \langle Y|\hat{p}|Y\rangle = \langle Z|\hat{p}|Z\rangle = 0$$

and

$$\langle X|\hat{p}|Y\rangle = \langle Y|\hat{p}|Z\rangle = \langle Z|\hat{p}|X\rangle = 0$$

There are finite interactions between the CB and the VB and we can obtain

$$\begin{aligned}\langle S|\hat{p}|X\rangle &= \hat{i}g\hat{p} = \hat{i}g\hat{p}_x\\ \langle S|\hat{p}|Y\rangle &= \hat{j}g\hat{p} = \hat{j}g\hat{p}_x\\ \langle S|\hat{p}|Z\rangle &= \hat{k}g\hat{p} = \hat{k}g\hat{p}_x\end{aligned}$$

where $\hat{i}, \hat{j}$, and $\hat{k}$ are the unit vectors along x, y, and z axes, respectively.

It is well known [49] that

$$\begin{bmatrix}\uparrow'\\ \downarrow'\end{bmatrix} = \begin{bmatrix} e^{-i\phi/2}\cos(\theta/2) & e^{i\phi/2}\sin(\theta/2)\\ -e^{-i\phi/2}\sin(\theta/2) & e^{i\phi/2}\cos(\theta/2)\end{bmatrix}\begin{bmatrix}\uparrow\\ \downarrow\end{bmatrix}$$

From above equation, we can write

$$\text{and}\begin{bmatrix}X'\\ Y'\\ Z'\end{bmatrix} = \begin{bmatrix}\cos\theta\cos\phi & \cos\theta\sin\phi & -\sin\theta\\ -\sin\phi & \cos\phi & 0\\ \sin\theta\cos\phi & \sin\theta\sin\phi & \cos\theta\end{bmatrix}\begin{bmatrix}X\\ Y\\ Z\end{bmatrix}$$

Besides, the spin vector can be written as

$$\vec{S} = \frac{\hbar}{2}\vec{\sigma},\ where,\ \sigma_x = \begin{bmatrix}0 & 1\\ 1 & 0\end{bmatrix},\ \sigma_y = \begin{bmatrix}0 & -i\\ i & 0\end{bmatrix},\ and\ \sigma_z = \begin{bmatrix}1 & 0\\ 0 & -1\end{bmatrix}$$

$$\hat{P}_{CV}(\vec{k}) = \langle u_1(\vec{k}, \vec{r})|\hat{p}|u_2(\vec{k}, r')\rangle$$

$$= \left\langle \left\{a_{k_+}\left[(iS)\downarrow'\right] + b_{k_+}\left[\left(\frac{X' - iY'}{\sqrt{2}}\right)\uparrow'\right] + c_{k_+}\left[Z'\downarrow'\right]\right\}|\hat{P}|\right.$$

$$\left.\left\{a_{k_-}(iS)\uparrow' - b_{k_-}\left[\left(\frac{X' + iY'}{\sqrt{2}}\right)\downarrow' + c_{k_-}\left[Z'\uparrow'\right]\right]\right\}\right\rangle$$

Using aforementioned relations, we obtain

$$\begin{aligned}\hat{p}_{CV}(\vec{k}) &= \left\langle u_1(\vec{k},\vec{r})|\hat{P}|u_2(\vec{k},r')\right\rangle \\ &= \frac{b_{k_+}a_{k_-}}{\sqrt{2}}\left\{\left\langle(X'-iY')|\hat{P}|iS\right\rangle\langle\uparrow'\mid\uparrow'\rangle\right\}+c_{k_+}a_{k_-}\left\{\left\langle Z'|\hat{P}|iS\right\rangle\langle\downarrow'\mid\uparrow'\rangle\right\} \\ &- \frac{a_{k_+}b_{k_-}}{\sqrt{2}}\left\{\left\langle iS|\hat{P}|(X'+iY')\right\rangle\langle\downarrow'\mid\downarrow'\rangle\right\}+a_{k_+}c_{k_-}\left\{\left\langle iS|\hat{P}|Z'\right\rangle\langle\downarrow'\mid\uparrow'\rangle\right\}\end{aligned} \tag{1.17}$$

We can also write

$$\begin{aligned}\left\langle(X'-iY')|\hat{P}|iS\right\rangle &= \left\langle(X')|\hat{P}|iS\right\rangle-\left\langle(iY')|\hat{P}|iS\right\rangle \\ &= i\int u^*_{X^+}\hat{P}S-\int -iu^*_{Y^+}\hat{P}iu_x = i\left\langle X'|\hat{P}|iS\right\rangle-\left\langle Y'|\hat{P}|iS\right\rangle\end{aligned}$$

From the above relations, for X', Y', and Z' we get

$$|X'\rangle = \cos\theta\cos\phi|X\rangle + \cos\theta\sin\phi|Y\rangle - \sin\theta|Z\rangle$$

Thus, $\langle X'|\hat{P}|S\rangle = \cos\theta\cos\phi\langle X'|\hat{P}|S\rangle + \cos\theta\sin\phi\langle Y|\hat{P}|S\rangle - \sin\theta\langle Z|\hat{P}|S\rangle = \hat{P}\hat{r}_1$
where $\hat{r}_1 = \hat{i}\cos\theta\cos\phi + \hat{j}\cos\theta\sin\phi - \hat{k}\sin\theta$

$$|Y'\rangle = -\sin\phi|X\rangle + \cos\phi|Y\rangle + 0|Z\rangle$$

Thus, $\langle Y'|\hat{P}|S\rangle = -\sin\phi\langle X|\hat{P}|S\rangle + \cos\phi\langle Y|\hat{P}|S\rangle + 0\langle Z|\hat{P}|S\rangle = \hat{P}\hat{r}_2$
where $\hat{r}_2 = -\hat{i}\sin\phi + \hat{j}\cos\phi$
so that $\langle(X'-iY')|\hat{P}|S\rangle = \hat{P}(i\hat{r}_1-\hat{r}_2)$
Thus,

$$\frac{a_{k_-}b_{k_+}}{\sqrt{2}}\langle(X'-iY')|\hat{P}|S\rangle\langle\uparrow'\mid\uparrow'\rangle = \frac{a_{k_-}b_{k_+}}{\sqrt{2}}(i\hat{r}_1-\hat{r}_2)\langle\uparrow'\mid\uparrow'\rangle \tag{1.18}$$

Now since

$$\langle iS|\hat{P}|(X'+iY')\rangle = i\langle S|\hat{P}|X'\rangle - \langle S|\hat{P}|Y'\rangle = \hat{P}(i\hat{r}_1-\hat{r}_2)$$

We can write,

$$-\left[\frac{a_{k_+}b_{k_-}}{\sqrt{2}}\left\{\langle iS|\hat{P}|(X'+iY')\rangle\langle\downarrow'\mid\downarrow'\rangle\right\}\right] = \left[\frac{a_{k_+}b_{k_-}}{\sqrt{2}}\hat{P}(i\hat{r}_1-\hat{r}_2)\langle\downarrow'\mid\downarrow'\rangle\right] \tag{1.19}$$

Similarly, we obtain

$$|Z'\rangle = \sin\theta\cos\phi|X\rangle + \sin\theta\sin\phi|Y\rangle + \cos\theta|Z\rangle$$

So that, $\langle Z'|\hat{P}|S\rangle = i\langle Z'|\hat{P}|S\rangle - i\hat{P}\{\sin\theta\cos\phi\hat{i} + \sin\theta\sin\phi\hat{j} + \cos\theta\hat{k}\} = i\hat{P}\hat{r}_3$
where $\hat{r}_3 = \hat{i}\sin\theta\cos\phi + \hat{j}\sin\theta\sin\phi + \hat{k}\cos\theta$
Thus,

$$c_{k_+}a_{k_-}\langle Z'|\hat{P}|S\rangle\langle\downarrow'|\uparrow'\rangle = c_{k_+}a_{k_-}i\hat{P}\hat{r}_3\langle\downarrow'|\uparrow'\rangle \tag{1.20}$$

similarly, we can write

$$c_{k_+}a_{k_-}\langle iS|\hat{P}|Z'\rangle\langle\downarrow'|\uparrow'\rangle = c_{k_-}a_{k_+}i\hat{P}\hat{r}_3\langle\downarrow'|\uparrow'\rangle \tag{1.21}$$

Therefore, we obtain

$$\begin{aligned}&\frac{a_{k_-}b_{k_+}}{\sqrt{2}}\left\{\left\langle (X'+iY')|\hat{P}|S\right\rangle\langle\uparrow'|\uparrow'\rangle\right\} - \frac{a_{k_-}b_{k_+}}{\sqrt{2}}\left\{\left\langle iS|\hat{P}|(X'+iY')\right\rangle\langle\downarrow'|\downarrow'\rangle\right\}\\ &= \frac{\hat{P}}{\sqrt{2}}\left(-a_{k_+}b_{k_-}\langle\downarrow'|\downarrow'\rangle + a_{k_-}b_{k_+}\langle\uparrow'|\uparrow'\rangle\right)(i\hat{r}_2 - \hat{r}_2)\end{aligned} \tag{1.22}$$

In addition, we can write

$$c_{k_+}a_{k_-}\langle Z'|\hat{P}|iS\rangle\langle\downarrow'|\uparrow'\rangle + c_{k_-}a_{k_+}\langle iS|\hat{P}|Z'\rangle\langle\downarrow'|\uparrow'\rangle = i\hat{P}(c_{k_+}a_{k_-} + c_{k_-}a_{k_+})\hat{r}_3\langle\downarrow'|\downarrow'\rangle \tag{1.23}$$

Combining eqs. (1.23) and (1.24), we find

$$\begin{aligned}\hat{p}_{CV}(\vec{k}) = &\frac{\hat{p}}{\sqrt{2}}(i\hat{r}_1 - \hat{r}_2)\{(b_{k_+}a_{k_-})\langle\uparrow'|\uparrow'\rangle - (b_{k_-}a_{k_+})\langle\downarrow'|\downarrow'\rangle\}\\ &+ i\hat{P}\hat{r}_3(c_{k_+}a_{k_-} - c_{k_-}c_{k_+})\langle\downarrow'|\uparrow'\rangle\end{aligned} \tag{1.24}$$

From the aforementioned relations, we obtain

$$\left.\begin{aligned}\uparrow' &= e^{-i\phi/2}\cos(\theta/2)\uparrow + e^{i\phi/2}\sin(\theta/2)\downarrow\\ \downarrow' &= e^{-i\phi/2}\sin(\theta/2)\uparrow + e^{i\phi/2}\cos(\theta/2)\downarrow\end{aligned}\right\} \tag{1.25}$$

Therefore,

$$\begin{aligned}\langle\downarrow'|\uparrow'\rangle_x = &-\sin(\theta/2)\cos(\theta/2)\langle\uparrow|\uparrow\rangle_x + e^{-i\theta}\cos^2(\theta/2)\langle\downarrow|\uparrow\rangle_x\\ &- e^{i\phi}\sin^2(\theta/2)\langle\uparrow|\downarrow\rangle_x + \sin(\theta/2)\cos(\theta/2)\langle\downarrow|\downarrow\rangle_x\end{aligned} \tag{1.26}$$

But we know from above that

$$\langle\uparrow|\uparrow\rangle_x = 0, \langle\downarrow|\uparrow\rangle = \frac{1}{2}, \ \langle\downarrow|\uparrow\rangle_x = \frac{1}{2} \text{and} \langle\downarrow|\downarrow\rangle_x = 0$$

Thus, we get

$$\begin{aligned}\langle\downarrow'|\uparrow'\rangle_x &= \frac{1}{2}\left[e^{-i\phi}\cos^2(\theta/2) - e^{i\phi}\sin^2(\theta/2)\right] \\ &= \frac{1}{2}\left[(\cos\phi - i\sin\phi)\cos^2(\theta/2) - (\cos\phi + i\sin\phi)\sin^2(\theta/2)\right] \\ &= \frac{1}{2}[\cos\phi\cos\theta - i\sin\phi]\end{aligned} \tag{1.27}$$

Similarly, we obtain

$$\langle\downarrow'|\uparrow\rangle_y = \frac{1}{2}[i\cos\phi + \sin\phi\cos\theta] and \langle\downarrow'|\uparrow\rangle_z = \frac{1}{2}[-\sin\theta]$$

$$\langle\downarrow'|\uparrow'\rangle = \hat{i}\langle\downarrow'|\uparrow'\rangle_x + \hat{j}\langle\downarrow'|\uparrow'\rangle_y + \hat{k}\langle\downarrow'|\uparrow'\rangle$$

Therefore

$$\begin{aligned}&\tfrac{1}{2}\{(\cos\theta\cos\phi - i\sin\phi)\hat{i} + (i\cos\phi + \sin\phi\cos\theta)\hat{j} - \sin\theta\hat{k}\} \\ &= \tfrac{1}{2}[\{(\cos\theta\cos\phi)\hat{i} + (\sin\phi\cos\theta)\hat{j} - \sin\theta\hat{k}\} + i\{-\hat{i}\sin\phi + \hat{j}\cos\phi\}] \\ &= \tfrac{1}{2}[\hat{r}_1 + i\hat{r}_2] = -\tfrac{1}{2}i[i\hat{r}_1 - \hat{r}_2]\end{aligned}$$

Similarly, we can write

$$\langle\uparrow'|\uparrow'\rangle = \tfrac{1}{2}[\hat{i}\sin\theta\cos\phi + \hat{j}\sin\theta\sin\phi + \hat{k}\cos\theta] = \tfrac{1}{2}\hat{r}_3 \text{ and } \langle\downarrow'|\downarrow'\rangle = -\tfrac{1}{2}\hat{r}_3$$

Using the above results, we can write

$$\begin{aligned}\hat{p}_{CV}(\vec{k}) &= \frac{\hat{P}}{\sqrt{2}}(i\hat{r}_2 - \hat{r}_2)\{(a_{k_-}b_{k_+})\langle\uparrow'|\uparrow'\rangle - (b_{k_-}a_{k_+})\langle\downarrow'|\downarrow'\rangle\} \\ &\quad + i\hat{P}\hat{r}_3\{(c_{k_+}a_{k_-} - c_{k_-}a_{k_+})\langle\downarrow'|\uparrow'\rangle\} \\ &= \frac{\hat{P}}{2}\hat{r}_3(i\hat{r}_1 - \hat{r}_2)\left\{\left(\frac{a_{k_-}b_{k_+}}{\sqrt{2}} + \frac{b_{k_-}a_{k_+}}{\sqrt{2}}\right)\right\} + \frac{\hat{P}}{2}\hat{r}_3(i\hat{r}_2 - \hat{r}_2)\{(c_{k_+}a_{k_-} + c_{k_-}c_{k_+})\}\end{aligned}$$

Thus,

$$\hat{p}_{CV}(\vec{k}) = \frac{\hat{p}}{2}\hat{r}_3(i\hat{r}_2 - \hat{r}_2)\left\{a_{k_+}\left(\frac{b_{k_-}}{\sqrt{2}} + c_{k_-}\right) + a_{k_-}\left(\frac{b_{k_+}}{\sqrt{2}} + c_{k_+}\right)\right\} \tag{1.28}$$

we can write that

$|\hat{r}_1| = |\hat{r}_2| = |\hat{r}_2| = 1$, and also, $\hat{P}\hat{r}_3 = \hat{P}_x \sin\theta\cos\phi\hat{i} + \hat{P}_y \sin\theta\sin\phi\hat{j} + \hat{P}_z\cos\theta\hat{k}$

where $\hat{P} = \langle S|P|X\rangle = \langle S|P|Y\rangle = \langle S|P|Z\rangle$,

$$\langle S|P'|X\rangle = \int u_C^\star(0,\vec{r})\hat{P}u_{VX}(0,\vec{r})d^3r = \hat{P}_{CVX}(0)$$

and $\langle S|\hat{P}|Z\rangle = \hat{P}_{CVZ}(0)$

Thus, $\hat{P} = \hat{P}_{CVX}(0) = \hat{P}_{CVY}(0) = \hat{P}_{CVZ}(0) = \hat{P}_{CV}(0)$

where $\hat{P}_{CV}(0) = \int u_C^\star(0,\vec{r})\hat{P}_{u_V}(0,\vec{r})d^3r \equiv \hat{P}$

For a plane polarized light wave, we have the polarization vector $\vec{\varepsilon}_s = \hat{k}$, when the light wave vector travels along the z-axis. Therefore, for a plane polarized light wave, we consider $\vec{\varepsilon}_s = \hat{k}$.

Then, from eq. (1.28) we get

$$(\vec{\varepsilon}\cdot\hat{p}_{CV}(\vec{k})) = \vec{k}\cdot\frac{\hat{P}}{2}\hat{r}_3(i\hat{r}_1 - \hat{r}_2)[A(\vec{k}) + B(\vec{k})]\cos\omega t \tag{1.29}$$

and

$$\left.\begin{aligned} A(\vec{k}) &= a_{k_-}\left(\frac{b_{k_+}}{\sqrt{2}} + c_{k_+}\right) \\ B(\vec{k}) &= a_{k_+}\left(\frac{b_{k_-}}{\sqrt{2}} + c_{k_-}\right) \end{aligned}\right\} \tag{1.30}$$

Thus,

$$\begin{aligned} \left|\vec{\varepsilon}\cdot\hat{p}_{CV}(\vec{k})\right|^2 &= \left|\vec{k}\cdot\frac{\hat{P}}{2}\hat{r}_3\right|^2 |i\hat{r}_1 - \hat{r}_2|^2[A(\vec{k}) + B(\vec{k})]^2\cos\omega t \\ &= \frac{1}{4}\left|\hat{P}_z\cos\theta\right|^2[A(\vec{k}) + B(\vec{k})]^2\cos^2\omega t \end{aligned} \tag{1.31}$$

Hence, the average value of $\left|\vec{\varepsilon}\cdot\hat{p}_{CV}\left(\vec{k}\right)\right|^2$ for a plane polarized light wave is given as

$$\frac{2\pi}{3}|\hat{P}_z|^2\left[A(\vec{k}) + B(\vec{k})\right]^2 \tag{1.32}$$

where $|\hat{P}_z|^2 = (\frac{1}{2})\left|\vec{k}\cdot\hat{p}_{cv}(0)\right|^2$
and

$$\left|\vec{k}\cdot\hat{p}_{cv}(0)\right|^2 = \frac{m^2}{4m_r}\frac{E_{g_0}(E_{g_0}+\Delta)}{(E_{g_0}+\frac{2}{3}\Delta)} \tag{1.33}$$

We can express $A(\vec{k})$ and $B(\vec{k})$ in terms of constants of the energy spectra as follows:
Substituting $a_{k_\pm}, b_{k_\pm}, c_{k_\pm}$, and $\gamma_{0k_\pm}$ in $A(\vec{k})$ and $B(\vec{k})$ in eq. (1.31), we get

$$A(\vec{k}) = \beta\left(t+\frac{\rho}{\sqrt{2}}\right)\left\{\left(\frac{E_{g_0}}{E_{g_0}+\delta'}\right)\gamma_{0k_+}^2 - \gamma_{0k_+}^2\gamma_{0k_-}^2\left(\frac{E_{g_0}-\delta'}{E_{g_0}+\delta'}\right)\right\}^{1/2} \tag{1.34}$$

$$B(\vec{k}) = \beta\left(t+\frac{\rho}{\sqrt{2}}\right)\left\{\left(\frac{E_{g_0}}{E_{g_0}+\delta'}\right)\gamma_{0k_-}^2 - \gamma_{0k_+}^2\gamma_{0k_-}^2\left(\frac{E_{g_0}-\delta'}{E_{g_0}+\delta'}\right)\right\}^{1/2} \tag{1.35}$$

where $\gamma_{0k_+}^2 = \frac{\xi_{1k}-E_{g_0}}{2(\xi_{1k}+\delta')} \equiv \frac{1}{2}\left[1-\left(\frac{E_{g_0}+\delta'}{\xi_{1k}+\delta'}\right)\right]$ and $\gamma_{0k_-}^2 = \frac{\xi_{1k}+E_{g_0}}{2(\xi_{1k}+\delta')} \equiv \frac{1}{2}\left[1-\left(\frac{E_{g_0}-\delta'}{\xi_{1k}+\delta'}\right)\right]$

Substituting $x \equiv \xi_{1k}+\delta'$ in $\gamma_{0k_\pm}^2$, we can write

$$A(\vec{k}) = \beta\left(t+\frac{\rho}{\sqrt{2}}\right)\left\{\left(\frac{E_{g_0}}{E_{g_0}+\delta'}\right)\frac{1}{2}\left(1-\frac{E_{g_0}+\delta'}{x}\right) - \frac{1}{4}\left(\frac{E_{g_0}-\delta'}{E_{g_0}+\delta'}\right)\left(1-\frac{E_{g_0}+\delta'}{x}\right)\left(1-\frac{E_{g_0}-\delta'}{x}\right)\right\}^{1/2}$$

Thus, $A(\vec{k}) = \frac{\beta}{2}(t+\frac{\rho}{\sqrt{2}})\left\{1-\frac{2a_0}{x}+\frac{a_1}{x^2}\right\}^{1/2}$

where $a_0 \equiv (E_{g_0}^2+\delta'^2)(E_{g_0}+\delta')^{-1}$ and $a_1 \equiv (E_{g_0}-\delta')^2$.

After calculation, one can show that

$$A(\vec{k}) = \frac{\beta}{2}\left(t+\frac{\rho}{\sqrt{2}}\right)(E_{g_0}-\delta')\left[\frac{1}{\xi_{1k}+\delta'}-\frac{1}{E_{g_0}+\delta'}\right]^{1/2}\left[\frac{1}{\xi_{1k}+\delta'}\frac{(E_{g_0}+\delta')}{(E_{g_0}-\delta')^2}\right]^{1/2} \tag{1.36}$$

Similarly, from eq. (1.36), we can write

$$B(\vec{k}) = \beta\left(t+\frac{\rho}{\sqrt{2}}\right)\left\{\left(\frac{E_{g_0}}{E_{g_0}+\delta'}\right)\frac{1}{2}\left(1+\frac{E_{g_0}-\delta'}{x}\right) - \frac{1}{4}\left(\frac{E_{g_0}-\delta'}{E_{g_0}+\delta'}\right)\left(1-\frac{E_{g_0}+\delta'}{x}\right)\left(1+\frac{E_{g_0}-\delta'}{x}\right)\right\}^{1/2}$$

So that, finally, we get

$$B(\vec{k}) = \frac{\beta}{2}\left(t + \frac{\rho}{\sqrt{2}}\right)\left(1 + \frac{E_{g_0} - \delta'}{\xi_{1k} + \delta'}\right) \tag{1.37}$$

Using eqs. (1.32), (1.33), (1.36), and (1.37), we can write

$$\left(\frac{|e|A_0}{2m}\right)^2 \frac{\left\langle \left|\vec{\varepsilon}\cdot\hat{p}_{CV}(\vec{k})\right|^2\right\rangle_{av}}{E_c(\vec{k}) - E_v(\vec{k})} = \left(\frac{|e|A_0}{2m}\right)^2 \frac{2\pi}{3}|\vec{k}\cdot\hat{p}_{CV}(0)|^2 \frac{\beta^2}{4}\left(t + \frac{\rho}{\sqrt{2}}\right)^2$$
$$\frac{1}{\xi_{1k}}\left\{\left(1 + \frac{E_{g_0} - \delta'}{\xi_{1k} + \delta'}\right) + (E_{g_0} - \delta')\left[\frac{1}{\xi_{1k} + \delta'} - \frac{1}{E_{g_0} + \delta'}\right]^{1/2}\left[\frac{1}{\xi_{1k} + \delta'} - \frac{E_{g_0} + \delta'}{\left(E_{g_0} - \delta'\right)^2}\right]^{1/2}\right\}^2 \tag{1.38}$$

Following Nag [5], it can be shown that

$$A_0^2 = \frac{I\lambda^2}{2\pi^2 c^3\sqrt{\varepsilon_{sc}\varepsilon_0}} \tag{1.39}$$

where I is the light intensity of wavelength λ, ε_0 is the permittivity of free space, and c is the velocity of light. Thus, the simplified electron energy spectrum in III–V, ternary, and quaternary materials in the presence of light waves can approximately be written as

$$\frac{\hbar^2 k^2}{2m_c} = \beta_0(E, \lambda) \tag{1.40}$$

Where

$$\beta_0(E, \lambda) \equiv [I_{11}(E) - \theta_0(E, \lambda)]$$

$$\theta_0(E, \lambda) \equiv \frac{|e|^2}{96 m_r \pi c^3}\frac{I\lambda^2}{\sqrt{\varepsilon_{sc}\varepsilon_0}}\frac{E_{g_0}(E_{g_0} + \Delta)}{(E_{g_0} + \frac{2}{3}\Delta)}\frac{\beta^2}{4}\left(t + \frac{\rho}{\sqrt{2}}\right)^2 \frac{1}{\phi_0(E)}$$

$$\left\{\left(1 + \frac{E_{g_0} - \delta'}{\phi_0(E) + \delta'}\right) + \left(E_{g_0} - \delta'\right)\left[\frac{1}{\phi_0(E) + \delta'} - \frac{1}{E_{g_0} + \delta'}\right]^{1/2}\left[\frac{1}{\phi_0(E) + \delta'} - \frac{E_{g_0} + \delta'}{(E_{g_0} - \delta')^2}\right]^{1/2}\right\}^2$$

and

$$\phi_0(E) \equiv E_{g_0}\left(1+2\left(1+\frac{m_c}{m_v}\right)\frac{I_{11}(E)}{E_{g_0}}\right)^{1/2}$$

Thus, under the limiting condition $\vec{k} \to 0$, from eq. (1.40), we observe that $E \neq 0$ and is positive. Therefore, in the presence of external light waves, the energy of the electron does not tend to zero when $\vec{k} \to 0$, whereas for the unperturbed three-band Kane model, $I_{11}(E) = \left[\hbar^2 k^2/(2m_c)\right]$ where $E \to 0$ for $\vec{k} \to 0$. As the CB is taken as the reference level of energy, the lowest positive value of E for $k' \to 0$ provides the increased band gap (ΔE_g) of the semiconductor due to photon excitation. The values of the increased band gap can be obtained by computer iteration processes for various values of I and λ, respectively.

Special cases:

1) For the two-band Kane model, we have $\Delta \to 0$. Under this condition, $I_{11}(E) \to E(1+aE) = \frac{\hbar^2 k^2}{2m_c}$. Since $\beta \to 1, t \to 1, \rho \to 0, \delta' \to 0$ for $\Delta \to 0$ from eq. (1.40), we can write the energy spectrum of III–V, ternary, and quaternary materials in the presence of external photoexcitation whose unperturbed conduction electrons obey the two-band Kane model as

$$\frac{\hbar^2 k^2}{2m_c} = \tau_0(E,\lambda) \tag{1.41}$$

Where

$$\tau_0(E,\lambda) \equiv E(1+aE) - B_0(E,\lambda),$$

$$B_0(E,\lambda) \equiv \frac{|e|^2}{384\pi c^3 m_r}\frac{I\lambda^2 E_{g_0}}{\sqrt{\varepsilon_{sc}\varepsilon_0}}\frac{1}{\phi_1(E)}\left\{\left(1+\frac{E_{g_0}}{\phi_0(E)}\right)+E_{g_0}\left[\frac{1}{\phi_1(E)}-\frac{1}{E_{g_0}}\right]\right\}^2,$$

$$\phi_1(E) \equiv E_{g_0}\left\{1+\frac{2m_c}{m_r}aE(1+aE)\right\}^{1/2}$$

For relatively wide band gap semiconductors, one can write $a \to 0$, $b \to 0$, $c \to 0$, and $I_{11}(E) \to E$ Thus, from eq. (1.41), we get

$$\frac{\hbar^2 k^2}{2m_c} = \rho_0(E,\lambda) \tag{1.42}$$

Where $\rho_0(E,\lambda) \equiv E - \frac{|e|^2 I\lambda^2}{96\pi C^3 m_r\sqrt{\varepsilon_{sc}\varepsilon_0}}\left[1+(2m_c m_r)aE\right]^{-3/2}$

Equations (1.40), (1.41), and (1.42) can approximately be written as

$$\frac{\hbar^2 k^2}{2m_c} = U_\lambda I_{11}(E) - P_\lambda \tag{1.43}$$

$$\frac{\hbar^2 k^2}{2m_c} = t_{1\lambda} E + t_{2\lambda} E^2 - \delta_\lambda \tag{1.44}$$

and

$$\frac{\hbar^2 k^2}{2m_c} = t_{1\lambda} E - \delta_\lambda \tag{1.45}$$

where

$$U_\lambda = (1+\theta_\lambda),\ \theta_\lambda = \frac{C_0}{A}\left(t_\lambda + \frac{BJ_\lambda}{A}\right),\ C_0 = \left[\frac{|e|^2}{96\pi c^3 m_r} \frac{I\lambda^2 E_{g_0}(E_{g_0}+\Delta)}{\sqrt{\varepsilon_{sc}\varepsilon_0}(E_{g_0}+\frac{2}{3}\Delta)} \frac{\beta^2}{4}\left(1+\frac{\rho}{\sqrt{2}}\right)^2\right],$$

$$A = E_{g_0},\ B = \left[1+\frac{m^*}{m_V}\right],\ G_\lambda = \left[\frac{2B}{(A+\delta')^3} - \frac{BC_\lambda}{(A+\delta')}\right],$$

$$C_\lambda = [(E_{g_0}+\delta')^{-1} + (E_{g_0}+\delta')(E_{g_0}-\delta')^{-2}](A+\delta')^{-1}$$

$$P_\lambda = \frac{C_0}{A} J_\lambda,\ J_\lambda = (D_\lambda + 2(E_{g_0}-\delta')\sqrt{f_\lambda}),\ D_\lambda = \left(1+\frac{2(E_{g_0}-\delta')}{(A+\delta')}\right),$$

$$f_\lambda = \left[\frac{1}{(A+\delta')^2} + \frac{1}{(E_{g_0}-\delta')^2} - C_\lambda\right],\ t_{1\lambda} = \left(1+\frac{3m_c}{m_r}\alpha\delta_\lambda\right),\ \alpha = \frac{1}{E_{g_0}}$$

$$\delta_\lambda = \frac{|e|^2 I\lambda^2}{96 m_r \pi c^3 \sqrt{\varepsilon_{sc}\varepsilon_0}} \text{ and } t_{2\lambda} = \alpha t_{1\lambda}$$

It is well known that the band tails are being formed in the forbidden zone of HDS and can be explained by the overlapping of the impurity band with the CB and VB [25]. Kane [26] and Bonch Bruevich [27] have independently derived the theory of band tailing for semiconductors having unperturbed parabolic energy bands. Kane's model [26] was used to explain the experimental results on tunneling [28] and the optical absorption edges [29, 30] in this context. Halperin and Lax [31] developed a model for band tailing applicable only to the deep tailing states. Although Kane's concept is often used in the literature for the investigation of band tailing [32, 33], it may be noted that this model [26, 34] suffers from serious assumptions in the sense that the local impurity potential is assumed to be small and slowly varying in space coordinates [33]. In this respect, the local impurity potential may be assumed to be a constant. To avoid these approximations, in this book, we have developed the electron energy spectra for HDS for studying the CEC based on the concept of the variation of the kinetic energy [25, 33] of the electron with the local point in space coordinates. This kinetic energy is then averaged over the entire region of variation using a Gaussian-type potential energy. It may be noted that a more general treatment of many-body theory for the DOS of HDS merges with one-electron theory under macroscopic conditions [25]. In addition, the experimental results for the Fermi energy and others are the average effect of this macroscopic case. Thus, the present treatment of the one-electron system is more applicable from experimental point of view and it is also easy to understand the overall effect in such a case [35]. In an HDS, each impurity atom is surrounded by the electrons, assuming a regular distribution of atoms, and it is screened independently [32, 34, 36]. The interaction energy between electrons and impurities is known as the impurity screening potential. This energy is determined by the interimpurity distance and the screening radius, which is known as the screening length. The screening radius changes with the electron concentration and the effective mass. Furthermore, these entities are important for HDS in characterizing semiconductor properties [37, 38] and modern electronic devices [32, 39].

Based on Kane's model, the works on Fermi energy and the screening length in an n-type GaAs have already been initiated [40, 41]. Incidentally, the limitations of Kane's model [26, 33], as mentioned earlier, are also present in their studies.

The Gaussian distribution $F(V)$ of the impurity potential is given by [26–27]

$$F(V) = (\pi\eta_g^2)^{-1/2} \exp\left(-V^2/\eta_g^2\right) \tag{1.46a}$$

where η_g is the impurity scattering potential. It appears from eq. (1.46a′) that the variance parameter η_g is not equal to zero, but the mean value is zero. Furthermore, the impurities are assumed to be uncorrelated and the band mixing effect has been neglected in this simplified theoretical formalism.

Under the condition of HD, using the method of averaging the kinetic energy of the electron, the HD dispersion relations in this case in the presence of light waves can be written as

$$\frac{\hbar^2 k^2}{2m_c} = T_1(E, \eta_g, \lambda) \tag{1.46b}$$

$$\frac{\hbar^2 k^2}{2m_c} = T_2(E, \eta_g, \lambda) \tag{1.47}$$

$$\frac{\hbar^2 k^2}{2m_c} = T_3(E, \eta_g, \lambda) \tag{1.48}$$

where $T_1(E, \eta_g, \lambda) = [U_\lambda[T_{31}(E, \eta_g) + iT_{32}(E, \eta_g)] - P_\lambda]$,

$$T_{31}\left(E,\eta_g\right)\equiv\left(\frac{2}{1+Erf(E/\eta_g)}\right)$$

$$\left[\frac{\alpha b}{c}\theta_0(E,\eta_g)+\left[\frac{\alpha c+bc-\alpha b}{c^2}\right]\gamma_0(E,\eta_g)+\frac{1}{c}\left(1-\frac{\alpha}{c}\right)\left(1-\frac{b}{c}\right)12[1+Erf\left(\frac{E}{\eta_g}\right)\right]$$

$$-\frac{1}{c}\left(1-\frac{\alpha}{c}\right)\left(1-\frac{b}{c}\right)\frac{2}{c\eta_g\sqrt{\pi}}\exp(-u_2^2)\left[\sum_{p=1}^{\infty}\frac{\exp(-p^2/4)}{p}sinh(pu_2)\right],$$

$$b\equiv\left(\frac{1}{E_g+\Delta}\right),c\equiv\left(\frac{1}{E_g+\frac{2}{3}\Delta}\right),u_2\equiv\frac{1+cE}{c\eta_g},$$

$$T_{32}(E,\eta_g)\equiv\left(\frac{2}{1+Erf(E/\eta_g)}\right)\frac{1}{c}\left(1-\frac{\alpha}{c}\right)\left(1-\frac{b}{c}\right)\frac{\sqrt{\pi}}{c\eta_g}\exp(-u_2^2).$$

$$\theta_0(E,\eta_g)=\frac{\eta_g E}{2\sqrt{\pi}}\exp\left(\frac{-E^2}{\eta_g^2}\right)+\frac{1}{4}(\eta_g^2+2E^2)\left[1+Erf\left(\frac{E}{\eta_g}\right)\right],$$

$$\gamma_0(E,\eta_g)=\eta_g\exp(-E^2/\eta_g^2)(2\sqrt{\pi})^{-1}+\frac{E}{2}(1+Erf(E/\eta_g)),$$

$$T_2(E,\eta_g,\lambda)=[U_{1\lambda}\gamma_3(E,\eta_g)+(t_{2\lambda})2\theta_0(E,\eta_g)[1+Erf(E/\eta_g)]^{-1}-\delta_\lambda]$$

$$\gamma_3(E,\eta_g)\equiv\left[\frac{2}{(1+Erf(E/\eta_g))}\right]\gamma_0(E,\eta_g)$$

and $T_3(E,\eta_g,\lambda)=[t_{1\lambda}\gamma_3(E,\eta_g)-\delta_\lambda]$

The HUP can be written as

$$\Delta p_i\Delta i{\sim}A\hbar \tag{1.49}$$

where $i=x$, y, and z,p is the momentum, $\Delta' s$ are the errors in measuring p_i, $i,\hbar$ is Dirac's constant, and A is the dimensionless constant.

Since $p_i=\hbar k_i$ where k_i is the electron wave vector, from eq. (1.50a) we can write

$$\Delta k_x\Delta k_y\,\Delta k_z=\frac{A^3}{\Delta V} \tag{1.50a}$$

where $\Delta V=\Delta x\,\Delta y\,\Delta z$

The use HUP tells us each electron occupies at least a volume ΔV and this electron must exists in either of the two possible spin orientations due to Pauli's exclusion principle. If n_{03D} is the electron statistics (ES) in unit volume, then $\Delta V = \frac{2}{n_{03D}}$ and the combination of this with eq. (1.50a) leads to

$$n_{03D} = \frac{2}{C_{3D}} \left(\Delta k_x \, \Delta k_y \, \Delta k_z \right) \text{ where } C_{3D} = A^3 \tag{1.50b}$$

If n_{03D} is the valley degeneracy, we can write

$$n_{03D} = \frac{2g_v}{C_{3D}} \left(\Delta k_x \Delta k_y \Delta k_z \right) \tag{1.50c}$$

For two and one dimensions, we get

$$n_{02D} = \frac{2g_v}{C_{2D}} \left(\Delta k_x \Delta k_y \right) \tag{1.50d}$$

and

$$n_{01D} = \frac{2g_v}{C_{1D}} \left(\Delta k_x \right) \tag{1.51a}$$

where n_{02D} and n_{01D} are ES per unit area per subband and per unit length per subband, respectively, and C_{2D} and C_{1D} are two dimensionless constants in the respective cases.

In accordance with HUP

$$V(E_F) = \left(\Delta k_x \Delta k_y \Delta k_z \right) \tag{1.51b}$$

Using eq. (1.46b) we get

$$V(E_{FHDL}) = \frac{4\pi}{3} \left[\frac{2m_c}{\hbar^2} \right]^{3/2} \left[T_1(E_{FHDL}, \eta_g \lambda) \right]^{3/2} \tag{1.51c}$$

where E_{FHDL} is the Fermi energy in this case.

Using eqs. (1.50b) and (1.51c), the ES can be written as

$$n_0 = \frac{8\pi g_v}{3C_{3D}} \left[\frac{2m_c}{\hbar^2} \right]^{3/2} \text{Real Part of} \left[T_1(E_{FHDL}, \eta_g \lambda) \right]^{3/2} \tag{1.51d}$$

By substituting $C_{3D} = (2\pi)^3$, eq. (1.51d) assumes the form

$$n_0 = \frac{g_v}{3\pi^2}\left[\frac{2m_c}{\hbar^2}\right]^{3/2} \text{Real Part of}\left[T_1(E_{FHDL}, \eta_g\lambda)\right]^{3/2} \tag{1.51e}$$

Equation (1.51e) is the expression of ES in HD III–V, ternary, and quaternary semiconductors in accordance with the three-band Kane model in the presence of intense light waves under the condition of extreme degeneracy with the formation of band tails without using the DOS function approach and by directly applying the HUP.

Similarly for perturbed two-band Kane model and for parabolic energy bands, we can write

$$n_0 = \frac{g_v}{3\pi^2}\left[\frac{2m_c}{\hbar^2}\right]^{3/2}\left[T_2(E_{FHDL}, \eta_g\lambda)\right]^{3/2} \tag{1.51f}$$

$$n_0 = \frac{g_v}{3\pi^2}\left[\frac{2m_c}{\hbar^2}\right]^{3/2}\left[T_3(E_{FHDL}, \eta_g\lambda)\right]^{3/2} \tag{1.51g}$$

In the absence of HD, the ES can be written as

$$n_0 = \frac{g_v}{3\pi^2}\left(\frac{2m_c}{\hbar^2}\right)^{3/2}\left[\beta_0(E_{FL}, \lambda)\right]^{3/2} \tag{1.52a}$$

$$n_0 = \frac{g_v}{3\pi^2}\left(\frac{2m_c}{\hbar^2}\right)^{3/2}\left[\tau_0(E_{FL}, \lambda)\right]^{3/2} \tag{1.52b}$$

$$n_0 = \frac{g_v}{3\pi^2}\left(\frac{2m_c}{\hbar^2}\right)^{3/2}\left[\rho_0(E_{FL}, \lambda)\right]^{3/2} \tag{1.52c}$$

In the absence of HD and photons, the ES in this case can be written as

$$n_0 = \frac{g_v}{3\pi^2}\left(\frac{2m_c}{\hbar^2}\right)^{3/2}\left[I_{11}(E_F)\right]^{3/2} \tag{1.53a}$$

$$n_0 = \frac{g_v}{3\pi^2}\left(\frac{2m_c}{\hbar^2}\right)^{3/2}\left[E_F(1+\alpha E_F)\right]^{3/2} \tag{1.53b}$$

$$n_0 = \frac{g_v}{3\pi^2}\left(\frac{2m_c}{\hbar^2}\right)^{3/2}\left[E_F\right]^{3/2} \tag{1.53c}$$

Equations (1.53b) and (1.53c) are well known in the literature [13].

The electronic contribution to the second- and third-order elastic constants for HD materials can be written as [94–106]

$$\Delta C_{44} = \frac{-G_0^2 Z_x}{9}\left[\frac{\partial n_0}{\partial(E_{FHD}-E_{0HD})}\right] \tag{1.54a}$$

and

$$\Delta C_{456} = \frac{G_0^3 Z_x}{27}\left[\frac{\partial^2 n_0}{\partial(E_{FHD}-E_{0HD})^2}\right] \tag{1.54b}$$

where G_0 is the deformation potential constant, $Z_x = (\text{Re al Part})^x$, E_{FHD} is the Fermi energy in HD materials in the presence of band tails, and E_{0HD} is obtained from the corresponding HD dispersion relation under the conditions $E = E_{0\,HD}$ when $k = 0$.

Using eqs. (1.54a), (1.54b), and (1.51e) and $x = 1$ successively, we can study ΔC_{44} and ΔC_{456} for HD materials in the presence of light waves whose energy band structures in the absence of any field are given by three-band Kane model. Using eqs. (1.54a), (1.54b), (1.51f), and (1.51g) together with $x = 0$, we can study the same for two-band Kane models along with respective parabolic energy bands in this case.

In the absence of band tails, we can write

$$\Delta C_{44} = \frac{-G_0^2}{9}\left[\frac{\partial n_0}{\partial E_{FL}}\right] \tag{1.54c}$$

and

$$\Delta C_{456} = \frac{G_0^3}{27}\left[\frac{\partial^2 n_0}{\partial E_{FL}{}^2}\right] \tag{1.54d}$$

where E_{FL} is the corresponding Fermi energy in this case.

Using eqs. (1.54c), (1.54d), (1.52a), (1.52b), and (1.52c) successively, the expressions for ΔC_{44} and ΔC_{456} for materials in the absence of band tails and in the presence of light waves whose energy band structures in the absence of any field are given by three- and two-band Kane models together with parabolic energy bands can, respectively, be written as

$$\Delta C_{44} = \frac{-G_0^2 g_v}{27\pi^2}\left(\frac{2m_c}{\hbar^2}\right)^{3/2}\left[\left[\left[\beta_0(E_{FL},\lambda)\right]^{3/2}\right]'\right] \tag{1.55a}$$

$$\Delta C_{44} = \frac{-G_0^2 g_v}{27\pi^2}\left(\frac{2m_c}{\hbar^2}\right)^{3/2}\left[\left[\left[\tau_0(E_{FL},\lambda)\right]^{3/2}\right]'\right] \tag{1.55b}$$

$$\Delta C_{44} = \frac{-G_0^2 g_v}{27\pi^2}\left(\frac{2m_c}{\hbar^2}\right)^{3/2}\left[\left[\left[\rho_0(E_{FL},\lambda)\right]^{3/2}\right]'\right] \tag{1.55c}$$

and

$$\Delta C_{456} = \frac{G_0^3 g_v}{81\pi^2}\left(\frac{2m_c}{\hbar^2}\right)^{3/2}\left[\left[[\beta_0(E_{FL},\lambda)]^{3/2}\right]''\right] \tag{1.56a}$$

$$\Delta C_{456} = \frac{G_0^3 g_v}{81\pi^2}\left(\frac{2m_c}{\hbar^2}\right)^{3/2}\left[\left[[\tau_0(E_{FL},\lambda)]^{3/2}\right]''\right] \tag{1.56b}$$

$$\Delta C_{456} = \frac{G_0^3 g_v}{81\pi^2}\left(\frac{2m_c}{\hbar^2}\right)^{3/2}\left[\left[[\rho_0(E_{FL},\lambda)]^{3/2}\right]''\right] \tag{1.56c}$$

where the primes denote the differentiation with respect to Fermi energy.

ΔC_{44} and ΔC_{456} for bulk materials in the absence of any field can be expressed as

$$\Delta C_{44} = \frac{-G_0^2}{9}\left[\frac{\partial n_0}{\partial E_F}\right] \tag{1.56d}$$

and

$$\Delta C_{456} = \frac{G_0^3}{27}\left[\frac{\partial^2 n_0}{\partial E_F{}^2}\right] \tag{1.56e}$$

where E_F is the corresponding Fermi energy.

Using eqs. (1.56d), (1.56e), (1.53a), and (1.53b) successively, the expressions for ΔC_{44} and ΔC_{456} for bulk materials in the absence of any field whose energy band structures in the absence of any field are given by three- and two-band Kane models can, respectively, be written as

$$\Delta C_{44} = \frac{-G_0^2 g_v}{27\pi^2}\left(\frac{2m_c}{\hbar^2}\right)^{3/2}\left[\left[[I_{11}(E_F)]^{3/2}\right]'\right] \tag{1.57a}$$

$$\Delta C_{44} = \frac{-G_0^2 g_v}{27\pi^2}\left(\frac{2m_c}{\hbar^2}\right)^{3/2}\left[\left[[E_F(1+\alpha E_F)]^{3/2}\right]'\right] \tag{1.57b}$$

and

$$\Delta C_{456} = \frac{G_0^3 g_v}{81\pi^2}\left(\frac{2m_c}{\hbar^2}\right)^{3/2}\left[\left[[I_{11}(E_F)]^{3/2}\right]''\right] \tag{1.58a}$$

$$\Delta C_{456} = \frac{G_0^3 g_v}{81\pi^2}\left(\frac{2m_c}{\hbar^2}\right)^{3/2}\left[\left[[E_F(1+\alpha E_F)]^{3/2}\right]''\right] \tag{1.58b}$$

Under the conditions $\Delta \gg E_g$ or $\Delta \ll E_g$ together with the inequality $\alpha E_F \ll 1$, the electron concentration in bulk specimens of III–V, ternary, and quaternary semiconductors whose energy band structures are defined by two-band Kane model can be expressed in the presence of finite temperature as

$$n_0 = N_c\left[F_{1/2}(\eta) + \left(\frac{15ak_BT}{4}\right)F_{3/2}(\eta)\right] \tag{1.59}$$

where $N_C \equiv g_v.2\left(\frac{2\pi m_c k_B T}{h^2}\right)^{3/2}$, $\eta \equiv \frac{E_F}{k_B T}$ and $F_j(\eta)$ is the one parameter Fermi–Dirac integral of order j, which can be written as [73]

$$F_j(\eta) = \left(\frac{1}{\Gamma(j+1)}\right)\int_0^{\infty} y^j(1+\exp(y-\eta)^{-1}dy,\ jgt-1 \tag{1.60a}$$

where $\Gamma(j+1)$ is the complete Gamma function or for all j, analytically continued as a complex integral around the negative axis

$$F_j(\eta) = A_j\int_0^{(0+)} y^j(1+\exp(-y-\eta)^{-1}dy, \tag{1.60b}$$

in which $A_j \equiv \frac{\Gamma(-j)}{2\pi\sqrt{-1}}$

Using eqs. (1.56d), (1.56e), and (1.59), ΔC_{44} and ΔC_{456} under the condition $\alpha E \ll 1$ can be written for two-band Kane model at a finite temperature as

$$\Delta C_{44} = \frac{-G_0^2}{9k_BT}N_c\left[F_{-1/2}(\eta) + \left(\frac{15ak_BT}{4}\right)F_{1/2}(\eta)\right] \tag{1.61a}$$

$$\Delta C_{456} = \frac{G_0^3}{27(k_BT)^2}N_c\left[F_{-3/2}(\eta) + \left(\frac{15ak_BT}{4}\right)F_{-1/2}(\eta)\right] \tag{1.61b}$$

Under the condition of nondegeneracy, eqs. (1.61a) and (1.61b) assume the form

$$\Delta C_{44} = \frac{-G_0^2}{9k_BT}n_0 \tag{1.62a}$$

$$\Delta C_{456} = \frac{G_0^3}{27(k_BT)^2}n_0 \tag{1.62b}$$

Equations (1.62a) and (1.62b) are well known in the literature.

1.2.2 The CECs under magnetic quantization in HD Kane-type semiconductors in the presence of light waves

(i) Using eq. (1.46b), the magnetodispersion law, in the absence of spin, for HD III–V, ternary, and quaternary semiconductors, in the presence of photoexcitation, whose unperturbed conduction electrons obey the three-band Kane model, is given by

$$T_1(E,\eta_g,\lambda)=\left(n+\frac{1}{2}\right)\hbar\omega_0+\frac{\hbar^2k_z^2}{2m_c} \tag{1.63}$$

The application of HUP leads to the expression of ES in this case as

$$n_0=\frac{4g_veB}{C_B\hbar}\sum_{n=0}^{n_{\max}}\Delta k_z \tag{1.64}$$

where C_B is a constant.

Using eq. (1.56) we can write

$$T_1(E_{FHDLB},\eta_g,\lambda)=\left(n+\frac{1}{2}\right)\hbar\omega_0+\frac{\hbar^2(\Delta k_z)^2}{2m_c} \tag{1.65}$$

where E_{FBHDLB} is the Fermi energy in the present case.

Using eqs. (1.64) and (1.65) together with the substitution $C_B=4\pi^2$ lead to the expression of electron concentration as

$$n_0=\frac{Bg_v|e|\sqrt{2m_c}}{\pi^2\hbar^2}\text{Real part of}\left[\sum_{n=0}^{n_{\max}}\left[\left\{T_1(E_{FHDLB},\eta_g,\lambda)-\left(n+\frac{1}{2}\right)\hbar\omega_0\right\}^{1/2}\right]\right] \tag{1.66a}$$

where E_{FHDLB} is the Fermi energy under quantizing magnetic field in the presence of light waves as measured from the edge of the CB in the vertically upward direction in the absence of any quantization.

The electronic contribution to the second- and third-order elastic constants for HD materials can be written as [94–106]

$$\Delta C_{44}=\frac{-G_0^2Z_x}{9}\,Real\ Part\ of\left[\frac{\partial n_0}{\partial(E_{FHDLB}-E_{0HDB2})}\right] \tag{1.66b}$$

and

$$\Delta C_{456}=\frac{G_0^3Z_x}{27}\,Real\ Part\ of\left[\frac{\partial^2 n_0}{\partial(E_{FHDLB}-E_{0HDB2})^2}\right] \tag{1.66c}$$

where E_{0HDB2} is obtained from the corresponding HD dispersion relation under the conditions $E = E_{0HDB2}$ when $k_z = 0$.

Using eqs. (1.66a), (1.66b), and (1.66c) together with $x = 1$ we can study ΔC_{44} and ΔC_{456} in this case.

Using eq. (1.41), the magneto-dispersion law, in the absence of spin and band tails for III–V, ternary, and quaternary semiconductors, in the presence of photo-excitation, whose unperturbed conduction electrons obey the two-band Kane model, is given by

$$\beta_0(E, \lambda) = \left(n + \frac{1}{2}\right)\hbar\omega_0 + \frac{\hbar^2 k_z^2}{2m_c} \tag{1.67}$$

The ES is given by

$$n_0 = \frac{Bg_v|e|\sqrt{2m_c}}{\pi^2\hbar^2} \sum_{n=0}^{n_{\max}} \left[\left\{\beta_0(E_{FLB}, \lambda) - \left(n + \frac{1}{2}\right)\hbar\omega_0\right\}^{1/2}\right] \tag{1.68}$$

where E_{FLB} is the Fermi energy under quantizing magnetic field in the presence of light waves and band tails as measured from the edge of the CB in the vertically upward direction in the absence of any quantization.

In the absence of band tails we can write

$$\Delta C_{44} = \frac{-G_0^2}{9}\left[\frac{\partial n_0}{\partial E_{FLB}}\right] \tag{1.69}$$

and

$$\Delta C_{456} = \frac{G_0^3}{27}\left[\frac{\partial^2 n_0}{\partial E_{FLB}^2}\right] \tag{1.70}$$

Using eqs. (1.68)–(1.70), we get

$$\Delta C_{44} = \frac{-G_0^2 Bg_v|e|\sqrt{2m_c}}{9\pi^2\hbar^2} \sum_{n=0}^{n_{\max}} \left[\left\{\beta_0(E_{FLB}, \lambda) - \left(n + \frac{1}{2}\right)\hbar\omega_0\right\}^{1/2}\right]' \tag{1.71}$$

$$\Delta C_{456} = \frac{G_0^3 Bg_v|e|\sqrt{2m_c}}{27\pi^2\hbar^2} \sum_{n=0}^{n_{\max}} \left[\left\{\beta_0(E_{FLB}, \lambda) - \left(n + \frac{1}{2}\right)\hbar\omega_0\right\}^{1/2}\right]'' \tag{1.72}$$

In the absence of light waves and HD, the ES can be written as

$$n_0 = \frac{Bg_v|e|\sqrt{2m_c}}{\pi^2\hbar^2} \sum_{n=0}^{n_{\max}} \left[\left\{I_{11}(E_{FB}) - \left(n + \frac{1}{2}\right)\hbar\omega_0\right\}^{1/2}\right] \tag{1.73}$$

where E_{FB} is the Fermi energy in this case.

In this case ΔC_{44} and ΔC_{456} can be written as

$$\Delta C_{44} = \frac{-G_0^2}{9}\left[\frac{\partial n_0}{\partial E_{FB}}\right] \tag{1.74}$$

and

$$\Delta C_{456} = \frac{G_0^3}{27}\left[\frac{\partial^2 n_0}{\partial E_{FB}^2}\right] \tag{1.75}$$

Using eqs. (1.73)–(1.75), we get

$$\Delta C_{44} = \frac{-G_0^2 B g_v |e| \sqrt{2m_c}}{9\pi^2\hbar^2} \sum_{n=0}^{n_{max}} \left[\left\{I_{11}(E_{FB}) - \left(n+\frac{1}{2}\right)\hbar\omega_0\right\}^{1/2}\right]' \tag{1.76}$$

$$\Delta C_{456} = \frac{G_0^3 B g_v |e| \sqrt{2m_c}}{27\pi^2\hbar^2} \sum_{n=0}^{n_{max}} \left[\left\{I_{11}(E_{FB}) - \left(n+\frac{1}{2}\right)\hbar\omega_0\right\}^{1/2}\right]'' \tag{1.77}$$

(ii) Using eq. (1.47), the magneto-dispersion law, in the absence of spin, for HD III–V, ternary, and quaternary semiconductors, in the presence of photoexcitation, whose unperturbed conduction electrons obey the two-band Kane model, is given by

$$T_2(E, \eta_g, \lambda) = \left(n+\frac{1}{2}\right)\hbar\omega_0 + \frac{\hbar^2 k_z^2}{2m_c} \tag{1.78}$$

The ES is given by

$$n_0 = \frac{B g_v |e| \sqrt{2m_c}}{\pi^2\hbar^2} \sum_{n=0}^{n_{max}} \left[\left\{T_2(E_{FHDLB}, \eta_g, \lambda) - \left(n+\frac{1}{2}\right)\hbar\omega_0\right\}^{1/2}\right] \tag{1.79}$$

Using eqs. (1.79), (1.66b), and (1.66c) together with $x = 0$ we can study ΔC_{44} and ΔC_{456} in this case.

Using eq. (1.42), the magneto-dispersion law, in the absence of spin and band tails for III–V, ternary, and quaternary semiconductors, in the presence of photoexcitation, whose unperturbed conduction electrons obey the two-band Kane model, is given by

$$\tau_0(E, \lambda) = \left(n+\frac{1}{2}\right)\hbar\omega_0 + \frac{\hbar^2 k_z^2}{2m_c} \tag{1.80}$$

The ES is given by

$$n_0 = \frac{Bg_v|e|\sqrt{2m_c}}{\pi^2\hbar^2}\sum_{n=0}^{n_{\max}}\left[\left\{\tau_0(E_{FLB},\lambda) - \left(n+\frac{1}{2}\right)\hbar\omega_0\right\}^{1/2}\right] \tag{1.81}$$

Using eqs. (1.69)–(1.81) we can write

$$\Delta C_{44} = \frac{-G_0^2 Bg_v|e|\sqrt{2m_c}}{9\pi^2\hbar^2}\sum_{n=0}^{n_{\max}}\left[\left\{\tau_0(E_{FLB},\lambda) - \left(n+\frac{1}{2}\right)\hbar\omega_0\right\}^{1/2}\right]' \tag{1.82}$$

$$\Delta C_{456} = \frac{G_0^3 Bg_v|e|\sqrt{2m_c}}{27\pi^2\hbar^2}\sum_{n=0}^{n_{\max}}\left[\left\{\tau_0(E_{FLB},\lambda) - \left(n+\frac{1}{2}\right)\hbar\omega_0\right\}^{1/2}\right]'' \tag{1.83}$$

In the absence of light waves and band tails, the ES for two-band Kane model in the presence of magnetic quantization can be written as

$$n_0 = \frac{Bg_v|e|\sqrt{2m_c}}{\pi^2\hbar^2}\sum_{n=0}^{n_{\max}}\left[\left\{E_{FB}(1+\alpha E_{FB}) - \left(n+\frac{1}{2}\right)\hbar\omega_0\right\}^{1/2}\right] \tag{1.84}$$

Using eqs. (1.74), (1.75) and (1.84) we can write

$$\Delta C_{44} = \frac{-G_0^3 Bg_v|e|\sqrt{2m_c}}{9\pi^2\hbar^2}\sum_{n=0}^{n_{\max}}\left[\left\{E_{FB}(1+\alpha E_{FB}) - \left(n+\frac{1}{2}\right)\hbar\omega_0\right\}^{1/2}\right]' \tag{1.85}$$

$$\Delta C_{456} = \frac{G_0^3 Bg_v|e|\sqrt{2m_c}}{27\pi^2\hbar^2}\sum_{n=0}^{n_{\max}}\left[\left\{E_{FB}(1+\alpha E_{FB}) - \left(n+\frac{1}{2}\right)\hbar\omega_0\right\}^{1/2}\right]'' \tag{1.86}$$

(iii) Using eq. (1.48a), the magneto-dispersion law, in the absence of spin, for HD III-V, ternary and quaternary semiconductors, in the presence of photo-excitation whose unperturbed conduction electrons obey the parabolic energy bands is given by

$$T_3(E,\eta_g,\lambda) = \left(n+\frac{1}{2}\right)\hbar\omega_0 + \frac{\hbar^2k_z^2}{2m_c} \tag{1.87}$$

The ES is given by

$$n_0 = \frac{Bg_v|e|\sqrt{2m_c}}{\pi^2\hbar^2}\sum_{n=0}^{n_{\max}}\left[\left\{T_3(E_{FHDLB},\eta_g,\lambda) - \left(n+\frac{1}{2}\right)\hbar\omega_0\right\}^{1/2}\right] \tag{1.88}$$

Using eqs. (1.88), (1.66b), and (1.66c) together with $x = 1$ we can study ΔC_{44} and ΔC_{456} in this case.

Using eq. (1.43), the magneto-dispersion law, in the absence of spin and band tails for III–V, ternary, and quaternary semiconductors, in the presence of photo-excitation whose unperturbed conduction electrons obey the parabolic energy bands is given by

$$\rho_0(E,\lambda) = \left(n+\frac{1}{2}\right)\hbar\omega_0 + \frac{\hbar^2 k_z^2}{2m_c} \tag{1.89}$$

The ES is given by

$$n_0 = \frac{Bg_v|e|\sqrt{2m_c}}{\pi^2\hbar^2}\sum_{n=0}^{n_{\max}}\left[\left\{\rho_0(E_{FLB},\lambda) - \left(n+\frac{1}{2}\right)\hbar\omega_0\right\}^{1/2}\right] \tag{1.90}$$

Using eqs. (1.69b), (1.69c), and (1.90) we can write

$$\Delta C_{44} = \frac{-G_0^2 Bg_v|e|\sqrt{2m_c}}{9\pi^2\hbar^2}\sum_{n=0}^{n_{\max}}\left[\left\{\rho_0(E_{FLB},\lambda) - \left(n+\frac{1}{2}\right)\hbar\omega_0\right\}^{1/2}\right]' \tag{1.91}$$

$$\Delta C_{456} = \frac{G_0^3 Bg_v|e|\sqrt{2m_c}}{27\pi^2\hbar^2}\sum_{n=0}^{n_{\max}}\left[\left\{\rho_0(E_{FLB},\lambda) - \left(n+\frac{1}{2}\right)\hbar\omega_0\right\}^{1/2}\right]'' \tag{1.92}$$

In the absence of light waves and band tails, the ES for isotropic parabolic energy bands can be written under magnetic quantization as

$$n_0 = \left(\frac{g_v eB\sqrt{2m_c}}{\pi^2\hbar^2}\right)\sum_{n=0}^{n_{\max}}\left[E_{FB} - \left(n+\frac{1}{2}\right)\hbar\omega_0\right]^{\frac{1}{2}} \tag{1.93}$$

Equation (1.79) is well known in the literature [13].

Using eqs. (1.69g), (1.69h), and (1.93) we get

$$\Delta C_{44} = \left(\frac{-G_0^2 g_v eB\sqrt{2m_c}}{9\pi^2\hbar^2}\right)\sum_{n=0}^{n_{\max}}\left[\left[E_{FB} - \left(n+\frac{1}{2}\right)\hbar\omega_0\right]^{\frac{1}{2}}\right]' \tag{1.94}$$

$$\Delta C_{456} = \left(\frac{G_0^3 g_v eB\sqrt{2m_c}}{27\pi^2\hbar^2}\right)\sum_{n=0}^{n_{\max}}\left[\left[E_{FB} - \left(n+\frac{1}{2}\right)\hbar\omega_0\right]^{\frac{1}{2}}\right]'' \tag{1.95}$$

Under the condition $\alpha E_{FB} \ll 1$, the electron concentration at finite temperature in this case can be expressed as

$$n_0 = N_c\theta_{B1}\left[\sum_{n=0}^{n_{\max}} \frac{1}{\sqrt{a_{01}}}\left[\left(1+\frac{3}{2}\alpha b_{01}\right)F_{\frac{-1}{2}}(\bar{\eta}_{B_1}) + \frac{3}{4}\alpha k_B T F_{\frac{1}{2}}(\bar{\eta}_{B1})\right]\right] \tag{1.96}$$

where $\theta_{B1} = \frac{\hbar\omega_0}{k_B T}$, $a_{01} = \left[1+\alpha\left(n+\frac{1}{2}\right)\hbar\omega_0\right]$, $b_{01} = \left[\left[1+\alpha\left(n+\frac{1}{2}\right)\hbar\omega_0\right]/a_{01}\right]$ *and* $\bar{\eta}_{B_1} = \frac{E_{FB}-b_{01}}{k_B T}$

The eq. (1.96) is well-known in the literature [13]

The influence of finite temperature leads us to the following expressions of ΔC_{44} *and* ΔC_{456} *for optoelectronic materials whose energy band structures are defined by the two-band Kane model under magnetic quantization under the condition* $\alpha E \ll 1$*as*

$$\Delta C_{44} = \frac{-G_0^2}{9k_B T}N_c\theta_{B1}\left[\sum_{n=0}^{n_{\max}} \frac{1}{\sqrt{a_{01}}}\left[\left(1+\frac{3}{2}\alpha b_{01}\right)F_{\frac{-3}{2}}(\bar{\eta}_{B_1}) + \frac{3}{4}\alpha k_B T F_{\frac{-1}{2}}(\bar{\eta}_{B1})\right]\right] \tag{1.97}$$

$$\Delta C_{456} = \frac{G_0^3}{27(k_B T)^2}N_c\theta_{B1}\left[\sum_{n=0}^{n_{\max}} \frac{1}{\sqrt{a_{01}}}\left[\left(1+\frac{3}{2}\alpha b_{01}\right)F_{\frac{-5}{2}}(\bar{\eta}_{B_1}) + \frac{3}{4}\alpha k_B T F_{\frac{-3}{2}}(\bar{\eta}_{B1})\right]\right] \tag{1.98}$$

In the absence of light waves and band tails, the electron concentration for isotropic parabolic energy bands can be written under magnetic quantization at a finite temperature as

$$n_0 = N_c\theta_{B1}\left[\sum_{n=0}^{n_{\max}} [F_{\frac{-1}{2}}(\bar{\eta}_{B_2})]\right] \text{ where } \bar{\eta}_{B_2} = \frac{E_{FB}-\left(n+\frac{1}{2}\right)\hbar\omega_0}{k_B T} \tag{1.99}$$

Using eqs. (1.99), (1.51g), and (1.51h) we get

$$\Delta C_{44} = \frac{-G_0^2}{9k_B T}N_c\theta_{B1}\left[\sum_{n=0}^{n_{\max}} [F_{\frac{-3}{2}}(\bar{\eta}_{B_2})]\right] \tag{1.100}$$

$$\Delta C_{456} = \frac{G_0^3}{27(k_B T)^2}N_c\theta_{B1}\left[\sum_{n=0}^{n_{\max}} [F_{\frac{-5}{2}}(\bar{\eta}_{B_2})]\right] \tag{1.101}$$

Equation (1.99) is well known in the literature [13].

1.2.3 The CECs under crossed electric and quantizing magnetic fields in HD Kane-type semiconductors in the presence of light waves

(i) The electron dispersion law in the present case is given by

$$T_1(E,\eta_g,\lambda)=\left(n+\frac{1}{2}\right)\hbar\omega_0+\frac{[\hbar k_z(E)]^2}{2m_c}-\frac{E_0}{B}\hbar k_y\{T_1(E,\eta_g,\lambda)\}'-\left\{\frac{m_cE_0^2[\{T_1(E,\eta_g,\lambda)\}']^2}{2B^2}\right\} \tag{1.102}$$

The electron concentration in this case can be expressed as

$$n_0=\frac{2g_vB\sqrt{2m_c}}{3L_x\pi^2\hbar^2E_0}\text{Real part of}\sum_{n=0}^{n_{max}}\left[M_{161}(E_{F_{BLHDC}},n,E_0,B,\lambda)\right] \tag{1.103}$$

where $E_{F_{BLHDC}}$ is the Fermi energy in this case,

$$M_{161}(n,E_{F_{BL}},\lambda)=\Bigg[\Bigg[[T_1(E_{F_{BLHDC}},n,E_0,B,\lambda)-\left(n+\frac{1}{2}\right)\hbar\omega_0-\frac{m_cE_0^2}{2B^2}$$

$$[\{T_1(E_{F_{BLHDC}},n,E_0,B,\lambda)\}']^2+|e|E_0L_x[\{T_1(E_{F_{BLHDC}},n,E_0,B,\lambda)\}']\Bigg]^{3/2}$$

$$-[T_1(E_{F_{BLHDC}},n,E_0,B,\lambda)-\left(n+\frac{1}{2}\right)\hbar\omega_0-\frac{m_cE_0^2}{2B^2}[\{T_1(E_{F_{BLHDC}},n,E_0,B,\lambda)\}']^2]^{3/2}]$$

$$\frac{1}{[\{T_1(E_{F_{BLHDC}},n,E_0,B,\lambda)\}']} \tag{1.104a}$$

The electronic contribution to the second- and third-order elastic constants for HD materials can be written as [94–106]

$$\Delta C_{44}=\frac{-G_0^2Z_x}{9}\left[\frac{\partial n_0}{\partial\left(E_{F_{BLHDC}}-E_{0HDB1}\right)}\right] \tag{1.104b}$$

and

$$\Delta C_{456}=\frac{G_0^3Z_x}{27}\left[\frac{\partial^2 n_0}{\partial\left(E_{F_{BLHDC}}-E_{0HDB1}\right)^2}\right] \tag{1.104c}$$

where E_{0HDB1} is obtained from the corresponding HD dispersion relation under the conditions $E=E_{0HDB1}$ when $k_z=0$ and $k_y=0$.

Using eqs. (1.104a)–(1.104c) with $x=1$ we can study ΔC_{44} and ΔC_{456} in this case.

The electron dispersion law in the present case in the absence of band tails is given by

$$\beta_0(E,\lambda) = \left(n+\frac{1}{2}\right)\hbar\omega_0 + \frac{[\hbar k_z(E)]^2}{2m_c} - \frac{E_0}{B}\hbar k_y\{\beta_0(E,\lambda)\}' - \left\{\frac{m_c E_0^2[\{\beta_0(E,\lambda)\}']^2}{2B^2}\right\} \tag{1.105}$$

The electron concentration in this case can be expressed as

$$n_0 = \frac{2g_v B\sqrt{2m_c}}{3L_x\pi^2\hbar^2 E_0}\sum_{n=0}^{n_{\max}}[M_{1612}(E_{F_{BLC}}, n, E_0, B, \lambda)], \tag{1.106a}$$

where

$$M_{1612}(n_{F_{BL}},\lambda) \equiv \Bigg[\Bigg[\beta_0(E_{F_{BLC}}, n, E_0, B, \lambda) - \left(n+\frac{1}{2}\right)\hbar\omega_0 - \frac{m_c E_0^2}{2B^2}$$

$$[\{\beta_0(E_{F_{BLC}}, n, E_0, B, \lambda)\}']^2 + |e|E_0 L_x[\{\beta_0(E_{F_{BLC}}, n, E_0, B, \lambda)\}']\Bigg]^{3/2}$$

$$- \left[\beta_0(E_{F_{BLC}}, n, E_0, B, \lambda) - \left(n+\frac{1}{2}\right)\hbar\omega_0 - \frac{m_c E_0^2}{2B^2}[\{\beta_0(E_{F_{BLC}}, n, E_0, B, \lambda)\}']^2\right]^{3/2}\Bigg]$$

$$\frac{1}{[\{\beta_0(E_{F_{BLC}}, n, E_0, B, \lambda)\}']}$$

The electronic contribution to the second- and third-order elastic constants in this case can be written as [94–106]

$$\Delta C_{44} = \frac{-G_0^2}{9}\left[\frac{\partial n_0}{\partial E_{F_{BLC}}}\right] \tag{1.106b}$$

and

$$\Delta C_{456} = \frac{G_0^3}{27}\left[\frac{\partial^2 n_0}{\partial E^2{}_{F_{BLC}}}\right] \tag{1.106c}$$

ΔC_{44} and ΔC_{456} in this case can, respectively, be expressed as

$$\Delta C_{44} = \frac{-2G_0^2 g_v B\sqrt{2m_c}}{27L_x\pi^2\hbar^2 E_0}\sum_{n=0}^{n_{\max}}[[M_{1612}(E_{F_{BLC}}, n, E_0, B, \lambda)]'], \tag{1.107}$$

$$\Delta C_{456} = \frac{2G_0^3 g_v B\sqrt{2m_c}}{81L_x\pi^2\hbar^2 E_0}\sum_{n=0}^{n_{max}}[[M_{1612}(E_{F_{BLC}}, n, E_0, B, \lambda)]''], \tag{1.108}$$

(ii) The electron dispersion law in the present case is given by

$$T_2(E,\eta_g,\lambda) = \left(n+\frac{1}{2}\right)\hbar\omega_0 + \frac{[\hbar k_z(E)]^2}{2m_c} - \frac{E_0}{B}\hbar k_y\{T_2(E,\eta_g,\lambda)\}' - \left\{\frac{m_c E_0^2[\{T_2(E,\eta_g,\lambda)\}']^2}{2B^2}\right\} \tag{1.109}$$

The electron concentration in this case can be expressed as

$$n_0 = \frac{2g_v B\sqrt{2m_c}}{3L_x\pi^2\hbar^2 E_0}\sum_{n=0}^{n_{max}}\left[M_{162}(E_{F_{BLC}}, n, E_0, B, \lambda)\right], \tag{1.110}$$

where

$$M_{162}(n, E_{F_{BLC}}, \lambda) = \left[\left[T_2(E_{F_{BLC}}, n, E_0, B, \lambda) - \left(n+\frac{1}{2}\right)\hbar\omega_0 - \frac{m_c E_0^2}{2B^2}\right.\right.$$
$$\left.[\{T_2(E_{F_{BLC}}, n, E_0, B, \lambda)\}']^2 + |e|E_0 L_x[\{T_2(E_{F_{BLC}}, n, E_0, B, \lambda)\}']\right]^{3/2}$$
$$\left.-\left[T_2(E_{F_{BLC}}, n, E_0, B, \lambda) - \left(n+\frac{1}{2}\right)\hbar\omega_0 - \frac{m_c E_0^2}{2B^2}[\{T_2(E_{F_{BLC}}, n, E_0, B, \lambda)\}']^2\right]^{3/2}\right]$$
$$\frac{1}{[\{T_2(E_{F_{BLC}}, n, E_0, B, \lambda)\}']} \tag{1.111}$$

Using eqs. (1.111), (1.104b), and (1.104c) with $x = 0$ we can study ΔC_{44} and ΔC_{456} in this case.

The electron dispersion law in the present case in the absence of band tails is given by

$$\tau_0(E,\lambda) = (n+12)\hbar\omega_0 + \frac{[\hbar k_z(E)]^2}{2m_c} - \frac{E_0}{B}\hbar k_y\{\tau_0(E,\lambda)\}' - \left\{\frac{m_c E_0^2[\{\tau_0(E,\lambda)\}']^2}{2B^2}\right\} \tag{1.112}$$

The electron concentration in this case can be expressed as

$$n_0 = \frac{2g_v B\sqrt{2m_c}}{3L_x\pi^2\hbar^2 E_0}\sum_{n=0}^{n_{max}}[M_{1614}(E_{F_{BLC}}, n, E_0, B, \lambda)],$$

$$M_{1614}(E_{F_{BLC}}, n, E_0, B, \lambda) = \left[\tau_0(E_{F_{BLC}}, n, E_0, B, \lambda) - \left(n+\frac{1}{2}\right)\hbar\omega_0\right.$$

$$\left.-\frac{m_c E_0^2}{2B^2}\{\tau_0(E_{F_{BLC}}, n, E_0, B, \lambda)\}'\right]^2 + |e|E_0 L_x[\{\tau_0(E_{F_{BLC}}, n, E_0, B, \lambda)\}']^{3/2}$$

$$-\left[\tau_0(E_{F_{BLC}}, n, E_0, B, \lambda) - \left(n+\frac{1}{2}\right)\hbar\omega_0 - \frac{m_c E_0^2}{2B^2}[\{\tau_0(E_{F_{BLC}}, n, E_0, B, \lambda)\}']^2\right]^{3/2}$$

$$\frac{1}{[\{\tau_0(E_{F_{BLC}}, n, E_0, B, \lambda)\}']} \tag{1.113}$$

Using eqs. (1.106b), (1.106c), and (1.113) leads to the expressions of ΔC_{44} and ΔC_{456} as

$$\Delta C_{44} = \frac{-2G_0^2 g_v B\sqrt{2m_c}}{27L_x\pi^2\hbar^2 E_0}\sum_{n=0}^{n_{max}}\left[\left[M_{1614}(E_{F_{BLC}}, n, E_0, B, \lambda)\right]'\right] \tag{1.114}$$

$$\Delta C_{456} = \frac{2G_0^3 g_v B\sqrt{2m_c}}{81L_x\pi^2\hbar^2 E_0}\sum_{n=0}^{n_{max}}\left[\left[M_{1614}(E_{F_{BLC}}, n, E_0, B, \lambda)\right]''\right] \tag{1.115}$$

(iii) The electron dispersion law in the present case is given by

$$T_3(E, \eta_g, \lambda) = \left(n+\frac{1}{2}\right)\hbar\omega_0 + \frac{[\hbar k_z(E)]^2}{2m_c} - \frac{E_0}{B}\hbar k_y\{T_3(E, \eta_g, \lambda)\}'$$

$$-\left\{\frac{m_c E_0^2[\{T_3(E, \eta_g, \lambda)\}']^2}{2B^2}\right\} \tag{1.116}$$

The electron concentration in this case can be expressed as

$$n_0 = \frac{2g_v B\sqrt{2m_c}}{3L_x\pi^2\hbar^2 E_0}\sum_{n=0}^{n_{max}}\left[M_{163}(E_{F_{BLHDC}}, n, E_0, B, \lambda)\right], \tag{1.117}$$

where

$$M_{163}(n, E_{F_{BL}}, \lambda) \equiv \left[\left[T_3(E_{F_{BLHDC}}, n, E_0, B, \lambda) - \left(n+\frac{1}{2}\right)\hbar\omega_0 - \frac{m_c E_0^2}{2B^2}\right.\right.$$

$$\left.[\{T_3(E_{F_{BLHDC}}, n, E_0, B, \lambda)\}']^2 + |e|E_0 L_x[\{T_3(E_{F_{BLHDC}}, n, E_0, B, \lambda)\}']\right]^{3/2}$$

$$\left.-\left[T_3(E_{F_{BLHDC}}, n, E_0, B, \lambda) - \left(\frac{n+1}{2}\right)\hbar\omega_0 - \frac{m_c E_0^2}{2B^2}[\{T_3(E_{F_{BLHDC}}, n, E_0, B, \lambda)\}']^2\right]^{3/2}\right]$$

$$\frac{1}{[\{T_3(E_{F_{BLHDC}}, n, E_0, B, \lambda)\}']} \tag{1.118}$$

Using eqs. (1.117), (1.104b), and (1.104c) with $x = 0$ we can study ΔC_{44} and ΔC_{456} in this case.

The electron dispersion law in the present case in the absence of band tails is given by

$$\rho_0(E,\lambda) = \left(n + \frac{1}{2}\right)\hbar\omega_0 + \frac{[\hbar k_z(E)]^2}{2m_c} - \frac{E_0}{B}\hbar k_y\{\rho_0(E,\lambda)\}' - \left\{\frac{m_c E_0^2[\{\rho_0(E,\lambda)\}']^2}{2B^2}\right\} \tag{1.119}$$

The electron concentration in this case can be expressed as

$$n_0 = \frac{2g_v B\sqrt{2m_c}}{3L_x\pi^2\hbar^2E_0}\sum_{n=0}^{n_{max}}\left[M_{1615}(E_{F_{BLC}}, n, E_0, B, \lambda)\right], \tag{1.120}$$

where $M_{1615}(n, E_{F_{BL}}, \lambda) \equiv \Big[[\rho_0(E_{F_{BLC}}, n, E_0, B, \lambda) - (n + \frac{1}{2})\hbar\omega_0 - \frac{m_c E_0^2}{2B^2}$

$$[\{\rho_0(E_{F_{BLC}}, n, E_0, B, \lambda)\}']^2 + |e|E_0L_x[\{\rho_0(E_{F_{BLC}}, n, E_0, B, \lambda)\}']]^{3/2}$$

$$-\left[\rho_0(E_{F_{BLC}}, n, E_0, B, \lambda) - \left(n + \frac{1}{2}\right)\hbar\omega_0 - \frac{m_c E_0^2}{2B^2}[\{\rho_0(E_{F_{BLC}}, n, E_0, B, \lambda)\}']^2\right]^{3/2}\Bigg]$$

$$\frac{1}{[\{\rho_0(E_{F_{BLC}}, n, E_0, B, \lambda)\}']}$$

Using eqs. (1.106b), (1.106c), and (1.113) leads to the expressions of ΔC_{44} and ΔC_{456} as

$$\Delta C_{44} = \frac{-2G_0^2 g_v B\sqrt{2m_c}}{27L_x\pi^2\hbar^2E_0}\sum_{n=0}^{n_{max}}\left[\left[M_{1615}(E_{F_{BLC}}, n, E_0, B, \lambda)\right]'\right] \tag{1.121}$$

$$\Delta C_{456} = \frac{2G_0^3 g_v B\sqrt{2m_c}}{81L_x\pi^2\hbar^2E_0}\sum_{n=0}^{n_{max}}\left[\left[M_{1615}(E_{F_{BLC}}, n, E_0, B, \lambda)\right]''\right] \tag{1.122}$$

(iv) In the absence of light waves and HD the dispersion relation in III–V semiconductors whose energy band structures are defined by the three-band Kane model can be written in the presence of cross fields configuration as

$$I_{11}(E) = \left(n + \frac{1}{2}\right)\hbar\omega_0 + \frac{[\hbar k_z(E)]^2}{2m_c} - \frac{E_0}{B}\hbar k_y\{I_{11}(E)\}' - \frac{m_c E_0^2\left[\{I_{11}(E)\}'\right]^2}{2B^2} \tag{1.123}$$

The electron concentration in this case assume the forms

$$n_0 = \frac{2g_v B\sqrt{2m_c}}{3L_x\pi^2\hbar^2 E_0}\sum_{n=0}^{n_{\max}}[T_{43}(n,\bar{E}_{FB})] \tag{1.124}$$

where $\bar{E}_{FB}$ is the Fermi energy in this case.

$$T_{43}(n,E_{FB}) \equiv \left[\left[I_{11}(\bar{E}_{FB}) - \left(n+\frac{1}{2}\right)\hbar\omega_0 - \frac{m_c E_0^2}{2B^2}[\{I_{11}(\bar{E}_{FB})\}']^2 + |e|E_0L_x[\{I_{11}(\bar{E}_{FB})\}']\right]^{3/2} - \left[I_{11}(\bar{E}_{FB}) - \left(n+\frac{1}{2}\right)\hbar\omega_0 - \frac{m_c E_0^2}{2B^2}[\{I_{11}(\bar{E}_{FB})\}']^2\right]^{3/2}\right]\frac{1}{[\{I_{11}(\bar{E}_{FB})\}']} \tag{1.125a}$$

The electronic contribution to the second- and third-order elastic constants in this case can be written as [94–106]

$$\Delta C_{44} = \frac{-G_0^2}{9}\left[\frac{\partial n_0}{\partial \bar{E}_{FB}}\right] \tag{1.125b}$$

and

$$\Delta C_{456} = \frac{G_0^3}{27}\left[\frac{\partial^2 n_0}{\partial \bar{E}^2_{FB}}\right] \tag{1.125c}$$

Using eqs. (1.125a), (1.125b), and (1.125c) we get

$$\Delta C_{44} = \frac{-2g_v G_0^2 B\sqrt{2m_c}}{27L_x\pi^2\hbar^2 E_0}\sum_{n=0}^{n_{\max}}[(T_{43}(n,\bar{E}_{FB}))'] \tag{1.126}$$

$$\Delta C_{456} = \frac{2g_v G_0^3 B\sqrt{2m_c}}{81L_x\pi^2\hbar^2 E_0}\sum_{n=0}^{n_{\max}}[(T_{43}(n,\bar{E}_{FB}))] \tag{1.127}$$

(a) Under the condition $\Delta \gg E_g$, eq. (1.123) assumes the well known from [1]

$$E(1+\alpha E) = \left(n+\frac{1}{2}\right)\hbar\omega_0 - E_0B\hbar k_y(1+2\alpha E) - m_cE_0^2 2B^2(1+2\alpha E)^2 + \frac{[\hbar k_z(E)]^2}{2m_c} \tag{1.128a}$$

The expression for n_0 in this case assumes the form

$$n_0 = \frac{2g_v B\sqrt{2m_c}}{3L_x \pi^2 \hbar^2 E_0} \sum_{n=0}^{n_{\max}} [T_{45}(n, \bar{E}_{FB})] \tag{1.128b}$$

where $T_{45}(n, \bar{E}_{FB}) \equiv \left[\left[\bar{E}_{FB}(1+\alpha\bar{E}_{FB}) - (n+\frac{1}{2})\hbar\omega_0 + |e|E_0 L_x(1+2\alpha\bar{E}_{FB}) - \frac{m_c E_0^2}{2B^2}\right.\right.$ $\left.(1+2\alpha\bar{E}_{FB})^2\right]^{3/2} - \left[\bar{E}_{FB}(1+\alpha\bar{E}_{FB}) - (n+\frac{1}{2})\hbar\omega_0 - \frac{m_c E_0^2}{2B^2}(1+2\alpha\bar{E}_{FB})^2\right]^{3/2}\Big][1+2\alpha\bar{E}_{FB}]^{-1}$

Using eqs. (1.125b), (1.125c), and (1.128) we get

$$\Delta C_{44} = \frac{-2g_v G_0^2 B\sqrt{2m_c}}{27L_x \pi^2 \hbar^2 E_0} \sum_{n=0}^{n_{\max}} [(T_{45}(n, \bar{E}_{FB}))'] \tag{1.129}$$

$$\Delta C_{456} = \frac{2g_v G_0^3 B\sqrt{2m_c}}{81L_x \pi^2 \hbar^2 E_0} \sum_{n=0}^{n_{\max}} [(T_{45}(n, \bar{E}_{FB}))''] \tag{1.130}$$

(b) For parbolic energy bands $\alpha \to 0$, and we can write

$$E = \left(n+\frac{1}{2}\right)\hbar\omega_0 - \frac{[(\hbar k_z E)]^2}{2m_c} - \frac{1}{2}m_c\left(\frac{E_0}{B}\right)^2 - \frac{E_0}{B}\hbar k_y \tag{1.131}$$

The electron concentration in this case can be expressed at a finite temperature as

$$n_0 = N_c \phi g_v \left[\frac{k_B T}{|e|E_0 L_x} \sum_{n=0}^{n_{\max}} \left[F_{\frac{1}{2}}(\eta_1) - F_{\frac{1}{2}}(\eta_2)\right]\right]. \tag{1.132}$$

where $\phi \equiv \frac{\hbar\omega_0}{k_B T}, \eta_1 \equiv \frac{\bar{E}_{FB} - \bar{\phi}_1}{k_B T}, \bar{\phi}_1 \equiv \left[\left(n+\frac{1}{2}\right)\hbar\omega_0 + \frac{1}{2}m^*\left(\frac{E_0}{B}\right)^2 - |e|E_0 L_x\right], \eta_2 \equiv \frac{(\bar{E}_{FB} - \bar{\phi}_2)}{k_B T}$ and $\left.\bar{\phi}_2 \equiv \bar{\phi}_1 + |e|E_0 L_x\right]$

Equation (1.119) is well known in the literature [74].
Using eqs. (1.125b), (1.125c), and (1.132) we get

$$\Delta C_{44} = -\frac{G_0^2}{9}\frac{N_c \phi g_v}{|e|E_0 L_x} \left[\sum_{n=0}^{n_{\max}} \left[F_{\frac{-1}{2}}(\eta_1) - F_{\frac{-1}{2}}(\eta_2)\right]\right] \tag{1.133}$$

$$\Delta C_{456} = \frac{G_0^3 N_c \phi g_v}{81 k_B T |e|E_0 L_x k_B T} \left[\sum_{n=0}^{n_{\max}} \left[F_{\frac{-3}{2}}(\eta_1) - F_{\frac{-3}{2}}(\eta_2)\right]\right] \tag{1.134}$$

1.2.4 The CECs in QWs of HD Kane-type semiconductors in the presence of light waves

(i) The 2D DR in QWs of HD III–V, ternary, and quaternary materials, whose unperturbed band structure is defined by the three-band Kane model, in the presence of light waves, can be expressed as

$$\frac{\hbar^2 k_s^2}{2m_c} + \frac{\hbar^2}{2m_c}\left(\frac{n_z \pi}{d_z}\right)^2 = T_1(E, \eta_g, \lambda) \tag{1.135}$$

Using eq. (1.50d) and summing over n_z together with the substitution of $C_{2D} = 1$, using HUP leads to the expression of ES as

$$n_0 = \left(\frac{m_c g_v}{\pi \hbar^2}\right) Real\,part\,of \sum_{n_z=1}^{n_{z\max}} \left[T_1(E_{F2DLHD}, n_z, \lambda) - \frac{\hbar^2}{2m_c}\left(\frac{\pi n_z}{d_z}\right)^2\right] \tag{1.136}$$

where E_{F2DLHD} is the Fermi energy in the present case as measured from the edge of the CB in the vertically upward direction in absence of any quantization.

The electronic contribution to the second- and third-order elastic constants for HD materials can be written as [94–106]

$$\Delta C_{44} = \frac{-G_0^2 Z_x}{9}\left[\frac{\partial n_0}{\partial(E_{FHD} - E_{0HD})}\right] \tag{1.137}$$

and

$$\Delta C_{456} = \frac{G_0^3 Z_x}{27 d_z}\left[\frac{\partial^2 n_0}{\partial(E_{F2DLHD} - E_{0H2D})^2}\right] \tag{1.138}$$

where E_{0H2D} is the subband energy in this case.

Using eqs. (1.136), (1.137), and (1.138) together with $x = 1$ we can study ΔC_{44} and ΔC_{456} in this case.

The 2D DR in QWs of III–V, ternary, and quaternary materials in the absence of band tails, whose unperturbed band structure is defined by the three-band Kane model, in the presence of light waves, can be expressed as

$$\frac{\hbar^2 k_s^2}{2m_c} + \frac{\hbar^2}{2m_c}\left(\frac{n_z \pi}{d_z}\right) = \beta_0(E, \lambda) \tag{1.139}$$

The ES can be written as

$$n_0 = \left(\frac{m_c g_v}{\pi \hbar^2}\right) \sum_{n_z=1}^{n_{z\max}} \left[\beta_0(E_{F2DL}, n_z, \lambda) - \frac{\hbar^2}{2m_c}\left(\frac{\pi n_z}{d_z}\right)^2\right] \tag{1.140}$$

where E_{F2DL} is the Fermi energy in the present case as measured from the edge of the CB in the vertically upward direction in the absence of any quantization.

ΔC_{44} and ΔC_{456} in this case can be written as

$$\Delta C_{44} = \frac{-G_0^2}{9d_z}\left[\frac{\partial n_0}{\partial E_{F2DL}}\right] \tag{1.141}$$

and

$$\Delta C_{456} = \frac{G_0^3}{27d_z}\left[\frac{\partial^2 n_0}{\partial E_{F2DL}{}^2}\right] \tag{1.142}$$

Using eqs. (1.140), (1.141), and (1.142) we can write

$$\Delta C_{44} = \left(\frac{-G_0^2 m_c g_v}{9\pi \hbar^2 d_z}\right) \sum_{n_z=1}^{n_{z\max}} \left[\beta_0(E_{F2DL}, n_z, \lambda) - \frac{\hbar^2}{2m_c}\left(\frac{\pi n_z}{d_z}\right)^2\right]' \tag{1.143}$$

$$\Delta C_{456} = \left(\frac{G_0^3 m_c g_v}{27\pi \hbar^2 d_z}\right) \sum_{n_z=1}^{n_{z\max}} \left[\beta_0(E_{F2DL}, n_z, \lambda) - \frac{\hbar^2}{2m_c}\left(\frac{\pi n_z}{d_z}\right)^2\right]'' \tag{1.144}$$

In the absence of band tails and light waves and for isotropic three-band Kane model, the 2D electron dispersion relation in this case can be written as

$$\frac{\hbar^2 k_s^2}{2m_c} + \frac{\hbar^2}{2m_c}(n_z \pi / d_z)^2 = I_{11}(E) \tag{1.145}$$

The carrier concentration assumes the form

$$n_0 = \frac{m_c g_v}{\pi \hbar^2} \sum_{n_z=1}^{n_{z\max}} [T_{53}(E_{Fs}, n_z)] \tag{1.146}$$

where E_{Fs} is the Fermi energy in this case and

$$T_{53}(E_{Fs}, n_z) \equiv \left[I_{11}(E_{Fs}) - \frac{\hbar^2}{2m_c}\left(\frac{n_z \pi}{d_z}\right)^2\right]$$

ΔC_{44} and ΔC_{456} in this case can be written as

$$\Delta C_{44} = \frac{-G_0^2}{9d_z}\left[\frac{\partial n_0}{\partial E_{Fs}}\right] \tag{1.147}$$

and

$$\Delta C_{456} = \frac{G_0^3}{27d_z}\left[\frac{\partial^2 n_0}{\partial E_{Fs}{}^2}\right] \tag{1.148}$$

Using eqs. (1.146), (1.147), and (1.148) we can write

$$\Delta C_{44} = \frac{-G_0^2 m_c g_v}{9\pi\hbar^2 d_z}\sum_{n_z=1}^{n_{z\max}}[T_{53}(E_{Fs}, n_z)]' \tag{1.149}$$

$$\Delta C_{456} = \frac{G_0^3 m_c g_v}{27\pi\hbar^2 d_z}\sum_{n_z=1}^{n_{z\max}}[T_{53}(E_{Fs}, n_z)]'' \tag{1.150}$$

(ii) The 2D DR in QWs of HD III–V, ternary, and quaternary materials, whose unperturbed band structure is defined by the two-band Kane model, in the presence of light waves, can be expressed as

$$\frac{\hbar^2 k_s^2}{2m_c} + \frac{\hbar^2}{2m_c}\left(\frac{n_z\pi}{d_z}\right)^2 = T_2(E, \eta_g, \lambda) \tag{1.151}$$

The ES can be written as

$$n_0 = \left(\frac{m_c g_v}{\pi\hbar^2}\right)\sum_{n_z=1}^{n_{z\max}}\left[T_2(E_{F2DLHD}, n_z, \lambda) - \frac{\hbar^2}{2m_c}\left(\frac{\pi n_z}{d_z}\right)^2\right] \tag{1.152}$$

where E_{F2DLHD} is the Fermi energy in the present case as measured from the edge of the CB in the vertically upward direction in the absence of any quantization.

Using eqs. (1.137), (1.138), and (1.152) we can study ΔC_{44} and ΔC_{456} in this case.

The 2D DR in QWs of III–V, ternary, and quaternary materials in the absence of band tails, whose unperturbed band structure is defined by the two-band Kane model, in the presence of light waves, can be expressed as

$$\frac{\hbar^2 k_s^2}{2m_c} + \frac{\hbar^2}{2m_c}\left(\frac{n_z\pi}{d_z}\right)^2 = \tau_0(E, \lambda) \tag{1.153}$$

The ES can be written as

$$n_0 = \left(\frac{m_c g_v}{\pi\hbar^2}\right)\sum_{n_z=1}^{n_{z\max}}\left[\tau_0(E_{F2DL}, n_z, \lambda) - \frac{\hbar^2}{2m_c}\left(\frac{\pi n_z}{d_z}\right)^2\right] \tag{1.154}$$

Using eqs. (1.141), (1.142), and (1.154) leads to the expressions of ΔC_{44} and ΔC_{456} as

$$\Delta C_{44} = \left(\frac{-G_0^2 m_c g_v}{9\pi\hbar^2 d_z}\right)\sum_{n_z=1}^{n_{z\max}}\left[\tau_0(E_{F2DL}, n_z, \lambda) - \frac{\hbar^2}{2m_c}\left(\frac{\pi n_z}{d_z}\right)^2\right]' \tag{1.155}$$

$$\Delta C_{456} = \left(\frac{G_0^3 m_c g_v}{27\pi\hbar^2 d_z}\right)\sum_{n_z=1}^{n_{z\max}}\left[\tau_0(E_{F2DL}, n_z, \lambda) - \hbar^2 2m_c\left(\frac{\pi n_z}{d_z}\right)^2\right]'' \tag{1.156}$$

In the absence of light waves and HD, the 2D electron dispersion relation for isotropic two-band Kane model can be written as

$$E(1+\alpha E) = \frac{\hbar^2 k_s^2}{2m_c} + \frac{\hbar^2}{2m_c}\left(\frac{n_z\pi}{d_z}\right)^2 \tag{1.157}$$

The ES can be written as

$$n_0 = \frac{m_c g_v}{\pi\hbar^2}\sum_{n_z=1}^{n_{z\max}}\left[E_{Fs}(1+\alpha E_{Fs}) - \frac{\hbar^2}{2m_c}\left(\frac{n_z\pi}{d_z}\right)^2\right] \tag{1.158}$$

Equation (1.158) is well known in the literature [74].

ΔC_{44} and ΔC_{456} in this case can be expressed as

$$\Delta C_{44} = \frac{-G_0^2 m_c g_v}{9\pi\hbar^2 d_z}\sum_{n_z=1}^{n_{z\max}}(1+2\alpha E_{Fs}) \tag{1.159a}$$

$$\Delta C_{456} = \frac{2\alpha G_0^3 m_c g_v}{27\pi\hbar^2 d_z}\sum_{n_z=1}^{n_{z\max}} 1 \tag{1.159b}$$

Thus, we observe that ΔC_{456} in this case is independent of ES.

The ES at a finite temperature can be written as

$$n_0 = \frac{m_c k_B T g_v}{\pi\hbar^2}\sum_{n_z=1}^{n_{z\max}}\left[(1+2\alpha E_{n_{z3}})F_0(\eta_{n_1}) + 2\alpha k_B T F_1(\eta_{n_1})\right] \tag{1.160}$$

where $\eta_{n_1} \equiv \left(E_{Fs} - E_{n_{z3}}\right)/k_B T$.

Equation (1.160) is well known in the literature [74].

Using eqs. (1.54c), (1.54d), and (1.160) we get

$$\Delta C_{44} = \frac{-m_c g_v G_0^2}{9 d_z \pi \hbar^2 k_B T} \sum_{n_z=1}^{n_{z\max}} [(1 + 2\alpha E_{n_{z_3}}) F_{-1}(\eta_{n_1}) + 2\alpha k_B T F_0(\eta_{n_1})] \tag{1.161}$$

$$\Delta C_{456} = \frac{m_c g_v G_0^3}{27 k_B T d_z \pi \hbar^2 (k_B T)^2} \sum_{n_z=1}^{n_{z\max}} [(1 + 2\alpha E_{n_{z_3}}) F_{-2}(\eta_{n_1}) + 2\alpha k_B T F_{-1}(\eta_{n_1})] \tag{1.162}$$

(iii) The 2D DR in QWs of HD III–V, ternary, and quaternary materials, whose unperturbed band structure is defined by the parabolic energy bands in the presence of light waves, can be expressed as

$$\frac{\hbar^2 k_s^2}{2m_c} + \frac{\hbar^2}{2m_c}\left(\frac{n_z \pi}{d_z}\right)^2 = T_3(E, \eta_g, \lambda) \tag{1.163}$$

The ES can be written as

$$n_0 = \left(\frac{m_c g_v}{\pi \hbar^2}\right) \sum_{n_z=1}^{n_{z\max}} \left[T_3(E_{F2DLHD}, n_z, \lambda) - \frac{\hbar^2}{2m_c}\left(\frac{\pi n_z}{d_z}\right)^2\right] \tag{1.164}$$

Using eqs. (1.137), (1.138), and (1.164) together with $x = 0$ we can study ΔC_{44} and ΔC_{456} in this case.

The 2D DR in QWs of III–V, ternary, and quaternary materials in the absence of band tails, whose unperturbed band structure is defined by the parabolic energy band in the presence of light waves, can be expressed as

$$\frac{\hbar^2 k_s^2}{2m_c} + \frac{\hbar^2}{2m_c}\left(\frac{n_z \pi}{d_z}\right)^2 = \rho_0(E, \lambda) \tag{1.165}$$

The ES can be written as

$$n_0 = \left(\frac{m_c g_v}{\pi \hbar^2}\right) \sum_{n_z=1}^{n_{z\max}} \left[\rho_0(E_{F2DL}, n_z, \lambda) - \frac{\hbar^2}{2m_c}\left(\frac{\pi n_z}{d_z}\right)^2\right] \tag{1.166}$$

Using eqs. (1.141), (1.142), and (1.166) ΔC_{44} and ΔC_{456} can be expressed as

$$\Delta C_{44} = \left(\frac{-G_0^2 m_c g_v}{9\pi \hbar^2 d_z}\right) \sum_{n_z=1}^{n_{z\max}} \left[\rho_0(E_{F2DL}, n_z, \lambda) - \frac{\hbar^2}{2m_c}\left(\frac{\pi n_z}{d_z}\right)^2\right]' \tag{1.167}$$

$$\Delta C_{456} = \left(\frac{G_0^3 m_c g_v}{27\pi \hbar^2 d_z}\right) \sum_{n_z=1}^{n_{z\max}} \left[\rho_0(E_{F2DL}, n_z, \lambda) - \frac{\hbar^2}{2m_c}\left(\frac{\pi n_z}{d_z}\right)^2\right]'' \tag{1.168}$$

In the absence of light waves and HD for isotropic parabolic energy band, the 2D electron dispersion relation can be written as

$$E = \frac{\hbar^2 k_s^2}{2m_c} + \frac{\hbar^2}{2m_c}\left(\frac{n_z \pi}{d_z}\right)^2 \tag{1.169}$$

The ES can be written as

$$n_0 = \frac{m_c g_v}{\pi \hbar^2} \sum_{n_z = 1}^{n_{z\max}} [(E_{Fs} - E_{n_{z33}})] \tag{1.170}$$

where E_{Fs} is the Fermi energy in this case

and $E_{n_{z33}} = \frac{\hbar^2}{2m_c}\left(\frac{n_z \pi}{d_z}\right)^2$

Equation (1.170) is well known in the literature [74].

ΔC_{44} and ΔC_{456} in this case can be expressed as

$$\Delta C_{44} = \frac{-G_0^2 m_c g_v}{9\pi \hbar^2 d_z} \sum_{n_z = 1}^{n_{z\max}} 1 \tag{1.171}$$

and

$$\Delta C_{456} = 0 \tag{1.172a}$$

Thus, we observe that ΔC_{44} for isotropic parabolic energy bands under the condition of extreme carrier degeneracy is independent of electron concentration and the corresponding ΔC_{456} vanishes.

The ES at a finite temperature can be written as

$$n_0 = \frac{m_c k_B T g_v}{\pi \hbar^2} \sum_{n_z = 1}^{n_{z\max}} [F_0(\eta_{n_{11}})] \tag{1.172b}$$

where $\eta_{n_{11}} \equiv \left(E_{Fs} - E_{n_{z33}}\right)/k_B T$.

Equation (1.172a) is well known in the literature [74].

Using eqs. (1.147), (1.148), and (1.172b) we get

$$\Delta C_{44} = \frac{-m_c g_v G_0^2}{9 d_z \pi \hbar^2 k_B T} \sum_{n_z = 1}^{n_{z\max}} [F_{-1}(\eta_{n_{11}})] \tag{1.172c}$$

$$\Delta C_{456} = \frac{m_c g_v G_0^3}{27 k_B T d_z \pi \hbar^2 (k_B T)^2} \sum_{n_z=1}^{n_{z\max}} [F_{-2}(\eta_{n_{11}})] \quad (1.172d)$$

1.2.5 The CECs in doping superlattices of HD Kane-type semiconductors in the presence of light waves

(i) The DR in doping superlattices of HD III–V, ternary, and quaternary materials in the presence of external photoexcitation whose unperturbed electrons are defined by the three-band Kane model can be expressed as

$$T_1(E, \eta_g, \lambda) = \left(n_i + \frac{1}{2}\right) \hbar \omega_{91HD}(E, \eta_g, \lambda) + \frac{\hbar^2 k_s^2}{2m_c} \quad (1.173)$$

where $\omega_{91HD}(E, \eta_g, \lambda) \equiv \left(\frac{n_0 |e|^2}{\varepsilon_{sc} T'_1(E, \eta_g, \lambda) m_c d_0}\right)^{1/2}$

The electron concentration can be written as

$$n_0 = \left(\frac{m_c g_v}{\pi \hbar^2}\right) \text{Real Part of} \sum_{n_i=0}^{n_{i\max}} \left[\left[M_{40HD}(E_{F2DLHDD}, \eta_g, \lambda)\right]\right] \quad (1.174a)$$

and $E_{F2DLHDD}$ is the Fermi energy in the present case as measured from the edge of the CB in the vertically upward direction in the absence of any quantization.

$$\Delta C_{44} = \frac{-G_0^2 Z_x}{9 d_0} \left[\frac{\partial n_0}{\partial(E_{F2DLHDD} - E_{0HD4})}\right] \quad (1.174b)$$

and

$$\Delta C_{456} = \frac{G_0^3 Z_x}{27 d_0} \left[\frac{\partial^2 n_0}{\partial(E_{F2DLHDD} - E_{0HD4})^2}\right] \quad (1.174c)$$

where G_0 is the deformation potential constant, $Z_x = (\text{Real Part})^x$, $E_{F2DLHDD}$ is the Fermi energy in HD materials in the presence of band tails, and E_{0HD4} is obtained from the corresponding HD dispersion relation under the conditions $E = E_{0HD4}$ when $k = 0$.

Using eqs. (1.174a)–(1.174c) together with $x = 1$ we can find ΔC_{44} and ΔC_{456} in this case.

In the absence of band tails DR in doping superlattices of III–V, ternary, and quaternary materials in the presence of external photoexcitation whose unperturbed electrons are defined by the two-band Kane model can be expressed as

$$\beta_0(E,\lambda) = \left(n_i + \tfrac{1}{2}\right)\hbar\omega_{911}(E,\lambda) + \frac{\hbar^2 k_s^2}{2m_c} \tag{1.175}$$

where $\omega_{911}(E,\lambda) \equiv \left(\frac{n_0|e|^2}{\varepsilon_{sc}\beta'_0(E,\lambda)m_c d_0}\right)^{1/2}$

The electron concentration can be written as

$$n_0 = \left(\frac{m_c g_v}{\pi\hbar^2}\right)\sum_{n_i=0}^{n_{i\max}}[[M_{401}(E_{F2DLD},\lambda)]] \tag{1.176}$$

where $M_{401}(E_{F2DLD},\lambda,n_i) = \{\beta_0(E_{F2DLD},\lambda) - (n_i + 12)\hbar\omega_{911}(E_{F2DLD},\lambda)\}$

and E_{F2DLD} is the Fermi energy in the present case as measured from the edge of the CB in the vertically upward direction in the absence of any quantization.

The electronic contribution to the second- and third-order elastic constants for HD materials can be written as [55–58]

$$\Delta C_{44} = \frac{-G_0^2 Z_x}{9d_0}\left[\frac{\partial n_0}{\partial(E_{F2DLD} - E_{n_{l101}})}\right] \tag{1.177}$$

and

$$\Delta C_{456} = \frac{G_0^3 Z_x}{27d_0}\left[\frac{\partial^2 n_0}{\partial\left(E_{F2DLD} - E_{n_{l101}}\right)^2}\right] \tag{1.178}$$

where $E_{n_{l101}}$ can be obtained by substituting $k_s = 0$ in the corresponding dispersion relation.

Using eqs. (1.177), (1.178), and (1.176) together with $x = 1$ we can find ΔC_{44} and ΔC_{456} in this case.

The electron energy spectrum in nipi structures of III–V, ternary, and quaternary materials can be expressed in the absence of both band tails and light waves as

$$I_{11}(E) = \left(n_i + \frac{1}{2}\right)\hbar\omega_9(E) + \frac{\hbar^2 k_s^2}{2m_c} \tag{1.179}$$

where $\omega_9(E,\lambda) \equiv \left(\frac{n_0|e|^2}{\varepsilon_{sc}I'_{11}(E)m_c d_0}\right)^{1/2}$

Using eq. (1.179) leads to the expression of the electron concentration as

$$n_0 = \frac{m_c g_v}{\pi\hbar^2}\sum_{n_i=0}^{n_{i\max}}\left[T_{83}(\bar{E}_{Fn_i}, n_i)\right] \tag{1.180}$$

where $\bar{E}_{Fn_i}$ is the Fermi energy in this case.

$$T_{83}(\bar{E}_{Fn_i}, n_i) \equiv \left[I_{11}(\bar{E}_{Fn_i}) - \left(n_i + \frac{1}{2}\right)\hbar\omega_9(\bar{E}_{Fn_i}) \right]$$

The electronic contribution to the second- and third-order elastic constants for HD materials can be written as

$$\Delta C_{44} = \frac{-G_0^2}{9d_0} \text{Real Part of} \left[\frac{\partial n_0}{\partial(\bar{E}_{Fn_i} - E_{2ni})} \right] \tag{1.181}$$

and

$$\Delta C_{456} = \frac{G_0^3}{27d_0} \text{Real Part of} \left[\frac{\partial^2 n_0}{\partial(\bar{E}_{Fn_i} - E_{2ni})^2} \right] \tag{1.182}$$

where E_{2ni} can be obtained by substituting $k_s = 0$ in the corresponding dispersion relation.

Using eqs. (1.182), (1.181), and (1.180) we can find ΔC_{44} and ΔC_{456} in this case.

(ii) The DR in doping superlattices of HD III–V, ternary, and quaternary materials in the presence of external photoexcitation whose unperturbed electrons are defined by the two-band Kane model can be expressed as

$$T_2(E, \eta_g, \lambda) = \left(n_i + \frac{1}{2}\right)\hbar\omega_{92HD}(E, \eta_g, \lambda) + \frac{\hbar^2 k_s^2}{2m_c} \tag{1.183}$$

where $\omega_{92HD}(E, \eta_g, \lambda) \equiv \left(\frac{n_0 |e|^2}{\varepsilon_{sc} T'_2(E, \eta_g, \lambda) m_c d_0} \right)^{1/2}$

The electron concentration can be written as

$$n_0 = \left(\frac{m_c g_v}{\pi\hbar^2}\right) \sum_{n_i=0}^{n_{i_{\max}}} \left[\left[M_{41HD}(E_{F2DLHDD}, \eta_g, \lambda)\right]\right] \tag{1.184}$$

where

$$M_{41HD}(E_{F2DLHDD}, \eta_g, \lambda, n_i) = \left\{ T_2(E_{F2DLHDD}, \eta_g, \lambda) - \left(n_i + \frac{1}{2}\right)\hbar\omega_{92HD}(E_{F2DLHDD}, \eta_g, \lambda) \right\}$$

and $E_{F2DLHDD}$ is the Fermi in this case.

Using eqs. (1.174b), (1.174c), and (1.184) together with $x = 0$ we can find ΔC_{44} and ΔC_{456} in this case.

In the absence of band tails, DR in doping superlattices of III–V, ternary, and quaternary materials in the presence of external photoexcitation whose unperturbed electrons are defined by the two-band Kane model can be expressed as

$$\tau_0(E,\lambda) = \left(n_i + \frac{1}{2}\right)\hbar\omega_{912}(E,\lambda) + \frac{\hbar^2 k_s^2}{2m_c} \tag{1.185}$$

$$\omega_{912}(E,\lambda) = \left(\frac{n_0|e|^2}{\varepsilon_{sc}\tau_0'(E,\lambda)m_c d_0}\right)^{1/2} \tag{1.186}$$

The electron concentration can be written as

$$n_0 = \left(\frac{m_c g_v}{\pi\hbar^2}\right)\sum_{n_i=0}^{n_{i\max}}[[M_{402}(E_{F2DLD},\lambda)]] \tag{1.187}$$

where,

$$M_{402}(E_{F2DLD},\lambda,n_i) = \left\{\tau_0(E_{F2DLD},\lambda) - \left(n_i + \frac{1}{2}\right)\hbar\omega_{912}(E_{F2DLD},\lambda)\right\}$$

and E_{F2DLD} is the Fermi energy in this case.

Using eqs. (1.177), (1.178), and (1.187) together with $x = 0$ we can find ΔC_{44} and ΔC_{456} in this case.

The electron energy spectrum in nipi structures of III–V, ternary, and quaternary materials can be expressed in the absence of both band tails and light waves and whose unperturbed dispersion relation is given by the two-band Kane model as

$$E(1+\alpha E) = \left(n_i + \frac{1}{2}\right)\hbar\omega_{10}(E) + \frac{\hbar^2 k_s^2}{2m_c} \tag{1.188}$$

where,

$$\omega_{10}(E) = \left(\frac{n_0|e|^2}{\varepsilon_{sc}(1+2\alpha E)m_c d_0}\right)^{1/2} \tag{1.189}$$

From eq. (1.188), we observe that the effective electron mass (EEM) in this case is a function of the Fermi energy, nipi subband index, and the other material constants, which is the characteristic feature of nipi structures of III–V, ternary, and quaternary compounds whose bulk dispersion relations are defined by the three-band Kane model.

Using eq. (1.188) leads to the expression of the electron concentration as

$$n_0 = \frac{m_c g_v}{\pi\hbar^2} \sum_{n_i=0}^{n_{i\max}} [T_{832}(\bar{E}_{Fn_i}, n_i)] \tag{1.190}$$

where $T_{832}(\bar{E}_{Fn_i}, n_i) \equiv \left[\bar{E}_{Fn_i}(1+\alpha\bar{E}_{Fn_i}) - (n_i + \frac{1}{2})\hbar\omega_{10}(\bar{E}_{Fn_i})\right]$

and $\bar{E}_{Fn_i}$ is the Fermi energy in this case.

Using eqs. (1.177), (1.178), and (1.190) together with $x = 0$ we can find ΔC_{44} and ΔC_{456} in this case.

(iii) The DR in doping superlattices of HD III–V, ternary, and quaternary materials in the presence of external photoexcitation whose unperturbed electrons are defined by the parabolic energy bands can be expressed as

$$T_3(E, \eta_g, \lambda) = \left(n_i + \frac{1}{2}\right)\omega_{93HD}(E, \eta_g, \lambda) + \frac{\hbar^2 k_s^2}{2m_c} \tag{1.191}$$

where $\omega_{93HD}(E, \eta_g, \lambda) \equiv \left(\frac{n_0|e|^2}{\varepsilon_{sc}T'_3(E, \eta_g, \lambda)m_c d_0}\right)^{1/2}$

The electron concentration can be written as

$$n_0 = \left(\frac{m_c g_v}{\pi\hbar^2}\right) \sum_{n_i=0}^{n_{i\max}} \left[\left[M_{42HD}(E_{F2DLHDD}, \eta_g, \lambda)\right]\right] \tag{1.192}$$

where

$$M_{42HD}(E_{F2DLHDD}, \eta_g, \lambda, n_i) = \left\{T_3(E_{F2DLHDD}, \eta_g, \lambda) - \left(n_i + \frac{1}{2}\right)\hbar\omega_{93HD}(E_{F2DLHDD}, \eta_g, \lambda)\right\}$$

and $E_{F2DLHDD}$ is the Fermi energy in this case.

Using eqs. (1.177), (1.178), and (1.192) together with $x = 0$ we can find ΔC_{44} and ΔC_{456} in this case.

In the absence of band tails DR in doping superlattices of III–V, ternary, and quaternary materials in the presence of external photoexcitation whose unperturbed electrons are defined by the parabolic energy band can be expressed as

$$\rho_0(E, \lambda) = \left(n_i + \frac{1}{2}\right)\hbar\omega_{913}(E, \lambda) + \frac{\hbar^2 k_s^2}{2m_c} \tag{1.193}$$

where $\omega_{913}(E, \lambda) \equiv \left(\frac{n_0|e|^2}{\varepsilon_{sc}\rho_0{}'(E, \lambda)m_c d_0}\right)^{1/2}$

The electron concentration can be written as

$$n_0 = \left(\frac{m_c g_v}{\pi \hbar^2}\right) \sum_{n_i=0}^{n_{i\max}} [[M_{403}(E_{F2DLD}, \lambda)]] \tag{1.194}$$

where

$$M_{403}(E_{F2DLD}, \lambda, n_i) = \left\{\rho_0(E_{F2DLD}, \lambda) - \left(n_i + \frac{1}{2}\right)\hbar\omega_{913}(E_{F2DLD}, \lambda)\right\}$$

and E_{F2DLD} is the Fermi energy in this case.

Using eqs. (1.177), (1.178), and (1.194) together with $x = 0$ we can find ΔC_{44} and ΔC_{456} in this case.

The electron energy spectrum in nipi structures of III–V, ternary, and quaternary materials whose energy band structures are defined by parabolic energy bands can be expressed in the absence of both band tails and light waves as

$$E = \left(n_i + \frac{1}{2}\right)\hbar\omega_{11} + \frac{\hbar^2 k_s^2}{2m_c} \tag{1.195}$$

where $\omega_{11} \equiv \left(\frac{n_0 |e|^2}{\varepsilon_{sc} m_c d_0}\right)^{1/2}$

The use of eq. (1.195) leads to the expression of the electron concentration at a finite temperature as

$$n_0 = \frac{m_c g_v k_B T}{\pi \hbar^2} \sum_{n_i=0}^{n_i \max} F_0(\eta_{4n_i}) \text{ where } \eta_{4n_i} = \frac{\bar{E}_{Fn_i} - E_{4ni}}{k_B T} \tag{1.196}$$

The electronic contribution to the second- and third-order elastic constants in this particular case can be expressed as

$$\Delta C_{44} = \frac{-G_0^2}{9d_0}\left[\frac{\partial n_0}{\partial(\bar{E}_{Fn_i} - E_{4ni})}\right] \tag{1.197}$$

and

$$\Delta C_{456} = \frac{G_0^3}{27d_0}\left[\frac{\partial^2 n_0}{\partial(\bar{E}_{Fn_i} - E_{4ni})^2}\right] \tag{1.198}$$

Using eq. (1.197), (1.198), and (1.196) we can find ΔC_{44} and ΔC_{456} in this case.

1.2.6 The CEC of QDs of HD Kane-type semiconductors in the presence of light waves

(i) The DR in QDs of HD III–V, ternary, and quaternary materials in the presence of external photoexcitation whose unperturbed electrons are defined by the three-band Kane model can be expressed as

$$\frac{\hbar^2(n_z\pi/d_z)^2}{2m_c}+\frac{\hbar^2(n_y\pi/d_y)^2}{2m_c}+\frac{\hbar^2(n_x\pi/d_x)^2}{2m_c}=T_1(E_{17,1},\eta_g,\lambda) \tag{1.199}$$

where $E_{17,1}$ is the totally quantized energy in this case.

The electron concentration can be written at a finite temperature as

$$n_0=\frac{2g_v}{d_xd_yd_z}\text{Real Part of}\sum_{n_x=1}^{n_{x\max}}\sum_{n_y=1}^{n_{y\max}}\sum_{n_z=1}^{n_{z\max}}F_{-1}(\eta_{126}) \tag{1.200}$$

where $\eta_{126}=\frac{E_{F126}-E_{17.1}}{k_BT}$ and E_{F126} is the Fermi energy in this case.

The electronic contribution to the second- and third-order elastic constants for HD materials in this case can be written as [55–58]

$$\Delta C_{44}=\frac{-G_0^2}{9}\text{Real part of}\left[\frac{\partial n_0}{\partial(E_{F126}-E_{17.1})}\right] \tag{1.201a}$$

and

$$\Delta C_{456}=\frac{G_0^3}{27}\text{Real part of}\left[\frac{\partial^2 n_0}{\partial(E_{F126}-E_{17.1})^2}\right] \tag{1.201b}$$

Using eqs. (1.201a), (1.201b), and (1.200) we can find ΔC_{44} and ΔC_{456} in this case.

In the absence of band tails, the totally quantized energy $E_{17,20}$ in this case is given by

$$\beta_0(E_{17,20},\lambda)=\frac{\hbar^2\pi^2}{2m_c}\left[\left(\frac{n_x}{d_x}\right)^2+\left(\frac{n_y}{d_y}\right)^2+\left(\frac{n_z}{d_z}\right)^2\right] \tag{1.202}$$

The electron concentration can be written at a finite temperature as

$$n_0=\frac{2g_v}{d_xd_yd_z}\sum_{n_x=1}^{n_{x\max}}\sum_{n_y=1}^{n_{y\max}}\sum_{n_z=1}^{n_{z\max}}F_{-1}(\eta_{1261}) \tag{1.203}$$

where $\eta_{1261}=\frac{E_{F1261}-E_{17,20}}{k_BT}$ and E_{F1261} is the Fermi energy in this case.

The electronic contribution to the second- and third-order elastic constants for HD materials in this case can be written as

$$\Delta C_{44} = \frac{-G_0^2}{9}\left[\frac{\partial n_0}{\partial(E_{F1261})}\right] \tag{1.204a}$$

and

$$\Delta C_{456} = \frac{G_0^3}{27}\left[\frac{\partial^2 n_0}{\partial(E_{F1261})^2}\right] \tag{1.204b}$$

Using eqs. (1.204a), (1.204b), and (1.203) we can find ΔC_{44} and ΔC_{456} in this case.

In the absence of band tails and external light waves, the totally quantized energy $E_{17,201}$ in this case is given by

$$I_{11}(E_{17,201}) = \frac{\hbar^2\pi^2}{2m_c}\left[\left(\frac{n_x}{d_x}\right)^2 + \left(\frac{n_y}{d_y}\right)^2 + \left(\frac{n_z}{d_z}\right)^2\right] \tag{1.205}$$

The electron concentration can be written at a finite temperature as

$$n_0 = \frac{2g_v}{d_x d_y d_z}\sum_{n_x=1}^{n_{x\max}}\sum_{n_y=1}^{n_{y\max}}\sum_{n_z=1}^{n_{z\max}} F_{-1}(\eta_{1262}) \tag{1.206}$$

where $\eta_{1262} = \frac{E_{F1262} - E_{17,201}}{k_B T}$ and E_{F1262} is the Fermi energy in this case.

The electronic contribution to the second- and third-order elastic constants for HD materials in this case can be written as

$$\Delta C_{44} = \frac{-G_0^2}{9}\left[\frac{\partial n_0}{\partial(E_{F1262})}\right] \tag{1.207a}$$

and

$$\Delta C_{456} = \frac{G_0^3}{27}\left[\frac{\partial^2 n_0}{\partial(E_{F1262})^2}\right] \tag{1.207b}$$

Using eqs. (1.207a), (1.207b), and (1.206) we can find ΔC_{44} and ΔC_{456} in this case.

(ii) The DR in QDs of HD III–V, ternary, and quaternary materials in the presence of external photoexcitation whose unperturbed electrons are defined by the two-band Kane model can be expressed as

$$\frac{\hbar^2(n_z\pi/d_z)^2}{2m_c}+\frac{\hbar^2(n_y\pi/d_y)^2}{2m_c}+\frac{\hbar^2(n_x\pi/d_x)^2}{2m_c}=T_2(E_{17,3},\eta_g,\lambda) \tag{1.208}$$

where $E_{17,3}$ is the totally quantized energy in this case.

The electron concentration can be written at a finite temperature as

$$n_0=\frac{2g_v}{d_xd_yd_z}\sum_{n_x=1}^{n_{x\max}}\sum_{n_y=1}^{n_{y\max}}\sum_{n_z=1}^{n_{z\max}}F_{-1}(\eta_{127}) \tag{1.209}$$

where $\eta_{127}=\frac{E_{F127}-E_{17.3}}{k_BT}$ and E_{F127} is the Fermi energy in this case.

The electronic contribution to the second- and third-order elastic constants for HD materials in this case can be written as

$$\Delta C_{44}=\frac{-G_0^2}{9}\left[\frac{\partial n_0}{\partial(E_{F127}-E_{17.3})}\right] \tag{1.210a}$$

and

$$\Delta C_{456}=\frac{G_0^3}{27}\left[\frac{\partial^2 n_0}{\partial(E_{F127}-E_{17.3})^2}\right] \tag{1.210b}$$

Using eqs. (1.291a), (1.291b), and (1.289) we can find ΔC_{44} and ΔC_{456} in this case.

In the absence of band tails, the totally quantized energy $E_{17,24}$ in this case is given by

$$\tau_0(E_{17,24},\lambda)=\frac{\hbar^2\pi^2}{2m_c}\left[\left(\frac{n_x}{d_x}\right)^2+\left(\frac{n_y}{d_y}\right)^2+\left(\frac{n_z}{d_z}\right)^2\right] \tag{1.211}$$

The electron concentration can be written at a finite temperature as

$$n_0=\frac{2g_v}{d_xd_yd_z}\sum_{n_x=1}^{n_{x\max}}\sum_{n_y=1}^{n_{y\max}}\sum_{n_z=1}^{n_{z\max}}F_{-1}(\eta_{1264}) \tag{1.212}$$

where $\eta_{1264}=\frac{E_{F1264}-E_{17,24}}{k_BT}$ and E_{F1264} is the Fermi energy in this case.

The electronic contribution to the second- and third-order elastic constants for HD materials in this case can be written as

$$\Delta C_{44}=\frac{-G_0^2}{9}\left[\frac{\partial n_0}{\partial(E_{F1264})}\right] \tag{1.213a}$$

and

$$\Delta C_{456} = \frac{G_0^3}{27}\left[\frac{\partial^2 n_0}{\partial (E_{F1264})^2}\right] \tag{1.213b}$$

Using eqs. (1.213a), (1.213b), and (1.212) we can find ΔC_{44} and ΔC_{456} in this case.

In the absence of band tails and light waves, the totally quantized energy $E_{17,205}$ in this case is given by

$$E_{17,205}(1+\alpha E_{17,205}) = \frac{\hbar^2\pi^2}{2m_c}\left[\left(\frac{n_x}{d_x}\right)^2 + \left(\frac{n_y}{d_y}\right)^2 + \left(\frac{n_z}{d_z}\right)^2\right] \tag{1.214}$$

The electron concentration can be written at a finite temperature as

$$n_0 = \frac{2g_v}{d_x d_y d_z}\sum_{n_x=1}^{n_{x\max}}\sum_{n_y=1}^{n_{y\max}}\sum_{n_z=1}^{n_{z\max}} F_{-1}(\eta_{1265}) \tag{1.215}$$

where $\eta_{1265} = \frac{E_{F1265} - E_{17,205}}{k_B T}$ and E_{F1265} is the Fermi energy in this case.

The electronic contribution to the second- and third-order elastic constants for HD materials in this case can be written as

$$\Delta C_{44} = \frac{-G_0^2}{9}\left[\frac{\partial n_0}{\partial (E_{F1265})}\right] \tag{1.216a}$$

and

$$\Delta C_{456} = \frac{G_0^3}{27}\left[\frac{\partial^2 n_0}{\partial (E_{F1265})^2}\right] \tag{1.216b}$$

Using eqs. (1.216a), (1.216b), and (1.215) we can find ΔC_{44} and ΔC_{456} in this case.

(iii) The DR in QDs of HD III–V, ternary, and quaternary materials in the presence of external photo-excitation whose unperturbed electrons are defined by the parabolic energy bands can be expressed as

$$\frac{\hbar^2(n_z\pi/d_z)^2}{2m_c} + \frac{\hbar^2(n_y\pi/d_y)^2}{2m_c} + \frac{\hbar^2(n_x\pi/d_x)^2}{2m_c} = T_3(E_{17,5}, \eta_g, \lambda) \tag{1.217}$$

where $E_{17,5}$ is the totally quantized energy in this case.

The electron concentration can be written at a finite temperature as

$$n_0 = \frac{2g_v}{d_x d_y d_z} \sum_{n_x=1}^{n_{x\max}} \sum_{n_y=1}^{n_{y\max}} \sum_{n_z=1}^{n_{z\max}} F_{-1}(\eta_{1275}) \tag{1.218}$$

where $\eta_{1275} = \frac{E_{F1275} - E_{17.5}}{k_B T}$ and E_{F1275} is the Fermi energy in this case.

The electronic contribution to the second- and third-order elastic constants for HD materials in this case can be written as

$$\Delta C_{44} = \frac{-G_0^2}{9}\left[\frac{\partial n_0}{\partial(E_{F1275} - E_{17.5})}\right] \tag{1.219a}$$

and

$$\Delta C_{456} = \frac{G_0^3}{27}\left[\frac{\partial^2 n_0}{\partial(E_{F1275} - E_{17.5})^2}\right] \tag{1.219b}$$

Using eqs. (1.219a), (1.219b), and (1.218) we can find ΔC_{44} and ΔC_{456} in this case.

In the absence of band tails, the totally quantized energy $E_{17,25}$ in this case is given by

$$\rho_0(E_{17,25}, \lambda) = \frac{\hbar^2 \pi^2}{2m_c}\left[\left(\frac{n_x}{d_x}\right)^2 + \left(\frac{n_y}{d_y}\right)^2 + \left(\frac{n_z}{d_z}\right)^2\right] \tag{1.220}$$

The electron concentration can be written at a finite temperature as

$$n_0 = \frac{2g_v}{d_x d_y d_z} \sum_{n_x=1}^{n_{x\max}} \sum_{n_y=1}^{n_{y\max}} \sum_{n_z=1}^{n_{z\max}} F_{-1}(\eta_{1265}) \tag{1.221}$$

where $\eta_{1265} = \frac{E_{F1265} - E_{17,25}}{k_B T}$ and E_{F1265} is the Fermi energy in this case.

The electronic contribution to the second- and third-order elastic constants for HD materials in this case can be written as

$$\Delta C_{44} = \frac{-G_0^2}{9}\left[\frac{\partial n_0}{\partial(E_{F1265})}\right] \tag{1.222a}$$

and

$$\Delta C_{456} = \frac{G_0^3}{27}\left[\frac{\partial^2 n_0}{\partial(E_{F1265})^2}\right] \tag{1.222b}$$

Using eqs. (1.222a), (1.222b), and (1.221) we can find ΔC_{44} and ΔC_{456} in this case.

In the absence of band tails and light waves, the totally quantized energy $E_{17,206}$ in this case is given by

$$E_{17,206} = \frac{\hbar^2\pi^2}{2m_c}\left[\left(\frac{n_x}{d_x}\right)^2 + \left(\frac{n_y}{d_y}\right)^2 + \left(\frac{n_z}{d_z}\right)^2\right] \tag{1.223}$$

The electron concentration can be written at a finite temperature as

$$n_0 = \frac{2g_v}{d_x d_y d_z}\sum_{n_x=1}^{n_{x\max}}\sum_{n_y=1}^{n_{y\max}}\sum_{n_z=1}^{n_{z\max}} F_{-1}(\eta_{1266}) \tag{1.224}$$

where $\eta_{1266} = \frac{E_{F1266} - E_{17,206}}{k_B T}$ and E_{F1266} is the Fermi energy in this case.

The electronic contribution to the second- and third-order elastic constants for materials in this case can be written as

$$\Delta C_{44} = \frac{-G_0^2}{9}\left[\frac{\partial n_0}{\partial(E_{F1266})}\right] \tag{1.225a}$$

and

$$\Delta C_{456} = \frac{G_0^3}{27}\left[\frac{\partial^2 n_0}{\partial(E_{F1266})^2}\right] \tag{1.225b}$$

Using eqs. (1.225a), (1.225b), and (1.224) we can find ΔC_{44} and ΔC_{456} in this case.

1.2.7 The magneto-CECs in QWs of HD Kane-type Semiconductors in the presence of light waves

(i) The magneto-DR in QWs of HD III–V, ternary, and quaternary materials, whose unperturbed band structure is defined by the three-band Kane model, in the presence of light waves, can be expressed as

$$\left(n + \frac{1}{2}\right)\hbar\omega_0 + \frac{\hbar^2}{2m_c}\left(\frac{n_z\pi}{d_z}\right)^2 = T_1(E_{17,8}, \eta_g, \lambda) \tag{1.226}$$

where $E_{17,8}$ is the totally quantized energy in this case.

The electron concentration can be written at a finite temperature as

$$n_0 = \frac{eBg_v}{\pi\hbar} \text{Real Part of} \sum_{n=0}^{n_{max}} \sum_{n_z=1}^{n_{z_{max}}} F_{-1}(\eta_{128}) \tag{1.227}$$

where $\eta_{128} = \frac{E_{F128} - E_{17.8}}{k_B T}$ and E_{F128} is the Fermi energy in this case.

The electronic contribution to the second- and third-order elastic constants for HD materials in this case can be written as

$$\Delta C_{44} = \frac{-G_0^2}{9d_z} \text{Real Part of} \left[\frac{\partial n_0}{\partial(E_{F128} - E_{17,8})}\right] \tag{1.228a}$$

and

$$\Delta C_{456} = \frac{G_0^3}{27d_z} \text{Real Part of} \left[\frac{\partial^2 n_0}{\partial(E_{F128} - E_{17,8})^2}\right] \tag{1.228b}$$

Using eqs. (1.228a), (1.228b), and (1.227) we can find ΔC_{44} and ΔC_{456} in this case.

The magneto-CEC in the absence of band tails in QWs of III–V, ternary, and quaternary materials, whose unperturbed band structure is defined by the three-band Kane model, in the presence of light waves can be expressed as

$$(n + \tfrac{1}{2})\hbar\omega_0 + \frac{\hbar^2}{2m_c}\left(\frac{n_z\pi}{d_z}\right)^2 = \beta_0(E_{17,9}, \lambda) \tag{1.229}$$

where $E_{17,9}$ is the totally quantized energy in this case

The electron concentration can be written at a finite temperature as

$$n_0 = \frac{eBg_v}{\pi\hbar} \sum_{n=0}^{n_{max}} \sum_{n_z=1}^{n_{z_{max}}} F_{-1}(\eta_{129}) \tag{1.230}$$

where $\eta_{129} = \frac{E_{F129} - E_{17.9}}{k_B T}$ and E_{F129} is the Fermi energy in this case.

The electronic contribution to the second- and third-order elastic constants for HD materials in this case can be written as

$$\Delta C_{44} = \frac{-G_0^2}{9d_z} \left[\frac{\partial n_0}{\partial E_{F129}}\right] \tag{1.231a}$$

and

$$\Delta C_{456} = \frac{G_0^3}{27d_z} \left[\frac{\partial^2 n_0}{\partial E_{F129}^2}\right] \tag{1.231b}$$

Using eqs. (1.231a), (1.231b), and (1.230) we can find ΔC_{44} and ΔC_{456} in this case.

The magneto-DR in the absence of band tails in QWs of III–V, ternary, and quaternary materials, whose unperturbed band structure is defined by the three-band Kane model, in the presence of light waves can be expressed as

$$\left(n+\frac{1}{2}\right)\hbar\omega_0+\frac{\hbar^2}{2m_c}\left(\frac{n_z\pi}{d_z}\right)^2=I_{11}(E_{17,10}) \tag{1.232}$$

where $E_{17,10}$ is the totally quantized energy in this case.

The electron concentration can be written at a finite temperature as

$$n_0=\frac{eBg_v}{\pi\hbar}\sum_{n=0}^{n_{\max}}\sum_{n_z=1}^{n_{z\max}}F_{-1}(\eta_{1210}) \tag{1.233}$$

where $\eta_{1210}=\frac{E_{F1210}-E_{17,10}}{k_BT}$ and E_{F1210} is the Fermi energy in this case.

The electronic contribution to the second- and third-order elastic constants for HD materials in this case can be written as

$$\Delta C_{44}=\frac{-G_0^2}{9d_z}\left[\frac{\partial n_0}{\partial E_{F1210}}\right] \tag{1.234a}$$

and

$$\Delta C_{456}=\frac{G_0^3}{27d_z}\left[\frac{\partial^2 n_0}{\partial E_{F1210}^2}\right] \tag{1.234b}$$

Using eqs. (1.234a), (1.234b), and (1.233) we can find ΔC_{44} and ΔC_{456} in this case.

(ii) The magneto-DR in QWs of HD III–V, ternary, and quaternary materials, whose unperturbed band structure is defined by the two-band Kane model, in the presence of light waves, can be expressed as

$$\left(n+\frac{1}{2}\right)\hbar\omega_0+\frac{\hbar^2}{2m_c}\left(\frac{n_z\pi}{d_z}\right)^2=T_2(E_{17,9},\eta_g,\lambda) \tag{1.235}$$

where $E_{17,9}$ is the totally quantized energy in this case.

The electron concentration can be written at a finite temperature as

$$n_0=\frac{eBg_v}{\pi\hbar}\sum_{n=0}^{n_{\max}}\sum_{n_z=1}^{n_{z\max}}F_{-1}(\eta_{129}) \tag{1.236}$$

where $\eta_{129}=\frac{E_{F129}-E_{17,9}}{k_BT}$ and E_{F129} is the Fermi energy in this case.

The electronic contribution to the second- and third-order elastic constants for HD materials in this case can be written as

$$\Delta C_{44} = \frac{-G_0^2}{9d_z}\left[\frac{\partial n_0}{\partial(E_{F129} - E_{17.9})}\right] \tag{1.237a}$$

and

$$\Delta C_{456} = \frac{G_0^3}{27d_z}\left[\frac{\partial^2 n_0}{\partial(E_{F129} - E_{17.9})^2}\right] \tag{1.237b}$$

Using eqs. (1.237a), (1.237b), and (1.236) we can find ΔC_{44} and ΔC_{456} in this case.

The magneto-DR in the absence of band tails in QWs of III–V, ternary, and quaternary materials, whose unperturbed band structure is defined by the two-band Kane model, in the presence of light waves, can be expressed as

$$\left(n + \frac{1}{2}\right)\hbar\omega_0 + \frac{\hbar^2}{2m_c}\left(\frac{n_z\pi}{d_z}\right)^2 = \tau_0(E_{17,10}, \lambda) \tag{1.238}$$

where $E_{17,10}$ is the totally quantized energy in this case.

The electron concentration can be written at a finite temperature as

$$n_0 = \frac{eBg_v}{\pi\hbar}\sum_{n=0}^{n_{\max}}\sum_{n_z=1}^{n_{z\max}} F_{-1}(\eta_{1210}) \tag{1.239}$$

where $\eta_{1210} = \frac{E_{F1210} - E_{17.10}}{k_B T}$ and E_{F1210} is the Fermi energy in this case.

The electronic contribution to the second- and third-order elastic constants for HD materials in this case can be written as

$$\Delta C_{44} = \frac{-G_0^2}{9d_z}\left[\frac{\partial n_0}{\partial E_{F1210}}\right] \tag{1.240a}$$

and

$$\Delta C_{456} = \frac{G_0^3}{27d_z}\left[\frac{\partial^2 n_0}{\partial E_{F1210}^2}\right] \tag{1.240b}$$

Using eqs. (1.240a), (1.240b), and (1.239) we can find ΔC_{44} and ΔC_{456} in this case.

The magneto-DR in the absence of band tails in QWs of III–V, ternary, and quaternary materials, whose unperturbed band structure is defined by the two-band Kane model, in the absence of light waves, can be expressed as

$$\left(n+\frac{1}{2}\right)\hbar\omega_0+\frac{\hbar^2}{2m_c}\left(\frac{n_z\pi}{d_z}\right)^2=E_{17,11}(1+\alpha E_{17,11}) \tag{1.241}$$

where, $E_{17,11}$ is the totally quantized energy in this case.

The electron concentration can be written at a finite temperature as

$$n_0=\frac{eBg_v}{\pi\hbar}\sum_{n=0}^{n_{\max}}\sum_{n_z=1}^{n_{z\max}}F_{-1}(\eta_{1211}) \tag{1.242}$$

where $\eta_{1211}=\frac{E_{F1211}-E_{17.11}}{k_BT}$ and E_{F1211} is the Fermi energy in this case.

The electronic contribution to the second- and third- order elastic constants for HD materials in this case can be written as

$$\Delta C_{44}=\frac{-G_0^2}{9d_z}\left[\frac{\partial n_0}{\partial E_{F1211}}\right] \tag{1.243a}$$

and

$$\Delta C_{456}=\frac{G_0^3}{27d_z}\left[\frac{\partial^2 n_0}{\partial E_{F1211}^2}\right] \tag{1.243b}$$

Using eqs. (1.243a), (1.243b) and (1.242) we can find ΔC_{44} and ΔC_{456} in this case.

(iii) The magneto DR in QWs of HD III-V, ternary and quaternary materials, whose unperturbed band structure is defined by the parabolic energy bands in the presence of light waves, can be expressed as

$$\left(n+\frac{1}{2}\right)\omega_0+\frac{\hbar^2}{2m_c}\left(\frac{n_z\pi}{d_z}\right)^2=T_3(E_{17,10},\eta_g,\lambda) \tag{1.244}$$

where, $E_{17,10}$ is the totally quantized energy in this case.

The electron concentration can be written at a finite temperature as

$$n_0=\frac{eBg_v}{\pi\hbar}\sum_{n=0}^{n_{\max}}\sum_{n_z=1}^{n_{z\max}}F_{-1}(\eta_{1210}) \tag{1.245}$$

where $\eta_{1210}=\frac{E_{F1210}-E_{17.10}}{k_BT}$ and E_{F1210} is the Fermi energy in this case.

The electronic contribution to the second- and third- order elastic constants for HD materials in this case can be written as

$$\Delta C_{44} = \frac{-G_0^2}{9d_z}\left[\frac{\partial n_0}{\partial(E_{F1210} - E_{17.10})}\right] \tag{1.246a}$$

and

$$\Delta C_{456} = \frac{G_0^3}{27d_z}\left[\frac{\partial^2 n_0}{\partial(E_{F1210} - E_{17.10})^2}\right] \tag{1.246b}$$

Using eqs. (1.246a), (1.246b) and (1.245) we can find ΔC_{44} and ΔC_{456} in this case.

The magneto-DR in the absence of band tails in QWs of III–V, ternary, and quaternary materials, whose unperturbed band structure is defined by the parabolic energy bands in the presence of light waves, can be expressed as

$$\left(n + \frac{1}{2}\right)\hbar\omega_0 + \frac{\hbar^2}{2m_c}\left(\frac{n_z\pi}{d_z}\right)^2 = \rho_0(E_{17,11}, \lambda) \tag{1.247}$$

where $E_{17,11}$ is the totally quantized energy in this case.

The electron concentration can be written at a finite temperature as

$$n_0 = \frac{eBg_v}{\pi\hbar}\sum_{n=0}^{n_{\max}}\sum_{n_z=1}^{n_{z\max}} F_{-1}(\eta_{1211}) \tag{1.248}$$

where $\eta_{1211} = \frac{E_{F1211} - E_{17.11}}{k_B T}$ and E_{F1211} is the Fermi energy in this case.

The electronic contribution to the second- and third-order elastic constants for HD materials in this case can be written as

$$\Delta C_{44} = \frac{-G_0^2}{9d_z}\left[\frac{\partial n_0}{\partial E_{F1211}}\right] \tag{1.249a}$$

and

$$\Delta C_{456} = \frac{G_0^3}{27d_z}\left[\frac{\partial^2 n_0}{\partial E_{F1211}^2}\right] \tag{1.249b}$$

Using eqs. (1.249a), (1.249b), and (1.248) we can find ΔC_{44} and ΔC_{456} in this case.

The magneto-DR in the absence of band tails in QWs of III–V, ternary, and quaternary materials, whose unperturbed band structure is defined by the parabolic energy bands, in the absence of light waves, can be expressed as

$$\left(n + \frac{1}{2}\right)\hbar\omega_0 + \frac{\hbar^2}{2m_c}\left(\frac{n_z\pi}{d_z}\right)^2 = E_{17,12} \tag{1.250}$$

where $E_{17,12}$ is the totally quantized energy in this case.

The electron concentration can be written at a finite temperature as

$$n_0 = \frac{eBg_v}{\pi\hbar} \sum_{n=0}^{n_{\max}} \sum_{n_z=1}^{n_{z\max}} F_{-1}(\eta_{1212}) \tag{1.251}$$

where $\eta_{1212} = \frac{E_{F1212} - E_{17.12}}{k_B T}$ and E_{F1212} is the Fermi energy in this case.

The electronic contribution to the second- and third-order elastic constants for HD materials in this case can be written as

$$\Delta C_{44} = \frac{-G_0^2}{9d_z}\left[\frac{\partial n_0}{\partial E_{F1211}}\right] \tag{1.252}$$

and

$$\Delta C_{456} = \frac{G_0^3}{27d_z}\left[\frac{\partial^2 n_0}{\partial E_{F1212}^2}\right] \tag{1.253}$$

Using eqs. (1.252), (1.253), and (1.251) we can find ΔC_{44} and ΔC_{456} in this case.

1.2.8 The CECs in accumulation and inversion layers of Kane-type Semiconductors in the presence of light waves

(a) In the presence of a surface electric field F_s along z direction and perpendicular to the surface, eq. (1.46a) assumes the form where, for this section, E represents the electron energy as measured from the edge of the CB at the surface in the vertically upward direction.

The quantization rule for inversion layers is given by

$$\int_0^{z_t} k_z dz = \frac{2}{3}(S_i)^{3/2} \tag{1.254}$$

where z_t is the classical turning point and S_i is the zeros of the Airy function $(Ai(-S_i) = 0)$.

Using eqs. (1.254) and (1.46b) under the weak electric field limit, the 2D DR in accumulation layers of HD III–V, ternary, and quaternary materials, whose unperturbed band structure is defined by the three-band Kane model, in the presence of light waves can be expressed as

$$T_1(E,\eta_g,\lambda) = \frac{\hbar^2 k_s^2}{2m_c} + S_i\left[\frac{\hbar|e|F_s\left[T_1(E,\eta_g,\lambda)\right]'}{\sqrt{2m_c}}\right]^{2/3} \tag{1.255}$$

Equation (1.255) represents the DR of the 2D electrons in accumulation layers of HD III–V, ternary, and quaternary materials under the weak electric field limit in the presence of light waves whose bulk electrons obey the HD three-band Kane model. Since the DR in accordance with the HD three-band Kane model is complex in nature, eq. (1.255) will also be complex. The both complexities occur due to the presence of poles in the finite complex plane of the dispersion relation of the materials in the presence of band tails.

The surface electron concentration is given by

$$n_S = g_v \, Real\,Part\,of\,the \sum_{i=0}^{i_{\max}}\left[\left[\frac{m_c}{\pi\hbar^2}P_{3HDL}(E'_{fL},i,\eta_g,\lambda)\right] + \frac{1}{3\pi^2}\left(\frac{2m_c}{\hbar^2}\right)^{3/2} t_i\left[T_1(E_{FB},\eta_g,\lambda)\right]^{3/2}\right] \tag{1.256}$$

where $P_{3HDL}(E'_{fL},i,\eta_g,\lambda) = \left[T_1(E'_{fL},\eta_g,\lambda) - S_i\left[\frac{\hbar|e|F_2\left[T_1(E'_{fL},\eta_g,\lambda)\right]'}{\sqrt{2m_c}}\right]^{2/3}\right]$, the Fermi energy in this case E'_{fL} that can be expressed as $E'_{fL} = eV_g - \frac{e^2 n_S d_{ex}}{\varepsilon_{OX}} + E_{FB}$, V_g is the gate voltage, n_s is the surface electron concentration, d_{ox} is the thickness of the oxide layer, ε_{ox} is the permittivity of the oxide layer, $F_s = \frac{en_s}{\varepsilon_{sc}}$, ε_{sc} is the semiconductor permittivity, and E_{FB} could be, under the condition extreme degeneracy, determined from the equation

$$n_B = \frac{4\pi}{3}\frac{2g_v}{(2\pi)^3}\left(\frac{2m_c}{\hbar^2}\right)^{3/2} Real\,Part\,of\left[T_1(E_{FB},\eta_g,\lambda)\right]^{3/2} \tag{1.257}$$

and n_B is the bulk electron concentration, $t_i = \frac{E_{Li\max}}{eF_S(1+i_{\max})}$, $E_{Li\max}$, is the root of the real part of the equation

$$0 = Real\,Part\,of\left[T_1(E_{Li_{\max}},\eta_g,\lambda) - S_{i_{\max}}\left[\frac{\hbar|e|F_s\left[T_1(E_{Li_{\max}},\eta_g,\lambda)\right]'}{\sqrt{2m_c}}\right]^{2/3}\right] \tag{1.258}$$

The electronic contribution to the second- and third-order elastic constants for HD materials in this case can be written as

$$\Delta C_{44} = \frac{-G_0^2}{9} \text{Real part of} \left[\frac{\partial n_0}{t_i \partial (E'_{fL} - E'_{i1L})} \right] \tag{1.259a}$$

and

$$\Delta C_{456} = \frac{G_0^3}{27} \text{Real Part of} \left[\frac{\partial^2 n_0}{t_i \partial (E'_{fL} - E'_{i1L})^2} \right] \tag{1.259b}$$

where t_i is the thickness.

Thus, using eqs. (1.259a), (1.259b), and (1.256) we can find ΔC_{44} and ΔC_{456} in this case.

Under the weak electric field limit, the 2D DR in accumulation layers of HD III–V, ternary, and quaternary materials, whose unperturbed band structure is defined by the three-band Kane model, in the absence of light waves, can be expressed as

$$\bar{T}_1(E, \eta_g) = \frac{\hbar^2 k_s^2}{2m_c} + S_i \left[\frac{\hbar |e| F_s \left[\bar{T}_1 \left(E, \eta_g \right) \right]'}{\sqrt{2m_c}} \right]^{2/3} \tag{1.260}$$

where $\bar{T}_1(E, \eta_g) = T_{31}(E, \eta_g) + iT_{32}(E, \eta_g)$

Equation (1.260) represents the DR of the 2D electrons in accumulation layers of HD III–V, ternary, and quaternary materials under the weak electric field limit in the absence of light waves whose bulk electrons obey the HD three-band Kane model.

The surface electron concentration is given by

$$n_S = g_v \text{Real Part of the} \sum_{i=0}^{i_{max}} \left[\left[\frac{m_c}{\pi \hbar^2} \bar{P}_{3HD} \left(E'_{f1}, i, \eta_g \right) \right] + \frac{1}{3\pi^2} \left(\frac{2m_c}{\hbar^2} \right)^{3/2} t_i \left[\bar{T}_1 \left(E_{FB1}, \eta_g \right) \right]^{3/2} \right] \tag{1.261}$$

$\bar{P}_{3HD}(E'_{f1}, i, \eta_g) = \left[\bar{T}_1 \left(E'_{f1}, \eta_g \right) - S_i \left[\frac{\hbar |e| F_2 [\bar{T}_1 (E'_{f1}, \eta_g)]'}{\sqrt{2m_c}} \right]^{2/3} \right]$, E'_{f1} is the Fermi energy in this case, which can be expressed as $E'_{f1} = eV_g - \frac{e^2 n_s d_{ex}}{\varepsilon_{ox}} + E_{FB1}$, and E_{FB1} could be, under the condition extreme degeneracy, determined from the equation

$$n_B = \frac{4\pi}{3} \frac{2g_v}{(2\pi)^3} \left(\frac{2m_c}{\hbar^2} \right)^{3/2} \text{Real Part of} \left[\bar{T}_1 \left(E_{FB}, \eta_g \right) \right]^{3/2} \tag{1.262}$$

where $t_i = \frac{E_{i\max}}{eF_S(1 + i_{\max})}$, $E_{i\max}$, is the root of the real part of the equation

$$0 = \text{Real Part of}\left[\bar{T}_1\left(E_{i_{\max}}, \eta_g\right) - S_{i_{\max}}\left[\frac{\hbar|e|F_s\left[\bar{T}_1\left(E_{i_{\max}}, \eta_g\right)\right]'}{\sqrt{2m_c}}\right]^{2/3}\right] \tag{1.263}$$

The electronic contribution to the second- and third-order elastic constants for HD materials in this case can be written as

$$\Delta C_{44} = \frac{-G_0^2}{9}\text{Real Part of}\left[\frac{\partial n_0}{t_i\partial(E'_{f1} - E'_{i11})}\right] \tag{1.264a}$$

and

$$\Delta C_{456} = \frac{G_0^3}{27}\text{Real Part of}\left[\frac{\partial^2 n_0}{t_i\partial(E'_{f1} - E'_{i11})^2}\right] \tag{1.264b}$$

Thus, using eqs. (1.264a), (1.264b), and (1.261) we can find ΔC_{44} and ΔC_{456} in this case.

In the absence of band tails and light waves, the 2D DR in III–V, ternary, and quaternary materials whose bulk electrons obey the three-band Kane model under the condition of weak electric field limit assumes the form

$$I_{11}(E) = \frac{\hbar^2 k_s^2}{2m_c} + S_i\left[\frac{\hbar|e|F_s[I_{11}(E)]'}{\sqrt{2m_c}}\right]^{2/3} \tag{1.265}$$

n_{2Dw} in the present case can be written as

$$n_{2Dw} = \frac{g_v m_c}{\pi\hbar^2}\sum_{i=0}^{i_{\max}}[P_{4w}(E_{Fiw}, i)] \tag{1.266}$$

Where

$$P_{4w}(E_{Fiw}, i) \equiv \left\{I_{11}(E_{Fiw}) - S_i\left[\frac{\hbar e F_s[I_{11}(E_{Fiw})]'}{\sqrt{2m_c}}\right]^{2/3}\right\}$$

and E_{Fiw} is the Fermi energy in this case.

The electronic contribution to the second- and third-order elastic constants for HD materials in this case can be written as

$$\Delta C_{44} = \frac{-G_0^2}{9}\left[\frac{\partial n_0}{t_i \partial\left(E_{Fiw} - E_{n_{iw2}}\right)}\right] \tag{1.267a}$$

and

$$\Delta C_{456} = \frac{G_0^3}{27}\left[\frac{\partial^2 n_0}{t_i \partial\left(E_{Fiw} - E_{n_{iw2}}\right)^2}\right] \tag{1.267b}$$

Thus, using eqs. (1.267a), (1.267b), and (1.266) we can find ΔC_{44} and ΔC_{456} in this case.

(b) In the presence of a surface electric field F_s along z direction and perpendicular to the surface, eq. (1.46a) assumes the form

$$T_2(E - |e|F_s z, \eta_g, \lambda) = \frac{\hbar^2 k^2}{2m_c} \tag{1.268}$$

Using eqs. (1.47) and (1.268) under the weak electric field limit, the 2D DR in accumulation layers of HD III–V, ternary, and quaternary materials, whose unperturbed band structure is defined by the two-band Kane model, in the presence of light waves can be expressed as

$$T_2(E, \eta_g, \lambda) = \frac{\hbar^2 k_s^2}{2m_c} + S_i\left[\frac{\hbar|e|F_s\left[T_2(E, \eta_g, \lambda)\right]'}{\sqrt{2m_c}}\right]^{2/3} \tag{1.269}$$

Equation (1.269) represents the DR of the 2D electrons in accumulation layers of HD III–V, ternary, and quaternary materials under the weak electric field limit in the presence of light waves whose bulk electrons obey the HD two-band Kane model. Since the DR in accordance with the HD two-band Kane model is real in nature, eq. (1.269) will also be real.

The surface electron concentration is given by

$$n_S = g_v \sum_{i=0}^{i_{\max}}\left[\left[\frac{m_c}{\pi\hbar^2} P_{3HDL2}(E'_{fL2}, i, \eta_g, \lambda)\right] + \frac{1}{3\pi^2}\left(\frac{2m_c}{\hbar^2}\right)^{3/2} t_i\left[T_2(E_{FB2}, \eta_g, \lambda)\right]^{3/2}\right] \tag{1.270}$$

where, $P_{3HDL}(E'_{fL}, i, \eta_g, \lambda) = \left[T_1(E'_{fL}, \eta_g, \lambda) - S_i\left[\frac{\hbar|e|F_2\left[T_1(E'_{fL}, \eta_g, \lambda)\right]'}{\sqrt{2m_c}}\right]^{2/3}\right]$, the Fermi energy in this case is E'_{fL2}, which can be expressed as $E'_{fL2} = eV_g - \frac{e^2 n_S d_{ex}}{\varepsilon_{ox}} + E_{FB2}$,, and E_{FB2} could be, under the condition extreme degeneracy, determined from the equation

$$n_B = \frac{4\pi}{3}\frac{2g_v}{(2\pi)^3}\left(\frac{2m_c}{\hbar^2}\right)^{3/2}\left[T_2(E_{FB2}, \eta_g, \lambda)\right]^{3/2} \tag{1.271}$$

$t_i = \frac{E_{Li\max}}{eF_S(1+i_{\max})}$, $E_{Li\max}$, is the root of the real part of the equation

$$0 = \left[T_2(E_{Li\max}, \eta_g, \lambda) - S_{i\max}\left[\frac{\hbar|e|F_s\left[T_2(E_{Li\max}, \eta_g, \lambda)\right]'}{\sqrt{2m_c}}\right]^{2/3}\right] \tag{1.272}$$

The electronic contribution to the second- and third-order elastic constants for HD materials in this case can be written as

$$\Delta C_{44} = \frac{-G_0^2}{9}\left[\frac{\partial n_0}{t_i\partial(E'_{fL2} - E'_{i1L2})}\right] \tag{1.273a}$$

and

$$\Delta C_{456} = \frac{G_0^3}{27}\left[\frac{\partial^2 n_0}{t_i\partial(E'_{fL2} - E'_{i1L2})^2}\right] \tag{1.273b}$$

Thus, using eqs. (1.273a), (1.273b), and (1.270) we can find ΔC_{44} and ΔC_{456} in this case.

Under the weak electric field limit, the 2D DR in accumulation layers of HD III–V, ternary, and quaternary materials, whose unperturbed band structure is defined by the three-band Kane model, in the absence of light waves, can be expressed as

$$\bar{T}_2(E, \eta_g) = \frac{\hbar^2 k_s^2}{2m_c} + S_i\left[\frac{\hbar|e|F_s\left[\bar{T}_2(E, \eta_g)\right]'}{\sqrt{2m_c}}\right]^{2/3} \tag{1.274}$$

where $\bar{T}_2(E, \eta_g) = \left[\frac{2}{1+Erf(E/\eta_g)}\right]\left[\gamma_0(E, \eta_g) + \alpha\theta_0(E, \eta_g)\right]$

Equation (1.274) represents the DR of the 2D electrons in accumulation layers of HD III–V, ternary, and quaternary materials under the weak electric field limit in the absence of light waves whose bulk electrons obey the HD three-band Kane model.

The surface electron concentration is given by

$$n_S = g_v\sum_{i=0}^{i_{\max}}\left[\left[\frac{m_c}{\pi\hbar^2}\bar{P}_{3HD}\left(E'_{f1}, i, \eta_g\right)\right] + \frac{1}{3\pi^2}\left(\frac{2m_c}{\hbar^2}\right)^{3/2} t_i\left[\bar{T}_2\left(E_{FB1}, \eta_g\right)\right]^{3/2}\right] \tag{1.275}$$

where $\bar{P}_{3HD}\left(E'_{f1}, i, \eta_g\right) = \left[\bar{T}_2\left(E'_{f1}, \eta_g\right) - S_i\left[\frac{\hbar|e|F_2[\bar{T}_2(E'_{f1}, \eta_g)]'}{\sqrt{2m_c}}\right]^{2/3}\right]$, the Fermi energy in this case E'_{f1}, which can be expressed as $E'_{f1} = eV_g - \frac{e^2 n_s d_{ex}}{\varepsilon_{ox}} + E_{FB1}$, and E_{FB1} could be, under the condition extreme degeneracy, determined from the equation

$$n_B = \frac{4\pi}{3}\frac{2g_v}{(2\pi)^3}\left(\frac{2m_c}{\hbar^2}\right)^{3/2}\left[\bar{T}_2\left(E_{FB}, \eta_g\right)\right]^{3/2} \tag{1.276}$$

$t_i = \frac{E_{i\,max}}{eF_s(1 + i_{max})}$, $E_{i\,max}$, is the root of the real part of the equation

$$0 = \left[\bar{T}_2\left(E_{i_{max}}, \eta_g\right) - S_{i_{max}}\left[\frac{\hbar|e|F_s\left[\bar{T}_2\left(E_{i_{max}}, \eta_g\right)\right]'}{\sqrt{2m_c}}\right]^{2/3}\right] \tag{1.277}$$

The electronic contribution to the second- and third-order elastic constants for HD materials in this case can be written as

$$\Delta C_{44} = \frac{-G_0^2}{9}\left[\frac{\partial n_0}{t_i \partial\left(E'_{f1} - E'_{i11}\right)}\right] \tag{1.278a}$$

and

$$\Delta C_{456} = \frac{G_0^3}{27}\left[\frac{\partial^2 n_0}{t_i \partial\left(E'_{f1} - E'_{i11}\right)^2}\right] \tag{1.278b}$$

Thus, using eq. (1.278a), (1.278b), and (1.275) we can find ΔC_{44} and ΔC_{456} in this case.

In the absence of band tails and light waves, the 2D DR in III–V, ternary, and quaternary materials whose bulk electrons obey the three-band Kane model under the condition of weak electric field limit assumes the form

$$E(1 + \alpha E) = \frac{\hbar^2 k_s^2}{2m_c} + S_i\left[\frac{\hbar|e|F_s(1 + 2\alpha E)}{\sqrt{2m_c}}\right]^{2/3} \tag{1.279}$$

n_{2Dw} in the present case can be written as

$$n_{2Dw} = \frac{g_v m_c}{\pi\hbar^2}\sum_{i=0}^{i_{max}}\left[P_{4w4}(E_{Fiw}, i)\right] \tag{1.280}$$

Where

$$P_{4w4}(E_{Fiw}, i) \equiv \left\{ E_{Fiw}(1+\alpha E_{Fiw}) - S_i \left[\frac{\hbar e F_s(1+\alpha E_{Fiw})}{\sqrt{2m_c}}\right]^{2/3} \right\}$$

and E_{Fiw} is the Fermi energy in this case.

The electronic contribution to the second- and third-order elastic constants for HD materials in this case can be written as

$$\Delta C_{44} = \frac{-G_0^2}{9}\left[\frac{\partial n_0}{t_i \partial\left(E_{Fiw} - E_{n_{iw24}}\right)}\right] \tag{1.281a}$$

and

$$\Delta C_{456} = \frac{G_0^3}{27}\left[\frac{\partial^2 n_0}{t_i \partial\left(E_{Fiw} - E_{n_{iw24}}\right)^2}\right] \tag{1.281b}$$

Thus, using eqs. (1.281a), (1.281b), and (1.280) we can find ΔC_{44} and ΔC_{456} in this case.

(c) In the presence of a surface electric field F_s along z direction and perpendicular to the surface, eq. (1.48) assumes the form

$$T_3(E - |e|F_s z, \eta_g, \lambda) = \frac{\hbar^2 k^2}{2m_c} \tag{1.282}$$

Using eqs. (1.254) and (1.282) under the weak electric field limit, the 2D DR in accumulation layers of HD III–V, ternary, and quaternary materials, whose unperturbed band structure is defined by the two-band Kane model, in the presence of light waves can be expressed as

$$T_3(E, \eta_g, \lambda) = \frac{\hbar^2 k_s^2}{2m_c} + S_i\left[\frac{\hbar|e|F_s\left[T_3(E, \eta_g, \lambda)\right]}{\sqrt{2m_c}}\right]^{2/3} \tag{1.283}$$

Equation (1.283) represents the DR of the 2D electrons in accumulation layers of HD III–V, ternary, and quaternary materials under the weak electric field limit in the presence of light waves whose bulk electrons obey the HD two-band Kane model. Since the DR in accordance with the HD two-band Kane model is real in nature, eq. (1.283) will also be real.

The surface electron concentration is given by

$$n_S = g_v \sum_{i=0}^{i_{\max}} \left[\left[\frac{m_c}{\pi \hbar^2} P_{3HDL23}(E'_{fL23}, i, \eta_g, \lambda) \right] + \frac{1}{3\pi^2} \left(\frac{2m_c}{\hbar^2} \right)^{3/2} t_i \left[T_3(E_{FB23}, \eta_g, \lambda) \right]^{3/2} \right] \tag{1.284}$$

where $P_{3HDL23}(E'_{fL23}, i, \eta_g, \lambda) = \left[T_3(E'_{fL23}, \eta_g, \lambda) - S_i \left[\frac{\hbar |e| F_2 [T_3(E'_{fL23}, \eta_g, \lambda)]'}{\sqrt{2m_c}} \right]^{2/3} \right]$, E'_{fL23} is the Fermi energy in this case, which can be expressed as $E'_{fL23} = eV_g - \frac{e^2 n_S d_{ex}}{\varepsilon_{ox}} + E_{FB23}$ and E_{FB2}, could be, under the condition extreme degeneracy, determined by the following equation:

$$n_B = \frac{4\pi}{3} \frac{2g_v}{(2\pi)^3} \left(\frac{2m_c}{\hbar^2} \right)^{3/2} \left[T_3(E_{FB23}, \eta_g, \lambda) \right]^{3/2} \tag{1.285}$$

$t_i = \frac{E_{Li\max}}{eF_S(1 + i_{\max})}$, $E_{Li\max}$, is the root of the real part of the equation

$$0 = \left[T_3(E_{Li_{\max}}, \eta_g, \lambda) - S_{i_{\max}} \left[\frac{\hbar |e| F_s \left[T_3(E_{Li_{\max}}, \eta_g, \lambda) \right]'}{\sqrt{2m_c}} \right]^{2/3} \right] \tag{1.286}$$

The electronic contribution to the second- and third-order elastic constants for HD materials in this case can be written as

$$\Delta C_{44} = \frac{-G_0^2}{9} \left[\frac{\partial n_0}{t_i \partial (E'_{fL23} - E'_{i1L23})} \right] \tag{1.287a}$$

and

$$\Delta C_{456} = \frac{G_0^3}{27} \left[\frac{\partial^2 n_0}{t_i \partial (E'_{fL23} - E'_{i1L23})^2} \right] \tag{1.287b}$$

Thus, using eqs. (1.287a), (1.287b), and (1.284) we can find ΔC_{44} and ΔC_{456} in this case.

Under the weak electric field limit, the 2D DR in accumulation layers of HD III–V, ternary, and quaternary materials, whose unperturbed band structure is defined by the three-band Kane model, in the absence of light waves, can be expressed as

$$\bar{T}_3(E, \eta_g) = \frac{\hbar^2 k_s^2}{2m_c} + S_i \left[\frac{\hbar |e| F_s \left[\bar{T}_3(E, \eta_g) \right]'}{\sqrt{2m_c}} \right]^{2/3} \tag{1.288}$$

where $\bar{T}_3(E, \eta_g) = \left[\frac{2}{1 + Erf(E/\eta_g [\gamma_0(E, \eta_g)]} \right]$

Equation (1.288) represents the DR of the 2D electrons in accumulation layers of HD III–V, ternary, and quaternary materials under the weak electric field limit in the absence of light waves whose bulk electrons obey the HD three-band Kane model.

The surface electron concentration is given by

$$n_S = g_v \sum_{i=0}^{i_{\max}} \left[\left[\frac{m_c}{\pi\hbar^2} \bar{P}_{3HD3}\left(E'_{f13}, i, \eta_g\right) \right] + \frac{1}{3\pi^2} \left(\frac{2m_c}{\hbar^2} \right)^{3/2} t_i \left[\bar{T}_3\left(E_{FB13}, \eta_g\right) \right]^{3/2} \right] \quad (1.289)$$

where $\bar{P}_{3HD3}(E'_{f13}, i, \eta_g) = \left[\bar{T}_3\left(E'_{f13}, \eta_g\right) - S_i \left[\frac{\hbar|e|F_2[\bar{T}_3(E'_{f13}, \eta_g)]'}{\sqrt{2m_c}} \right]^{2/3} \right]$, E'_{f13} is the Fermi energy in this case, which can be expressed as $E'_{f13} = eV_g - \frac{e^2 n_s d_{ex}}{\varepsilon_{ox}} + E_{FB13}$, and E_{FB13} could be, under the condition extreme degeneracy, determined by the following equation:

$$n_B = \frac{4\pi}{3} \frac{2g_v}{(2\pi)^3} \left(\frac{2m_c}{\hbar^2} \right)^{3/2} \left[\bar{T}_3\left(E_{FB13}, \eta_g\right) \right]^{3/2} \quad (1.290)$$

$t_i = \frac{E_{i\max}}{eF_s(1 + i_{\max})}$, $E_{i\max}$, is the root of the real part of the equation

$$0 = \left[\bar{T}_3\left(E_{i_{\max}}, \eta_g\right) - S_{i_{\max}} \left[\frac{\hbar|e|F_s\left[\bar{T}_3\left(E_{i_{\max}}, \eta_g\right)\right]'}{\sqrt{2m_c}} \right]^{2/3} \right] \quad (1.291)$$

The electronic contribution to the second- and third-order elastic constants for HD materials in this case can be written as

$$\Delta C_{44} = \frac{-G_0^2}{9} \left[\frac{\partial n_0}{t_i \partial\left(E'_{f13} - E'_{i113}\right)} \right] \quad (1.292a)$$

and

$$\Delta C_{456} = \frac{G_0^3}{27} \left[\frac{\partial^2 n_0}{t_i \partial\left(E'_{f13} - E'_{i113}\right)^2} \right] \quad (1.292b)$$

Thus, using eqs. (1.292a), (1.292b), and (1.289) we can find ΔC_{44} and ΔC_{456} in this case.

In the absence of band tails and light waves, the 2D DR in III–V, ternary, and quaternary materials whose bulk electrons obey the three-band Kane model under the condition of weak electric field limit assumes the form

$$E = \frac{\hbar^2 k_s^2}{2m_c} + S_i \left[\frac{\hbar|e|F_s}{\sqrt{2m_c}}\right]^{2/3} \tag{1.293}$$

The expression of n_{2Di} at a finite temperature can be written as

$$n_{2Di} = \frac{g_v m_c k_B T}{\pi \hbar^2} \sum_{i=0}^{i_{max}} F_0(\eta_{i44}) \tag{1.294}$$

where $\eta_{i44} \equiv (k_B T)^{-1} \left[E_{Fi44} - S_i \left[\frac{\hbar|e|F_s}{\sqrt{2m_c}}\right]^{2/3}\right]$, and E_{Fi44} is the Fermi energy as measured from the edge of the CB at the surface.

The electronic contribution to the second- and third-order elastic constants for HD materials in this case can be written as

$$\Delta C_{44} = \frac{-G_0^2}{9} \left[\frac{\partial n_0}{t_i \partial\left(E_{Fi44} - E_{n_{i45}}\right)}\right] \tag{1.295a}$$

and

$$\Delta C_{456} = \frac{G_0^3}{27} \left[\frac{\partial^2 n_0}{t_i \partial\left(E_{Fi44} - E_{n_{i45}}\right)^2}\right] \tag{1.295b}$$

Thus, using eqs. (1.295a), (1.295b), and (1.294) we can find ΔC_{44} and ΔC_{456} in this case.

1.2.9 The CECs in NWs of HD Kane-type semiconductors in the presence of light waves

(a) The 1D DR in NWs of HD III–V, ternary, and quaternary materials, whose unperturbed band structure is defined by the three-band Kane model in the absence of any field, in the presence of light waves can be expressed as

$$\frac{\hbar^2 (n_z \pi/d_z)^2}{2m_c} + \frac{\hbar^2 (n_y \pi/d_y)^2}{2m_c} + \frac{\hbar^2 k_x^2}{2m_c} = T_1(E, \eta_g, \lambda) \tag{1.296a}$$

Using eq. (1.51a) and summing over both n_y and n_z together with the substitution of $C_{1D} = \pi$, the use of HUP leads to the expression of ES as

$$n_0 = \frac{2g_v}{\pi} \text{Real Part of} \sum_{n_y=1}^{n_{y\max}} \sum_{n_z=1}^{n_{z\max}} \left[T_{3L1}(E_{F1HDNWL1}, n_y, n_z, \eta_g, \lambda)\right] \tag{1.296b}$$

where $T_{3L1}(E, n_y, n_z, \eta_g, \lambda) = \left[\left[T_1(E, \eta_g, \lambda) - \left[\frac{\hbar^2(n_z\pi/d_z)^2}{2m_c} + \frac{\hbar^2(n_y\pi/d_y)^2}{2m_c}\right]\right]\frac{2m_c}{\hbar^2}\right]^{1/2}$

and $E_{F1HDNWL1}$ is the Fermi energy in this case.

The electronic contribution to the second- and third-order elastic constants for HD materials in this case can be written as

$$\Delta C_{44} = \frac{-G_0^2}{9d_y d_z} \text{Real Part of} \left[\frac{\partial n_0}{\partial\left(E_{F1HDNWL1} - E'_{3HDNWL1}\right)}\right] \tag{1.297a}$$

and

$$\Delta C_{456} = \frac{G_0^3}{27d_y d_z} \text{Real part of} \left[\frac{\partial^2 n_0}{\partial\left(E_{F1HDNWL1} - E'_{3HDNWL1}\right)^2}\right] \tag{1.297b}$$

Thus, using eqs. (1.297a), (1.297b), and (1.296b) we can find ΔC_{44} and ΔC_{456} in this case.

The 1D DR for NWs of III–V materials whose energy band structures are defined by the three-band Kane model in the absence of band tailing assumes the form

$$\frac{\hbar^2(n_z\pi/d_z)^2}{2m_c} + \frac{\hbar^2(n_y\pi/d_y)^2}{2m_c} + \frac{\hbar^2 k_x^2}{2m_c} = \beta_0(E, \lambda) \tag{1.298}$$

$$f_{12L1}(E, n_y, n_z, \lambda) = \left[\left[\beta_0(E, \lambda) - \left[\frac{\hbar^2(n_z\pi/d_z)^2}{2m_c} + \frac{\hbar^2(n_y\pi/d_y)^2}{2m_c}\right]\right]\frac{2m_c}{\hbar^2}\right]^{1/2}$$

The ES per unit length can be written as

$$n_0 = \frac{2g_v}{\pi} \sum_{n_y=1}^{n_{y\max}} \sum_{n_z=1}^{n_{z\max}} \left[f_{12L1}(E_{F1NWL2}, n_y, n_z, \lambda)\right] \tag{1.299}$$

where $f_{12L1}(E_{F1NWL2}, n_y, n_z, \lambda) = \left[\left[\beta_0(E_{F1NWL2}, \lambda) - \left[\frac{\hbar^2(n_z\pi/d_z)^2}{2m_c} + \frac{\hbar^2(n_y\pi/d_y)^2}{2m_c}\right]\right]\frac{2m_c}{\hbar^2}\right]^{1/2}$ and E_{F1NWL2} is the Fermi energy in this case.

The electronic contribution to the second- and third-order elastic constants for HD materials in this case can be written as

$$\Delta C_{44} = \frac{-G_0^2}{9d_y d_z}\left[\frac{\partial n_0}{\partial E_{F1NWL2}}\right] \quad (1.300a)$$

and

$$\Delta C_{456} = \frac{G_0^3}{27d_y d_z}\left[\frac{\partial^2 n_0}{\partial E_{F1NWL2}^2}\right] \quad (1.300b)$$

Thus, using eqs. (1.300a), (1.300b), and (1.299) we can find ΔC_{44} and ΔC_{456} in this case.

(b) The 1D DR in NWs of HD III-V, ternary, and quaternary materials, whose unperturbed band structure is defined by the two-band Kane model in the absence of any field, in the presence of light waves can be expressed as

$$\frac{\hbar^2(n_z\pi/d_z)^2}{2m_c} + \frac{\hbar^2\left(n_y\pi/d_y\right)^2}{2m_c} + \frac{\hbar^2k_x^2}{2m_c} = T_2(E, \eta_g, \lambda) \quad (1.301)$$

The ES per unit length can be written as

$$n_0 = \frac{2g_v}{\pi}\sum_{n_y=1}^{n_{y\max}}\sum_{n_z=1}^{n_{z\max}}\left[T_{3L2}(E_{F1HDNWL2}, n_y, n_z, \eta_g, \lambda)\right] \quad (1.302)$$

The electronic contribution to the second- and third-order elastic constants for HD materials in this case can be written as

$$\Delta C_{44} = \frac{-G_0^2}{9d_y d_z}\left[\frac{\partial n_0}{\partial\left(E_{F1HDNWL2} - E'_{3HDNWL2}\right)}\right] \quad (1.303a)$$

and

$$\Delta C_{456} = \frac{G_0^3}{27d_y d_z}\left[\frac{\partial^2 n_0}{\partial\left(E_{F1HDNWL2} - E'_{3HDNWL2}\right)^2}\right] \quad (1.303b)$$

Thus, using eqs. (1.303a), (1.303b), and (1.302) we can find ΔC_{44} and ΔC_{456} in this case.

The 1D DR, for NWs of III–V materials whose energy band structures are defined by the two-band Kane model in the absence of band tailing assumes the form

$$\frac{\hbar^2(n_z\pi/d_z)^2}{2m_c} + \frac{\hbar^2\left(n_y\pi/d_y\right)^2}{2m_c} + \frac{\hbar^2k_x^2}{2m_c} = \tau_0(E, \lambda) \quad (1.304)$$

The ES per unit length can be written as

$$n_0 = \frac{2g_v}{\pi} \sum_{n_y=1}^{n_{y\max}} \sum_{n_z=1}^{n_{z\max}} \left[f_{12L2}(E_{F1NWL21}, n_y, n_z, \lambda) \right] \tag{1.305}$$

Where

$$f_{12L2}(E_{F1NWL21}, n_y, n_z, \lambda) = \left[\left[\tau_0(E_{F1NWL21}, \lambda) - \left[\frac{\hbar^2 (n_z \pi / d_z)^2}{2m_c} + \frac{\hbar^2 (n_y \pi / d_y)^2}{2m_c} \right] \right] \frac{2m_c}{\hbar^2} \right]^{1/2}$$

and $E_{F1NWL21}$ is the Fermi energy in this case.

The electronic contribution to the second- and third-order elastic constants for HD materials in this case can be written as

$$\Delta C_{44} = \frac{-G_0^2}{9d_y d_z} \left[\frac{\partial n_0}{\partial E_{F1NWL21}} \right] \tag{1.306a}$$

and

$$\Delta C_{456} = \frac{G_0^3}{27 d_y d_z} \left[\frac{\partial^2 n_0}{\partial E_{F1NWL21}^2} \right] \tag{1.306b}$$

Thus, using eqs. (1.306a), (1.306b), and (1.305) we can find ΔC_{44} and ΔC_{456} in this case.

(c) The 1D DR in NWs of HD III–V, ternary, and quaternary materials, whose unperturbed band structure is defined by the parabolic energy bands in the absence of any field, in the presence of light waves can be expressed as

$$\frac{\hbar^2 (n_z \pi / d_z)^2}{2m_c} + \frac{\hbar^2 (n_y \pi / d_y)^2}{2m_c} + \frac{\hbar^2 k_x^2}{2m_c} = T_3(E, \eta_g, \lambda) \tag{1.307}$$

The ES per unit length can be written as

$$n_0 = \frac{2g_v}{\pi} \sum_{n_y=1}^{n_{y\max}} \sum_{n_z=1}^{n_{z\max}} \left[T_{3L3}(E_{F1HDNWL3}, n_y, n_z, \eta_g, \lambda) \right] \tag{1.308}$$

Where

$$T_{3L3}(E_{F1HDNWL3}, n_y, n_z, \eta_g, \lambda) = \left[\left[T_3(E_{F1HDNWL3}, \eta_g, \lambda) - \left[\frac{\hbar^2 (n_z \pi / d_z)^2}{2m_c} + \frac{\hbar^2 (n_y \pi / d_y)^2}{2m_c} \right] \right] \frac{2m_c}{\hbar^2} \right]^{1/2}$$

and $E_{F1HDNWL3}$ is the Fermi energy in this case.

The electronic contribution to the second- and third-order elastic constants for HD materials in this case can be written as

$$\Delta C_{44} = \frac{-G_0^2}{9d_y d_z}\left[\frac{\partial n_0}{\partial\left(E_{F1HDNWL3} - E'_{3HDNWL3}\right)}\right] \quad (1.309a)$$

and

$$\Delta C_{456} = \frac{G_0^3}{27d_y d_z}\left[\frac{\partial^2 n_0}{\partial\left(E_{F1HDNWL3} - E'_{3HDNWL3}\right)^2}\right] \quad (1.309b)$$

Thus, using eqs. (1.309a), (1. 309b), and (1.308) we can find ΔC_{44} and ΔC_{456} in this case.

The 1D DR for NWs of III–V materials whose energy band structures are defined by the parabolic energy bands in the absence of band tailing assumes the form

$$\frac{\hbar^2(n_z\pi/d_z)^2}{2m_c} + \frac{\hbar^2(n_y\pi/d_y)^2}{2m_c} + \frac{\hbar^2 k_x^2}{2m_c} = \rho_0(E,\lambda) \quad (1.310)$$

The ES per unit length can be written as

$$n_0 = \frac{2g_v}{\pi}\sum_{n_y=1}^{n_{y\max}}\sum_{n_z=1}^{n_{z\max}}\left[f_{12L3}(E_{F1NWL3}, n_y, n_z, \lambda)\right] \quad (1.311)$$

where

$$f_{12L3}(E_{F1NWL3}, n_y, n_z, \lambda) = \left[\left[\rho_0(E_{F1NWL3}, \lambda) - \left[\frac{\hbar^2(n_z\pi/d_z)^2}{2m_c} + \frac{\hbar^2(n_y\pi/d_y)^2}{2m_c}\right]\right]\frac{2m_c}{\hbar^2}\right]^{1/2}$$

and E_{F1NWL3} is the Fermi energy in this case.

The electronic contribution to the second- and third-order elastic constants for HD materials in this case can be written as

$$\Delta C_{44} = \frac{-G_0^2}{9d_y d_z}\left[\frac{\partial n_0}{\partial E_{F1NWL3}}\right] \quad (1.312a)$$

and

$$\Delta C_{456} = \frac{G_0^3}{27d_y d_z}\left[\frac{\partial^2 n_0}{\partial E_{F1NWL3}^2}\right] \quad (1.312b)$$

Thus, using eqs. (1.312a), (1.312b), and (1.311) we can find ΔC_{44} and ΔC_{456} in this case.

1.2.10 The magneto-CECs in accumulation and inversion layers of Kane-type Semiconductors in the presence of light waves

(a) The 2D DR in accumulation layers of HD III–V, ternary, and quaternary materials, whose unperturbed band structure is defined by the three-band Kane model, in the presence of light waves under magnetic quantization, can be expressed as

$$T_1(E_{17,30}, \eta_g, \lambda) = \left(n + \frac{1}{2}\right)\hbar\omega_0 + S_i \left[\frac{\hbar|e|F_s\left[T_1(E_{17,30}, \eta_g, \lambda)\right]'}{\sqrt{2m_c}}\right]^{2/3} \tag{1.313}$$

where $E_{17,30}$ is the totally quantized energy in this case. Equation (1.313) represents the magneto-DR of the 2D electrons in accumulation layers of HD III–V, ternary, and quaternary materials under the weak electric field limit in the presence of light waves whose bulk electrons obey the HD three-band Kane model. Since the DR in accordance with the HD three-band Kane model is complex in nature, eq. (1.313) will also be complex in the energy plane. The total energy is quantized since the wave vector space is totally quantized.

The electron concentration can be written at a finite temperature as

$$n_0 = \frac{g_v eB}{\pi\hbar} \text{ Real Part of } \sum_{n=0}^{n_{max}} \sum_{i=0}^{i_{max}} F_{-1}(\eta_{17,30}) \tag{1.314}$$

Where $\eta_{17,30} = \frac{E_{F17,30} - E_{17,30}}{k_B T}$ and $E_{F17,30}$ is the Fermi energy in this case.

The electronic contribution to the second- and third-order elastic constants for HD materials in this case can be written as

$$\Delta C_{44} = \frac{-G_0^2}{9} \text{ Real Part of } \left[\frac{\partial n_0}{t_i \partial(E_{F17,30} - E_{17,30})}\right] \tag{1.315a}$$

and

$$\Delta C_{456} = \frac{G_0^3}{27} \text{ Real Part of } \left[\frac{\partial^2 n_0}{t_i \partial(E_{F17,30} - E_{17,30})^2}\right] \tag{1.315b}$$

Thus, using eqs. (1.315a), (1.315b), and (1.314) we can find ΔC_{44} and ΔC_{456} in this case.

In the absence of band tails and in the presence of light waves under the condition of weak electric field limit, the 2D electron dispersion relation under magnetic quantization assumes the form

$$\beta_0(E_{17,31},\lambda) = \left(n+\frac{1}{2}\right)\hbar\omega_0 + S_i\left[\frac{\hbar|e|F_s\left[\beta_0(E_{17,31},\lambda)\right]'}{\sqrt{2m_c}}\right]^{2/3} \tag{1.316}$$

where $E_{17,31}$ is the totally quantized energy in this case.

The electron concentration can be written at a finite temperature as

$$n_0 = \frac{g_v eB}{\pi\hbar}\sum_{n=0}^{n_{max}}\sum_{i=0}^{i_{max}} F_{-1}(\eta_{17,31}) \tag{1.317}$$

Where $\eta_{17,31} = \frac{E_{F17,31} - E_{17,31}}{k_B T}$ and $E_{F17,31}$ is the Fermi energy in this case.

The electronic contribution to the second- and third-order elastic constants for HD materials in this case can be written as

$$\Delta C_{44} = \frac{-G_0^2}{9}\left[\frac{\partial n_0}{t_i \partial E_{F17,31}}\right] \tag{1.318a}$$

and

$$\Delta C_{456} = \frac{G_0^3}{27}\left[\frac{\partial^2 n_0}{t_i \partial E_{F17,31}^2}\right] \tag{1.318b}$$

Thus, using eqs. (1.318a), (1.318b), and (1.317) we can find ΔC_{44} and ΔC_{456} in this case.

(b) The 2D DR in accumulation layers of HD III–V, ternary, and quaternary materials, whose unperturbed band structure is defined by the two-band Kane model, in the presence of light waves under magnetic quantization, can be expressed as

$$T_2(E_{17,32},\eta_g,\lambda) = \left(n+\frac{1}{2}\right)\hbar\omega_0 + S_i\left[\frac{\hbar|e|F_s\left[T_2(E_{17,32},\eta_g,\lambda)\right]'}{\sqrt{2m_c}}\right]^{2/3} \tag{1.319}$$

where $E_{17,32}$ is the totally quantized energy in this case.

Equation (1.319) represents the magneto-DR of the 2D electrons in accumulation layers of HD III–V, ternary, and quaternary materials under the weak electric field limit in the presence of light waves whose bulk electrons obey the HD two-band Kane model. Since the DR in accordance with the HD two-band Kane model is real in nature, eq. (1.319) will also be real in the energy plane. The total energy is quantized since the wave vector space is totally quantized.

The electron concentration can be written at a finite temperature as

$$n_0 = \frac{g_v eB}{\pi\hbar} \sum_{n=0}^{n_{\max}} \sum_{i=0}^{i_{\max}} F_{-1}(\eta_{17,32}). \quad (1.320)$$

where $\eta_{17,32} = \frac{E_{F17,32} - E_{17,32}}{k_B T}$ and $E_{F17,32}$ is the Fermi energy in this case.

The electronic contribution to the second- and third-order elastic constants for HD materials in this case can be written as

$$\Delta C_{44} = \frac{-G_0^2}{9} \left[\frac{\partial n_0}{t_i \partial(E_{F17,32} - E_{17,32})} \right] \quad (1.321a)$$

and

$$\Delta C_{456} = \frac{G_0^3}{27} \left[\frac{\partial^2 n_0}{t_i \partial(E_{F17,32} - E_{17,32})^2} \right] \quad (1.321b)$$

Thus, using eqs. (1.321a), (1.321b), and (1.320) we can find ΔC_{44} and ΔC_{456} in this case.

In the absence of band tails and under the condition of weak electric field limit, the 2D dispersion relation in the presence of light waves assumes the form

$$\tau_0(E_{17,33}, \lambda) = \left(n + \frac{1}{2}\right)\hbar\omega_0 + S_i \left[\frac{\hbar|e|F_s[\tau_0(E_{17,33}, \lambda)]'}{\sqrt{2m_c}} \right]^{2/3} \quad (1.322)$$

where $E_{17,33}$ is the totally quantized energy in this case.

The electron concentration can be written at a finite temperature as

$$n_0 = \frac{g_v eB}{\pi\hbar} \sum_{n=0}^{n_{\max}} \sum_{i=0}^{i_{\max}} F_{-1}(\eta_{17,33}) \quad (1.323)$$

Where

$$\eta_{17,33} = \frac{E_{F17,33} - E_{17,33}}{k_B T}$$

and $E_{F17,33}$ is the Fermi energy in this case.

The electronic contribution to the second- and third-order elastic constants for HD materials in this case can be written as

$$\Delta C_{44} = \frac{-G_0^2}{9}\left[\frac{\partial n_0}{t_i \partial(E_{F17,33} - E_{17,33})}\right] \tag{1.324a}$$

and

$$\Delta C_{456} = \frac{G_0^3}{27}\left[\frac{\partial^2 n_0}{t_i \partial(E_{F17,33} - E_{17,33})^2}\right] \tag{1.324b}$$

Thus, using eqs. (1.324a), (1.324b), and (1.323) we can find ΔC_{44} and ΔC_{456} in this case.

(c) The 2D DR in accumulation layers of HD III-V, ternary, and quaternary materials, whose unperturbed band structure is defined by the parabolic energy bands, in the presence of light waves under magnetic quantization, can be expressed as

$$T_3(E_{17,323}, \eta_g, \lambda) = \left(n + \frac{1}{2}\right)\hbar\omega_0 + S_i\left[\frac{\hbar|e|F_s\left[T_3(E_{17,323}, \eta_g, \lambda)\right]'}{\sqrt{2m_c}}\right]^{2/3} \tag{1.325}$$

where $E_{17,323}$ is the totally quantized energy in this case.

The electron concentration can be written at a finite temperature as

$$n_0 = \frac{g_v eB}{\pi\hbar}\sum_{n=0}^{n_{max}}\sum_{i=0}^{i_{max}} F_{-1}(\eta_{17,323}) \tag{1.326}$$

where $\eta_{17,323} = \frac{E_{F17,323} - E_{17,323}}{k_B T}$ and $E_{F17,323}$. is the Fermi energy in this case.

The electronic contribution to the second- and third-order elastic constants for HD materials in this case can be written as

$$\Delta C_{44} = \frac{-G_0^2}{9}\left[\frac{\partial n_0}{t_i \partial(E_{F17,323} - E_{17,323})}\right] \tag{1.327a}$$

and

$$\Delta C_{456} = \frac{G_0^3}{27}\left[\frac{\partial^2 n_0}{t_i \partial(E_{F17,323} - E_{17,323})^2}\right] \tag{1.327b}$$

Thus, using eqs. (1.327a), (1. 327b), and (1.326) we can find ΔC_{44} and ΔC_{456} in this case.

In the absence of band tails and under the condition of weak electric field limit, the 2D dispersion relation in the presence of light waves assumes the form

$$\rho_0(E_{17,333},\lambda) = \left(n+\frac{1}{2}\right)\hbar\omega_0 + S_i\left[\frac{\hbar|e|F_s[\rho_0(E_{17,333},\lambda)]}{\sqrt{2m_c}}\right]^{2/3} \quad (1.328)$$

where $E_{17,333}$ is the totally quantized energy in this case.

The electron concentration can be written at a finite temperature as

$$n_0 = \frac{g_v eB}{\pi\hbar}\sum_{n=0}^{n_{\max}}\sum_{i=0}^{i_{\max}} F_{-1}(\eta_{17,333}) \quad (1.329)$$

Where $\eta_{17,333} = \frac{E_{F17,333} - E_{17,333}}{k_B T}$ and $E_{F17,333}$ is the Fermi energy in this case.

The electronic contribution to the second- and third-order elastic constants for HD materials in this case can be written as

$$\Delta C_{44} = \frac{-G_0^2}{9}\left[\frac{\partial n_0}{t_i\partial(E_{F17,333} - E_{17,333})}\right] \quad (1.330a)$$

and

$$\Delta C_{456} = \frac{G_0^3}{27}\left[\frac{\partial^2 n_0}{t_i\partial(E_{F17,333} - E_{17,333})^2}\right] \quad (1.330b)$$

Thus, using eqs. (1.330a), (1.330b), and (1.329) we can find ΔC_{44} and ΔC_{456} in this case.

1.2.11 The magneto-CECs in doping superlattices of HD Kane-type Semiconductors in the presence of light waves

(a) The magneto-DR in doping superlattices of HD III–V, ternary, and quaternary materials in the presence of external photoexcitation whose unperturbed electrons are defined by the three-band Kane model can be expressed as

$$T_1(E_{17,40},\eta_g,\lambda) = \left(n_i+\frac{1}{2}\right)\hbar\omega_{91HD}(E_{17,40},\eta_g,\lambda) + \left(n+\frac{1}{2}\right)\hbar\omega_0 \quad (1.331)$$

where $E_{17,40}$ is the total energy in this case.

The electron concentration can be expressed at a finite temperature as

$$n_0 = \frac{g_v eB}{\pi\hbar}\,\text{Real part of}\sum_{n=0}^{n_{\max}}\sum_{i=0}^{i_{\max}} F_{-1}(\eta_{17,40}) \quad (1.332)$$

where $\eta_{17,40} = \frac{E_{F17,40} - E_{17,40}}{k_B T}$ and $E_{F17,40}$ is the Fermi energy in this case.

The electronic contribution to the second- and third-order elastic constants for HD materials in this case can be written as

$$\Delta C_{44} = \frac{-G_0^2}{9d_0} \text{Real } part\ of \left[\frac{\partial n_0}{\partial(E_{F17,40} - E_{17,40})}\right] \tag{1.333a}$$

and

$$\Delta C_{456} = \frac{G_0^3}{27d_0} \text{Real part of} \left[\frac{\partial^2 n_0}{\partial(E_{F17,40} - E_{17,40})^2}\right] \tag{1.333b}$$

Thus, using eqs. (1.333a), (1.333b), and (1.332) we can find ΔC_{44} and ΔC_{456} in this case.

(b) The magneto-DR in doping superlattices of HD III–V, ternary, and quaternary materials in the presence of external photoexcitation whose unperturbed electrons are defined by the two-band Kane model can be expressed as

$$T_2(E_{17,41}, \eta_g, \lambda) = \left(n_i + \frac{1}{2}\right)\hbar\omega_{92HD}(E_{17,41}, \eta_g, \lambda) + \left(n + \frac{1}{2}\right)\hbar\omega_0 \tag{1.334}$$

where $E_{17,41}$ is the total energy in this case.

The electron concentration can be expressed at a finite temperature as

$$n_0 = \frac{g_v eB}{\pi\hbar} \sum_{n=0}^{n_{\max}} \sum_{i=0}^{i_{\max}} F_{-1}(\eta_{17,41}) \tag{1.335}$$

where $\eta_{17,41} = \frac{E_{F17,41} - E_{17,41}}{k_B T}$ and $E_{F17,41}$ is the Fermi energy in this case.

The electronic contribution to the second- and third-order elastic constants for HD materials in this case can be written as

$$\Delta C_{44} = \frac{-G_0^2}{9d_0} \left[\frac{\partial n_0}{\partial(E_{F17,41} - E_{17,41})}\right] \tag{1.336a}$$

and

$$\Delta C_{456} = \frac{G_0^3}{27d_0} \left[\frac{\partial^2 n_0}{\partial(E_{F17,41} - E_{17,41})^2}\right] \tag{1.336b}$$

Thus, using eqs. (1.336a), (1.336b), and (1.335) we can find ΔC_{44} and ΔC_{456} in this case.

(c) The magneto-DR in doping superlattices of HD III–V, ternary, and quaternary materials in the presence of external phot-excitation whose unperturbed electrons are defined by the parabolic energy bands can be expressed as

$$T_3(E_{17,42}, \eta_g, \lambda) = \left(n_i + \frac{1}{2}\right)\hbar\omega_{93HD}(E_{17,42}, \eta_g, \lambda) + \left(n + \frac{1}{2}\right)\hbar\omega_0 \tag{1.337}$$

where $E_{17,42}$ is the total energy in this case.

The electron concentration can be expressed at a finite temperature as

$$n_0 = \frac{g_v eB}{\pi\hbar} \sum_{n=0}^{n_{max}} \sum_{i=0}^{i_{max}} F_{-1}(\eta_{17,42}) \tag{1.338}$$

where $\eta_{17,42} = \frac{E_{F17,42} - E_{17,42}}{k_B T}$ and $E_{F17,42}$ is the Fermi energy in this case.

The electronic contribution to the second- and third-order elastic constants for HD materials in this case can be written as

$$\Delta C_{44} = \frac{-G_0^2}{9d_0}\left[\frac{\partial n_0}{\partial(E_{F17,42} - E_{17,42})}\right] \tag{1.339a}$$

and

$$\Delta C_{456} = \frac{G_0^3}{27d_0}\left[\frac{\partial^2 n_0}{\partial(E_{F17,42} - E_{17,42})^2}\right] \tag{1.339b}$$

Thus, using eqs. (1.339a), (1.339b), and (1338) we can find ΔC_{44} and ΔC_{456} in this case.

1.2.12 The CECs in QWHD EMSLs of Kane-type semiconductors in the presence of light waves

(a) Following Sasaki [76], the electron dispersion law in HD III–V EMSLs in the presence of light waves, the dispersion relations of whose constituent materials in the absence of any perturbation are defined by the three-band Kane model, can be written as

$$k_x^2 = \left[\frac{1}{L_0^2}\{\cos^{-1}(f_{HD1}(E, k_y, k_z\lambda))\}^{-2} - k_\perp^2\right] \tag{1.340}$$

In which,

$$f_{HD1}(E, k_y, k_z\lambda) = a_{1HD1}\cos[a_0 C_{1HD1}(E, k_\perp) + b_0 D_{1HD1}(E, k_\perp)] - a_{2HD1}\cos[a_0 C_{1HD1}(E, k_\perp) - b_0 D_{1HD1}(E, k_\perp)],\ k_\perp^2 = k_y^2 + k_z^2,\ \mathrm{L}_0 = a_0 + b_0,$$

$$a_{1HD1} = \left[\sqrt{\frac{m_{c2}\text{Real part of }\left[T_1(0,\eta_{g2},\lambda)\right]}{m_{c1}\text{Real part of }\left[T_1(0,\eta_{g1},\lambda)\right]}}+1\right]^2$$

$$\left[4\left(\frac{m_{c2}\text{Real part of}\left[T_1(0,\eta_{g2},\lambda)\right]}{m_{c1}\text{Real part of}\left[T_1(0,\eta_{g1},\lambda)\right]}\right)^{1/2}\right]^{-1}$$

$$a_{2HD1} = \left[-1+\sqrt{\frac{m_{c2}\text{Real part of}\left[T_1(0,\eta_{g2},\lambda)\right]}{m_{c1}\text{Real part of}\left[T_1(0,\eta_{g1},\lambda)\right]}}\right]^2$$

$$\left[4\left(\frac{m_{c2}\text{Real part of}\left[T_1(0,\eta_{g2},\lambda)\right]}{m_{c1}\text{Real part of}\left[T_1(0,\eta_{g1},\lambda)\right]}\right)^{1/2}\right]^{-1}$$

$$C_{1HD1}(E,k_\perp,\lambda) = \left[\left(\frac{2m_{c1}}{\hbar^2}\right)T_1(E,\eta_{g_1},\lambda) - k_\perp^2\right]^{1/2}$$

$$\text{and } D_{1HD1}(E,k_\perp,\lambda) = \left[\left(\frac{2m_{c2}}{\hbar^2}\right)T_1(E,\eta_{g_1},\lambda) - k_\perp^2\right]^{1/2}$$

The DR in in QWHD EMSLs of Kane-type semiconductors in the presence of light waves, the dispersion relations of whose constituent materials in the absence of any perturbation are defined by the three-band Kane model, can be written as

$$\left(\frac{n_x\pi}{d_x}\right)^2 = \left[\frac{1}{L_0^2}\left\{\cos^{-1}(f_{HD1}(E,k_y,k_z,\lambda))\right\}^2 - k_\perp^2\right] \tag{1.341}$$

The electron concentration per unit area, ΔC_{44} and ΔC_{456} have to be evaluated numerically in this case.

(b) Following Sasaki [76], the electron dispersion law in HD III–V EMSLs in the presence of light waves, the dispersion relations of whose constituent materials in the absence of any perturbation are defined by the two-band Kane model, can be written as

$$k_x^2 = \left[\frac{1}{L_0^2}\left\{\cos^{-1}(f_{HD2}(E,k_y,k_z\lambda))\right\}^{-2} - k_\perp^2\right] \tag{1.342}$$

in which

$$f_{HD2}(E,k_y,k_z\lambda)=a_{1HD2}\cos[a_0C_{1HD2}(E,k_\perp)+b_0D_{1HD2}(E,k_\perp)]$$
$$-a_{2HD2}\cos[a_0C_{1HD2}(E,k_\perp)-b_0D_{1HD2}(E,k_\perp)],$$

$$a_{1HD2}=\left[\left[\sqrt{\frac{m_{c2}\left[T_2(0,\eta_{g2},\lambda)\right]}{m_{c1}\left[T_2(0,\eta_{gl},\lambda)\right]}}+1\right]^2\left[4\left(\frac{m_{c2}\left[T_2(0,\eta_{g2},\lambda)\right]}{m_{c1}\left[T_2(0,\eta_{gl},\lambda)\right]}\right)\right]^{1/2}\right]^{-1}$$

$$a_{2HD1}=\left[\left[-1+\sqrt{\frac{m_{c2}\left[T_2(0,\eta_{g2},\lambda)\right]}{m_{c1}\left[T_2(0,\eta_{g1},\lambda)\right]}}\right]^2\left[4\left(\frac{m_{c2}\left[T_2(0,\eta_{g2},\lambda)\right]}{m_{c1}\left[T_2(0,\eta_{g1},\lambda)\right]}\right)\right]^{1/2}\right]^{-1}$$

$$C_{1HD1}(E,k_\perp,\lambda)=\left[\left(\frac{2m_{c1}}{\hbar^2}\right)T_2(E,\eta_{g_1},\lambda)-k_\perp^2\right]^{1/2}$$

and

$$D_{1HD1}(E,k_\perp,\lambda)=\left[\left(\frac{2m_{c2}}{\hbar^2}\right)T_2(E,\eta_{g_1},\lambda)-k_\perp^2\right]^{1/2}$$

The DR in QWHD EMSLs of Kane-type semiconductors in the presence of light waves, the dispersion relations of whose constituent materials in the absence of any perturbation are defined by the two-band Kane model, can be written as

$$\left(\frac{n_x\pi}{d_x}\right)^2=\left[\frac{1}{L_0^2}\{\cos^{-1}(f_{HD2}(E,k_y,k_z,\lambda))\}^2-k_\perp^2\right] \tag{1.343}$$

The electron concentration per unit area, ΔC_{44} and ΔC_{456}, has to be evaluated numerically in this case.

(c) Following Sasaki [76], the electron dispersion law in HD III–V EMSLs in the presence of light waves, the dispersion relations of whose constituent materials in the absence of any perturbation are defined by the parabolic energy bands, can be written as

$$k_x^2=\left[\frac{1}{L_0^2}\{\cos^{-1}(f_{HD3}(E,k_y,k_z\lambda))\}^{-2}-k_\perp^2\right] \tag{1.344}$$

in which

$$f_{HD3}(E,k_y,k_z\lambda)=a_{2HD3}\cos[a_0C_{1HD3}(E,k_\perp)+b_0D_{1HD3}(E,k_\perp)]-a_{2HD3}\cos[a_0C_{1HD3}(E,k_\perp)$$
$$-b_0D_{1HD3}(E,k_\perp)],$$

$$a_{1HD2} = \left[\left[\sqrt{\frac{m_{c2}\left[T_3(0,\eta_{g2},\lambda)\right]}{m_{c1}\left[T_3(0,\eta_{g2},\lambda)\right]}}+1\right]^2\left[4\left(\frac{m_{c2}\left[T_3(0,\eta_{g2},\lambda)\right]}{m_{c1}\left[T_3(0,\eta_{g1},\lambda)\right]}\right)\right]^{1/2}\right]^{-1}$$

$$a_{1HD3} = \left[\left[\sqrt{\frac{m_{c2}\left[T_3(0,\eta_{g2},\lambda)\right]}{m_{c1}\left[T_3(0,\eta_{g1},\lambda)\right]}}\right]^2\left[4\left(\frac{m_{c2}\left[T_3(0,\eta_{g2},\lambda)\right]}{m_{c1}\left[T_3(0,\eta_{g1},\lambda)\right]}\right)\right]^{1/2}\right]^{-1}$$

$$C_{1HD1}(E,k_\perp,\lambda) = \left[\left(\frac{2m_{c1}}{\hbar^2}\right)T_3(E,\eta_{g_1},\lambda) - k_\perp^2\right]^{1/2}$$

and

$$D_{1HD3}(E,k_\perp,\lambda) = \left[\left(\frac{2m_{c2}}{\hbar^2}\right)T_3(E,\eta_{g_1},\lambda) - k_\perp^2\right]^{1/2}$$

The DR in QWHD EMSLs of Kane-type semiconductors in the presence of light waves, the dispersion relations of whose constituent materials in the absence of any perturbation are defined by the parabolic energy bands, can be written as

$$\left(\frac{n_x\pi}{d_x}\right)^2 = \left[\frac{1}{L_0^2}\{\cos^{-1}(f_{HD3}(E,k_y,k_z,\lambda))\}^2 - k_\perp^2\right] \tag{1.345}$$

The electron concentration per unit area,ΔC_{44} and ΔC_{456}, has to be evaluated numerically in this case.

1.2.13 The CECs in NWHD EMSLs of Kane-Type semiconductors in the presence of light waves

(a) The DR in NWHD EMSLs of Kane-type semiconductors in the presence of light waves, the dispersion relations of whose constituent materials in the absence of any perturbation are defined by the three-band Kane model, can be written as

$$k_x^2 = \left[\frac{1}{L_0^2}\left\{\cos^{-1}\left(f_{HD1}\left(E,\frac{n_y\pi}{d_y},\frac{n_z\pi}{d_z},\lambda\right)\right)\right\}^2 - \left[\left(\frac{n_y\pi}{d_y}\right)^2+\left(\frac{n_z\pi}{d_z}\right)^2\right]\right] \tag{1.346}$$

The electron concentration can be written as

$$n_0 = \frac{2}{\pi}\mathit{Real\,part\,of}$$

$$\sum_{n_y=1}^{n_{y\max}} \sum_{n_z=1}^{n_{z\max}} \left[\left[\frac{1}{L_0^2}\left\{\cos^{-1}\left(f_{HD1}\left(E_{F1.2.13}, \frac{n_y\pi}{d_y}, \frac{n_z\pi}{d_z}, \lambda\right)\right)\right\}^2 - \left[\left(\frac{n_y\pi}{d_y}\right)^2 + \left(\frac{n_z\pi}{d_z}\right)^2\right]\right]^{1/2}\right] \tag{1.347}$$

where $E_{F1.2.13}$ is the Fermi energy in this case.

The electronic contribution to the second- and third-order elastic constants for HD materials in this case can be written as

$$\Delta C_{44} = \frac{-G_0^2}{9d_y d_z} \text{Real part of} \left[\frac{\partial n_0}{\partial(E_{F1.2.13} - E_{nSL5HD4,})}\right] \tag{1.348a}$$

and

$$\Delta C_{456} = \frac{G_0^3}{27d_y d_z} \text{Real part of} \left[\frac{\partial^2 n_0}{\partial(E_{F1.2.13} - E_{nSL5HD4,})^2}\right] \tag{1.348b}$$

Thus, using eqs. (1.348a), (1.348b), and (1. 347) we can find ΔC_{44} and ΔC_{456} in this case.

(b) The DR in in NWHD EMSLs of Kane-type semiconductors in the presence of light waves, the dispersion relations of whose constituent materials in the absence of any perturbation are defined by the two-band Kane model, can be written as

$$k_x^2 = \left[\frac{1}{L_0^2}\left\{\cos^{-1}\left(f_{HD2}\left(E, \frac{n_y\pi}{d_y}, \frac{n_z\pi}{d_z}, \lambda\right)\right)\right\}^2 - \left[\left(\frac{n_y\pi}{d_y}\right)^2 + \left(\frac{n_z\pi}{d_z}\right)^2\right]\right] \tag{1.349}$$

The electron concentration can be written as

$$n_0 = \frac{2}{\pi} \sum_{n_y=1}^{n_{y\max}} \sum_{n_z=1}^{n_{z\max}}$$

$$\left[\left[\frac{1}{L_0^2}\left\{\cos^{-1}\left(f_{HD2}\left(E_{F1.2.131}, \frac{n_y\pi}{d_y}, \frac{n_z\pi}{d_z}, \lambda\right)\right)\right\}^2 - \left[\left(\frac{n_y\pi}{d_y}\right)^2 + \left(\frac{n_z\pi}{d_z}\right)^2\right]\right]^{1/2}\right] \tag{1.350}$$

where $E_{F1.2.131}$ is the Fermi energy in this case.

The electronic contribution to the second- and third-order elastic constants for HD materials in this case can be written as

$$\Delta C_{44} = \frac{-G_0^2}{9d_y d_z}\left[\frac{\partial n_0}{\partial(E_{F1.2.131} - E_{nSL5HD5})}\right] \tag{1.351a}$$

and

$$\Delta C_{456} = \frac{G_0^3}{27d_y d_z}\left[\frac{\partial^2 n_0}{\partial(E_{F1.2.131} - E_{nSL5HD5},)^2}\right] \tag{1.351b}$$

Thus, using eqs. (1.351a), (1.351b), and (1. 350) we can find ΔC_{44} and ΔC_{456} in this case.

(c) The DR in NWHD EMSLs of Kane-type semiconductors in the presence of light waves, the dispersion relations of whose constituent materials in the absence of any perturbation are defined by the parabolic energy bands, can be written as

$$k_x^2 = \left[\frac{1}{L_0^2}\left\{\cos^{-1}\left(f_{HD3}\left(E, \frac{n_y\pi}{d_y}, \frac{n_z\pi}{d_z}, \lambda\right)\right)\right\}^2 - \left[\left(\frac{n_y\pi}{d_y}\right)^2 + \left(\frac{n_z\pi}{d_z}\right)^2\right]\right] \tag{1.352}$$

The electron concentration can be written as

$$n_0 = \frac{2}{\pi}\sum_{n_y=1}^{n_{y\max}}\sum_{n_z=1}^{n_{z\max}}\left[\left[\frac{1}{L_0^2}\left\{\cos^{-1}(f_{HD3}(E_{F1.2.132}, \frac{n_y\pi}{d_y}, \frac{n_z\pi}{d_z}, \lambda))\right\}^2 - \left[\left(\frac{n_y\pi}{d_y}\right)^2 + \left(\frac{n_z\pi}{d_z}\right)^2\right]\right]^{1/2}\right] \tag{1.353}$$

where $E_{F1.2.132}$ is the Fermi energy in this case.

The electronic contribution to the second- and third-order elastic constants for HD materials in this case can be written as

$$\Delta C_{44} = \frac{-G_0^2}{9d_y d_z}\left[\frac{\partial n_0}{\partial(E_{F1.2.132} - E_{nSL5HD6})}\right] \tag{1.354a}$$

and

$$\Delta C_{456} = \frac{G_0^3}{27d_y d_z}\left[\frac{\partial^2 n_0}{\partial(E_{F1.2.132} - E_{nSL5HD6},)^2}\right] \tag{1.354b}$$

Thus, using eqs. (1.354a), (1.354b), and (1.353) we can find ΔC_{44} and ΔC_{456} in this case.

1.2.14 The magneto-CECs in HD EMSLs of Kane-type semiconductors in the presence of light waves

(a) In the presence of an external magnetic field along x-direction, the simplified magneto-dispersion law in this case can be written as

$$k_x^2 = [\rho_{4HD1}(n, E, \lambda)] \tag{1.355}$$

in which

$$\rho_{4HD1}(n, E, \lambda) = \frac{1}{L_0^2}\left[\cos^{-1}(f_{HD1}(n, E, \lambda))\right]^2 - \left\{\frac{2|e|B}{\hbar}\left(n + \frac{1}{2}\right)\right\}$$

$$f_{HD1}(E, n, \lambda) = a_{1HD1}\cos[a_0 C_{1HD1}(E, n, \lambda) + b_0 D_{1HD1}(E, n, \lambda)]$$
$$- a_{2HD1}\cos[a_0 C_{1HD1}(E, n, \lambda) - b_0 D_{1HD1}(E, n, \lambda)]$$

$$C_{1HD1}(E, n, \lambda) \equiv \left[\left(\frac{2m_{c1}}{\hbar^2}\right) T_1(E, \eta_{g_1}, \lambda) - \left\{\frac{2|e|B}{\hbar}\left(n + \frac{1}{2}\right)\right\}\right]^{1/2}$$

and

$$D_{1HD1}(E, n, \lambda) \equiv \left[\left(\frac{2m_{c2}}{\hbar^2}\right) T_1\left(E, \eta_{g_2}, \lambda\right) - \left\{\frac{2|e|B}{\hbar}\left(n + \frac{1}{2}\right)\right\}\right]^{1/2}$$

The electron concentration can be expressed as

$$n_0 = \frac{eB}{\pi^2\hbar} \,\text{Real part of} \sum_{n=0}^{n_{\max}} \left[[\rho_{4HD1}(n, E_{fSLHDB}, \lambda)]^{1/2}\right] \tag{1.356}$$

where E_{fSLHDB} is the Fermi energy in this case.

The electronic contribution to the second- and third-order elastic constants for HD materials in this case can be written as

$$\Delta C_{44} = \frac{-G_0^2}{9} \,\text{Real part of} \left[\frac{\partial n_0}{\partial(E_{f\,SLHDB} - E_{nSL5HD})}\right] \tag{1.357a}$$

and

$$\Delta C_{456} = \frac{G_0^3}{27} \,\text{Real part of} \left[\frac{\partial^2 n_0}{\partial(E_{f\,SLHDB} - E_{n\,SL5HD})^2}\right] \tag{1.357b}$$

Thus, using eqs. (1.357a), (1.357b), and (1.356) we can find ΔC_{44} and ΔC_{456} in this case.

(b) In the presence of an external magnetic field along x-direction, the simplified magneto-dispersion law in this case can be written as

$$k_x^2 = [\rho_{4HD2}(n, E, \lambda)] \tag{1.358}$$

in which

$$\rho_{4HD2}(n, E, \lambda) = \frac{1}{L_0^2}\left[\cos^{-1}(f_{HD2}(n, E, \lambda))\right]^2 - \left\{\frac{2|e|B}{\hbar}\left(n + \frac{1}{2}\right)\right\}$$

$$f_{HD2}(E, n, \lambda) = a_{1HD2}\cos[a_0 C_{1HD2}(E, n, \lambda) + b_0 D_{1HD2}(E, n, \lambda)]$$

$$- a_{2HD2}\cos[a_0 C_{1HD2}(E, n, \lambda) - b_0 D_{1HD2}(E, n, \lambda)]$$

$$C_{1HD2}(E, n, \lambda) \equiv \left[\left(\frac{2m_{c1}}{\hbar^2}\right)T_2(E, \eta_{g_1}, \lambda) - \left\{\frac{2|e|B}{\hbar}\left(n + \frac{1}{2}\right)\right\}\right]^{1/2}$$

and

$$D_{1HD2}(E, n, \lambda) \equiv \left[\left(\frac{2m_{c2}}{\hbar^2}\right)T_2(E, \eta_{g_2}, \lambda) - \left\{\frac{2|e|B}{\hbar}\left(n + \frac{1}{2}\right)\right\}\right]^{1/2}$$

The electron concentration can be expressed as

$$n_0 = \frac{eB}{\pi^2\hbar}\sum_{n=0}^{n_{max}}\left[[\rho_{4HD2}(n, E_{f\,SLHDB}, \lambda)]^{1/2}\right] \tag{1.359}$$

where E_{fSLHDB} is the Fermi energy in this case.

The electronic contribution to the second- and third-order elastic constants for HD materials in this case can be written as

$$\Delta C_{44} = \frac{-G_0^2}{9}\left[\frac{\partial n_0}{\partial(E_{f\,SLHDB} - E_{n\,SL5HD2})}\right] \tag{1.360a}$$

and

$$\Delta C_{456} = \frac{G_0^3}{27}\left[\frac{\partial^2 n_0}{\partial(E_{f\,SLHDB} - E_{n\,SL5HD2})^2}\right] \tag{1.360b}$$

Thus, using eqs. (1.360a), (1.360b), and (1.359) we can find ΔC_{44} and ΔC_{456} in this case.

(c) In the presence of an external magnetic field along x-direction, the simplified magneto-dispersion law in this case can be written as

$$k_x^2 = [\rho_{4HD3}(n, E, \lambda)] \tag{1.361}$$

in which

$$\rho_{4HD3}(n, E, \lambda) = \frac{1}{L_0^2}\left[\cos^{-1}(f_{HD3}(n, E, \lambda))\right]^2 - \left\{\frac{2|e|B}{\hbar}\left(n + \frac{1}{2}\right)\right\}$$

$$f_{HD3}(E, n, \lambda) = a_{1HD3}\cos[a_0 C_{1HD3}(E, n, \lambda) + b_0 D_{1HD3}(E, n, \lambda)]$$

$$- a_{2HD3}\cos[a_0 C_{1HD3}(E, n, \lambda) - b_0 D_{1HD3}(E, n, \lambda)]$$

$$C_{1HD3}(E, n, \lambda) \equiv \left[\left(\frac{2m_{c1}}{\hbar^2}\right) T_3(E, \eta_{g_1}, \lambda) - \left\{\frac{2|e|B}{\hbar}\left(n + \frac{1}{2}\right)\right\}\right]^{1/2}$$

and

$$D_{1HD3}(E, n, \lambda) \equiv \left[\left(\frac{2m_{c2}}{\hbar^2}\right) T_3(E, \eta_{g_2}, \lambda) - \left\{\frac{2|e|B}{\hbar}\left(n + \frac{1}{2}\right)\right\}\right]^{1/2}$$

The electron concentration can be expressed as

$$n_0 = \frac{eB}{\pi^2\hbar}\sum_{n=0}^{n_{\max}}\left[[\rho_{4HD3}(n, E_{fSLHDB}, \lambda)]^{1/2}\right] \tag{1.362}$$

where $E_{f\,SLHDB}$ is the Fermi energy in this case.

The electronic contribution to the second- and third-order elastic constants for HD materials in this case can be written as

$$\Delta C_{44} = \frac{-G_0^2}{9}\left[\frac{\partial n_0}{\partial(E_{f\,SLHDB} - E_{n\,SL5HD3})}\right] \tag{1.363a}$$

and

$$\Delta C_{456} = \frac{G_0^3}{27}\left[\frac{\partial^2 n_0}{\partial(E_{fSLHDB} - E_{nSL5HD3})^2}\right] \tag{1.363b}$$

Thus, using eqs. (1.363a), (1.363b,) and (1.362) we can find ΔC_{44} and ΔC_{456} in this case.

1.2.15 The magneto-CECs in QWHD EMSLs of Kane-type semiconductors in the presence of light waves

(a) In the presence of an external magnetic field along x-direction, the simplified magneto-dispersion law in QWHD III–V EMSLs in the presence of light waves, the dispersion relations of whose constituent materials in the absence of any perturbation are defined by the three-band Kane model, can be written as

$$\left(\frac{n_x\pi}{d_x}\right)^2 = \left[\rho_{4HGD1}(n, E_{17.50}, \lambda)\right] \tag{1.364}$$

where $E_{17.50}$ is the totally quantized energy in this case.

The electron concentration can be written at a finite temperature as

$$n_0 = \frac{g_v eB}{\pi\hbar} \text{Real part of} \sum_{n=0}^{n_{\max}} \sum_{n_x=0}^{n_{x\max}} F_{-1}(\eta_{17.50}) \tag{1.365}$$

where $\eta_{17.50} = \frac{E_{F17.50} - E_{17.50}}{k_B T}$ and $E_{F17.50}$ is the Fermi energy in this case.

The electronic contribution to the second- and third-order elastic constants for HD materials in this case can be written as

$$\Delta C_{44} = \frac{-G_0^2}{9d_x} \text{Real part of} \left[\frac{\partial n_0}{\partial(E_{F17.50} - E_{17.50})}\right] \tag{1.366a}$$

and

$$\Delta C_{456} = \frac{G_0^3}{27d_x} \text{Real part of} \left[\frac{\partial^2 n_0}{\partial(E_{F17.50} - E_{17.50})^2}\right] \tag{1.366b}$$

Thus, using eqs. (1.366a), (1.366b), and (1.365) we can find ΔC_{44} and ΔC_{456} in this case.

(b) In the presence of an external magnetic field along x-direction, the simplified magneto-dispersion law in QWHD III–V EMSLs in the presence of light waves, the dispersion relations of whose constituent materials in the absence of any perturbation are defined by the two-band Kane model, can be written as

$$\left(\frac{n_x\pi}{d_x}\right)^2 = \left[\rho_{4HGD2}(n, E_{17.51}, \lambda)\right] \tag{1.367}$$

where $E_{17.51}$ is the totally quantized energy in this case.

The electron concentration can be written at a finite temperature as

$$n_0 = \frac{g_v eB}{\pi\hbar} \sum_{n=0}^{n_{\max}} \sum_{n_x=0}^{n_{x\max}} F_{-1}(\eta_{17.51}) \tag{1.368}$$

where $\eta_{17.51} = \frac{E_{F17.51} - E_{17.51}}{k_B T}$ and $E_{F17.51}$ is the Fermi energy in this case.

Thus, using eqs. (1.366a), (1.366b), and (1.368) we can find ΔC_{44} and ΔC_{456} in this case.

(c) In the presence of an external magnetic field along x-direction, the simplified magneto-dispersion law in QWHD III-V EMSLs in the presence of light waves, the dispersion relations of whose constituent materials in the absence of any perturbation are defined by the isotropic parabolic energy bands, can be written as

$$\left(\frac{n_x \pi}{d_x}\right)^2 = [\rho_{4HGD3}(n, E_{17.52}, \lambda)] \tag{1.369}$$

where $E_{17.52}$ is the totally quantized energy in this case.

The electron concentration can be written at a finite temperature as

$$n_0 = \frac{g_v eB}{\pi\hbar} \sum_{n=0}^{n_{\max}} \sum_{n_x=0}^{n_{x\max}} F_{-1}(\eta_{17.52}) \tag{1.370}$$

where $\eta_{17.52} = \frac{E_{F17.52} - E_{17.52}}{k_B T}$ and $E_{F17.52}$ is the Fermi energy in this case.

The electronic contribution to the second- and third-order elastic constants for HD materials in this case can be written as

$$\Delta C_{44} = \frac{-G_0^2}{9d_x} \left[\frac{\partial n_0}{\partial (E_{F17.52} - E_{17.52})}\right] \tag{1.371}$$

and

$$\Delta C_{456} = \frac{G_0^3}{27d_x} \left[\frac{\partial^2 n_0}{\partial (E_{F17.52} - E_{17.52})^2}\right] \tag{1.372}$$

Thus, using eqs. (1.371), (1.372), and (1.370) we can find ΔC_{44} and ΔC_{456} in this case.

1.2.16 The CECs in QWHD superlattices of Kane-type semiconductors with graded interfaces in the presence of light waves

The electron dispersion law in bulk specimens of the heavily doped constituent materials of III–V SLs whose energy band structures are defined can by expressed as

$$\frac{\hbar^2 k^2}{2m_{cj}} = V_{1j}(E, \eta_{gj}, \lambda, \Delta_j, E_{g0j}) + iV_{2j}(E, \eta_{gj}, \lambda, \Delta_j, E_{g0j}) \tag{1.373}$$

where

$$j = 1, 2, \quad V_{1j}(E, \eta_{gj}, \lambda, \Delta_j, E_{g0j}) = \left[U_{\lambda j} T_{1j}(E, \Delta_j, E_{gj}, \eta_{gj}) - P_{\lambda j}\right]$$

$$U_{\lambda j} = (1 + \theta_{\lambda j}), \quad \theta_{\lambda j} = \frac{C_{0j}}{A_j}\left(t_{\lambda j} + \frac{B_j J_{\lambda j}}{A_j}\right), \quad C_{0j} = \left[\frac{|e|^2 I\lambda^2 E_{g0j}(E_{gj} + \Delta_j)\beta_j^2 (1 + \frac{\rho_j}{\sqrt{2}})^2}{384 m_{rj} \pi c^3 \sqrt{\varepsilon_{scj}\varepsilon_0}\left(E_{g0j} + \frac{2}{3}\Delta_j\right)}\right],$$

$$\beta_j = \left[6\left(E_{g0j} + \frac{2}{3}\Delta_j\right)\frac{(E_{g0j} + \Delta_j)}{\chi_j}\right]^{1/2}, \chi_j = (6E_{g0}{}^2 + 9E_{g0j}\Delta_j + 4\Delta_j{}^2), \quad \rho_j = \left[\frac{4\Delta_j{}^2}{3\chi_j}\right]^{1/2},$$

$$A_j = E_{g0j},$$

$$t_{\lambda j} = \left[E_{\lambda j} - \frac{G_{\lambda j}(E_{g0j} - \delta'_j)}{\sqrt{t_{\lambda j}}}\right], \quad E_{\lambda j} = \frac{2B_j(E_{g0j} - \delta'_j)}{\left(A_j + \delta'_j\right)^2}, \quad B_j = \left[1 + \frac{m_{cj}}{m_{vj}}\right],$$

$$G_{\lambda j} = \left[\frac{2B_j}{\left(A_j + \delta'_j\right)^3} - \frac{B_j C_{\lambda j}}{\left(A_j + \delta'_j\right)}\right],$$

$$C_{\lambda j} = \left[\left(E_{g0j} + \delta'_j\right)^{-1} + \left(E_{g0j} + \delta'_j\right)\left(E_{g0j} - \delta'_j\right)^{-2}\right]\left(A_j + \delta'_j\right)^{-1}, \quad P_{\lambda j} = \frac{C_{0j}}{A_j} J_{\lambda j},$$

$$J_{\lambda j} = \left(D_{\lambda j} + 2(E_{g0j} - \delta'_j)\sqrt{f_{\lambda j}}\right), D_{\lambda j} = \left(1 + \frac{2(E_{g0j} - \delta'_j)}{\left(A_j + \delta'_j\right)}\right),$$

$$f_{\lambda j} = \left[\frac{1}{\left(A_j + \delta'_j\right)^2} + \frac{1}{\left(E_{g0j} - \delta'_j\right)^2} - C_{\lambda j}\right], V_{2j}(E, \eta_{gj}, \lambda, \Delta_j, E_{g0j}) = \left[U_{\lambda j} T_{2j}(E, \Delta_j, E_{gj}, \eta_{gj})\right]$$

$$\text{and } T_{2j}(E, \Delta_j, E_{gj}, \eta_{gj}) \equiv \left(\frac{2}{1 + Erf(E/\eta_{gj})}\right)\frac{1}{c_j}\left(1 - \frac{\alpha_j}{c_j}\right)\left(1 - \frac{b_j}{c_j}\right)\frac{\sqrt{\pi}}{c_j \eta_{gj}}\exp(-u_j^2)$$

Therefore, the DR in HD III–V SLs with graded interfaces in the presence of light waves can be expressed as [77a]

$$k_z^2 = G_8 + iH_8 \tag{1.374}$$

where the notations are given in [77b].

The simplified DR of HDQWs of III–V superlattices with graded interfaces can be expressed as

$$\left(\frac{n_z\pi}{d_z}\right)^2 = G_8 + iH_8 \tag{1.375}$$

The electron concentration, ΔC_{44} and ΔC_{456}, has to be evaluated numerically.

1.2.17 The CECs in NWHD superlattices of Kane-type semiconductors with graded interfaces in the presence of light waves

The DR in NWHD superlattices of Kane-type semiconductors with graded interfaces in the presence of light waves can be expressed as

$$k_z^2 = G_{8,17,50} + iH_{8,17,50} \tag{1.376}$$

where the notations are given in [77b].

The electron concentration can be written as

$$n_0 = \frac{2g_v}{\pi} \text{Real part of} \sum_{n_x=1}^{n_{x\max}} \sum_{n_y=1}^{n_{y\max}} \left[[G_{8,17,50} + iH_{8,17,50}]|_{E=E_{F8.17,51}} \right]^{1/2} \tag{1.377}$$

where $E_{F8.17,51}$ is the Fermi energy in this case.

The electronic contribution to the second- and third-order elastic constants for HD materials in this case can be written as

$$\Delta C_{44} = \frac{-G_0^2}{9d_xd_y} \text{Real part of} \left[\frac{\partial n_0}{\partial(E_{F8.17.51} - E_{8.17.52})} \right] \tag{1.378a}$$

and

$$\Delta C_{456} = \frac{G_0^3}{27d_xd_y} \text{Real part of} \left[\frac{\partial^2 n_0}{\partial(E_{F8.17.51} - E_{8.17.52})^2} \right] \tag{1.378b}$$

$(E_{8.17.52})$ can be written as

$$\left[[G_{8,17,50} + iH_{8,17,50}]|_{E=E_{F8.17,52}} \right] = 0 \tag{1.378c}$$

Thus, using eqs. (1.378a), (1.378b), (1.378c), and (1.377) we can find ΔC_{44} and ΔC_{456} in this case.

1.2.18 The CECs in Quantum dot HD superlattices of Kane-type semiconductors with graded interfaces in the presence of light waves

The DR in QDHD superlattices of Kane-type semiconductors with graded interfaces in the presence of light waves can be expressed as

$$\left(\frac{n_z \pi}{d_z}\right)^2 = [G_{8,17,50} + iH_{8,17,50}]|_{E=E_{17,52}} \tag{1.379}$$

where $E_{17,52}$ is the totally quantized energy in this case.

The electron concentration can be expressed at a finite temperature as

$$n_0 = \frac{2g_v}{d_x d_y d_z} \text{Real part of} \sum_{n_x=1}^{n_{x\max}} \sum_{n_y=1}^{n_{y\max}} \sum_{n_z=1}^{n_{z\max}} F_{-1}(\eta_{17,52}) \tag{1.380}$$

where, $\eta_{17,52} = \frac{E_{F17.52} - E_{17.52}}{k_B T}$ and $E_{F17.52}$ is the Fermi energy in this case.

The electronic contribution to the second- and third- order elastic constants for HD materials in this case can be written as

$$\Delta C_{44} = \frac{-G_0^2}{9} \text{Real part of} \left[\frac{\partial n_0}{\partial (E_{F17.52} - E_{17.52})}\right] \tag{1.381a}$$

and

$$\Delta C_{456} = \frac{G_0^3}{27} \text{Real part of} \left[\frac{\partial^2 n_0}{\partial (E_{F17.52} - E_{17.52})^2}\right] \tag{1.381b}$$

Thus, using eqs. (1.381a), (1.381b) and (1.380) we can find ΔC_{44} and ΔC_{456} in this case.

1.2.19 The magneto-CECs in HD superlattices of Kane-type Semiconductors with graded interfaces in the presence of light waves

The magneto DR in HD superlattices of Kane-type semiconductors with graded interfaces in the presence of light waves can be expressed as

$$k_z^2 = G_{8,17,54} + iH_{8,17,54} \tag{1.382}$$

where the notations are given in [77b]

The electron concentration can be written as

$$n_0 = \frac{g_v eB}{\pi^2 \hbar} \text{Real part of} \sum_{n=0}^{n_{\max}} \left[\left[G_{8,17,54} + iH_{8,17,54} \right] \Big|_{E=E_{F8.17,54}}{}^{1/2} \right] \tag{1.383}$$

where $E_{F8.17,54}$ is the Fermi energy in this case.

The electronic contribution to the second- and third-order elastic constants for HD materials in this case can be written as

$$\Delta C_{44} = \frac{-G_0^2}{9} \text{Real part of} \left[\frac{\partial n_0}{\partial (E_{F8.17.54} - E_{17.55})} \right] \tag{1.384a}$$

and

$$\Delta C_{456} = \frac{G_0^3}{27} \text{Real part of} \left[\frac{\partial^2 n_0}{\partial (E_{F8.17.54} - E_{17.55})^2} \right] \tag{1.384b}$$

Thus, using eqs. (1.384a), (1.384b), and (1.383) we can find ΔC_{44} and ΔC_{456} in this case.

1.2.20 The magneto CEC in QWHD superlattices of Kane-type semiconductors with graded interfaces in the presence of light waves

The magneto DR in QWHD superlattices of Kane-type semiconductors with graded interfaces in the presence of light waves can be expressed as

$$\left(\frac{n_z \pi}{d_z} \right)^2 = \left[G_{8,17,54} + iH_{8,17,54} \right] \Big|_{E=E_{17,55}} \tag{1.385}$$

where $E_{17,55}$ is the totally quantized energy in this case.

The electron concentration can be written at a finite temperature as

$$n_0 = \frac{g_v eB}{\pi \hbar} \text{Real part of} \sum_{n_z=1}^{n_{z\max}} \sum_{n=0}^{n_{\max}} F_{-1}(\eta_{17,55}) \tag{1.386}$$

where $\eta_{17,55} = \frac{E_{F17,55} - E_{17,55}}{k_B T}$ and E_{F1755} is the Fermi energy in the present case.

The electronic contribution to the second- and third-order elastic constants for HD materials in this case can be written as

$$\Delta C_{44} = \frac{-G_0^2}{9 d_z} \text{Real part of} \left[\frac{\partial n_0}{\partial (E_{F17.55} - E_{17.55})} \right] \tag{1.387a}$$

and

$$\Delta C_{456} = \frac{G_0^3}{27d_z} \text{Real part of} \left[\frac{\partial^2 n_0}{\partial (E_{F17.55} - E_{17.55})^2} \right] \tag{1.387b}$$

where $\eta_{17,55} = \frac{E_{F17,55} - E_{17,55}}{k_B T}$ and E_{F1755} is the Fermi energy in the present case.

Thus, using eqs. (1.387a), (1.387b), and (1.386) we can find ΔC_{44} and ΔC_{456} in this case.

1.3 Suggestion for the experimental determination of CECs

In recent years, with the advent of quantum hall effect [78], there has been considerable interest in studying the thermoelectric power under strong magnetic field (TPSM) in various types of nanostructured materials having quantum confinement of their charge carriers in one, two, and three dimensions of the respective wave vector space leading to different carrier energy spectra [79–108]. The classical TPSM equation $G = \pi^2 k_B / 3e$ (which is a function of three fundamental constants of nature) is valid only under the condition of carrier nondegeneracy, being independent of carrier concentration and reflects the fact that the signature of the band structure of any material is totally absent in the same.

Zawadzki [84] demonstrated that the TPSM for electronic materials having degenerate electron concentration is essentially determined by their respective energy band structures. It has, therefore, different values in different materials and changes with the doping, magnitude of the reciprocal quantizing magnetic field under magnetic quantization, quantizing electric field as in inversion layers, nanothickness as in quantum wells, wires and dots, with superlattice period as in quantum confined semiconductor superlattices with graded interfaces having various carrier energy spectra and also in other types of field assisted nanostructured materials.

The magnitude of the thermoelectric power G can be written as [85]

$$G = \frac{1}{|e|Tn_0} \int_{-\infty}^{\infty} (E - E_F) R(E) \left[-\frac{\partial f_0}{\partial E} \right] dE \tag{1.388}$$

where $R(E)$ is the total number of states.

Equation (1.388) can be written under the condition of carrier degeneracy [80] as

$$G = \left(\frac{\pi k_B^2 T}{3|e|n_0} \right) \left(\frac{\partial n_0}{\partial E_F} \right) \tag{1.389}$$

For inversion layers and NIPI structures, under the condition of electric quantum limit, eq. (1.389) assumes the form

$$G = \frac{\pi^2 k_B^2 T}{3en_{02D}} \left[\frac{\partial n_{02D}}{\partial (E_{F2D} - E_{02D})} \right] \tag{1.390}$$

where n_{02D}, E_{F2D}, andE_{02D} are the surface electron concentration, the Fermi energy, and the subband energy for the said 2D systems at the electric quantum limit, respectively.

For heavily doped semiconductors, eq. (1.388) assumes the form

$$G = \frac{\pi^2 k_B^2 T}{3en_{OHD}} \left[\frac{\partial n_{OHD}}{\partial (E_{FHD} - E_{OHD})} \right] \tag{1.391}$$

where n_{02D}, E_{F2D}, and E_{02D} are the electron concentration, the Fermi energy, and the band tail energy for the heavily doped semiconductors; here E_{OHD} could be obtained from the heavily doped dispersion relation of the semiconductor under the conditions $E = E_{0\,HD}$ and $k = 0$.

The knowledge of the carrier contribution to the elastic constants is important in studying the mechanical properties of the materials and has been investigated in the literature [109–112]. The electronic contribution to the second- and third-order elastic constants for HD materials can be written as [109–112]

$$\Delta C_{44} = \frac{-G_0^2}{9} \left[\frac{\partial n_{OHD}}{\partial (E_{FHD} - E_{OHD})} \right] \tag{1.392}$$

and

$$\Delta C_{456} = \frac{G_0^3}{27} \left[\frac{\partial^2 n_{OHD}}{\partial (E_{FHD} - E_{OHD})^2} \right] \tag{1.393}$$

where G_0 is the deformation potential constant. Thus, using eqs. (1.391), (1.392), and (1.393), we can write

$$\Delta C_{44} = \left[\frac{-n_0 (\bar{G}_0)^2 |e| G}{(3\pi^2 k_B^2 T)} \right] \tag{1.394}$$

and

$$\Delta C_{456} = \left(\frac{n_0 |e| (\bar{G}_0)^3 G^2}{(3\pi^4 k_B^3 T)} \right) \left(1 + \frac{n_0}{G} \frac{\partial G}{\partial n_0} \right) \tag{1.395}$$

Therefore, by using eqs. (1.394) and (1.395) we can investigate ΔC_{44} and ΔC_{456} for all the cases of this monograph. Besides, the experimental graph of G versus n_0 allows us to determine the electronic contribution to the elastic constants for materials having arbitrary spectra.

1.4 Results and discussion

Using the appropriate equations we have plotted the normalized ΔC_{44} and ΔC_{456} as functions of electron concentration n_0 (for a given wave to 1.4 and 1.5 to 1.8 by taking HD specimens of n-InAs($E_{g_0} = 0.36eV$, $\Delta = 0.43eV$, $m^* = 0.026m_0$, $g_v = 1$, $\varepsilon_{sc} = 12.25\varepsilon_0$ [13]), n-InSb ($E_{g_0} = 0.2352eV$, $\Delta = 0.81eV$, $m^* = 0.01359m_0$, $g_v = 1$, length where we have considered red light for which $\lambda = 660nm$ and $I = 1nWm^{-}2$) at $T = 4.2\,K$ in Figure 1.1 $\varepsilon_{sc} = 15.56\varepsilon_0$ [13]), $n-Hg_{1-x}Cd_xTe(E_{g_0} = (-0.302 + 1.93x + 5.35\times10^{-4}(1-2x)T - 0.810x^2 + 0.832x^3)eV$, $\Delta = (0.63 + 0.24x - 0.27x^2)eV$, $m^* = 0.1m_0E_{g_0}(eV)^{-1}$, $g_v = 1$ ε_{sc}, $= [20.262 - 14.812x + 5.22795x^2]\varepsilon_0$ [13]) and $n-In_{1-}xGa_xAs_yP_{1-y}$ lattice matched to $InP(E_{g_0} = 1.337$ $(-0.73y + 0.13y^2)eV$, $\Delta = (0.114 + 0.26y - 0.22y^2)eV$, $m^* = (0.08 - 0.039y)m_0$ $y = (0.1896 - 0.4052x)/(0.1896 - 0.0123x)$, $g_v = 1$, $\varepsilon_{sc} = [10.65 + 0.1320y]\varepsilon_0)$ [13] in accordance with the perturbed three and two band models of Kane and that of perturbed parabolic energy bands respectively. In Figures 1.1–1.16 we have plotted the normalized ΔC_{44} and ΔC_{456} as functions of I, taking $\lambda = 660$ nm and $n_0 = 1\times10^{25}\,m^{-3}$ for the purpose of numerical computations. In Figures 1.17–1.24 we have plotted the normalized ΔC_{44} and ΔC_{456} as functions of the wavelength. In Figures 1.25 and 1.26, we have plotted the normalized ΔC_{44} and ΔC_{456} as functions of alloy composition for HD n-$Hg_{1-x}Cd_xTe$ while Figures 1.27 and 1.28 exhibit the same as functions of y for HD n-$In_{1-x}Ga_xAs_yP_{1-y}$ lattice matched to InP, respectively.

From Figures 1.1–1.4, we observe that both ΔC_{44} and ΔC_{456} increase with increasing electron concentration and their numerical values in the presence of light waves for all the materials are large when compared with $I = 0$.

The combined influence of the energy band constants on ΔC_{44} and ΔC_{456} for HD n-InAs and HD n- InSb can easily be assessed from Figures 1.1 and 1.2. For the purpose of relative assessment, all the plots in the absence of light waves have further been drawn. From Figures 1.10–1.12 and 1.13–1.16, we can see that both ΔC_{44} and ΔC_{456} increase with increasing light intensity for all the materials. It appears from Figures 1.17–1.20 and 1.21–1.24 that ΔC_{44} and ΔC_{456} increase as the wavelength shifts from red to violet color. The plots of Figures 1.25 and 1.26 are valid for x>0.17, since for x<0.17 the band gap becomes negative in HD n-$Hg_{1-x}Cd_xTe$, leading to semimetallic state. The plots of Figures 1.27 and 1.28 exhibit the variation of the normalized ΔC_{44} and ΔC_{456} with the alloy composition y for HD n-$In_{1-x}Ga_xAs_yP_{1-y}$ lattice matched to InP.

The influence of light is immediately apparent from the plots in Figures 1.9–1.24, since ΔC_{44} and ΔC_{456} depend strongly on I and λ, which is in direct contrast as compared with $I = 0$. The variations of ΔC_{44} and ΔC_{456} in Figures 1.9–1.24 reflect the direct signature of the light wave on the electronic, optic, and the other band structure-dependent properties of HD semiconducting materials in the presence of light waves and the photon-assisted transport for the corresponding semiconductor devices. Although both ΔC_{44} and ΔC_{456} tend to increase with the intensity and wavelength, but the rate of increase totally depends on the band structure. Note that in view of large changes of the elastic constants with n_0, detailed experimental work on second- and third-order elastic constants as functions of n_0 would be interesting for this case. It may be suggested that the experiments on the velocity of sound involving the shear mode as a function of n_0 may exhibit the carrier contribution to the elastic constants for materials having arbitrary carrier energy spectra. This fact again suggests another experimental determination of ΔC_{44} and ΔC_{456} beside the suggested methods of determining them as given by eqs. (1.394) and (1.395), respectively.

From all the Figures of (1.1) to (1.28), we observe that both the elastic constants increases from the light off case to the light on case, since the value of the Fermi energy in the presence of light waves becomes larger due to the increase in the carrier concentration as compared with the same as in the absence of photo-excitation. Therefore, the numerical magnitude of ΔC_{44} and ΔC_{456} in the presence of light is larger when compared with the same in the absence of light for the whole range of the concentration considered, although the same increases with an increase in carrier degeneracy.

It is worth mentioning that our basic eqs. (1.46b)–(1.48a) cover various materials having different energy band structures. In this chapter, the concentration, light intensity, wavelength, and the dependence of alloy composition of the CECs for bulk specimens of HD n-InAs, HD n-InSb, HD n-$Hg_{1-x}Cd_xTe$, and HD n-$In_{1-x}Ga_xAs_yP_{1-y}$ lattice matched to InP have been studied, as shown in Figures (1.1)–(1.28). Thus, we have covered a wide class of compounds whose energy band structures are defined by the three- and two-band Kane model in the absence of photon field. Under certain limiting conditions, all the results of ΔC_{44} and ΔC_{456} lead to the well-known expressions of ΔC_{44} and ΔC_{456} for nondegenerate compounds having parabolic energy bands as given by eqs. (1.65a) and (1.65b), respectively. This indirect test not only exhibits the mathematical compatibility of our formulation but also shows the fact that our simple analysis is a more generalized one, since one can obtain the corresponding results for the relatively wide gap materials having parabolic energy bands under certain limiting conditions from our present derivation.

Our experimental suggestion for the determination of the said elastic constants is valid for materials having arbitrary dispersion relations. Since the experimental curve of n_0 versus G for the present generalized systems is not available in the literature to the best of our knowledge, we cannot compare our theoretical formulation with the proposed experiment, although the generalized analysis as presented in this context can be checked when the experimental investigation of G for the present system would appear in the literature.

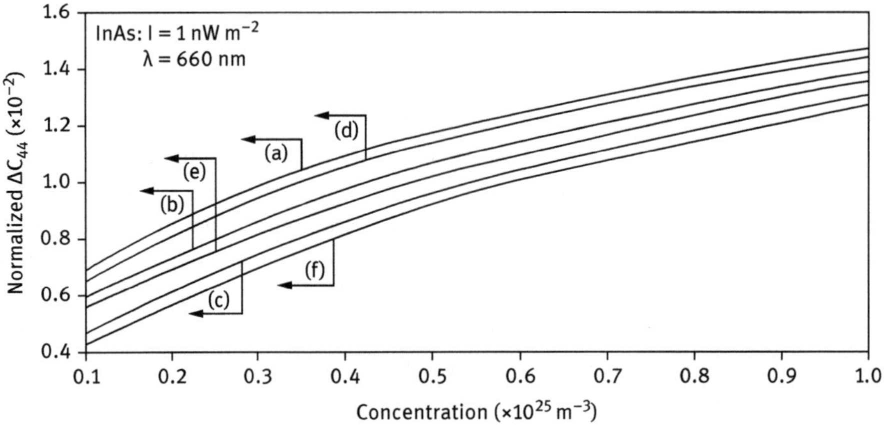

Figure 1.1: Plot of the normalized ΔC_{44} as a function of electron concentration for HD n-InAs in the presence of light waves, where the curves (a), (b), and (C) represent the perturbed three- and two-band Kane models and that of the perturbed parabolic energy bands, respectively. The curves (d), (e), and (f) represent the same in the absence of external photoexcitation.

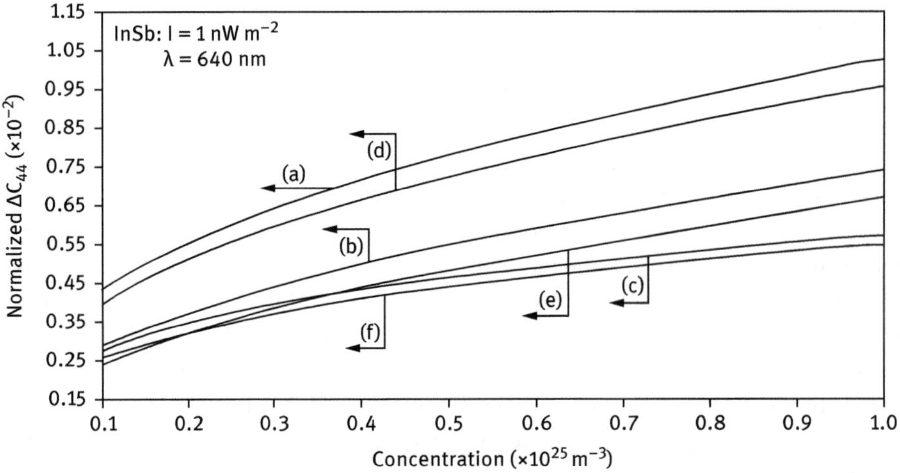

Figure 1.2: The plot of the normalized ΔC_{44} as a function of electron concentration for HD n-InSb in the presence of light waves, where the curves (a), (b), and (C) represent the perturbed three- and two-band Kane models and that of the perturbed parabolic energy bands, respectively. The curves (d), (e), and (f) represent the same in the absence of external photoexcitation.

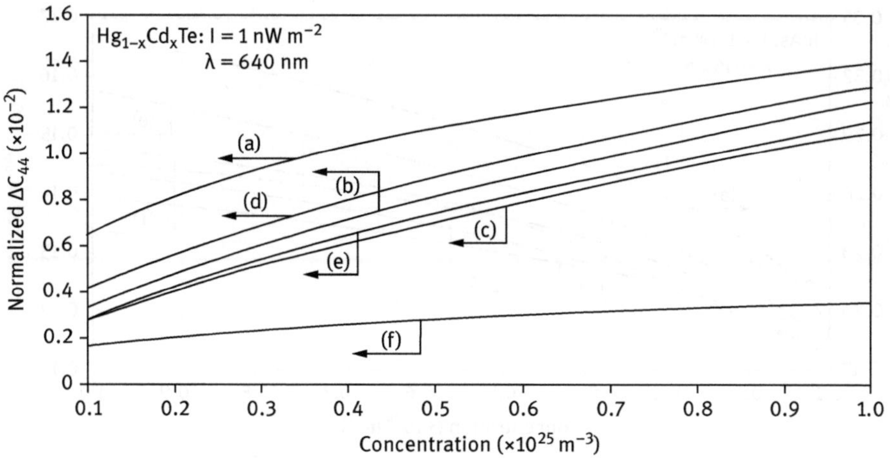

Figure 1.3: The plot of the normalized ΔC_{44} as a function of electron concentration for HD n-$Hg_{1-x}Cd_xTe$ in the presence of light waves, where the curves (a), (b), and (C) represent the perturbed three- and two-band Kane models and that of the perturbed parabolic energy bands, respectively. The curves (d), (e), and (f) represent the same in the absence of external photoexcitation.

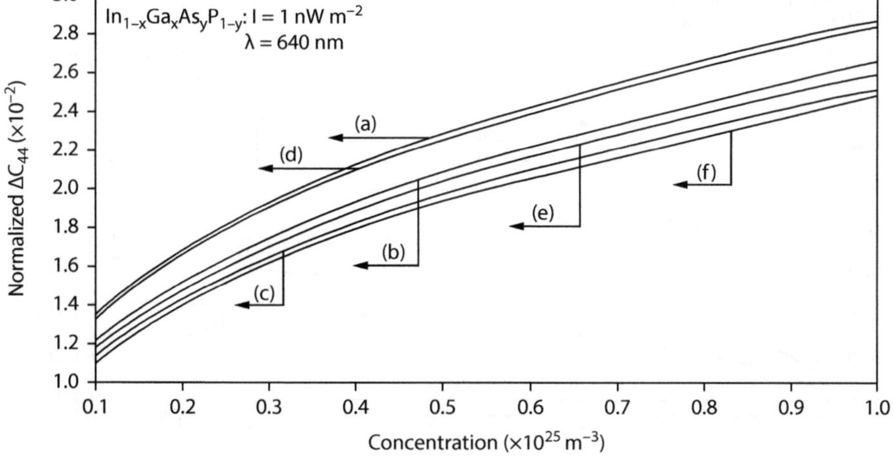

Figure 1.4: The plot of the normalized ΔC_{44} as a function of electron concentration for HD n-$In_{1-x}Ga_x\,As_yP_{1-y}$ in the presence of light waves, where the curves (a), (b), and (C) represent the perturbed three- and two-band Kane models and that of the perturbed parabolic energy bands, respectively. The curves (d), (e), and (f) represent the same in the absence of external photoexcitation.

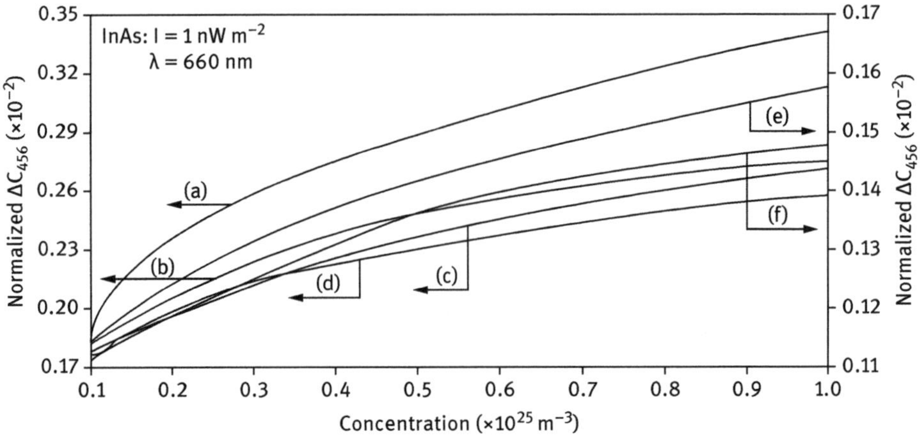

Figure 1.5: The plot of the normalized ΔC_{456} as a function of electron concentration for HD n-InAs in the presence of light waves, where the curves (a), (b), and (C) represent the perturbed three- and two-band Kane models and that of the perturbed parabolic energy bands, respectively. The curves (d), (e), and (f) represent the same in the absence of external photoexcitation.

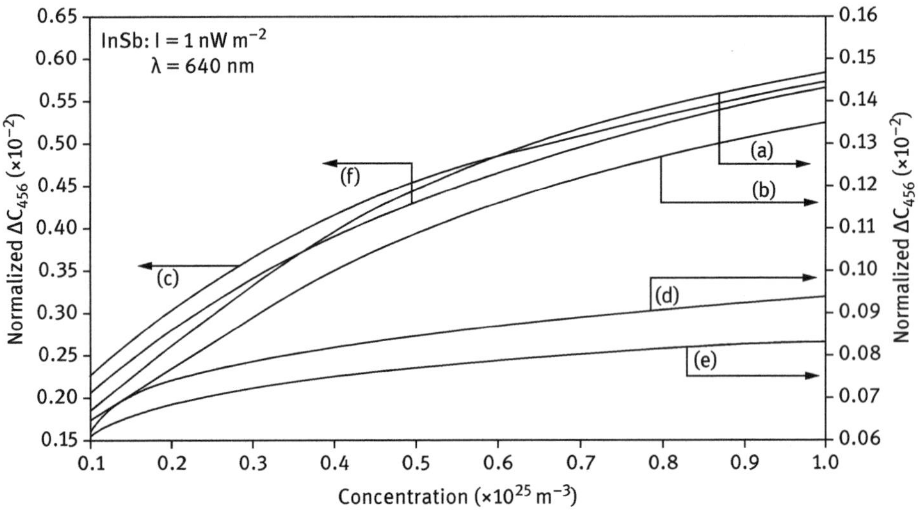

Figure 1.6: The plot of the normalized ΔC_{456} as a function of electron concentration for HD n-InSb in the presence of light waves, where the curves (a), (b), and (C) represent the perturbed three- and two-band Kane models and that of the perturbed parabolic energy bands, respectively. The curves (d), (e), and (f) represent the same in the absence of external photoexcitation.

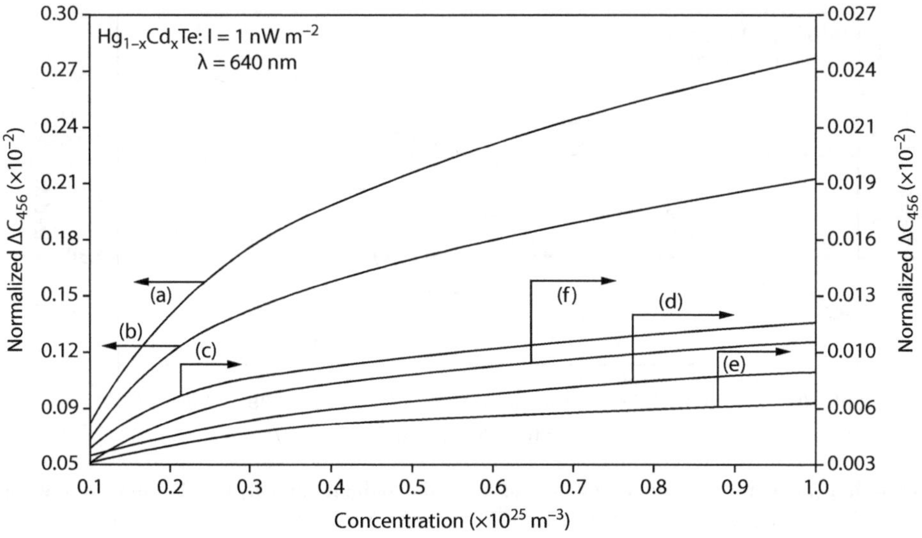

Figure 1.7: The plot of the normalized ΔC_{456} as a function of electron concentration for HD n-$Hg_{1-x}Cd_xTe$ in the presence of light waves, where the curves (a), (b), and (C) represent the perturbed three- and two-band Kane models and that of the perturbed parabolic energy bands, respectively. The curves (d), (e), and (f) represent the same in the absence of external photoexcitation.

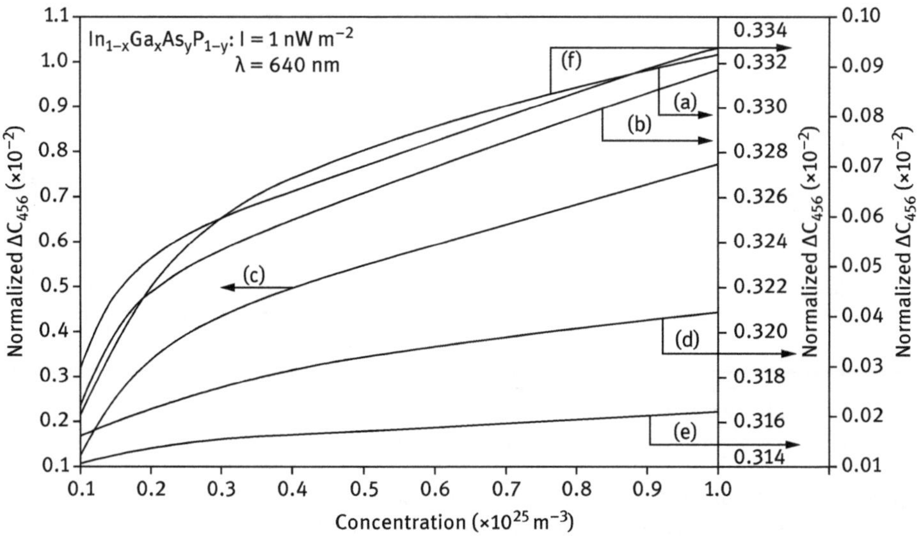

Figure 1.8: The plot of the normalized ΔC_{456} as a function of electron concentration for HD n-$In_{1-x}Ga_xAs_yP_{1-y}$ in the presence of light waves, where the curves (a), (b), and (C) represent the perturbed three- and two-band Kane models and that of the perturbed parabolic energy bands, respectively. The curves (d), (e), and (f) represent the same in the absence of external photoexcitation.

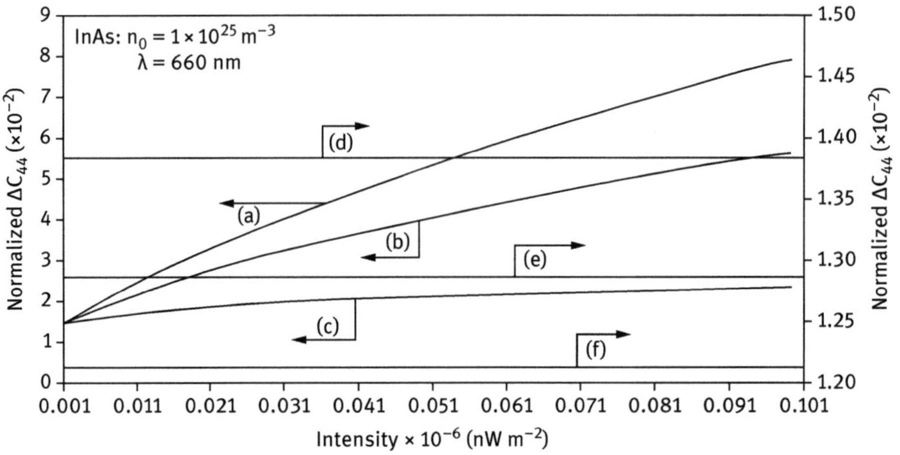

Figure 1.9: The plot of the normalized ΔC_{44} as a function of light intensity for HD n-InAs, where the curves (a), (b), and (C) represent the perturbed three- and two-band Kane models and that of the perturbed parabolic energy bands, respectively. The curves (d), (e), and (f) represent the same in the absence of external photoexcitation.

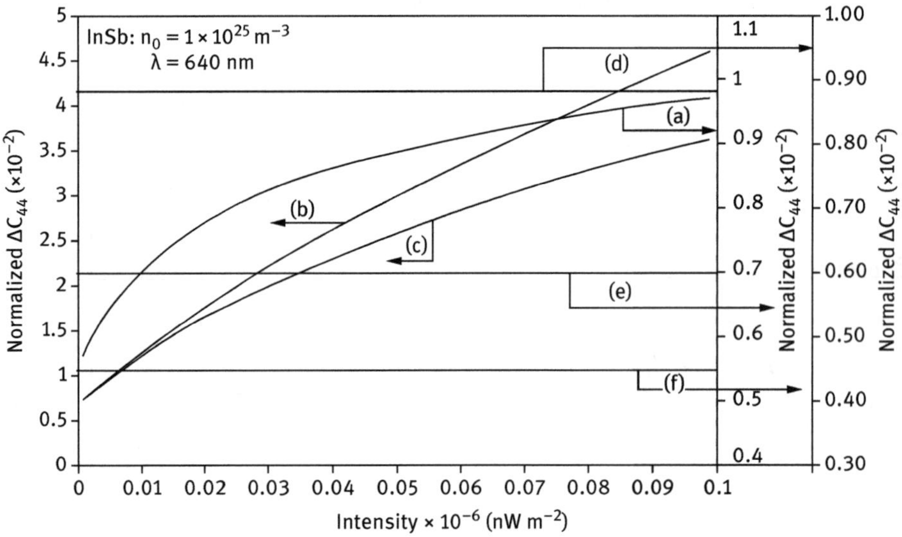

Figure 1.10: The plot of the normalized ΔC_{44} as a function of light intensity for HD n-InSb, where the curves (a), (b), and (C) represent the perturbed three- and two-band Kane models and that of the perturbed parabolic energy bands, respectively. The curves (d), (e), and (f) represent the same in the absence of external photoexcitation.

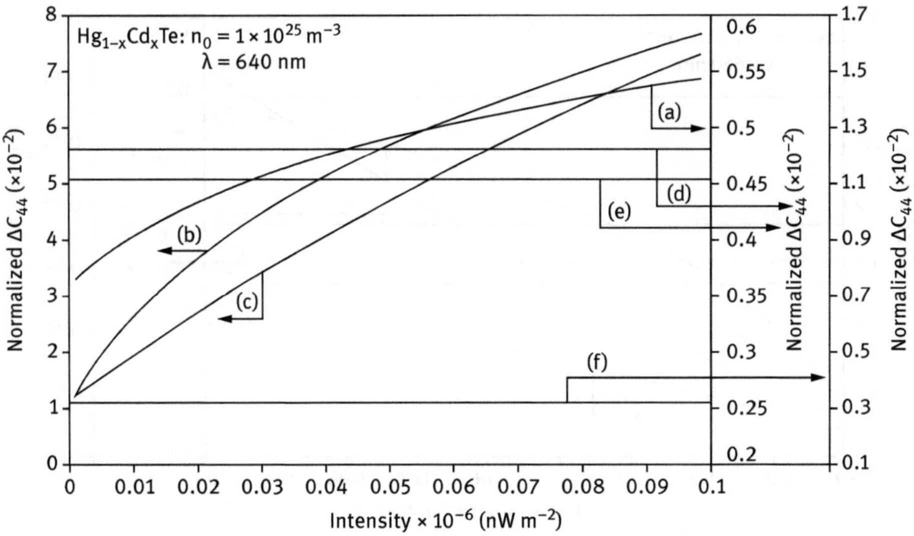

Figure 1.11: The plot of the normalized ΔC_{44} as a function of light intensity for HD n-$Hg_{1-x}Cd_xTe$, where the curves (a), (b), and (C) represent the perturbed three- and two-band Kane models and that of the perturbed parabolic energy bands, respectively. The curves (d), (e), and (f) represent the same in the absence of external photoexcitation.

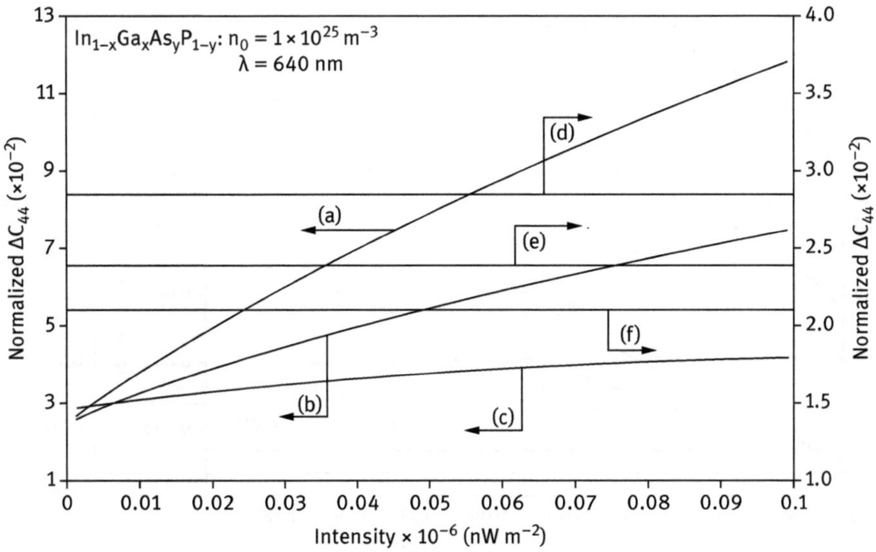

Figure 1.12: The plot of the normalized ΔC_{44} as a function of light intensity for HD n-$In_{1-x}Ga_xAs_yP_{1-y}$, where the curves (a), (b), and (C) represent the perturbed three- and two-band Kane models and that of the perturbed parabolic energy bands, respectively. The curves (d), (e), and (f) represent the same in the absence of external photoexcitation.

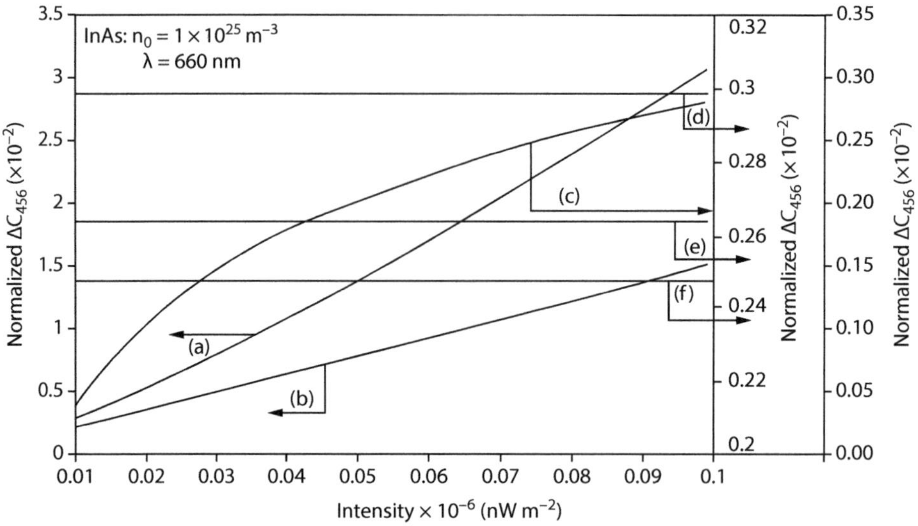

Figure 1.13: The plot of the normalized ΔC_{456} as a function of light intensity for HD n-InAs, where the curves (a), (b), and (C) represent the perturbed three- and two-band Kane models and that of the perturbed parabolic energy bands, respectively. The curves (d), (e), and (f) represent the same in the absence of external photoexcitation.

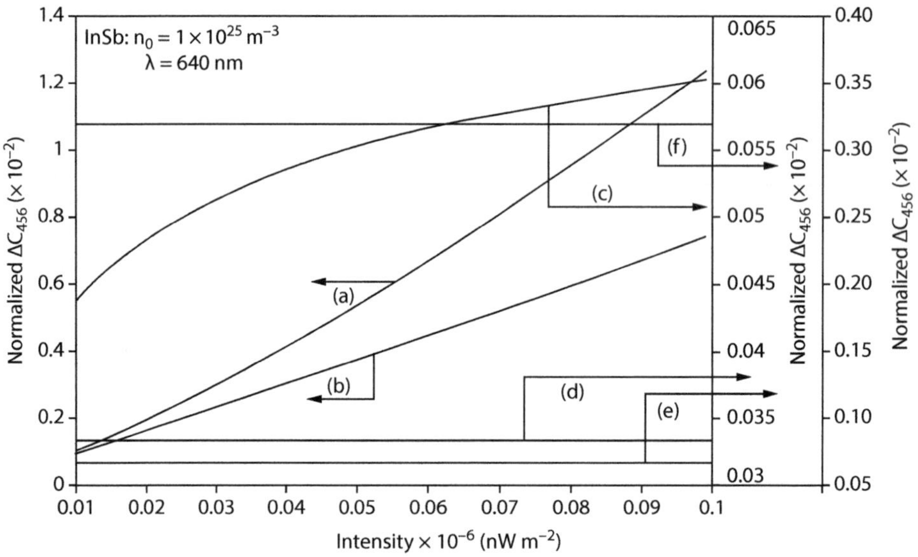

Figure 1.14: The plot of the normalized ΔC_{456} as a function of light intensity for HD n-InSb, where the curves (a), (b), and (C) represent the perturbed three- and two-band Kane models and that of the perturbed parabolic energy bands, respectively. The curves (d), (e), and (f) represent the same in the absence of external photoexcitation.

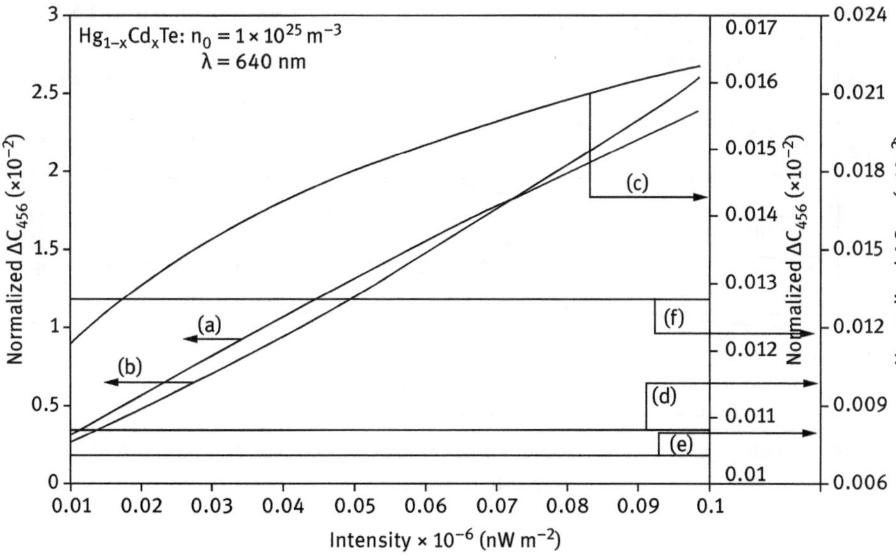

Figure 1.15: The plot of the normalized ΔC_{456} as a function of light intensity for HD n-$Hg_{1-x}Cd_xTe$, where the curves (a), (b), and (C) represent the perturbed three- and two-band Kane models and that of the perturbed parabolic energy bands, respectively. The curves (d), (e), and (f) represent the same in the absence of external photoexcitation.

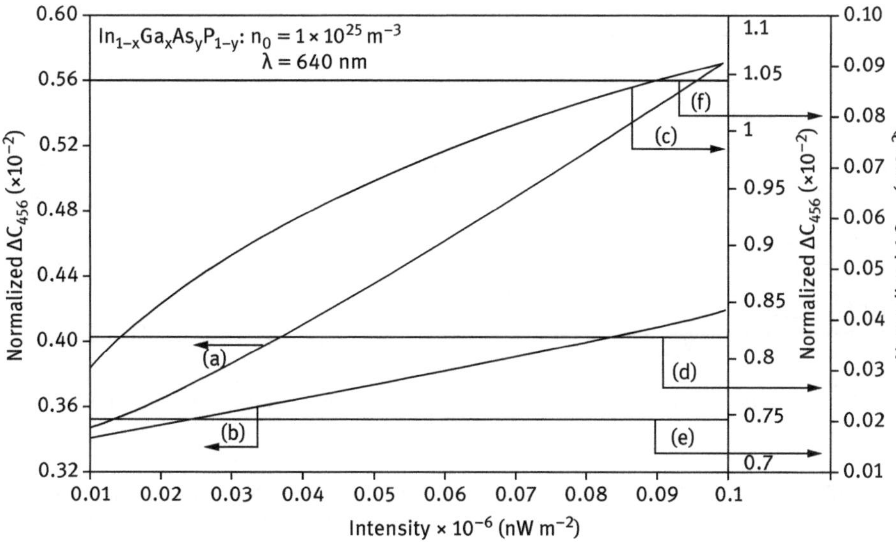

Figure 1.16: The plot of the normalized ΔC_{456} as a function of light intensity for HD n-$In_{1-x}Ga_xAs_yP_{1-y}$, where the curves (a), (b), and (C) represent the perturbed three- and two-band Kane models and that of the perturbed parabolic energy bands, respectively. The curves (d), (e), and (f) represent the same in the absence of external photoexcitation.

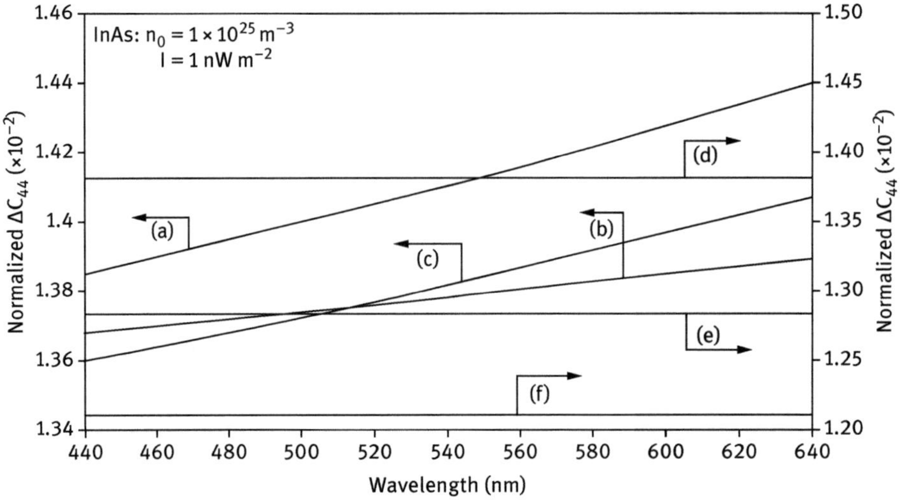

Figure 1.17: The plot of the normalized ΔC_{44} as a function of wavelength for HD n-InAs, where the curves (a), (b), and (C) represent the perturbed three- and two-band Kane models and that of the perturbed parabolic energy bands, respectively. The curves (d), (e), and (f) represent the same in the absence of external photoexcitation.

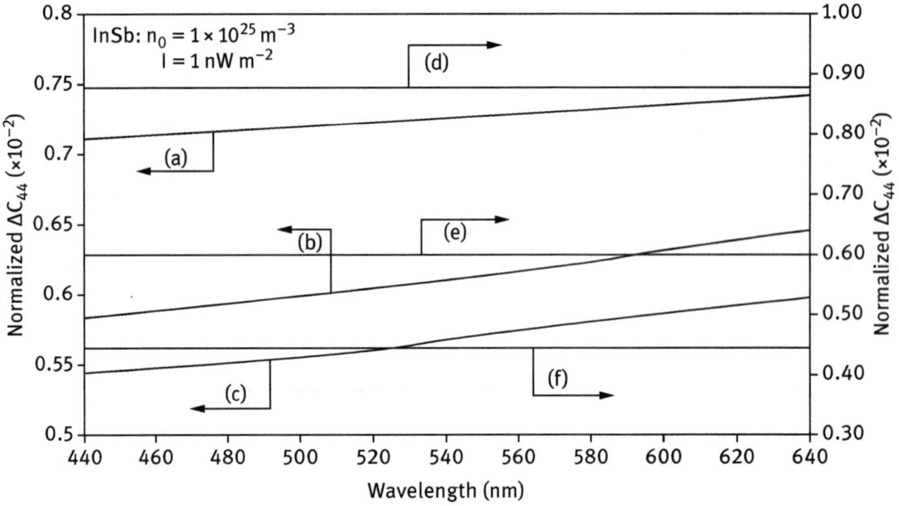

Figure 1.18: The plot of the normalized ΔC_{44} as a function of wavelength for HD n-InSb, where the curves (a), (b), and (C) represent the perturbed three- and two-band Kane models and that of the perturbed parabolic energy bands, respectively. The curves (d), (e), and (f) represent the same in the absence of external photoexcitation.

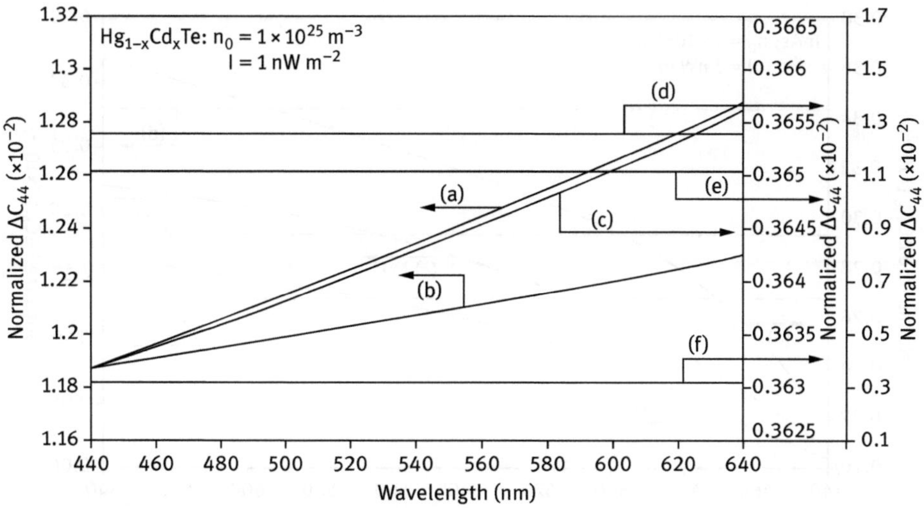

Figure 1.19: The plot of the normalized ΔC_{44} as a function of wavelength for HD n-$Hg_{1-x}Cd_xTe$, where the curves (a), (b), and (C) represent the perturbed three- and two-band Kane models and that of the perturbed parabolic energy bands, respectively. The curves (d), (e), and (f) represent the same in the absence of external photoexcitation.

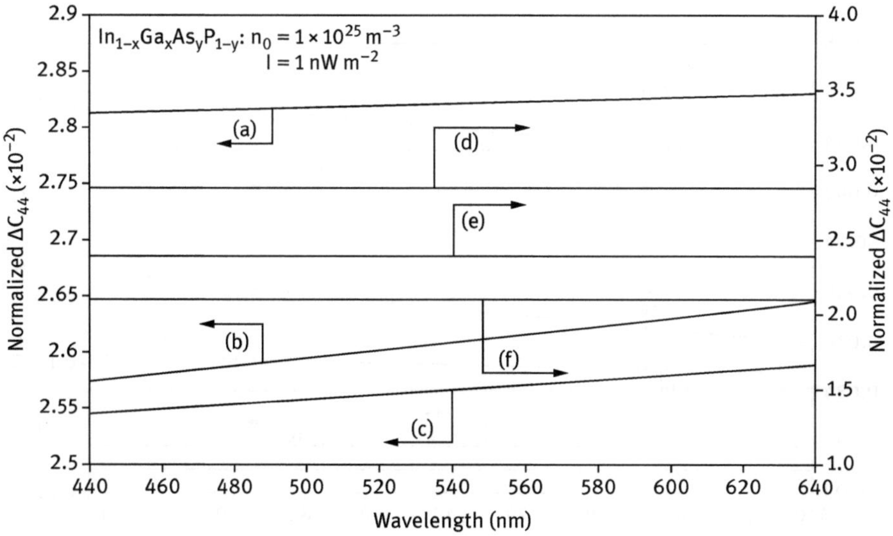

Figure 1.20: The plot of the normalized ΔC_{44} as a function of wavelength for HD n-$In_{1-x}Ga_xAs_yP_{1-y}$, where the curves (a), (b), and (C) represent the perturbed three- and two-band Kane models and that of the perturbed parabolic energy bands, respectively. The curves (d), (e), and (f) represent the same in the absence of external photoexcitation.

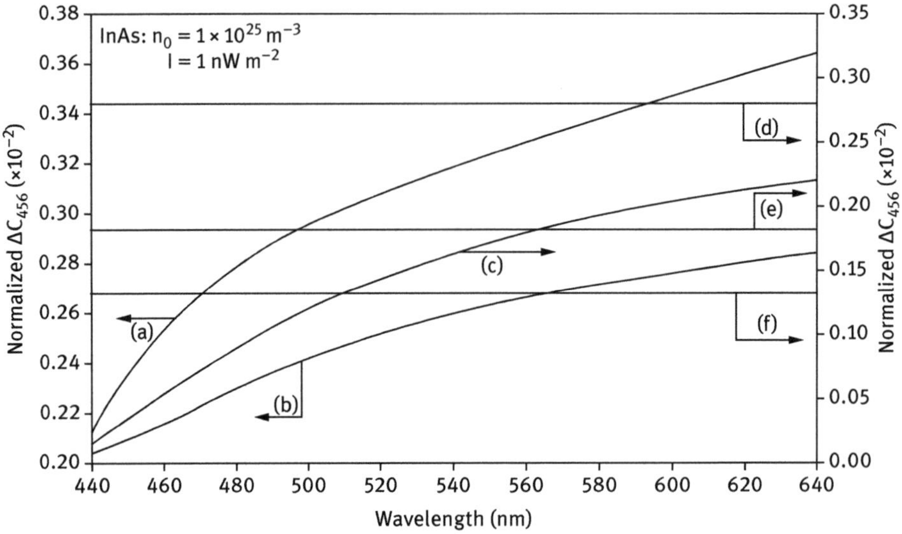

Figure 1.21: The plot of the normalized ΔC_{456} as a function of wavelength for HD n-InAs in the presence of light waves, where the curves (a), (b), and (C) represent the perturbed three- and two-band Kane models and that of the perturbed parabolic energy bands, respectively. The curves (d), (e), and (f) represent the same in the absence of external photoexcitation.

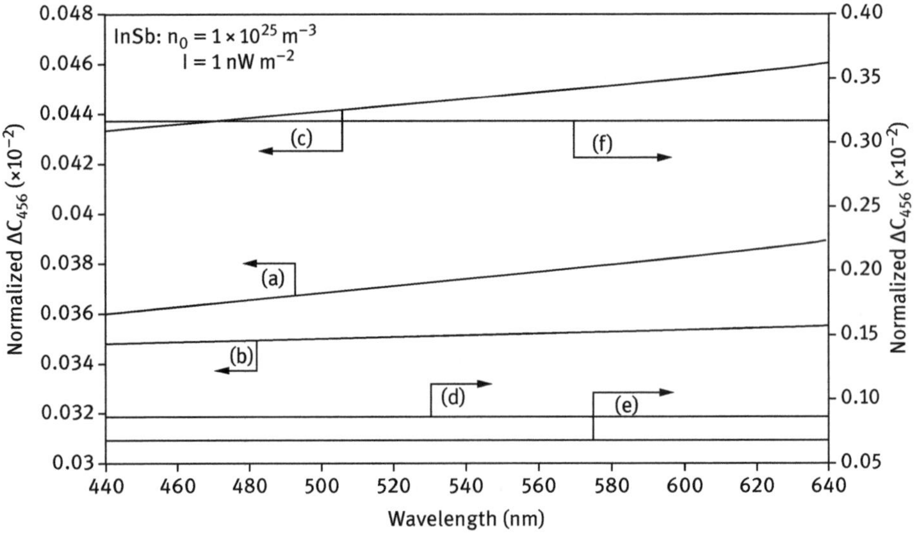

Figure 1.22: The plot of the normalized ΔC_{456} as a function of wavelength for HD n-InSb in the presence of light waves, where the curves (a), (b), and (C) represent the perturbed three- and two-band Kane models and that of the perturbed parabolic energy bands, respectively. The curves (d), (e), and (f) represent the same in the absence of external photoexcitation.

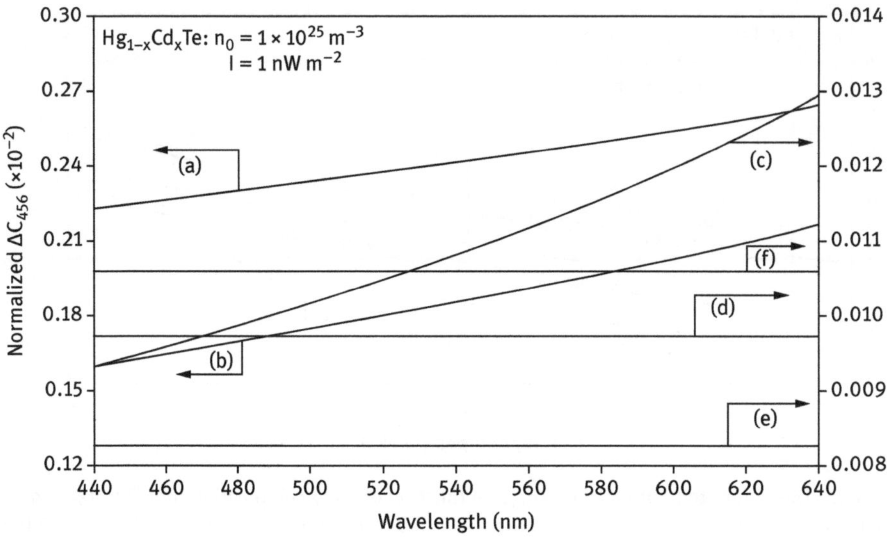

Figure 1.23: The plot of the normalized ΔC_{456} as a function of wavelength for HD n-$Hg_{1-x}Cd_xTe$ in the presence of light waves, where the curves (a), (b), and (C) represent the perturbed three and two-band Kane model and that of the perturbed parabolic energy bands, respectively. The curves (d), (e), and (f) represent the same in the absence of external photoexcitation.

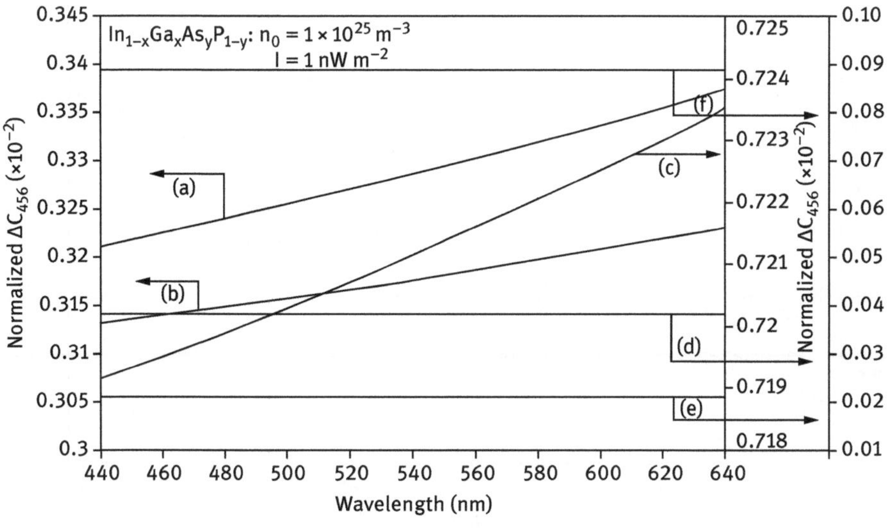

Figure 1.24: The plot of the normalized ΔC_{456} as a function of wavelength for HD n-$In_{1-x}Ga_xAs_yP_{1-y}$ in the presence of light waves, where the curves (a), (b), and (C) represent the perturbed three- and two-band Kane models and that of the perturbed parabolic energy bands, respectively. The curves (d), (e), and (f) represent the same in the absence of external photoexcitation.

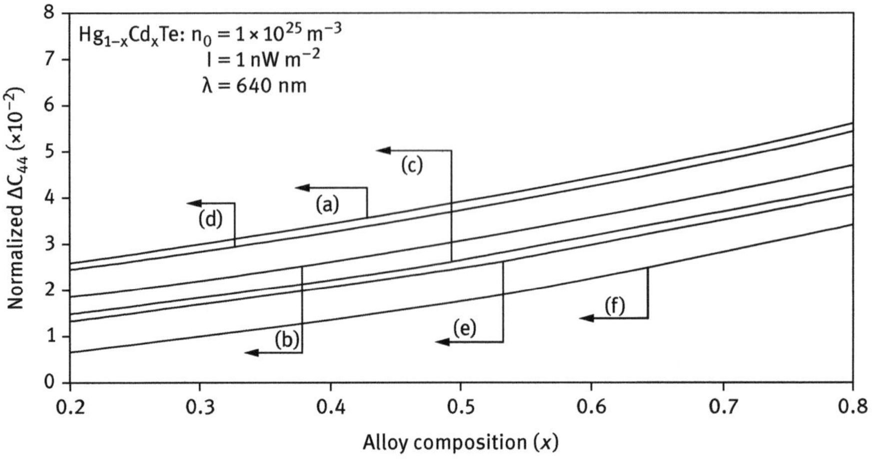

Figure 1.25: The plot of the normalized ΔC_{44} as a function of alloy composition for HD n-$Hg_{1-x}Cd_xTe$, where the curves (a), (b), and (C) represent the perturbed three- and two-band Kane models and that of the perturbed parabolic energy bands, respectively. The curves (d), (e), and (f) represent the same in the absence of external photoexcitation.

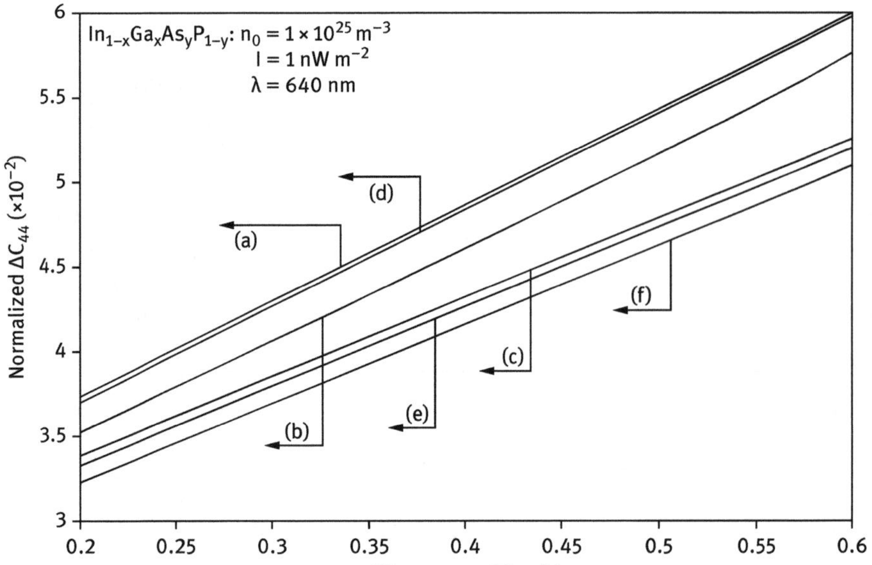

Figure 1.26: The plot of the normalized ΔC_{44} as a function of wavelength for HD n-$In_{1-x}Ga_xAs_yP_{1-y}$, where the curves (a), (b), and (C) represent the perturbed three- and two-band Kane models and that of the perturbed parabolic energy bands, respectively. The curves (d), (e), and (f) represent the same in the absence of external photoexcitation.

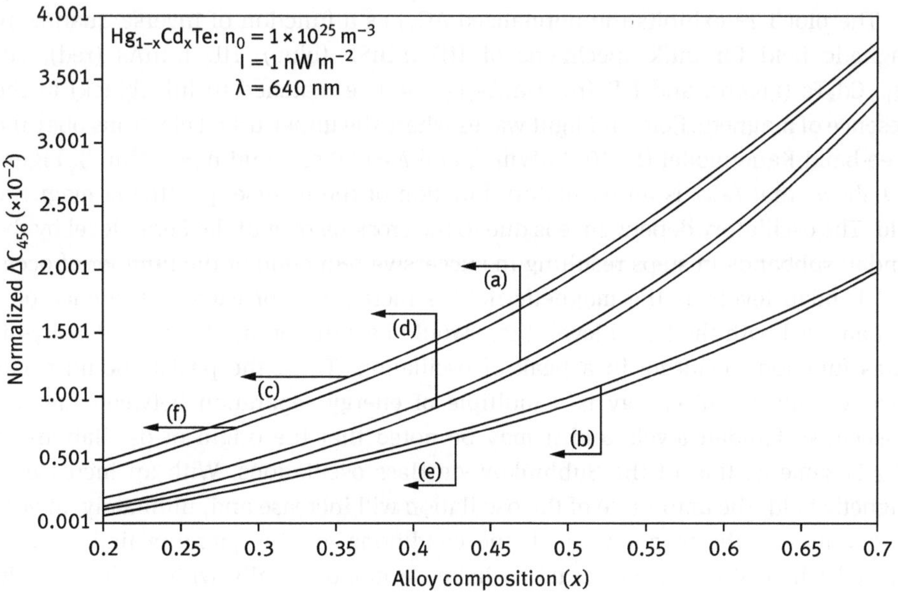

Figure 1.27: The plot of the normalized ΔC_{456} as a function of alloy composition for HD n-$Hg_{1-x}Cd_xTe$ in the presence of light waves, where the curves (a), (b), and (C) represent the perturbed three and two-band Kane model and that of the perturbed parabolic energy bands, respectively. The curves (d), (e), and (f) represent the same in the absence of external photoexcitation.

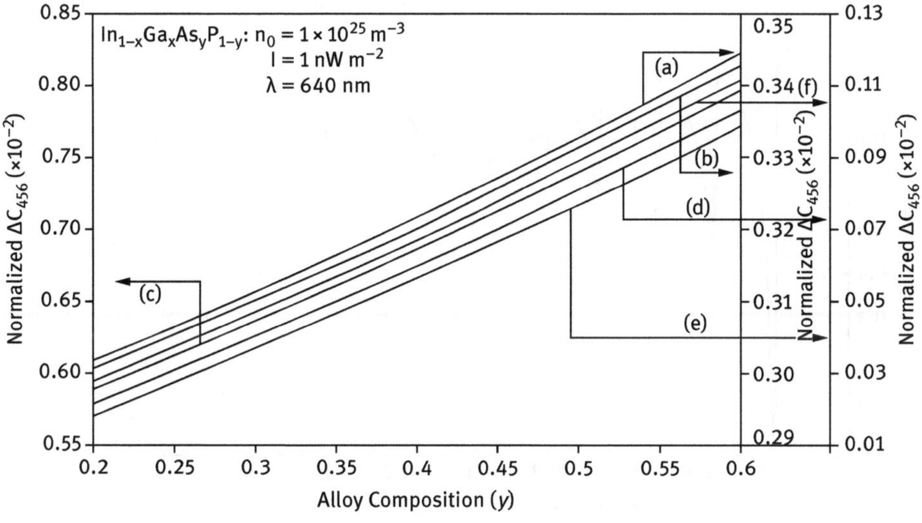

Figure 1.28: The plot of the normalized ΔC_{456} as a function of alloy composition for HD n-$In_{1-x}Ga_xAs_yP_{1-y}$ in the presence of light waves, where the curves (a), (b), and (C) represent the perturbed three and two-band Kane model and that of the perturbed parabolic energy bands, respectively. The curves (d), (e), and (f) represent the same in the absence of external photoexcitation.

The plot 1.29 exhibits the normalized ΔC_{44} as a function of inverse quantizing magnetic field for bulk specimens of HD n-InSb (blue), HD n-InAs (red), HD $Hg_{1-x}Cd_xTe$ (green), and HD $In_{1-x}Ga_xAs_yP_{1-y}$ lattice matched to InP (black) in the presence of magnetic field and light waves where the unperturbed electrons obey the three-band Kane model ($I = 10^{-4}\ nWm^{-2}$, and $\lambda = 660\ nm$, and $n_o = 10^{25} m^{-3}$). Figure 1.29 shows that ΔC_{44} is an oscillatory function of the inverse quantizing magnetic field. The oscillatory dependence is due to the crossing over of the Fermi level by the Landau subbands in steps resulting in successive reduction of the number of occupied Landau levels as the magnetic field is increased. For each coincidence of a Landau level with the Fermi level, there would be a discontinuity in the density-of-states function, resulting in a peak of oscillation. Thus, the peaks should occur whenever the Fermi energy is a multiple of energy separation between the two consecutive Landau levels and it may be noted that the origin of oscillations in ΔC_{44} is same as that of the Subhnikov–de Hass oscillations. With an increase in magnetic field, the amplitude of the oscillation will increase and, ultimately, at very large values of the magnetic field, the conditions for the quantum limit will be reached when ΔC_{44} will be found to increase monotonically with an increase in magnetic field. Figure 1.30 shows the normalized ΔC_{44} as a function of wave length in the presence of magnetic field and light waves for bulk specimens of HD n-InSb (blue), HD n-InAs (red), HD $Hg_{1-x}Cd_xTe$ (green), and HD $In_{1-x}Ga_xAs_yP_{1-y}$ lattice matched to InP (black) where the unperturbed electrons obey the three-band Kane

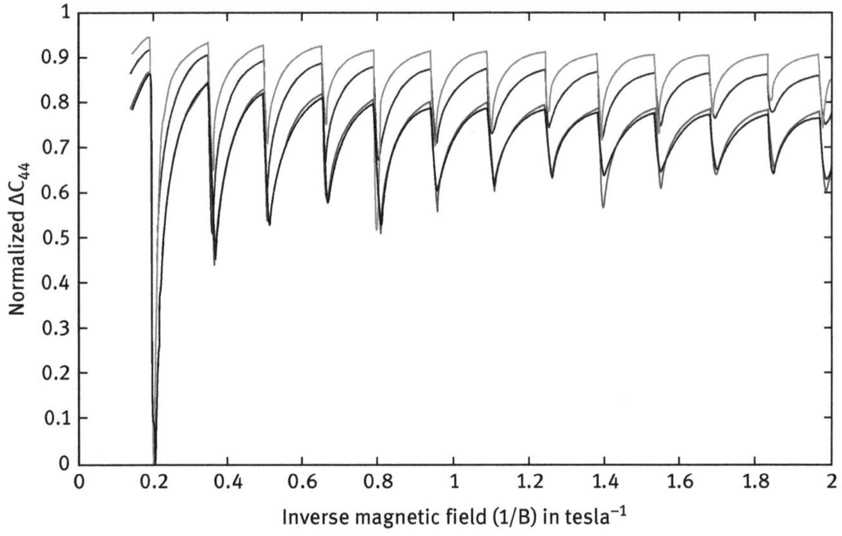

Figure 1.29: The plot of the normalized ΔC_{44} as a function of inverse quantizing magnetic field for bulk specimens of HD n-InSb (blue), HD n-InAs (red), HD $Hg_{1-x}Cd_xTe$ (green), and HD $In_{1-x}Ga_xAs_yP_{1-y}$ lattice matched to InP (black) in the presence of magnetic field and light waves where the unperturbed electrons obey the three-band Kane model ($I = 10^{-4}\ nWm^{-2}$ and $\lambda = 660$ nm, and $n_o = 10^{25} m^{-3}$).

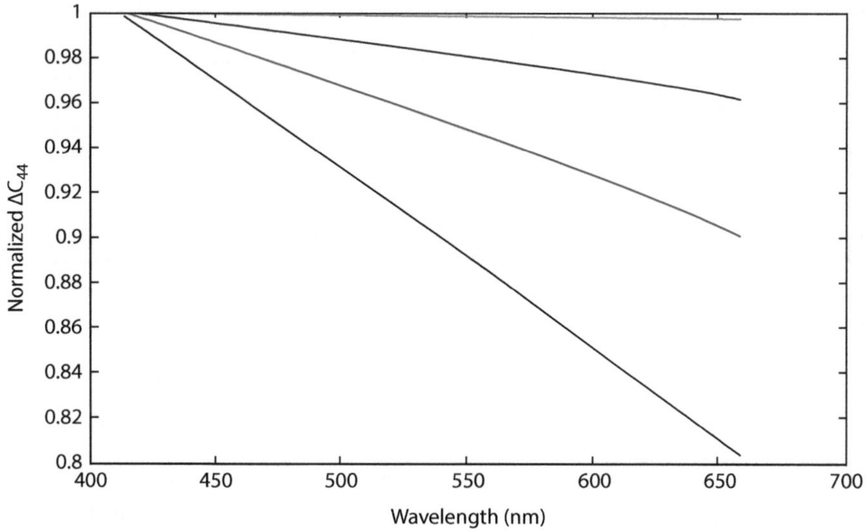

Figure 1.30: The plot of the normalized ΔC_{44} as a function of wavelength in the presence of magnetic field and light waves for bulk specimens of HD n-InSb (blue), HD n-InAs (red), HD $Hg_{1-x}Cd_xTe$ (green), and HD $In_{1-x}Ga_xAs_yP_{1-y}$ lattice matched to InP (black) where the unperturbed electrons obey the three-band Kane model ($1/B = 0.1\,\text{tesla}^{-1}$, $n_o = 10^{25}\,m^{-3}$, and $I = 10^{-4}\,nWm^{-2}$).

model ($1/B = 0.1\,\text{tesla}^{-1}$, $n_o = 10^{25}\,m^{-3}$, and $I = 10^{-4}\,nWm^{-2}$). It appears that ΔC_{44} increases with decreasing alloy composition for both the cases. Figure 1.31 shows the normalized ΔC_{44} as a function of light intensity in the presence of magnetic field and light waves for bulk specimens of HD n-InSb (blue), HD n-InAs (red), HD $Hg_{1-x}Cd_xTe$ (green), and HD $In_{1-x}Ga_xAs_yP_{1-y}$ lattice matched to InP (black) where the unperturbed electrons obey the three-band Kane model ($1/B = 0.1\,\text{tesla}^{-1}$, $n_o = 10^{25}\,m^{-3}$, and $\lambda = 660$ nm). Figure 1.31 shows that ΔC_{44} increases with decreasing light intensity for all the cases. Figure 1.32 exhibits the normalized ΔC_{44} as a function of electron concentration in the presence of magnetic field and light waves for bulk specimens of HD n-InSb (blue), HD n-InAs (red), HD $Hg_{1-x}Cd_xTe$ (green), and HD $In_{1-x}Ga_xAs_yP_{1-y}$ lattice matched to InP (black) where the unperturbed electrons obey the three-band Kane model ($1/B = 0.1\,\text{tesla}^{-1}$, $I = 10^{-4}\,nWm^{-2}$, and $\lambda = 660$ nm). Note that ΔC44 exhibits sharp oscillations at specified values of the electron concentration when the Fermi energy reaches the Landau energy for the whole range of the concentration as considered here. Figure 1.33 exhibits the normalized ΔC_{44} as a function of alloy composition in the presence of magnetic field and light waves for bulk specimens of HD $Hg_{1-x}Cd_xTe$ (blue) and HD $In_{1-x}Ga_xAs_yP_{1-y}$ lattice matched to InP (black) where the unperturbed electrons obey the three-band Kane model ($1/B = 0.1\,\text{tesla}^{-1}$, $I = 10^{-4}\,Wm^{-2}$, $\lambda = 660$ nm, and $n_o = 10^{25}\,m^{-3}$). We observed that ΔC_{44} under magnetic quantization decreases with an increase in alloy composition for all the cases where the numerical values depend on the values of the energy band constants. Figure 1.34 shows the normalized ΔC_{44} as a

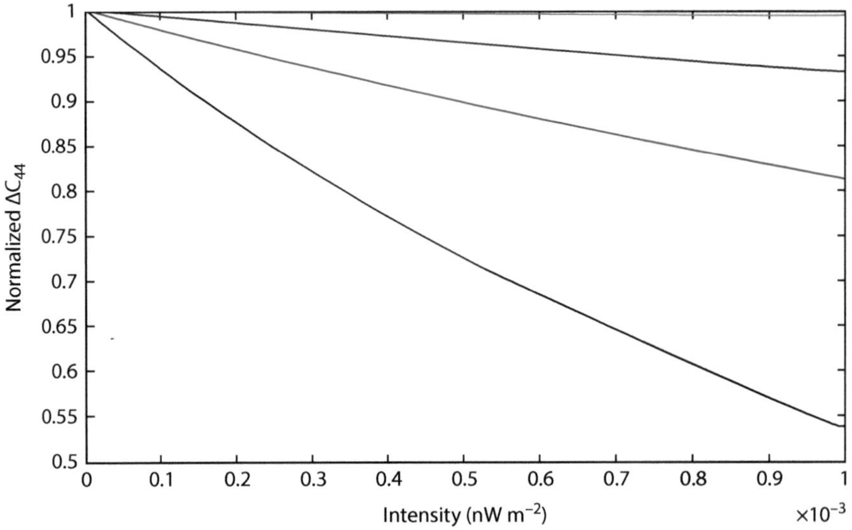

Figure 1.31: The plot of the normalized ΔC_{44} as a function of light intensity in the presence of magnetic field and light waves for bulk specimens of HD n-InSb (blue), HD n-InAs (red), HD $Hg_{1-x}Cd_xTe$ (green), and HD $In_{1-x}Ga_xAs_yP_{1-y}$ lattice matched to InP (black) where the unperturbed electrons obey the three-band Kane model ($1/B = 0.1\,\text{tesla}^{-1}$, $n_o = 10^{25}\text{m}^{-3}$ and $\lambda = 660$ nm).

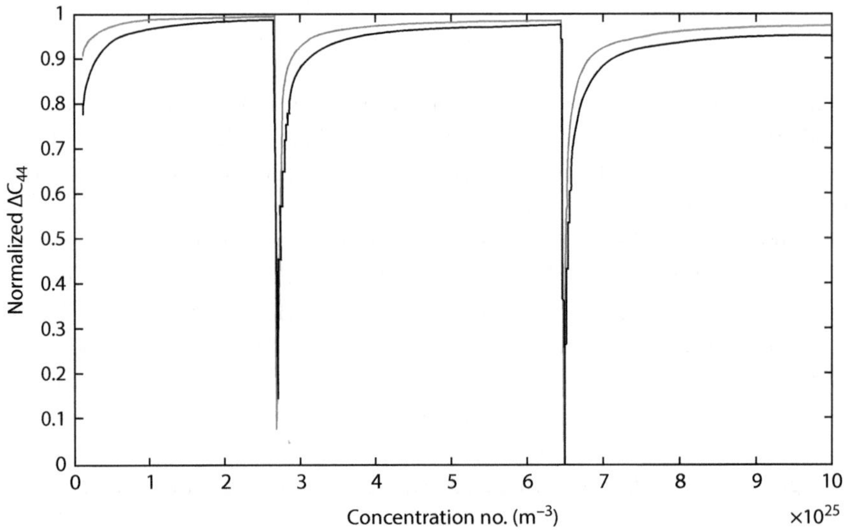

Figure 1.32: Plot of the normalized ΔC_{44} as a function of electron concentration in the presence of magnetic field and light waves for bulk specimens of HD $Hg_{1-x}Cd_xTe$ (green) and HD $In_{1-x}Ga_xAs_yP_{1-y}$ lattice matched to InP (black) where the unperturbed electrons obey the three band model of Kane ($1/B = 0.1\,\text{tesla}^{-1}$, $I = 10^{-4}\,\text{nWm}^{-2}$ and $\lambda = 660$ nm).

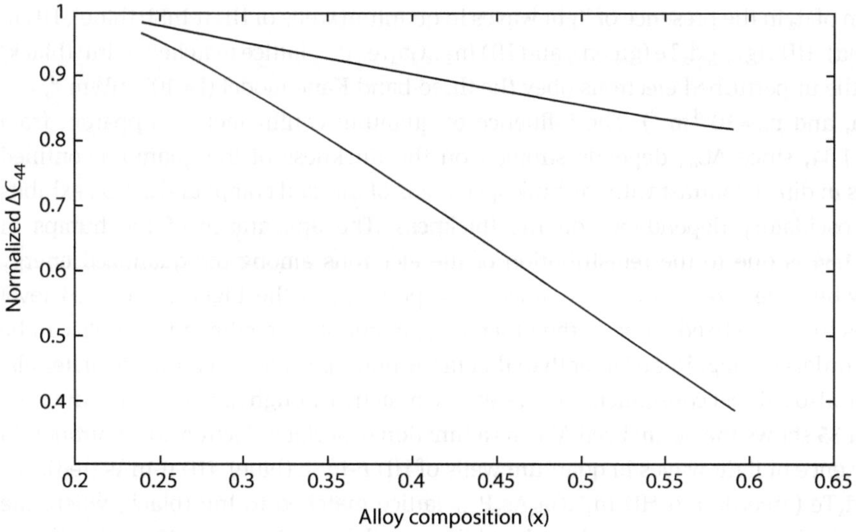

Figure 1.33: The plot of the normalized ΔC_{44} as a function of alloy composition in the presence of magnetic field and light waves for bulk specimens of HD $Hg_{1-x}Cd_xTe$ (blue) and HD $In_{1-x}Ga_xAs_yP_{1-y}$ lattice matched to InP (black) where the unperturbed electrons obey the three-band Kane model. ($1/B = 0.1$ tesla, $^{-1}I = 10^{-4}\,Wm^{-2}$, $\lambda = 660$ nm, and $n_o = 10^{25}\,m^{-3}$).

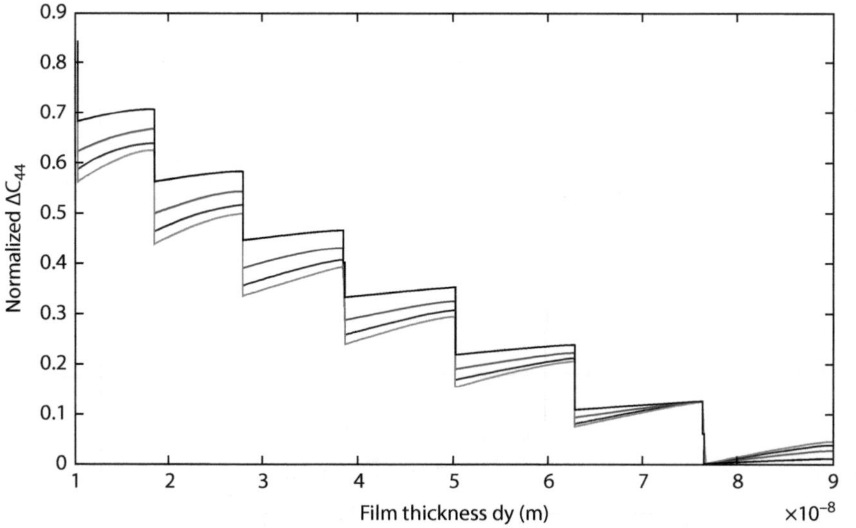

Figure 1.34: The plot of the normalized ΔC_{44}as a function of d_z in the presence of light waves in quantum wells of HD n-InSb (blue), HD n-InAs (red), HD $Hg_{1-x}Cd_xTe$ (green), and HD $In_{1-x}Ga_xAs_yP_{1-y}$ lattice matched to InP (black) where the unperturbed electrons obey the three-band Kane model ($I = 10^{-4}\,nWm^{-2}$, $\lambda = 660$ nm, and $n_o = 10^{18}m^{-2}$).

function of d_z in the presence of light waves in quantum wells of HD n-InSb (blue), HD n-InAs (red), HD $Hg_{1-x}Cd_xTe$ (green), and HD $In_{1-x}Ga_xAs_yP_{1-y}$ lattice matched to InP (black) where the unperturbed electrons obey the three-band Kane model ($I = 10^{-4}$ nWm^{-2}, $\lambda =$ 660 nm, and $n_o = 10^{18}$ m^{-2}). The influence of quantum confinement is apparent from Figure 1.34, since ΔC_{44} depends strongly on the thickness of the quantum-confined systems in direct contrast with the bulk specimens of the said compounds. ΔC_{44} exhibits strong oscillatory dependence on the thickness. The appearance of the humps in Figure 1.34 is due to the redistribution of the electrons among the quantized energy levels when the size quantum number corresponding to the highest occupied level changes from one fixed value to the other. ΔC_{44} in quantum-confined materials can be several orders of magnitude larger than that in the bulk specimens of the same materials, which is also a direct consequence of system asymmetry through dimensional reduction. Figure 1.35 shows the normalized ΔC_{44} as a function of surface electron concentration in the presence of light waves in quantum wells of HD n-InSb (blue), HD n-InAs (red), HD $Hg_{1-x}Cd_xTe$ (green), and HD $In_{1-x}Ga_xAs_yP_{1-y}$ lattice matched to InP (black) where the unperturbed electrons obey the three-band Kane model ($d_z = 10$ nm, $I = 10^{-4}$ nWm^{-2}, and $\lambda = 660$ nm). Figure 1.35 shows that ΔC_{44} decreases with increasing surface electron concentration and at a specified value of the same; the ΔC_{44} exhibits the step decrement attending a lower value and again decreases linearly with increasing surface ES. Figure 1.36 exhibits normalized ΔC_{44} as a function of light intensity in quantum wells of HD n-InSb (blue), HD n-InAs (red), HD $Hg_{1-x}Cd_xTe$ (green), and HD $In_{1-x}Ga_xAs_yP_{1-y}$

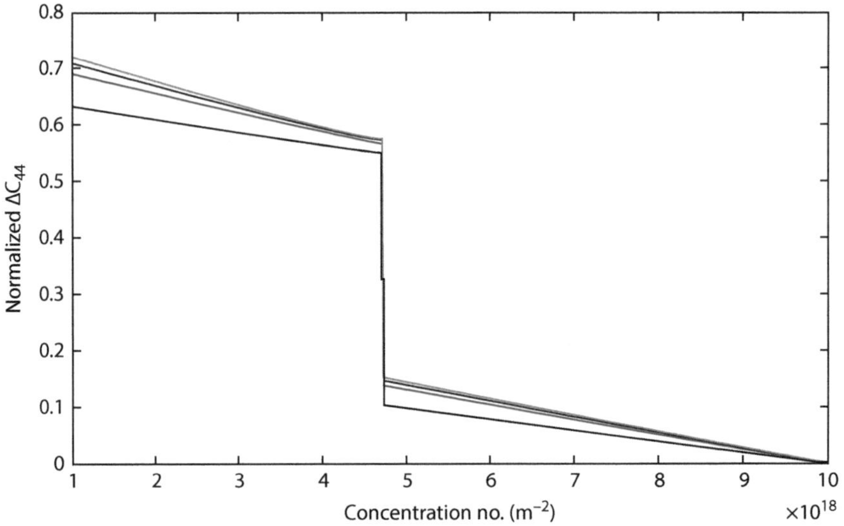

Figure 1.35: The plot of the normalized ΔC_{44} as a function of surface electron concentration in the presence of light waves in quantum wells of HD n-InSb (blue), HD n-InAs (red), HD $Hg_{1-x}Cd_xTe$ (green), and HD $In_{1-x}Ga_xAs_yP_{1-y}$ lattice matched to InP (black) where the unperturbed electrons obey the three-band Kane model ($d_z = 10$ nm, $I = 10^{-4}$ nWm^{-2}, and $\lambda = 660$ nm).

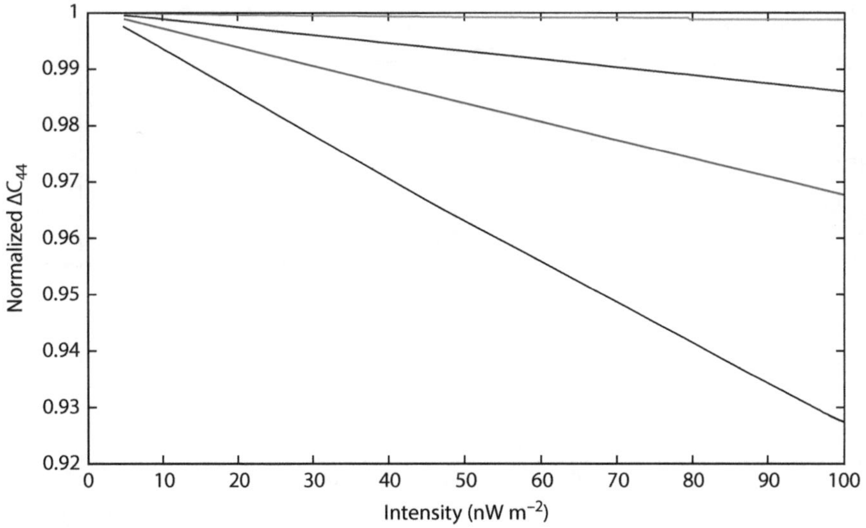

Figure 1.36: The plot of the normalized ΔC_{44} as a function of light intensity in quantum wells of HD n-InSb (blue), HD n-InAs (red), HD $Hg_{1-x}Cd_xTe$ (green), and HD $In_{1-x}Ga_xAs_yP_{1-y}$ lattice matched to InP (black) where the unperturbed electrons obey the three-band Kane model ($d_z = 10$ nm, $n_o = 10^{18}$ m^{-2}, and $\lambda = 660$ nm).

lattice matched to InP (black) where the unperturbed electrons obey the three-band Kane model ($d_z = 10$ nm, $n_o = 10^{18}$ m^{-2}, and $\lambda = 660$ nm). It appears that ΔC_{44} decreases with increasing light intensity in all the cases for all the HD quantum wells considered here. Figure 1.37 exhibits the plot of the normalized ΔC_{44} as a function of wavelength in quantum wells of HD n-InSb (blue), HD n-InAs (red), HD $Hg_{1-x}Cd_xTe$ (green), and HD $In_{1-x}Ga_xAs_yP_{1-y}$ lattice matched to InP (black) where the unperturbed electrons obey the three-band Kane model ($d_z = 10$ nm, $n_o = 10^{18}$ m^{-2}, and $I = 10^{-4}$ nWm^{-2}). From Figure 1.37 we conclude that ΔC_{44} decreases with increasing wavelength for all the heavily doped materials as considered for numerical examples in this case. Figure 1.38 exhibits the plot of the normalized ΔC_{44} as a function of alloy composition in the presence of light wave in quantum wells of HD $Hg_{1-x}Cd_xTe$ (blue) and HD $In_{1-x}Ga_xAs_yP_{1-y}$ lattice matched to InP (black) where the unperturbed electrons obey the three-band Kane model ($d_z = 10$ nm, $n_o = 10^{18}$m^{-2}, $\lambda = 660$ nm, and $I = 10^{-4}$ nWm^{-2}). It appears that ΔC_{44} increases with decreasing alloy composition for all the range of x as considered here. Figure 1.39 exhibits the normalized ΔC_{44} as a function of d_z in the presence of light waves in nanowires of HD n-InSb (blue), HD n-InAs (red), HD $Hg_{1-x}Cd_xTe$ (green), and HD $In_{1-x}Ga_xAs_yP_{1-y}$ lattice matched to InP (black) where the unperturbed electrons obey the three-band Kane model ($\lambda = 660$ nm, $I = 1$ nWm^{-2}, $n_o = 10^{11}$ m^{-1}, and $d_y = 10$ nm). It appears that in nanowires ΔC_{44} becomes invariant of the film thickness for the small value of film thickness, exhibits sharp fall at a particular value of thickness manifesting the quantum size effects in ΔC_{44} in Nano Wires and again increases rapidly with

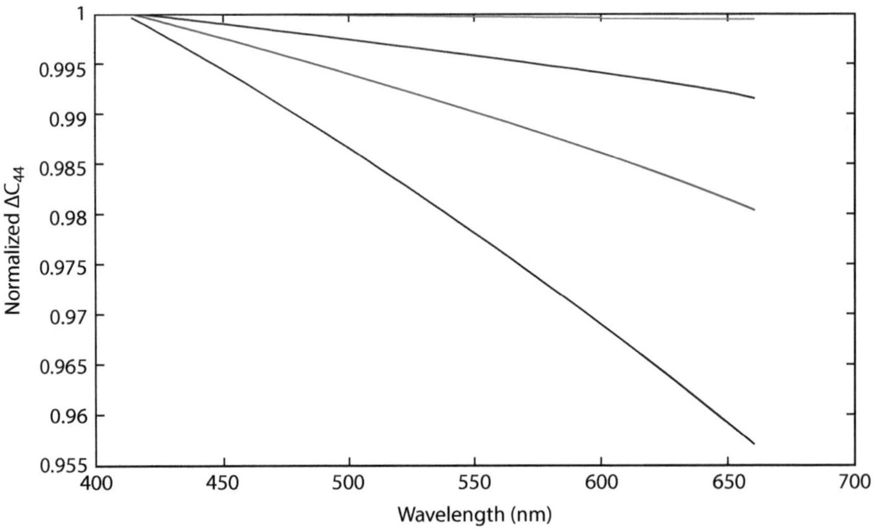

Figure 1.37: The plot of the normalized ΔC_{44} as a function of wavelength in quantum wells of HD n-InSb (blue), HD n-InAs (red), HD $Hg_{1-x}Cd_xTe$ (green), and HD $In_{1-x}Ga_xAs_yP_{1-y}$ lattice matched to InP (black) where the unperturbed electrons obey the three-band Kane model ($d_z = 10$ nm, $n_o = 10^{18}\,m^{-2}$, and $I = 10^{-4}\,nWm^{-2}$).

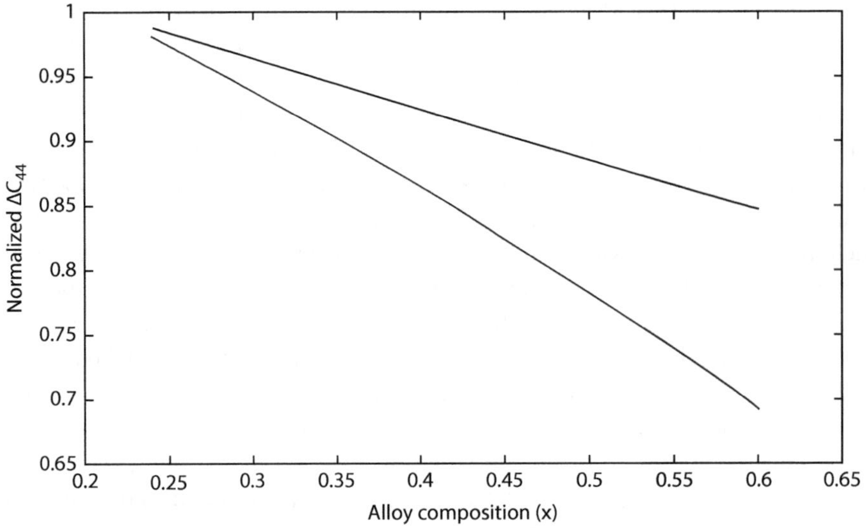

Figure 1.38: The plot of the normalized ΔC_{44} as a function of alloy composition in the presence of light wave in quantum wells of HD $Hg_{1-x}Cd_xTe$ (blue) and HD $In_{1-x}Ga_xAs_yP_{1-y}$ lattice matched to InP (black) where the unperturbed electrons obey the three-band Kane model ($d_z = 10$ nm, $n_o = 10^{18} m^{-2}$, $\lambda = 660$ nm, and $I = 10^{-4}\,nWm^{-2}$).

increasing thickness. The normalized ΔC_{44} as a function of electron concentration per unit length has been plotted in Figure 1.40 in the presence of light waves in nanowires of HD n-InSb (blue), HD n-InAs (red), HD $Hg_{1-x}Cd_xTe$ (green), and HD $In_{1-x}Ga_xAs_yP_{1-y}$ lattice matched to InP (black) where the unperturbed electrons obey the three-band Kane model ($\lambda = 660$ nm, $I = 1$ nWm^{-2}, $d_z = 10$ nm, and $d_y = 10$ nm). It is interesting to note that ΔC_{44} exhibits more or less the same type of variation with ES as observed earlier in Figure 1.39 in the case of the quantized variation of ΔC_{44} with d_z. Figure 1.45 shows the plot of the normalized ΔC_{44} as a function of light intensity in nanowires of HD n-InSb (blue), HD n-InAs (red), HD $Hg_{1-x}Cd_xTe$ (green), and HD $In_{1-x}Ga_xAs_yP_{1-y}$ lattice matched to InP (black) where the unperturbed electrons obey the three-band Kane model ($\lambda = 660$ nm, $d_z = 10$ nm, $d_y = 10$ nm, and $n_o = 10^{11}$ m^{-1}). The light intensity attenuates ΔC_{44} that decreases with an increase in intensity, although the amount of attenuation is different for different materials (Figure 1.41). Figure 1.42 shows the normalized ΔC_{44} as a function of wavelength in nanowires of HD n-InSb (blue), HD n-InAs (red), HD $Hg_{1-x}Cd_xTe$ (green), and HD $In_{1-x}Ga_xAs_yP_{1-y}$ lattice matched to InP (black) where the unperturbed electrons obey the three-band Kane model ($I = 1$ nWm^{-2}, $d_z = 10$ nm, $d_y = 10$ nm, and $n_o = 10^{11}$ m^{-1}). The influence of wavelength on ΔC_{44} decreases with the increment of wavelength of the external photoexcitation for the whole range of wavelengths. Figure 1.43 shows the plot of the normalized ΔC_{44} as a function of alloy composition in nanowires of HD $Hg_{1-x}Cd_xTe$ (green) and HD $In_{1-x}Ga_xAs_yP_{1-y}$ lattice matched to InP (black) where the unperturbed electrons obey the three-band Kane model ($n_o = 10^{11}$ m^{-1}, $d_y = 20$ nm, and

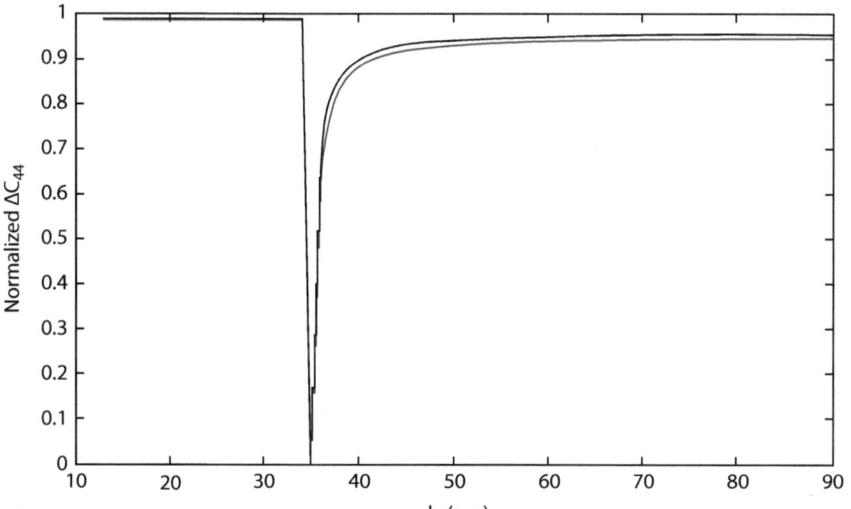

Figure 1.39: Plot of the normalized ΔC_{44}as a function of d_z in the presence of light waves in Nano Wires of HD n-InAs (red) and HD In $_{1-x}Ga_xAs_yP_{1-y}$ lattice matched to InP (black) where the unperturbed electrons obey the three band model of Kane(λ= 660 nm, I = 1nWm^{-2}, n_o= 10^{11} m^{-1} and d_y = 10 nm)

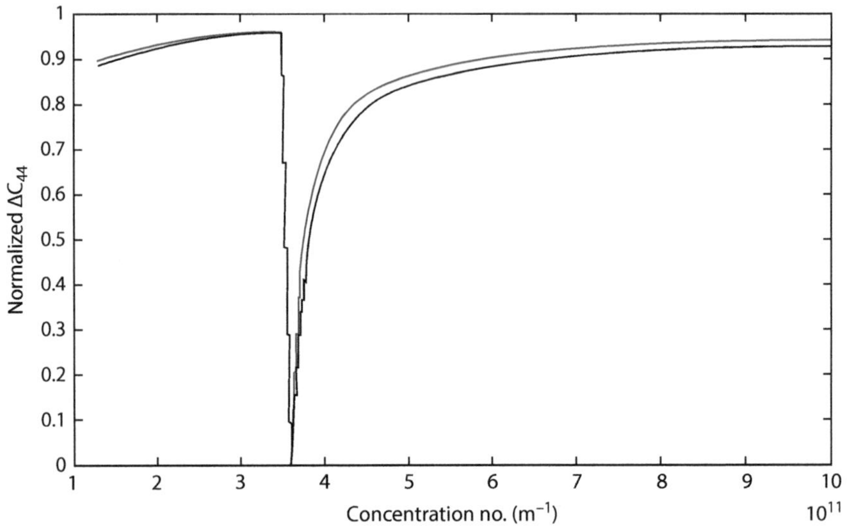

Figure 1.40: The plot of the normalized ΔC_{44} as a function of electron concentration in the presence of light waves in nanowires of HD n-InSb (blue), HD n-InAs (red), HD $Hg_{1-x}Cd_xTe$ (green), and HD $In_{1-x}Ga_xAs_yP_{1-y}$ lattice matched to InP (black) where the unperturbed electrons obey the three-band Kane model ($\lambda = 660$ nm, $I = 1$ nW m^{-2}, $d_z = 10$ nm, and $d_y = 10$ nm).

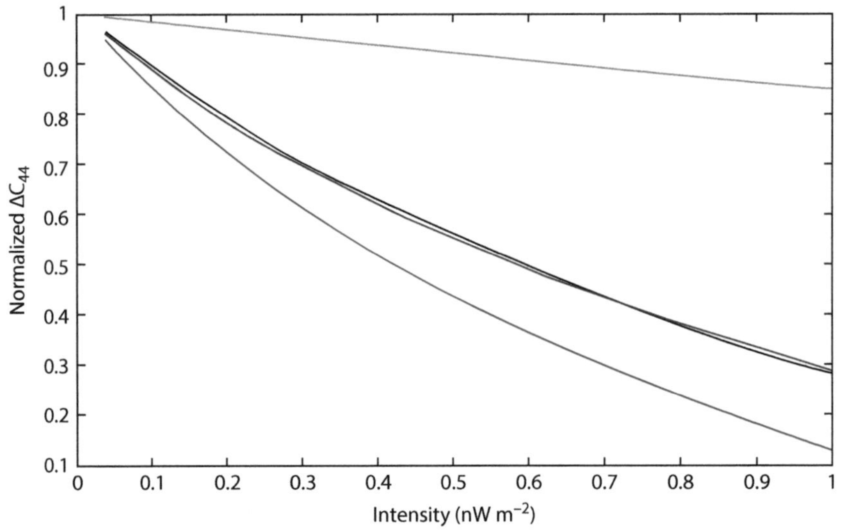

Figure 1.41: The plot of the normalized ΔC_{44} as a function of light intensity in nanowires of HD n-InSb (blue), HD n-InAs (red), HD $Hg_{1-x}Cd_xTe$ (green), and HD In $_{1-}xGa_xAs_yP_{1-y}$ lattice matched to InP (black) where the unperturbed electrons obey the three-band Kane model ($\lambda = 660$ nm, $d_z = 10$ nm, $d_y = 10$ nm, and $n_o = 10^{11} m^{-1}$).

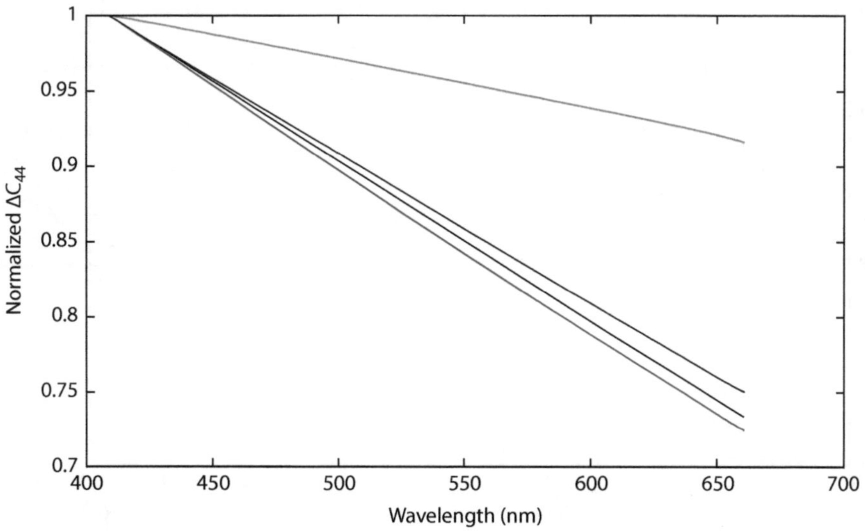

Figure 1.42: The plot of the normalized ΔC_{44} as a function of wavelength in nanowires of HD n-InSb (blue), HD n-InAs (red), HD $Hg_{1-x}Cd_xTe$ (green), and HD $In_{1-x}Ga_xAs_yP_{1-y}$ lattice matched to InP (black) where the unperturbed electrons obey the three-band Kane model ($I = 1 nWm^{-2}$, $d_z = 10$ nm, $d_y = 10$ nm, and $n_o = 10^{11}$ m^{-1})

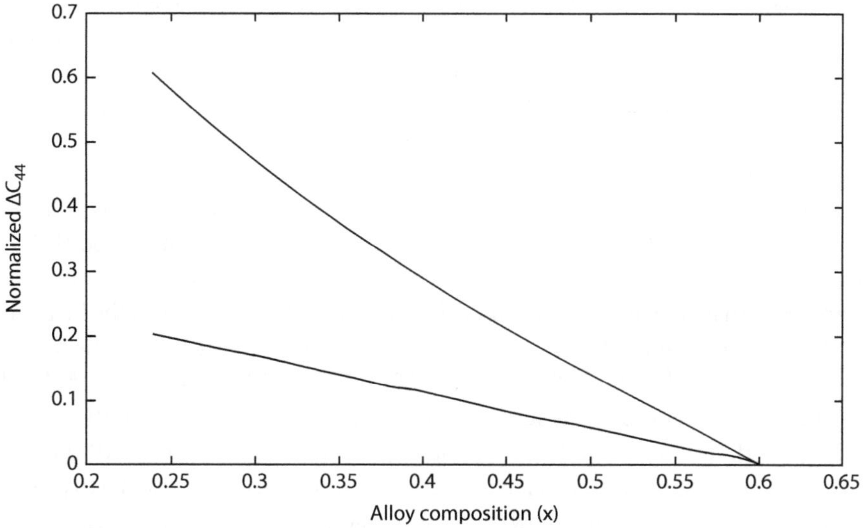

Figure 1.43: The plot of the normalized ΔC_{44}as a function of alloy composition in nanowires of HD $Hg_{1-x}Cd_xTe$ (green), and HD $In_{1-x}Ga_xAs_yP_{1-y}$ lattice matched to InP (black) where the unperturbed electrons obey the three-band Kane model ($n_o = 10^{11}m^{-1}$, $d_y = 20$ nm, and $d_z = 15$ nm).

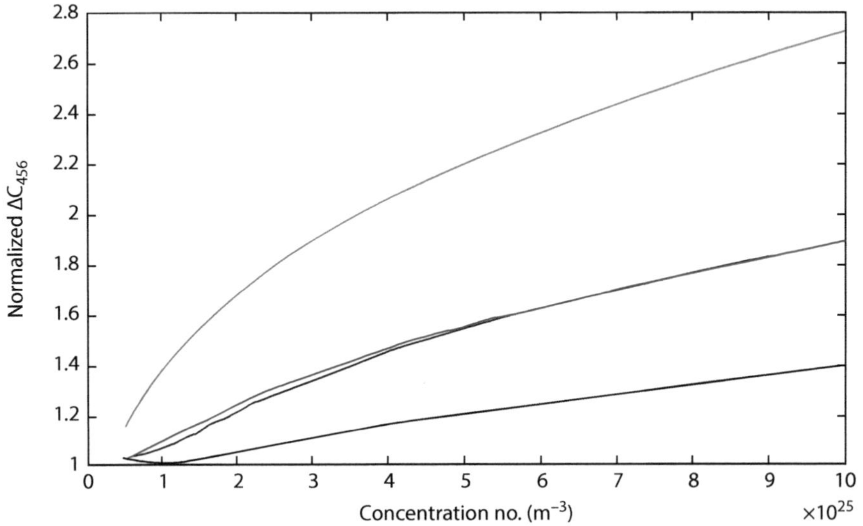

Figure 1.44: The plot of the normalized ΔC_{456} as a function of electron concentration per unit volume in the presence of light waves for bulk specimens of HD n-InSb (blue), HD n-InAs (red), HD $Hg_{1-x}Cd_xTe$ (green), and HD $In_{1-x}Ga_xAs_yP_{1-y}$ lattice matched to InP (black), respectively, where the unperturbed electrons obey the three-band Kane model. ($I = 10^{-4}$ nWm^{-2} and $\lambda = 660$ nm).

$d_z = 15$ nm). As usual, ΔC_{44} decreases with increasing alloy composition for all the quantized materials. Figure 1.44 shows the normalized ΔC_{456} as a function of electron concentration per unit volume in the presence of light waves for bulk specimens of HD n-InSb (blue), HD n-InAs (red), HD $Hg_{1-x}Cd_xTe$ (green), and HD $In_{1-x}Ga_xAs_yP_{1-y}$ lattice matched to InP (black), respectively, where the unperturbed electrons obey the three-band Kane model ($I = 10^{-4}$ nWm^{-2} and $\lambda = 660$ nm). From the figure it appears that ΔC_{456} increases with increasing n_0 for all the four materials considered here. From eq. (1.395) we can infer that ΔC_{456} will increase with an increase in n_0 if the rate of decrease of G^2 is smaller than the rate of increase of n_0. In this context from Figure 1.1 we can infer that ΔC_{44} decreases with increasing ES. This radically different behavior is due to the fact that from eq. (1.394) one can infer that ΔC_{44} will decrease with an increase in n_0 if the rate of decrease of G is greater than the rate of increase of n_0. Figure 1.17 shows the normalized ΔC_{456} as a function of light intensity for bulk specimens of HD n-InSb (blue), HD n-InAs (red), HD $Hg_{1-x}Cd_xTe$ (green), and HD $In_{1-x}Ga_xAs_yP_{1-y}$ lattice matched to InP (black), respectively, where the unperturbed electrons obey the three-band Kane model ($n_o = 10^{25}$ m^{-3} and $\lambda = 660$ nm). Figure 1.45 shows that ΔC_{456} increases with an increase in light intensity and there are considerable differences in the variation of ΔC_{456} among the materials considered here, reflecting the influence of the energy band constants of the different compounds. Figure 1.46 exhibits the variations of the normalized ΔC_{456} as a function of wavelength for bulk specimens of HD n-InSb (blue), HD n-InAs (red), HD $Hg_{1-x}Cd_xTe$ (green), and HD $In_{1-x}Ga_xAs_yP_{1-y}$ lattice matched to InP (black), respectively,

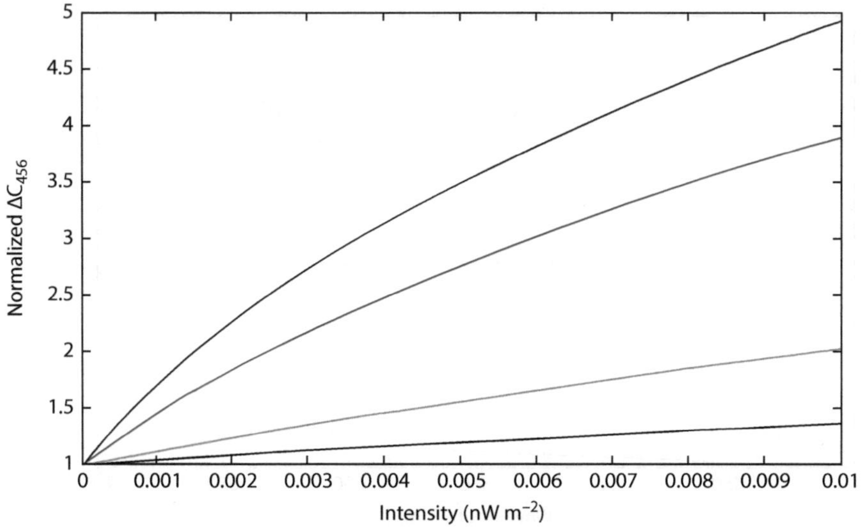

Figure 1.45: The plot of the normalized ΔC_{456} as a function of light intensity for bulk specimens of HD n-InSb (blue), HD n-InAs (red), HD $Hg_{1-x}Cd_xTe$ (green), and HD $In_{1-x}Ga_xAs_yP_{1-y}$ lattice matched to InP (black), respectively, where the unperturbed electrons obey the three-band Kane model ($n_o = 10^{25} m^{-3}$ and $\lambda = 660$ nm).

where the unperturbed electrons obey the three-band Kane model ($n_o = 10^{25}$ m^{-3} and $I = 10^{-4}$ nWm^{-2}). It appears from Figure 1.50 that ΔC_{456} increases with increasing wavelength for all the cases and the numerical magnitude differ widely from each other. Figure 1.47 shows the normalized ΔC_{456} as a function of alloy composition x in the presence of light wave for bulk specimens of HD $Hg_{1-x}Cd_xTe$ (blue) and HD $In_{1-x}Ga_xAs_yP_{1-y}$ lattice matched to InP (black) where the unperturbed electrons obey the three-band Kane model ($n_o = 10^{25}m^{-3}$, $I = 10^{-4}$ nWm^{-2}, and $\lambda = 660$ nm). It appears from the figure that ΔC_{456} increases with increasing alloy composition. In both the cases Figure 1.48 shows the normalized ΔC_{456} as a function of inverse quantizing magnetic field for bulk specimens of HD n-InSb (blue), HD n-InAs (red), HD $Hg_{1-x}Cd_xTe$ (green), and HD $In_{1-x}Ga_xAs_yP_{1-y}$ lattice matched to InP (black) in the presence of magnetic field and light waves where the unperturbed electrons obey the three-band Kane model ($I = 10^{-4}$ nWm^{-2}, $\lambda = 660$ nm, and $n_o = 10^{25}$ m^{-3}). ΔC_{456} exhibits oscillatory dependence with inverse quantizing magnetic field due to SdH effect. Figure 1.49 shows the normalized ΔC_{456} as a function of wavelength in the presence of magnetic field and light waves for bulk specimens of HD n-InSb (blue), HD n-InAs (red), HD $Hg_{1-x}Cd_xTe$ (green), and HD $In_{1-x}Ga_xAs_yP_{1-y}$ lattice matched to InP (black) where the unperturbed electrons obey the three-band Kane model ($1/B = 0.1$ tesla $^{-1}$, $n_o = 10^{25}$ m^{-3}, and $I = 10^{-4}$ nWm^{-2}). It appears from the figure that ΔC_{456} decreases with an increase in wavelength for the whole range of wavelengths considered in the plot. Figure 1.50 exhibits the normalized ΔC_{456} as a function of light intensity in the presence of magnetic field and light waves for

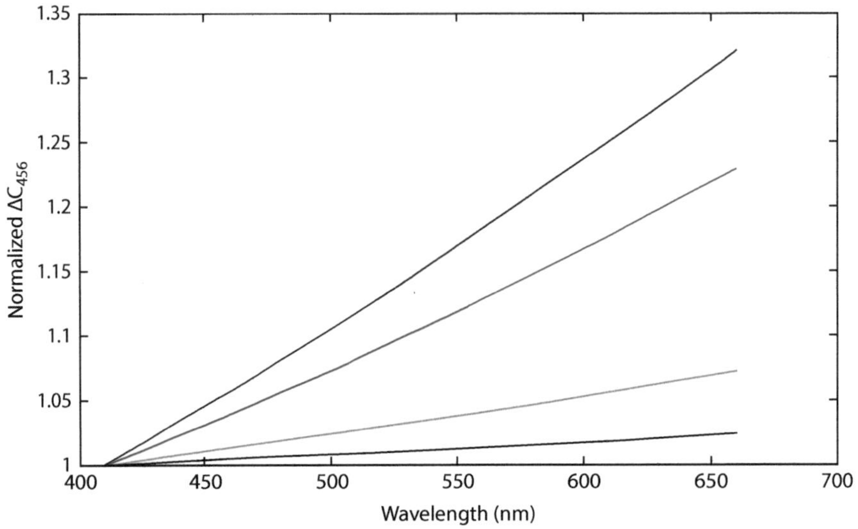

Figure 1.46: The plot of the normalized ΔC_{456} as a function of wavelength for bulk specimens of HD n-InSb (blue), HD n-InAs (red), HD $Hg_{1-x}Cd_xTe$ (green), and HD $In_{1-x}Ga_xAs_yP_{1-y}$ lattice matched to InP (black), respectively, where the unperturbed electrons obey the three-band Kane model ($n_o = 10^{25}$ m^{-3} and $I = 10^{-4}$ nWm^{-2}).

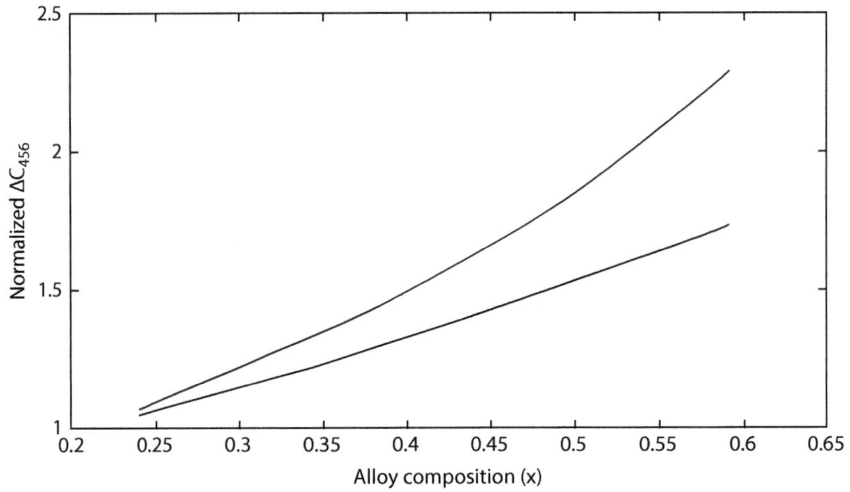

Figure 1.47: The plot of the normalized ΔC_{456} as a function of alloy composition x in the presence of light wave for bulk specimens of HD $Hg_{1-x}Cd_xTe$ (blue) and HD $In_{1-x}Ga_xAs_yP_{1-y}$ lattice matched to InP (black) where the unperturbed electrons obey the three-band Kane model ($n_o = 10^{25}m^{-3}$, $I = 10^{-4}$ nWm^{-2}, and $\lambda = 660$ nm).

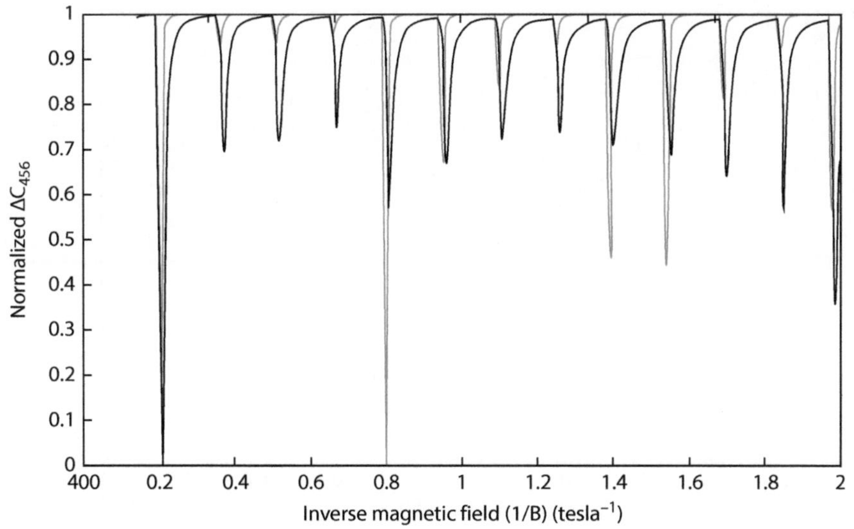

Figure 1.48: Plot of the normalized ΔC_{456} as a function of inverse quantizing magnetic field for bulk specimens of HD $Hg_{1-x}Cd_xTe$ (green) and HD $In_{1-x}Ga_xAs_yP_{1-y}$ lattice matched to InP (black) in the presence of magnetic field and light waves where the unperturbed electrons obey the three band model of Kane. ($I = 10^{-4}\ nWm^{-2}$ and $\lambda = 660$ nm and $n_o = 10^{25} m^{-3}$)

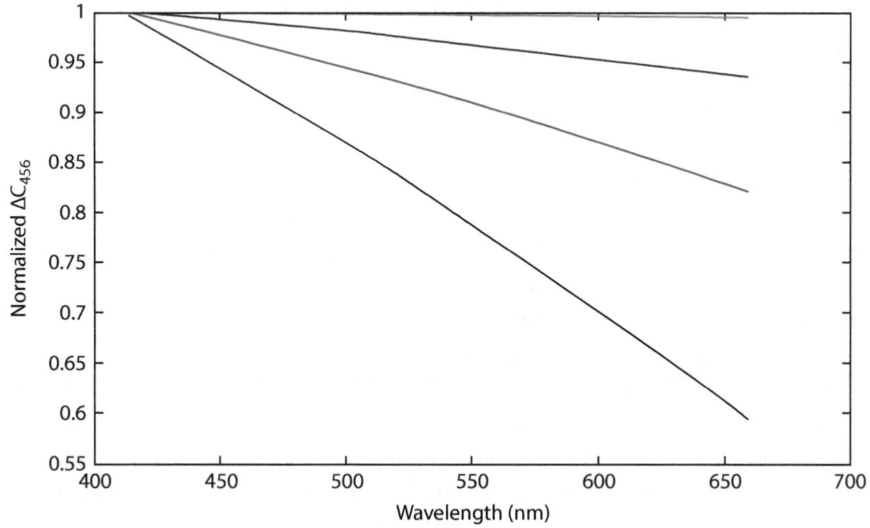

Figure 1.49: The plot of the normalized ΔC_{456} as a function of wavelength in the presence of magnetic field and light waves for bulk specimens of HD n-InSb (blue), HD n-InAs (red), HD $Hg_{1-x}Cd_xTe$ (green), and HD $In_{1-x}Ga_xAs_yP_{1-y}$ lattice matched to InP (black) where the unperturbed electrons obey the three-band Kane model ($1/B = 0.1\ tesla^{-1}$, $n_o = 10^{25} m^{-3}$, and $I = 10^{-4}\ nWm^{-2}$).

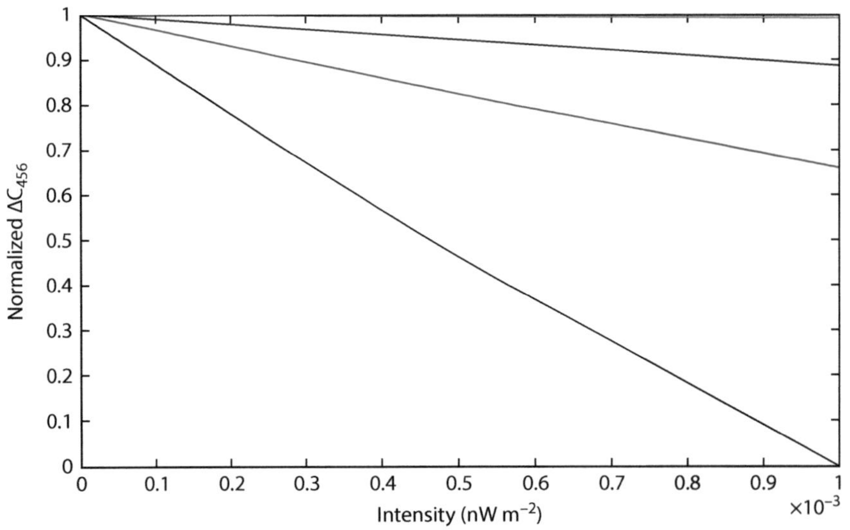

Figure 1.50: The plot of the normalized ΔC_{456} as a function of light intensity in the presence of magnetic field and light waves for bulk specimens of HD n-InSb (blue), HD n-InAs (red), HD $Hg_{1-x}Cd_xTe$ (green), and HD $In_{1-x}Ga_xAs_yP_{1-y}$ lattice matched to InP (black) where the unperturbed electrons obey the three-band Kane model ($1/B = 0.1$ tesla $^{-1}$, $n_o = 10^{25}$ m^{-3}, and $\lambda = 660$ nm).

bulk specimens of HD n-InSb (blue), HD n-InAs (red), HD $Hg_{1-x}Cd_xTe$ (green), and HD $In_{1-x}Ga_xAs_yP_{1-y}$ lattice matched to InP (black) where the unperturbed electrons obey the three-band Kane model ($1/B = 0.1$ tesla $^{-1}$, $n_o = 10^{25}$ m^{-3}, and $\lambda = 660$ nm). It appears that ΔC_{456} increases with a decrease in light intensity for all the cases. Figure 1.55 shows the normalized ΔC_{456} as a function of electron concentration in the presence of magnetic field and light waves for bulk specimens of HD n-InSb (blue), HD n-InAs (red), HD $Hg_{1-x}Cd_xTe$ (green), and HD $In_{1-x}Ga_xAs_yP_{1-y}$ lattice matched to InP (black) where the unperturbed electrons obey the three-band Kane model ($1/B = 0.1$ tesla $^{-1}$, $I = 10^{-4}$ nWm^{-2} and $\lambda = 660$ nm; Figure 1.51). It appears from the figure that ΔC_{456} exhibits stiky oscillatory dependence on electron concentration due to SdH effects.

Figure 1.52 exhibits the plot of the normalized ΔC_{456} as a function of alloy composition in the presence of magnetic field and light waves for bulk specimens of HD $Hg_{1-x}Cd_xTe$ (blue) and HD $In_{1-x}Ga_xAs_yP_{1-y}$ lattice matched to InP (black) where the unperturbed electrons obey the three-band Kane model ($1/B = 0.1$ tesla $^{-1}$ $I = 10^{-4}$ Wm^{-2}, and $\lambda = 660$ nm and $n_o = 10^{25}$ m^{-3}). It appears that ΔC_{456} decreases an increase in alloy composition and the rate of decrement is different for $Hg_{1-x}Cd_xTe$ and $In_{1-x}Ga_xAs_yP_{1-y}$ lattice matched to InP, respectively. Figure 1.53 shows the normalized ΔC_{456} as a function of d_z in the presence of light waves in quantum wells of HD n-InSb (blue), HD n-InAs (red), HD $Hg_{1-x}Cd_xTe$ (green), and HD $In_{1-x}Ga_xAs_yP_{1-y}$ lattice matched to InP (black) where the unperturbed electrons obey the three-band Kane model ($I = 10^{-4}$ nWm^{-2}, $\lambda = 660$ nm, and $n_o = 10^{17}$ m^{-2}) .

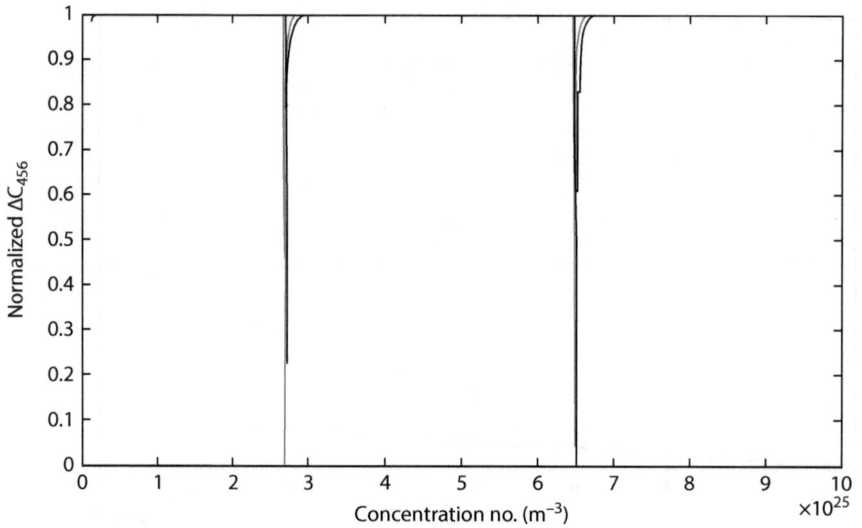

Figure 1.51: Plot of the normalized ΔC_{456}as a function of electron concentration in the presence of magnetic field and light waves for bulk specimens of HD $Hg_{1-x}Cd_xTe$ (green) and HD $In_{1-x}Ga_xAs_yP_{1-y}$ lattice matched to InP (black) where the unperturbed electrons obey the three band model of Kane ($1/B = 0.1$ tesla $^{-1}$, $I = 10^{-4}$ nWm^{-2} and $\lambda = 660$ nm) .

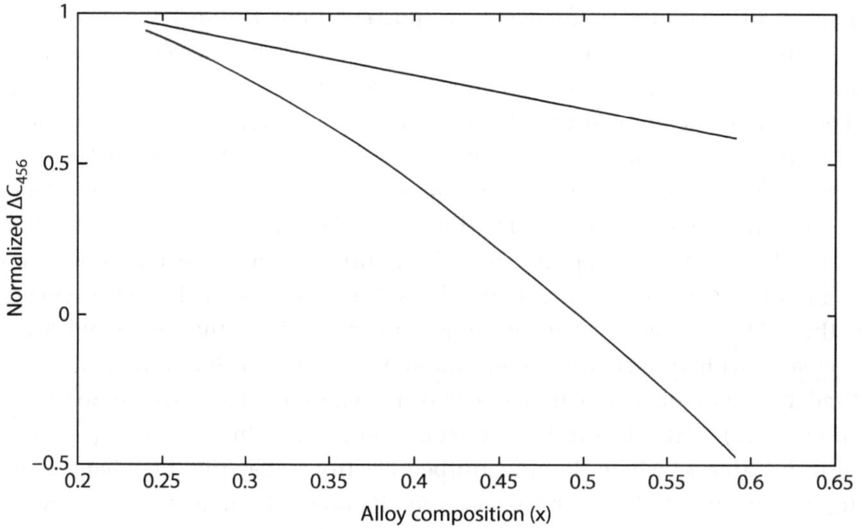

Figure 1.52: The plot of the normalized ΔC_{456} as a function of alloy composition in the presence of magnetic field and light waves for bulk specimens of HD $Hg_{1-x}Cd_xTe$ (blue) and HD $In_{1-x}Ga_xAs_yP_{1-y}$ lattice matched to InP (black) where the unperturbed electrons obey the three-band Kane model. ($1/B = 0.1$ tesla $^{-1}$ $I = 10^{-4}$ Wm^{-2}, $\lambda = 660$ nm, and $n_o = 10^{25}$ m^{-3}).

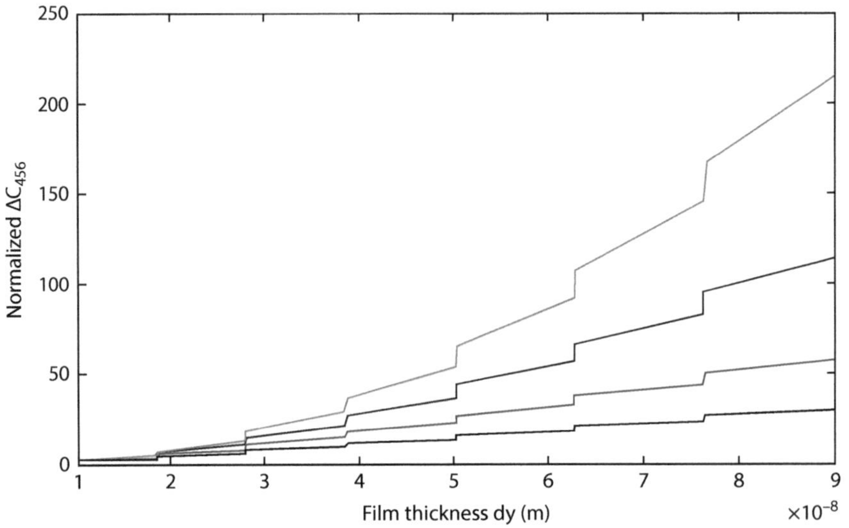

Figure 1.53: The plot of the normalized ΔC_{456} as a function of d_z in the presence of light waves in quantum wells of HD n-InSb (blue), HD n-InAs (red), HD $Hg_{1-x}Cd_xTe$ (green), and HD $In_{1-x}Ga_xAs_yP_{1-y}$ lattice matched to InP (black) where the unperturbed electrons obey the three-band Kane model ($I = 10^{-4}$ nWm^{-2}, $\lambda = 660$ nm, and $n_o = 10^{18}$ m^{-2}).

The influence of quantum confinement is apparent from Figure 1.57 since ΔC_{456} increases with an increase in the film thickness in an oscillatory way in quantum steps and the rate of oscillations are different for different materials as considered here. Figure 1.54 exhibits the normalized ΔC_{456} as a function of surface electron concentration in the presence of light waves in quantum wells of HD n-InSb (blue), HD n-InAs (red), HD $Hg_{1-x}Cd_xTe$ (green), and HD $In_{1-x}Ga_xAs_yP_{1-y}$ lattice matched to InP (black) where the unperturbed electrons obey the three-band Kane model ($d_z = 10$ nm, $I = 10^{-4}$ nWm^{-2}, and $\lambda = 660$ nm). It appears from the figure that ΔC_{456} decreases with an increase in concentration and at a particular value of ES when the Fermi energy touches the subband energy, an quantum jump is reflected in the figure and ΔC_{456} again decreases with an increase in concentration. Figure 1.55 shows the plot of the normalized ΔC_{456} as a function of light intensity in quantum wells of HD n-InSb (blue), HD n-InAs (red), HD $Hg_{1-x}Cd_xTe$ (green), and HD $In_{1-x}Ga_xAs_yP_{1-y}$ lattice matched to InP (black) where the unperturbed electrons obey the three-band Kane model ($d_z = 10$ nm, $n_o = 10^{17}$ m^{-2}, and $\lambda = 660$ nm). It appears from the figure that ΔC_{456} increases with the increasing intensity of the incident light and rate of increment is rather insensitive with the light intensity for the low value of the intensity whereas for larger value of the same the rate of change differs for different materials as considered here. Figure 1.56 shows the normalized ΔC_{456} as a function of wavelength in quantum wells of HD n-InSb (blue), HD n-InAs (red), HD $Hg_{1-x}Cd_xTe$ (green), and HD $In_{1-x}Ga_xAs_yP_{1-y}$ lattice matched to InP (black) where the unperturbed electrons obey the

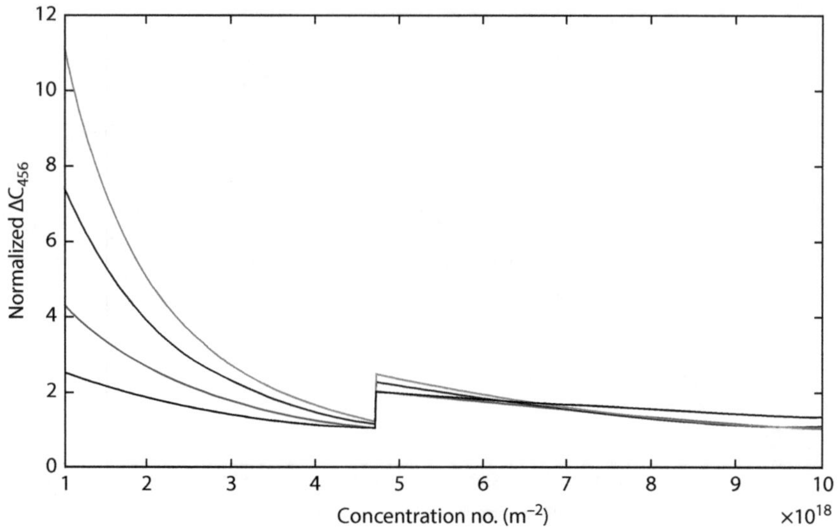

Figure 1.54: The plot of the normalized ΔC_{456} as a function of surface electron concentration in the presence of light waves in quantum wells of HD n-InSb (blue), HD n-InAs (red), HD $Hg_{1-x}Cd_xTe$ (green), and HD $In_{1-x}Ga_xAs_yP_{1-y}$ lattice matched to InP (black) where the unperturbed electrons obey the three-band Kane model ($d_z = 10$ nm, $I = 10^{-4}$ nWm^{-2}, and $\lambda = 660$ nm).

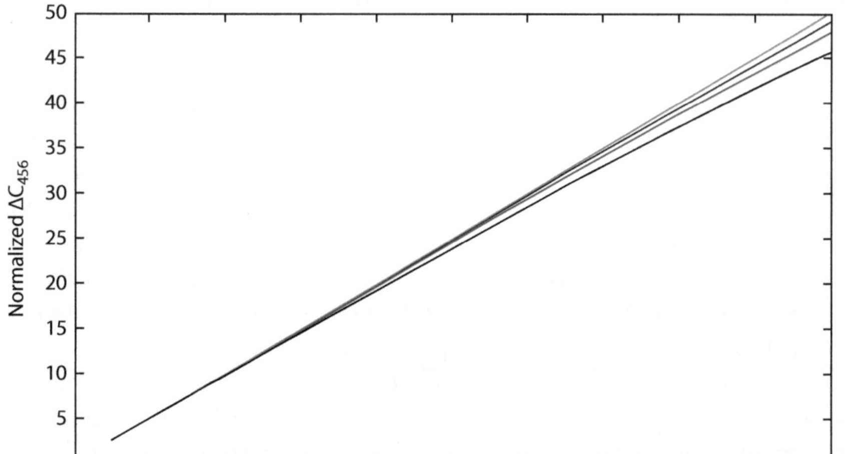

Figure 1.55: The plot of the normalized ΔC_{456} as a function of light intensity in quantum wells of HD n-InSb (blue), HD n-InAs (red), HD $Hg_{1-x}Cd_xTe$ (green), and HD $In_{1-x}Ga_xAs_yP_{1-y}$ lattice matched to InP (black) where the unperturbed electrons obey the three-band Kane model ($d_z = 10$ nm, $n_o = 10^{18}$ m^{-2}, and $\lambda = 660$ nm).

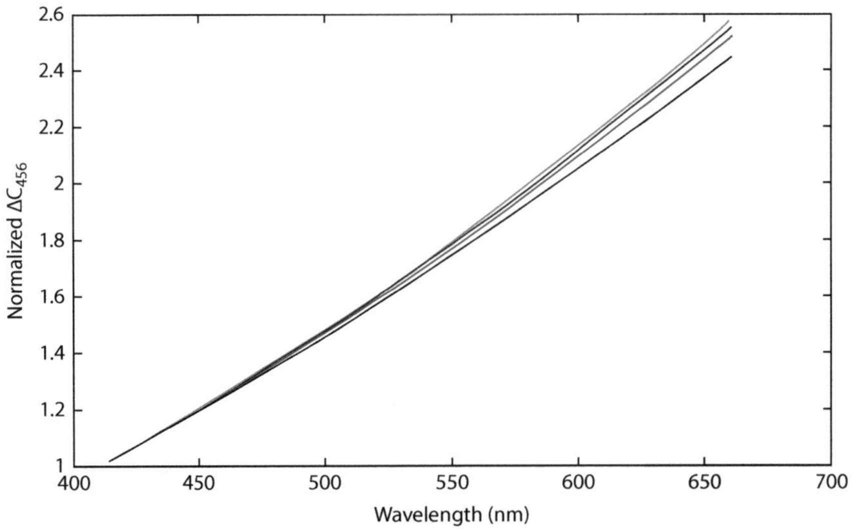

Figure 1.56: The plot of the normalized ΔC_{456} as a function of wavelength in quantum wells of HD n-InSb (blue), HD n-InAs (red), HD $Hg_{1-x}Cd_xTe$ (green), and HD $In_{1-x}Ga_xAs_yP_{1-y}$ lattice matched to InP (black) where the unperturbed electrons obey the three-band Kane model ($d_z = 10$ nm, $n_o = 10^{18}\ m^{-2}$, and $I = 10^{-4}\ nWm^{-2}$).

three-band Kane model ($d_z = 10$ nm, $n_o = 10^{17}\ m^{-2}$, and $I = 10^{-4}\ nWm^{-2}$). It appears from the figure that ΔC_{456} increases with an increase in wavelength. Figure 1.57 exhibits the normalized ΔC_{456} as a function of alloy composition in the presence of light wave in quantum wells of HD $Hg_{1-x}Cd_xTe$ (blue) and HD $In_{1-x}Ga_xAs_yP_{1-y}$ lattice matched to InP (black) where the unperturbed electrons obey the three-band Kane model ($d_z = 10$ nm, $n_o = 10^{18}\ m^{-2}$, $\lambda = 660$ nm, and $I = 10^{-4}\ nWm^{-2}$). Figure 1.57 reveals the fact that for $Hg_{1-x}Cd_xTe$, ΔC_{456} increases with an increase in the alloy composition, whereas the same for $In_{1-x}Ga_xAs_yP_{1-y}$ lattice matched to InP is alloy composition invariant. Figure 1.58 shows the normalized ΔC_{456} as a function of d_z in the presence of light waves in nanowires of HD n-InSb (blue), HD n-InAs (red), HD $Hg_{1-x}Cd_xTe$ (green), and HD $In_{1-x}Ga_xAs_yP_{1-y}$ lattice matched to InP (black) where the unperturbed electrons obey the three-band Kane model ($\lambda = 660$ nm, $I = 1\ nWm^{-2}$, $n_o = 10^{11}\ m^{-1}$, and $d_y = 10$ nm). It appears from the figure that ΔC_{456} exhibits extreme spiky oscillation with respect to film thickness in nanowires and influences of energy band constants are immaterial in this case. Figure 1.59 shows the normalized ΔC_{456} as a function of electron concentration in the presence of light waves in nanowires of HD n-InSb (blue), HD n-InAs (red), HD $Hg_{1-x}Cd_xTe$ (green), and HD $In_{1-x}Ga_xAs_yP_{1-y}$ lattice matched to InP (black) where the unperturbed electrons obey the three-band Kane model ($\lambda = 660$ nm, $I = 1\ nWm^{-2}$, $d_z = 10$ nm, and $d_y = 10$ nm). It appears from the figure that ΔC_{456} exhibits extreme spiky oscillation with respect to ES per unit length in nanowires and influences of energy band constants are immaterial in this case. Figure 1.60 exhibits the plot of the normalized ΔC_{456} as a

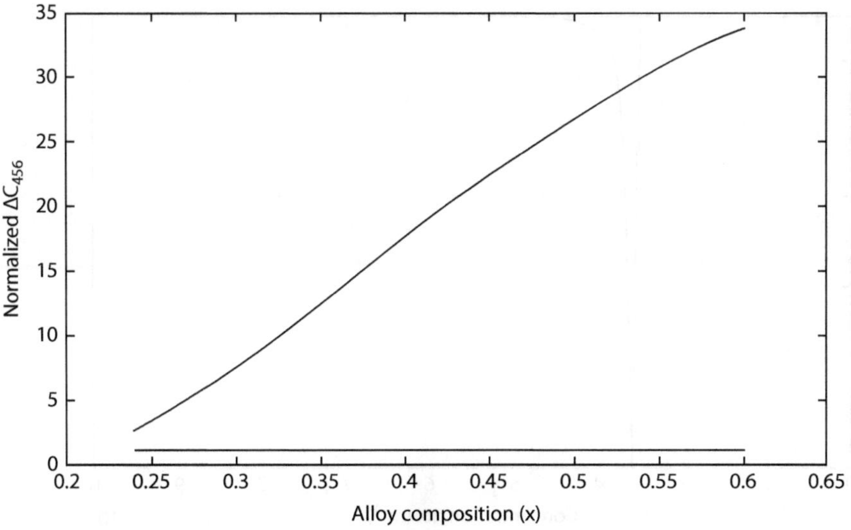

Figure 1.57: The plot of the normalized ΔC_{456} as a function of alloy composition in the presence of light wave in quantum wells of HD $Hg_{1-x}Cd_xTe$ (blue) and HD $In_{1-x}Ga_xAs_yP_{1-y}$ lattice matched to InP (black) where the unperturbed electrons obey the three-band Kane model ($d_z = 10$ nm, $n_o = 10^{18}\ m^{-2}$, $\lambda = 660$ nm, and $I = 10^{-4}\ nWm^{-2}$).

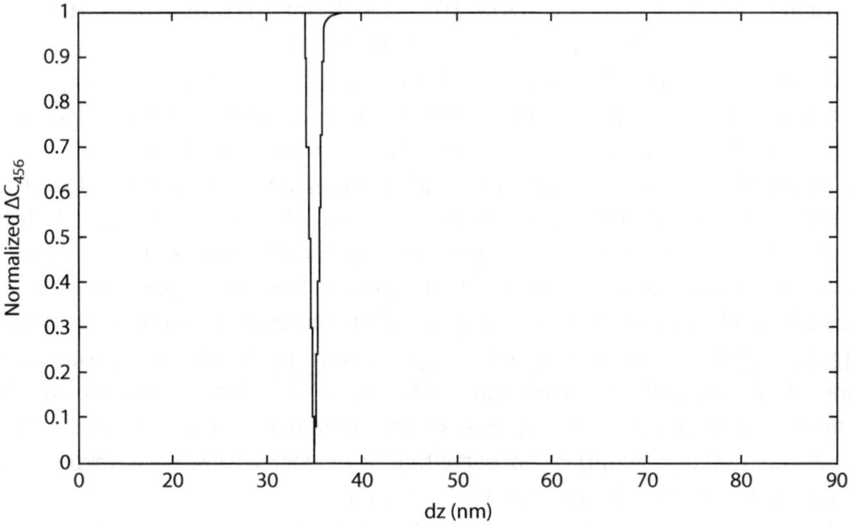

Figure 1.58: Plot of the normalized ΔC_{456} as a function of $d_z$in the presence of light waves in Nano Wires of HD $In_{1-x}Ga_xAs_yP_{1-y}$ lattice matched to InP (black) where the unperturbed electrons obey the three band model of Kane ($\lambda = 660$ nm, $I = 1nWm^{-2}$, $n_o = 10^{11}\ m^{-1}$ and $d_y = 10$ nm).

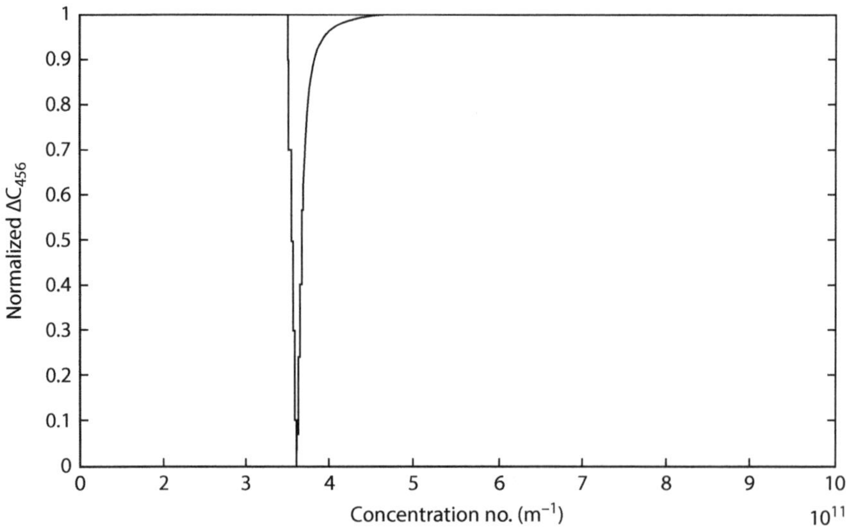

Figure 1.59: Fig.1.59. Plot of the normalized ΔC_{456}as a function of electron concentration in the presence of light waves in Nano Wires of HD $In_{1-x}Ga_xAs_yP_{1-y}$ lattice matched to InP (black) where the unperturbed electrons obey the three band model of Kane (λ= 660 nm, I = 1nWm^{-2}, d_z = 10 nm, d_y = 10 nm)

function of light intensity in nanowires of HD n-InSb (blue), HD n-InAs (red), HD $Hg_{1-x}Cd_xTe$ (green), and HD $In_{1-x}Ga_xAs_yP_{1-y}$ lattice matched to InP (black) where the unperturbed electrons obey the three-band Kane model (λ = 660 nm, d_z = 10 nm, d_y = 10 nm, and $n_o = 10^8\ m^{-1}$). It appears from the figure that ΔC_{456} increases with a decrease in light intensity for all the nanowire materials as considered here. Figure 1.61 shows the normalized ΔC_{456} as a function of wavelength in nanowires of HD n-InSb (blue), HD n-InAs (red), HD $Hg_{1-x}Cd_xTe$ (green), and HD $In_{1-x}Ga_xAs_yP_{1-y}$ lattice matched to InP (black) where the unperturbed electrons obey the three-band Kane model (I = 1 nWm^{-2}, d_z = 10 nm, d_y = 10 nm, and $n_o = 10^{11}\ m^{-1}$). It appears from the figure that ΔC_{456} decreases with an increase in light intensity for all the nanowire materials considered here. Figure 1.62 shows the normalized ΔC_{456} as a function of alloy composition in nanowires of HD $Hg_{1-x}Cd_xTe$ (green) and HD $In_{1-x}Ga_xAs_yP_{1-y}$ lattice matched to InP (black) where the unperturbed electrons obey the three-band Kane model ($n_o = 10^{11}$ m^{-1}, d_y = 20 nm, and d_z = 15 nm). We observe that ΔC_{456} decreases with an increase in alloy composition for both the optoelectronic materials.

For the purpose of condensation we have plotted very few cases with the hope that the readers will perform all the computer programming for all the quantized materials under different physical conditions for the purpose of creating new physics for CECs, effective electron mass, sub band energies and other important transport quantities which are totally band structure dependent. The numerical results presented in this chapter would be different for different materials, but the nature of variation would

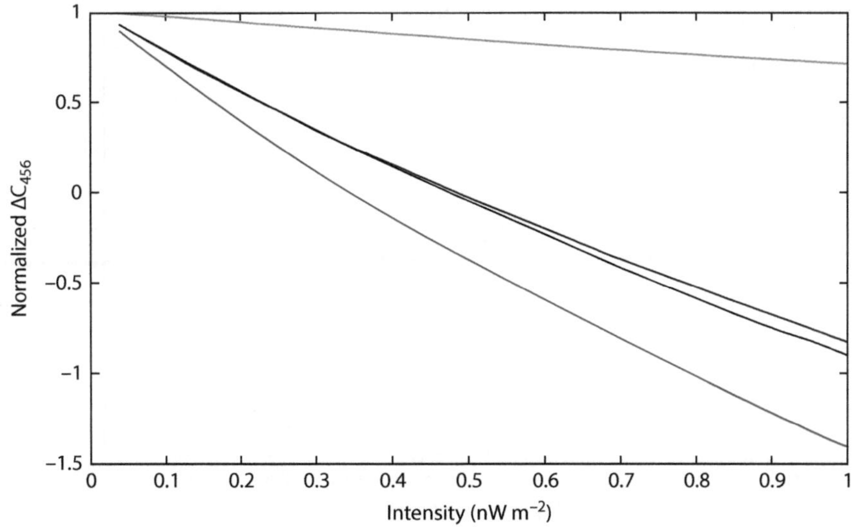

Figure 1.60: The plot of the normalizedΔC_{456} as a function of light intensity in nanowires of HD n-InSb (blue), HD n-InAs (red), HD $Hg_{1-x}Cd_xTe$ (green), and HD $In_{1-x}Ga_xAs_yP_{1-y}$ lattice matched to InP (black) where the unperturbed electrons obey the three-band Kane model ($\lambda = 660$ nm, $d_z = 10$ nm, $d_y = 10$ nm, and $n_o = 10^{11}\ m^{-1}$).

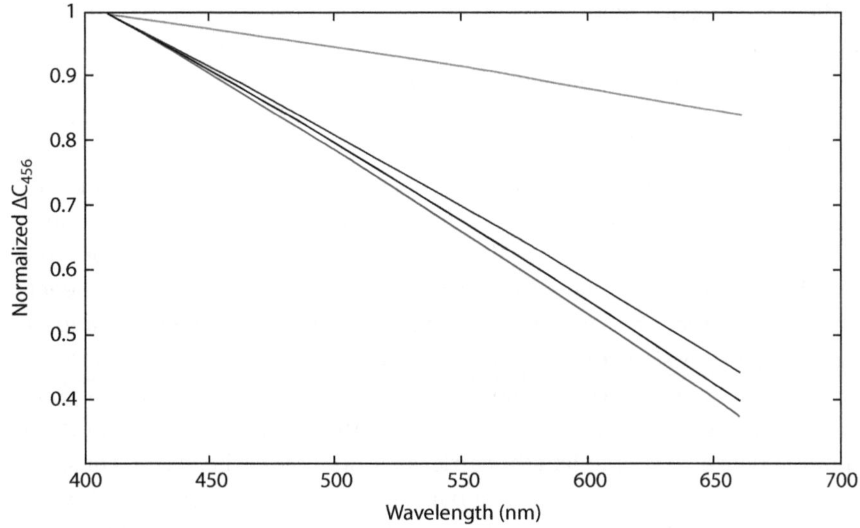

Figure 1.61: The plot of the normalized ΔC_{456} as a function of wavelength in nanowires of HD n-InSb (blue), HD n-InAs (red), HD $Hg_{1-x}Cd_xTe$ (green), and HD $In_{1-x}Ga_xAs_yP_{1-y}$ lattice matched to InP (black) where the unperturbed electrons obey the three-band Kane model ($I = 1nWm^{-2}$, $d_z = 10$ nm, $d_y = 10$ nm, and $n_o = 10^{11}\ m^{-1}$).

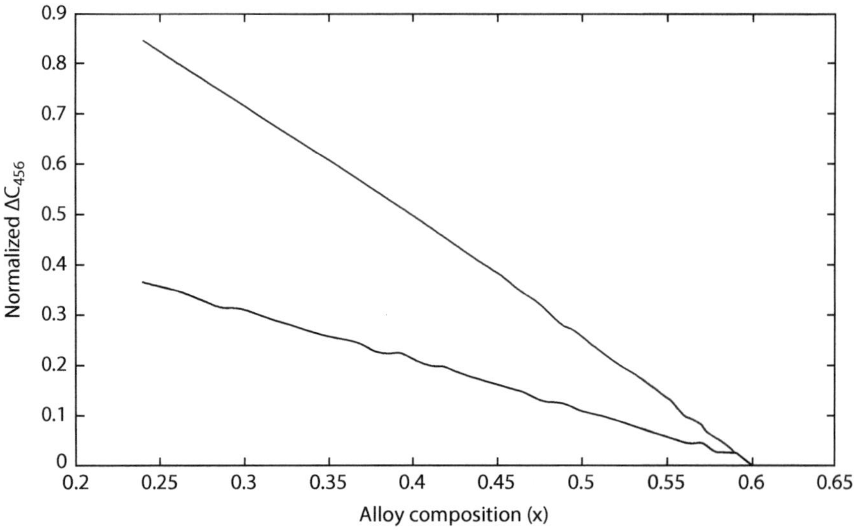

Figure 1.62: The plot of the normalized ΔC_{456} as a function of alloy composition in nanowires of HD $Hg_{1-x}Cd_xTe$ (green) and HD $In_{1-x}Ga_xAs_yP_{1-y}$ lattice matched to InP (black) where the unperturbed electrons obey the three-band Kane model ($n_o = 10^{11}\ m^{-1}$, $d_y = 20$ nm, and $d_z = 15$ nm).

remain unaltered. The theoretical results given here would be useful in analyzing various other experimental data related to the transport phenomenon. We must note that the study of transport phenomenon and the formulation of the electronic properties of low-dimensional field-aided HD compounds are based on the dispersion relations for such materials. It is worth remarking that this simplified formulation exhibits the basic qualitative features of the CECs for field-assisted low-dimensional materials. The basic objective of this chapter is not solely to demonstrate the influence of quantum confinement on the CECs for different quantum-confined HD nonparabolic materials in the presence of external fields but also to formulate the appropriate electron statistics in the most generalized form, since the transport and other phenomena in such nanostructured materials having different band structures and the derivation of the expressions of many important electronic properties are based on the electron statistics in such compounds. It is worth remarking that the analysis as presented in this chapter can be used to investigate the Burstein Moss shift, thermoelectric power, Debye screening length, carrier contribution to the elastic constants, diffusivity–mobility ratio, measurement of band -gap in the presence of light waves, diffusion coefficient of the minority carriers, nonlinear optical response, third-order nonlinear optical susceptibility, generalized Raman gain, the plasma frequency, the activity coefficient, magnetothermal effect in quantized structures, normalized Hall coefficient, reflection coefficient, heat capacity, magnetic susceptibilities, Faraday rotation, Fowler–Nordheim field emission, optical effective mass, Einstein's photoemission, Righi–Leduc coefficient, electric susceptibility, electric susceptibility mass, electron diffusion thermopower, hydrostatic Piezo-resistance coefficient, relaxation time for acoustic mode scattering, and gate capacitance, and other different transport coefficients of modern HD nonparabolic quantum-confined field-aided HD devices operated under different external conditions having varying band structures.

It is imperative to state that the present investigation excludes the many-body, hot electron, broadening, and the allied effects in the simplified theoretical formalism due to the absence of proper analytical techniques for including them for generalized systems as considered here. Our simplified approach will be appropriate for the purpose of comparisons when the methods of tackling the formidable problems after inclusion of the said effects for the generalized systems emerge. The results of this simplified approach get transformed to the well-known formulation of the CECs for wide gap materials having parabolic energy bands. This indirect test not only exhibits the mathematical compatibility of the formulation but also shows the fact that this simple analysis is more generalized one, since one can obtain the corresponding results for materials having parabolic energy bands under certain limiting conditions from the present derivation. For the purpose of computer simulations for obtaining the plots of CECs versus various external variables, we have taken very low temperatures since the quantization effects are basically low-temperature phenomena together with the fact that the temperature dependence of all the energy band constants of all the semiconductors and their nanostructures as considered in this chapter are not available in the literature. Our results as formulated in this chapter are valid for finite temperatures and are useful in comparing the results for temperature variations of CECs after the availability of the temperature dependences of such constants of various dispersion relations in this context. The inclusion of the said effects would certainly increase the accuracy of the results, although the qualitative features of CECs would not change in the presence of the aforementioned relations of the various quantized structures as discussed earlier. It may finally be noted that the basic aim of this chapter is not solely to demonstrate the influence of quantum confinement on the CECs for a wide class of quantized materials but also to formulate the appropriate carrier statistics in the most generalized form, since the transport and other phenomena in modern nanostructured devices having different band structures and the derivation of the expressions of many important carrier properties are based on the temperature-dependent carrier statistics in such systems.

1.5 Open research problems

The problems under this section of the book are by far the most important part for the readers, and few open research problems are listed in all chapters. The numerical values of the energy band constants for various semiconductors [42–48, 50–54, 59–72, 75, 113–231, 238–242, 245–249, 252–256, 258–264, 267–285] *are given in Appendix for the related computer simulations.*

(R.1.1) Investigate the CECs for the quantum-confined HD semiconductors whose respective DRs of the carriers in the absence of band tails are given below:

(a) The electron dispersion law in n-GaP can be written as [232]

$$E = \frac{\hbar^2 k_s^2}{2m_p^*} + \frac{\hbar^2 k_s^2}{2m_\perp^*} m \frac{\bar{\Delta}}{2} \pm \left[\left(\frac{\bar{\Delta}}{2} \right)^2 + P_1 k_z^2 + D_1 k_x^2 k_y^2 \right] \tag{R1.1}$$

where $\bar{\Delta} = 335meV, P_1 = 2 \times 10^{-10}\text{eVm}, D_1 = P_1 a_1$ *and* $a_1 = 5.4 \times 10^{10} m$

(b) The dispersion relation for the conduction electrons for IV–VI semiconductors can also be described by the models of Cohen [233], McClure and Choi, [234], Bangert and Kastner [235], Foley and Langenberg [236], and Takaoka et al. [237], respectively.

(i) In accordance with Cohen [233], the dispersion law of the carriers is given by

$$E(1+\alpha E) = \frac{p_x^2}{2m_1} + \frac{p_z^2}{2m_3} - \frac{\alpha E p_y^2}{2m'_2} + \left(\frac{\alpha p_y^4}{4m_2 m'_2} \right) + \frac{p_y^2}{2m_2}(1+\alpha E) \tag{R1.2}$$

where m_1, m_2, and m_3 are the effective carrier masses at the band-edge along x, y, and z directions, respectively, and m'_2 is the effective-mass tensor component at the top of the valence band (for electrons) or at the bottom of the CB (for holes).

(ii) The carrier energy spectra can be written, following McClure and Choi [230], as

$$E(1+\alpha E) = \frac{p_x^2}{2m_1} + \frac{p_y^2}{2m_2} + \frac{p_z^2}{2m_3} + \frac{p_y^2}{2m_2} \alpha E \left\{ 1 - \left(\frac{m_2}{m'_2} \right) \right\} + \frac{p_y^4 \alpha}{4m_2 m'_2} - \frac{\alpha p_x^2 p_y^2}{4m_1 m_2} - \frac{\alpha p_y^2 p_z^2}{4m_2 m_3} \tag{R1.3}$$

(iii) The carrier energy spectrum of IV–VI semiconductors in accordance with Foley and Langenberg [236] can be written as

$$E + \frac{E_g}{2} = E_-(k) + \left[\left[E_+(k) + \frac{E_g}{2} \right]^2 + P_\perp^2 k_s^2 + P_p^2 k_z^2 \right]^{1/2} \tag{R1.4}$$

where $E_+ = (k) = \frac{\hbar^2 k_s^2}{2m_\perp^+} + \frac{\hbar^2 k_z^2}{2m_p^+}$, $E_- = (k) = \frac{\hbar^2 k_s^2}{2m_\perp^-} + \frac{\hbar^2 k_z^2}{2m_p^-}$ represent the contribution from the interaction of the CB and the VB edge states with the more distant bands and the free electron term, $\frac{1}{m_\perp^\pm} = \frac{1}{2}\left[\frac{1}{m_{tc}} \pm \frac{1}{m_{tv}}\right]$, $\frac{1}{m_p^\pm} = \frac{1}{2}\left[\frac{1}{m_{1c}} \pm \frac{1}{m_{1v}}\right]$.

For n-PbTe

$$P_\perp = 4.61 \times 10^{-10}\ \text{eVm},\ P_p = 4.61 \times 10^{-10}\ \text{eVm},\ \frac{m_0}{m_{tv}} = 10.36,\ \frac{m_0}{m_{tv}} = 0.75,\ \frac{m_0}{m_{tc}} = 11.36,$$

$$\frac{m_0}{m_{1c}} = 1.20 \text{ and } g_v = 4$$

(iv) The carrier DR in accordance with the model of Takaoka et al. [237] can be written as

$$E2\left(1+\frac{E}{E_g}\right) - \frac{\beta\hbar^2 k_y^2}{2M_2} - \frac{\gamma\hbar^4 k_y^4}{4M_2^2 E_g} = \frac{\hbar^2 k_x^2}{2m_1} + \frac{\hbar^2 k_z^2}{2m_3}$$

where $\beta = 1 + \frac{E}{E_g}(1-\gamma) + \delta$, $\gamma = \frac{M_2}{M'_2}$, $\delta = \frac{M_2}{m_2}$ and the notations are defined in [237].

(c) The conduction electrons of n-GaSb obey the following two dispersion relations:

(i) In accordance with the model of Seiler et al. [250]

$$E = \left[-\frac{E_g}{2} + \frac{E_g}{2}[1+\alpha_4 k^2]^{1/2} + \frac{\bar{\varsigma}_0\hbar^2 k^2}{2m_0} + \frac{\bar{v}_0 f_1(k)\hbar^2}{2m_0} \pm \frac{\bar{\omega}_0 f_2(k)\hbar^2}{2m_0}\right] \tag{R1.5}$$

where $\alpha_4 \equiv 4P^2(E_g + \frac{2}{3}\Delta)\left[E_g^2(E_g+\Delta)\right]^{-1}$, P is the isotropic momentum matrix element, $f_1(k) \equiv k^{-2}\left[k_x^2 k_y^2 + k_y^2 k_z^2 + k_z^2 k_x^2\right]$ represents the warping of the Fermi surface, $f_2(k) \equiv \left[\left\{k^2(k_x^2 k_y^2 + k_y^2 k_z^2 + k_z^2 k_x^2) - 9k_x^2 k_y^2 k_z^2\right\}^{1/2} k^{-1}\right]$ represents the inversion asymmetry splitting of the CB, and $\bar{\varsigma}_0$, $\bar{v}_0$, and $\bar{\omega}_0$ represent the constants of the electron spectrum in this case.

(ii) In accordance with the model of Zhang et al. [251]

$$E = \left[E_2^{(1)} + E_2^{(2)} K_{4,1}\right]k^2 + \left[E_4^{(1)} + E_4^{(2)} K_{4,1}\right]k^4 + k^6\left[E_6^{(1)} + E_6^{(2)} K_{4,1} + E_6^{(3)} K_{6,1}\right] \tag{R1.6}$$

where $K_{4,1} \equiv \frac{5}{4}\sqrt{21}\left[\frac{k_x^4 + k_y^4 + k_z^4}{k^4} - \frac{3}{5}\right]$, $K_{6,1} \equiv \sqrt{\frac{639639}{32}}\left[\frac{k_x^2 k_y^2 k_z^2}{k^6} + \frac{1}{22}\left(\frac{k_x^4 + k_y^4 + k_z^4}{k^4} - \frac{3}{5}\right) - \frac{1}{105}\right]$, the coefficients are in eV, the values of k are $10(\frac{a}{2\pi})$ times those of k in atomic units (a is the lattice constant),

$$E_2^{(1)} = 1.0239620,\ E_2^{(2)} = 0,\ E_4^{(1)} = -1.1320772,\ E_4^{(2)} = 0.05658,\ E_6^{(1)} = 1.1072073$$

$$E_6^{(2)} = -0.1134024 \text{ and } E_6^{(3)} = -0.0072275.$$

(d) In addition to the well-known band models of III–V semiconductors as discussed in this book, the conduction electrons of such compounds obey the following three dispersion relations:

(i) In accordance with the model of Rossler [243]

$$E = \frac{\hbar^2 k^2}{2m_c} + (\alpha_{11} + \alpha_{12}k)k^4 + (\beta_{11} + \beta_{12}k)[k_x^2 k_y^2 + k_y^2 k_z^2 + k_z^2 k_x^2] \pm (\gamma_{11} + \gamma_{12}k)\left[k^2(k_x^2 k_y^2 + k_y^2 k_z^2 + k_z^2 k_x^2) - 9k_x^2 k_y^2 k_z^2\right]^{1/2} \quad \text{(R1.7)}$$

where $\alpha_{11} = -2132 \times 10^{-40}\,\text{eVm}^4$, $\alpha_{12} = 9030 \times 10^{-50}\,\text{eVm}^5$, $\beta_{11} = -2493 \times 10^{-40}\,\text{eVm}^4$, $\beta_{12} = 12594 \times 10^{-50}\,\text{eVm}^5$, $\gamma_{11} = 30 \times 10^{-30}\,\text{eVm}^3$, and $\gamma_{12} = -154 \times 10^{-42}\,\text{eVm}^4$.

(ii) In accordance with Johnson and Dicley [289], the electron energy spectrum assumes the form

$$E = \frac{E_g}{2} + \frac{\hbar^2 k^2}{2}\left[\frac{1}{m_0} + \frac{1}{m_{yb}}\right] + \frac{E_g}{2}\left[1 + 4\frac{\hbar^2 k^2 f_1(E)}{2m'_c\, E_g}\right] \quad \text{(R1.8)}$$

where $\frac{m_0}{m'_c} \equiv P^2\left[\frac{(E_g + \frac{2\Delta}{3})}{E_g(E_g + \Delta)}\right]$, $\bar{f}_1(E) \equiv \frac{(E_g+\Delta)(E + E_g + \frac{2\Delta}{3})}{(E_g+\frac{2\Delta}{3})(E + E_g + \Delta)}$, $m'_c = 0.139m_0$, and $m_{yb} = \left[\frac{1}{m'_c} - \frac{2}{m_0}\right]^{-1}$

(iii) In accordance with Agafonov et al. [289], the electron energy spectrum can be written as

$$E = \frac{\bar{\eta} - E_g}{2}\left[1 - \frac{\hbar^2 k^2}{2\bar{\eta}m^*}\left\{\frac{D\sqrt{3} - 3B}{2(\frac{\hbar^2}{2m^*})}\right\}\left[\frac{k_x^4 + k_y^4 + k_z^4}{k^4}\right]\right] \quad \text{(R1.9a)}$$

where $\bar{\eta} \equiv (E_g^2 + \frac{8}{3}P^2k^2)^{1/2}$, $\bar{B} \equiv -21\frac{\hbar^2}{2m_0}$, and $D \equiv -40\left(\frac{\hbar^2}{2m_0}\right)$

(iv) In accordance with the model of Kolodziejczak et al [243], the electron energy spectrum of III–V compounds can be expressed, taking into account the interaction of the higher bands, as

$$E = \frac{\hbar^2 k^2}{2m_0} + \frac{\chi_0 - E_g}{2} + a\left[\frac{\chi_0 + E_g}{2\chi_0}\right]\frac{\hbar^2 k^2}{2m_0} + \frac{\chi_0 - E_g}{2\chi_0}\left]\frac{\hbar^2 k^2}{2m_0}\right[b + \frac{c}{5}\right] \quad \text{(R1.9b)}$$

where $\chi_0 \equiv \left[E_g^2 - 4FE_g\left(\frac{E_g + \frac{2}{3}\Delta}{E_g + \Delta}\right)\left(\frac{\hbar^2 k^2}{2m_0}\right)\right]^{1/2}$

F represents the interaction between the $\Gamma_{25'}$ and Γ_{15} states,

$a \equiv \frac{-p^2\left[E(\Gamma_{15c}) - E(\Gamma_{15y})\right] - \left[E(\Gamma_{15}) - E(\Gamma_{25'})\right]}{2\left[E(\Gamma_{15c}) - E(\Gamma_{15y})\right]\left[E(\Gamma_{15c}) - E(\Gamma_{15y}) - E_g\right]}$, $b \equiv \frac{1}{3}[M + 4G]$, M, represents the interaction between $\Gamma_{25'}$ and Γ_{15} states, G represents the interaction between $\Gamma_{25'}$ and $\Gamma_{12'}$ states, $c \equiv \frac{1}{2}\left[\frac{(F - G + M)^2 - (F + 2G - M)^2}{F + 2G - M}\right]$, and the other notations are the same as in the above reference.

(e) The dispersion relation of the carriers in n-type $Pb_{1-x}Ga_xTe$ with x = 0.01 can be written following Vassilev [265] as

$$\left[E - 0.606k_s^2 - 0.0722k_z^2\right]\left[E + \bar{E}_g + 0.411k_s^2 + 0.0377k_z^2\right] = 0.23k_s^2 + 0.02k_z^2 \pm \left[0.06\bar{E}_g + 0.061k_s^2 + 0.0066k_z^2\right]k_s \tag{R1.10}$$

where $\bar{E}_g(=0.21eV)$ is the energy gap for the transition point, the zero of the energy E is at the edge of the CB of the Γpoint of the Brillouin zone and is measured positively upward, and k_x, k_y , and k_z are in the units of $10^9 m^{-1}$.

(f) The energy spectrum of the carriers in the two higher VBs and the single lower VB of Te can, respectively, be expressed as [250]

$$\bar{E} = A_{10}k_z^2 + B_{10}k_s^2 \pm \left[\Delta_{10}^2 + (\beta_{10}\mathrm{k_z})^2\right]^{1/2} \text{ and } \bar{E} = \Delta_{\|} + A_{10}k_z^2 + B_{10}k_s^2 \pm \beta_{10}k_z \tag{R1.11a}$$

where $\bar{E}$ is the energy of the hole as measured from the top of the valance and within it, $A_{10} = 3.77 \times 10^{-19}\mathrm{eVm}^2$, $B_{10} = 3.57 \times 10^{-19}\mathrm{eVm}^2$, $\Delta_{10} = 0.628\,\mathrm{eV}$, $(B_{10})^2 = 6 \times 10^{-20}$ $(\mathrm{eVm})^2$, and $\Delta_{\|} = 1004 \times 10^{-5}eV$ are the spectrum constants.

The dispersion relation of the conduction electrons of tellurium can be written in accordance with the model of Ortenberg and Button as [266]

$$E = t_1 + t_2k_z^2 + t_3k_s^2 + t_4k_s^4 + t_5k_s^2k_z^2 \pm \left[(t_1 + t_6k_s^2)^2 + t_7k_z^2\right]^{1/2} \tag{R1.11b}$$

where t_1, t_2,t_3,t_4,t_5, t_6, and t_7 are the energy band constants.

(g) The dispersion relation of the holes in p-InSb can be written in accordance with Cunningham [285] as

$$\bar{E} = c_4(1 + \gamma_4 f_4)k^2 \pm \frac{1}{3}\left[2\sqrt{2}\sqrt{c_4}\sqrt{16 + 5\gamma_4}\sqrt{E_4}g_4k\right] \tag{R1.12}$$

where $c_4 \equiv \frac{\hbar^2}{2m_0} + \theta_4, \theta_4 \equiv 4.7\frac{\hbar^2}{2m_0}, \gamma_4 \equiv \frac{b_4}{c_4}, b_4 \equiv \frac{3}{2}b_5 + 2\theta_4$, $b_5 \equiv 2.4\frac{\hbar^2}{2m_0}$, $f_4 \equiv \frac{1}{4}\left[\sin^2 2\theta + \sin^4\theta\sin^2 2\phi\right]$, $\theta$is measured from the positive z-axis, $\phi$is measured from positive x-axis,$g_4 \equiv \sin\theta\left[\cos^2\theta + \frac{1}{4}\sin^4\theta\sin^2 2\phi\right]$, and $E_4 = 5 \times 10^{-4}eV$

(h) The energy spectrum of the VBs of CuCl in accordance with Yekimov et al. [286] can be written as

$$E_h = (\gamma_6 - 2\gamma_7)\frac{\hbar^2k^2}{2m_0} \tag{R1.13}$$

and

$$E_{l,s} = (\gamma_6 + \gamma_7)\frac{\hbar^2 k^2}{2m_0} - \frac{\Delta_1}{2} \pm \left[\frac{\Delta_1^2}{4} + \gamma_7 \Delta_1 \frac{\hbar^2 k^2}{2m_0} + 9\left(\frac{\gamma_7 \hbar^2 k^2}{2m_0}\right)^2\right]^{1/2} \tag{R1.14}$$

where $\gamma_6 = 0.53,\ \gamma_7 = 0.07,\ \Delta_1 = 70\,\text{meV}$.

(i) In the presence of stress, χ_6 along <001> and <111> directions, the energy spectra of the holes in semiconductors having diamond structure VBs can be, respectively, expressed following Roman and Ewald [287] as

$$E = A_6 k^2 \pm \left[\bar{B}_7^2 + \delta_6^2 + B_7 \delta_6 (2k_z^2 - k_s^2)\right]^{1/2} \tag{R1.15}$$

and

$$E = A_6 k^2 \pm \left[\bar{B}_7^2 k^4 + \delta_7^2 + \frac{D_6}{\sqrt{3}} \delta_7 (2k_z^2 - k_s^2)\right]^{1/2} \tag{R1.16}$$

where, A_6, B_7, D_6, and C_6 are inverse mass band parameters in which $\delta_6 \equiv l_7(\bar{S}_{11} - \bar{S}_{12})\chi_6$, are the usual elastic compliance constants, $\bar{B}_7^2 \equiv \left(B_7^2 + \frac{c_6^2}{5}\right)$, and $\delta_7 \equiv \left(\frac{d_8 S_{44}}{2\sqrt{3}}\right)\chi_6$. For gray tin, $d_8 = -4.1\,\text{eV}$, $l_7 = -2.3eV$, $A_6 = 19.2\frac{\hbar^2}{2m_0}$, $B_7 = 26.3\frac{\hbar^2}{2m_0}$, $D_6 = 31\frac{\hbar^2}{2m_0}$, and $c_6^2 = -1112\frac{\hbar^2}{2m_0}$.

(j) The DR of the carriers of cadmium and zinc diphosphides are given by [235]

$$E \equiv \left[\beta_1 + \frac{\beta_2 \beta_3(k)}{8\beta_4}\right]k^2 \pm \left\{\left[\beta_4 \beta_3(k) x(\beta_5 + \frac{\beta_2 \beta_3(k)}{8\beta_4} k^2\right] + 8\beta_4^2\left(1 - \frac{\beta_3^2}{4}\right) - \beta_2\left(1 - \frac{\beta_3^2(k)}{4}\right)k^2\right\}^{1/2} \tag{R1.17}$$

where $\beta_1, \beta_2, \beta_4$, and β_5 are system constants and $\beta_3(k) = \frac{k_x^2 + k_y^2 - 2k_z^2}{k^2}$.

(k) The E–k relation of the conduction electrons in semiconductors in the presence of electron–phonon interaction assumes the form [257]

$$E = \frac{\hbar^2 k^2}{2m_c} - \alpha_c \hbar\omega_0 \frac{p_0}{\hbar k}\tan^{-1}\left[\frac{\hbar^2 k^2}{2m_c(\hbar\omega_0 - E)}\right]^{1/2} \tag{R1.18}$$

where α_c is the dimensionless coupling constant, $p_0 = (2m_c\hbar\omega_0)^{1/2}$, and ω_0 is the angular frequency of the optical phonon.

(R1.2) Investigate the CECs for bulk specimens of the HD semiconductors in the presences of exponential, Kane, Halperian, Lax, and Bonch–Burevich types of band tails [37] for all systems whose unperturbed carrier energy spectra are defined in (R1.1).

(R1.3) Investigate the CECs for QWs of all the HD semiconductors as considered in (R1.2).

(R1.4) Investigate the CECs for HD bulk specimens of the negative refractive index, organic, magnetic, and other advanced optical materials in the presence of an arbitrarily oriented alternating electric field.

(R1.5) Investigate the CECs for the QWs of HD negative refractive index, organic, magnetic, and other advanced optical materials in the presence of an arbitrarily oriented alternating electric field.

(R1.6) Investigate the CECs for the multiple QWs of HD materials whose unperturbed carrier energy spectra are defined in (R1.1).

(R1.7) Investigate the CECs for all the appropriate HD low-dimensional systems of this chapter in the presence of finite potential wells.

(R1.8) Investigate the CECs for all the appropriate HD low-dimensional systems of this chapter in the presence of parabolic potential wells.

(R1.9) Investigate the CECs for all the appropriate HD systems of this chapter forming quantum rings.

(R1.10) Investigate the CECs for all the aforementioned appropriate problems in the presence of elliptical hill and quantum square rings.

(R1.11) Investigate the CECs for triangular 2D systems in the presence of an arbitrarily oriented alternating electric field for all the HD materials whose unperturbed carrier energy spectra are defined in (R1.1).

(R1.12) Investigate the CECs for HD 2D systems of the negative refractive index and other advanced optical materials in the presence of an arbitrarily oriented alternating electric field and nonuniform light waves.

(R1.13) Investigate the CECs for triangular HD 2D systems of the negative refractive index, organic, magnetic, and other advanced optical materials in the presence of an arbitrarily oriented alternating electric field in the presence of strain.

(R1.14) (a) Investigate the CECs for HD 2D systems of the negative refractive index, organic, magnetic, and other advanced optical materials in the presence of many body effects. (b) Investigate all the appropriate problems of this chapter for a Dirac electron.

(R1.15) Investigate all the appropriate problems of this chapter by including the many body, image force, broadening, and hot carrier effects, respectively.

(R1.16) Investigate all the appropriate problems of this chapter by removing all the mathematical approximations + and establishing the respective appropriate uniqueness conditions.

References

[1] P.K. Basu, *Theory of Optical Process in Semiconductors, Bulk and Microstructures* (Oxford University Press, Oxford, 1997).

[2] K.P. Ghatak, S. Bhattacharya, S. Bhowmik, R. Benedictus, and S. Chowdhury, *J. Appl. Phys*, 103, 094314 (2008); K.P. Ghatak and S. Bhattacharya, *J. Appl. Phys*,.102, 073704 (2007); K.P. Ghatak, S. Bhattacharya, S.K. Biswas, A. De, and A.K. Dasgupta, *Phys. Scr.*, 75, 820 (2007); P.K. Bose, N. Paitya, S. Bhattacharya, D. De, S. Saha, K.M. Chatterjee, S. Pahari, and K.P. Ghatak, *Quantum Matter*, 1, 89 (2012); K.P. Ghatak, S. Bhattacharya, A. Mondal, S. Debbarma, P. Ghorai, and A. Bhattacharjee, *Quantum Matter*, 2, 25 (2013); S. Bhattacharya, D. De, S. Ghosh, J. P. Bannerje, M. Mitra, B. Nag, S. Saha, S. K. Bishwas, and M. Paul, *J. Comp. Theor. Nanosci*, 7, 1066 (2010).
[3] K.P. Ghatak, S. Bhattacharya, S. Pahari, S.N. Mitra, and P.K. Bose, *J. Phys. Chem. Solids*, 70, 122 (2009), S. Bhattacharya, D. De, R. Sarkar, S. Pahari, A. De, A.K. Dasgupta, S. N. Biswas, and K.P. Ghatak, *J. Comp. Theor. Nanosci.*, 5, 1345 (2008); S. Mukherjee, D. De, D. Mukherjee, S. Bhattacharya, A. Sinha, and K.P. Ghatak, *Physica B*, 393, 347 (2007).
[4] K. Seeger, *Semiconductor Physics* (Springer-Verlag, 7th edn, Berlin, Germany, 2006).
[5] B.R. Nag, *Physics of Quantum Well Devices* (Kluwer Academic Publishers, The Netherlands, 2000).
[6] A. Rogalski, *J. Alloys Comp.*, 371, 53 (2004).
[7] A. Baumgartner, A. Chaggar, A. Patanè, L. Eaves, and M. Henini, *Appl. Phys. Lett.*, 92, 091121 (2008).
[8] J. Devenson, R. Teissier, O. Cathabard, and A.N. Baranov, *Proc. SPIE.*, 6909, 69090U (2008).
[9] B.S. Passmore, J. Wu, M.O. Manasreh, and G.J. Salamo, *Appl. Phys. Lett.*, 91, 233508 (2007).
[10] M. Mikhailova, N. Stoyanov, I. Andreev, B. Zhurtanov, S. Kizhaev, E. Kunitsyna, K. Salikhov, and Y. Yakovlev, *Proc. SPIE*, 6585, 658526 (2007).
[11] W. Kruppa, J.B. Boos, B.R. Bennett, N.A. Papanicolaou, D. Park, and R. Bass, *Electron. Lett*, 42, 688 (2006).
[12] E.O. Kane, *Semiconductors and Semimetals*, R.K. Willardson and A.C. Beer (eds.) (Academic Press, New York, Vol. 1, 1966).
[13] B.R. Nag, *Electron Transport in Compound Semiconductors* (Springer, Heidelberg, 1980).
[14] G.E. Stillman, C.M. Wolfe, and J.O. Dimmock, *Semiconductors and Semimetals*, R.K. Willardon and A.C. Beer (eds) (Academic Press, New York, Vol. 12, 1977).
[15] D.J. Newson and A. Karobe, *Semicond. Sci. Tech.*, 3, 786 (1988).
[16] E.D. Palik, G.S. Picus, S. Teither, and R.E. Wallis, *Phys. Rev.*, 122, 475 (1961).
[17] P.Y. Lu, C.H. Wung, C.M. Williams, S.N.G. Chu, and C.M. Stiles, *Appl. Phys. Lett*, 49, 1372 (1986).
[18] N.R. Taskar, I.B. Bhat, K.K. Prat, D. Terry, H. Ehasani, and S.K. Ghandhi, *J. Vac. Sci. Tech*, 7A, 281 (1989).
[19] F. Koch, *Springer Series in Solid States Sciences* (Springer, Germany, 1984).
[20] L.R. Tomasetta, H.D. Law, R.C. Eden, I. Reyhimy, and K. Nakano, *IEEE J. Quant. Electron.*, 14, 800 (1978).
[21] T. Yamato, K. Sakai, S. Akiba, and Y. Suematsu, *IEEE J. Quantum Electron.*, 14, 95 (1978).
[22] T.P. Pearsall, B.I. Miller, and R.J. Capik, *Appl. Phys. Lett.*, 28, 499 (1976).
[23] M.A. Washington, R.E. Nahory, M.A. Pollack, and E.D. Beeke, *Appl. Phys. Lett.*, 33, 854 (1978).
[24] M.I. Timmons, S.M. Bedair, R.J. Markunas, and J.A. Hutchby, (1982), *Proceedings of the 16th IEEE Photovoltaic Specialist Conference* (IEEE, San Diego, California, 666).
[25] R.K. Willardson, ed. by A. C. Beer, 1, *Semiconductors and Semimetals* (Academic Press, New York, 1966) p. 102.
[26] E.O. Kane, *Phys. Rev*, 131, 79 (1963), *Phys. Rev. B.*, 139, 343 (1965).
[27] V.L. Bonch Bruevich, *Sov. Phys. Sol. Stat.*, 4, 53, (1963).
[28] R.A. Logan and A.G. Chynoweth, *Phys. Rev.*, 131, 89 (1963).
[29] C.J. Hwang, *J. Appl. Phys.*, 40, 3731 (1969).
[30] J.I. Pankove, *Phys. Rev. A*, 130, 2059 (1965).

[31] B.I. Halperin, M. Lax, *Phys. Rev.*, 148, 722 (1966).
[32] R.A. Abram, G.J. Rees, and B.L.H. Wilson, *Adv. Phys*, 27, 799, (1978).
[33] B.I. Shklovskii, and A.L. Efros, *Electronic Properties of Doped Semiconductors*, Springer series in Solid State Sciences, Vol. 45 (Springer, Berlin, 1984).
[34] E.O. Kane, *Solid State Electron.*, 28, 3 (1985).
[35] P.K. Chakraborty and J.C. Biswas, *J. Appl. Phys.*, 82, 3328 (1997).
[36] B.R. Nag, *Electron Transport in Compound Semiconductors*, Springer Series in Solid state Sciences, Vol. 11 (Springer, Heidelberg, 1980).
[37] P.E. Schmid, *Phys. Rev. B*, 23, 5531 (1981).
[38] Jr. G.E. Jellison, F.A. Modine, C.W. White, R.F. Wood, and R.T. Young, *Phys. Rev. Lett.*, 46, 1414 (1981).
[39] V.I. Fistul, *Heavily Doped Semiconductors* (Plenum, New York, 1969) ch 7.
[40] C.J. Hwang, *J. Appl. Phys*, 41, 2668 (1970); W. Sritrakool, H.R. Glyde, V. Sa Yakanit Can. *J. Phys*, 60, 373 (1982); H. Ikoma, *J. Phys. Soc. Jap.*, 27, 514 (1969).
[41] M.H. Chen, C.C. Wu, and C.J. Lin, *J. Low Temp. Phys.*, 55, 127 (1984).
[42] K.P. Ghatak and B. Mitra, *Physica Scripta*, 46, 182 (1992).
[43] B. Nag and K.P. Ghatak, *J. Phys. Chem. Solids*, 58, 427 (1997).
[44] K.P. Ghatak and B. Nag, *Physica status solidi (b)*, 205, 519 (1998).
[45] K.P. Ghatak, J.Y. Siddiqui, and B. Nag, Phys. Lett. A, 282, 428 (2001).
[46] K.P. Ghatak and S.N. Biswas, *MRS Proc.*, 308, 445 (1993).
[47] K.P. Ghatak, J.P. Banerjee, P.K. Chakrabarty, and B. Nag, *J. Wave Mater. Inter.*, 11, 166 (1996).
[48] L.J. Singh, S. Choudhary, A. Mallik, and K.P. Ghatak, J. Comput. Theor. Nanosci., 2, 287 (2005).
[49] A. Mallik, K.P. Ghatak, S. Choudhary, and L.J. Singh, *J. Comput. Theor. Nanosci.*, 2, 287 (2005).
[50] K.P. Ghatak, *Int. J. Electr*, 71, 239 (1991).
[51] K.P. Ghatak, *Acta Physica Hungarica*, 68, 253 (1990).
[52] B. Mitra and K.P. Ghatak, *physica status solidi (b)*, 154, K35 (1989).
[53] K.P. Ghatak and P.K. Bose, *J. Wave Mater. Inter.*, 12, 53 (1997).
[54] S.M. Adhikari and K.P. Ghatak, *J. Nanoeng. Nanomanuf.*, 3, 48 (2013).
[55] B. Chatterjee, S. Chakrabarti, S.K. Sen, M. Mitra and K.P. Ghatak, *Quantum Matter*, 5, 85 (2016).
[56] K.P. Ghatak, S. Bhattacharya, S.S. Roy, and L.J. Singh, *Nonlinear Optics*, 32, 307 (2010).
[57] D. Baruah, S. Choudhury, K.M. Singh, and K.P. Ghatak, *J. Phys. Confer. Ser*, 61, 80 (2007).
[58] K.P. Ghatak, and S.N. Biswas, *Acta Physica Slovaca*, 43, 425 (1993).
[59] K.P. Ghatak, and M. Mondal, *physica status solidi (b)*, 170, 57 (1992).
[60] K.P. Ghatak, D.K. Basu, and B. Nag, *J. Phys. Chem. Solids*, 58, 133 (1997).
[61] K.P. Ghatak, *Physica Status Solidi (b)*, 154, K29 (1989).
[62] K.P. Ghatak, J.P. Banerjee, B. Goswami, and B. Nag, *Nonlinear Opt. Read.*, 16, 241 (1996).
[63] B. Nag, and K.P. Ghatak, *Nonlinear Opt. Read.*, 19, 1 (1998).
[64] M. Mitra, M. Chakraborty, S. Debbarma, S. Chakraborty, and S.K. Sen, *Quantum Matter*, 5, 58 (2016).
[65] K.P. Ghatak and B. Nag, *Nanostruct. Mater*, 10, 923 (1998).
[66] M. Mitra, M. Chakraborty, S. Debbarma, S. Chakraborty, and S.K. Sen, *Quantum Matter*, 5, 58 (2016).
[67] B. Nag and K.P. Ghatak, *Physica Scripta*, 54, 657 (1996).
[68] K.P. Ghatak, S. Dutta, D.K. Basu, and B. Nag, *Il Nuovo Cimento D*, 20, 227 (1998).
[69] S. Bhattacharya, S. Chowdhury, S. Ghoshal, S.K. Biswas, D. De, and K.P. Ghatak, *J. Comput. Theor. Nanosci.*, 3, 423 (2006); K.P. Ghatak, S. Bhattacharya, S. Pahari, S.N. Mitra, P.K. Bose, and D. De, *J. Phys. Chem. Solids*, 70, 122 (2009); K.P. Ghatak, S. Karmakar, D. De, S. Pahari, S.K. Charaborty, and S.K. Biswas, *J. Comput. Theor. Nanosci*, 3, 153 (2006); K.P. Ghatak, J.P.

Banerjee, and B. Nag, *J. Appl. Phys.*, 83, 1420 (1998); K.P. Ghatak, S.K. Biswas, D. De, S. Ghosal, and S. Chatterjee, *Physica B*, 353, 127–149 (2004); K.P. Ghatak, J.P. Banerjee, D. Bhattacharyya, and B. Nag, *Nanotechnol*, 7, 110 (1996); K.P. Ghatak, B. Nag, and G. Mazumder, *MRS Proc*, 379, 109 (1995)

[70] R.W. Keyes, *IBM J. RES Dev.*, 5, 266 (1961).

[71] A.K. Sreedhar, S.C. Gupta, *Phys. Rev*, 5B, 1360 (1972).

[72] M. Abramowitz and I.A. Stegun, *Handbook of Mathematical Functions* (Dover Publications, New York, USA, 1965).

[73] J.S. Blakemore, *Semiconductor Statistics* (Dover Publications, New York, 1987); K.P. Ghatak, S. Bhattacharya, S.K. Biswas, A. Dey and A.K. Dasgupta, *Phys. Scr.*, 75, 820 (2007).

[74] K.P. Ghatak, S. Bhattacharya, and D. De, *Einstein Relation in Compound Semiconductors and their Nanostructures*. Springer series in materials science, 116 (2009).

[75] G. Paasch, T. Fiedler, M. Kolar and I. Bartos, *Phys. Stat. Sol. (b)*, 118, 641 (1983).

[76] H. Sasaki, *Phys. Rev. B*, 30, 7016 (1984).

[77a] H.X. Jiang, J.Y. Lin, *J. Appl. Phys*, 61, 624 (1987).

[77b] K.P. Ghatak, *Quantum Effects, Heavy Doping, and the Effective Mass (Series on the foundation of Natural Science and Technology-Vol. 8), World Scientific*, 2017.

[78] K.v. Klitzing, G. Dorda, and M. Pepper, *Phys. Rev. Lett*,.45, 494 (1980); K.v. Klitzing, *Rev. Mod. Phys*. 58, 519 (1986).

[79] J. Hajdu and G. Landwehr, *Strong and Ultrastrong Magnetic Fields and Their Applications*, F. Herlach (ed.), pp. 17 (Springer 1985).

[80] I.M. Tsidilkovskii, *Band of Semiconductors* (Pergamon Press, Oxford, 1982).

[81] E.A. Arushanov, A.F. Knyazev, A.N. Natepov, S.I. Radautsan, *Sov. Phys. Semi.*, 15, 828 (1981).

[82] S.P. Zelenim, A.S. Kondratev, and A.E. Kuchma, *Sov. Phys. Semi.* 16, 355 (1982).

[83] F.M. Peeters and P. Vasilopoulos, *Phys. Rev*,. *B*46, 4667 (1992).

[84] W. Zawadzki, *Two-Dimensional Systems, Heterostructures and Superlattices*, Springer Series in Solid State Sciences, Vol. 53, edited by G. Bauer, F. Kuchar, H. Heinrich (Springer, Berlin, Heidelberg, 1984).

[85] B.M. Askerov, N.F. Gashimzcede, and M.M. Panakhov, *Sov. Phys. Sol. State*, 29, 465 (1987).

[86] G.P. Chuiko, *Sov. Phys. Semi.*, 19, 1279 (1985).

[87] S. Pahari, S. Bhattacharya, and K.P. Ghatak, *J. Comput. Theor. Nanosci.*, Invited review article, 6, 2088 (2009).

[88] K.P. Ghatak, S. Bhattacharya, S. Bhowmik, R. Benedictus, and S. Choudhury, *J. Appl. Phys.*, 103, 034303 (2008).

[89] K.P. Ghatak, S. Bhattacharya, S. Pahari, D. De, S. Ghosh, and M. Mitra, *Annalen der Physik*, 17, 195 (2008).

[90] K.P. Ghatak and S.N. Biswas, *J. Appl. Phys*.70, 299 (1991).

[91] K.P. Ghatak and S.N. Biswas, *J. Low Temp. Phys.*, 78, 219 (1990).

[92] K.P. Ghatak and M. Mondal, *J. Appl. Phys.*, 65, 3480 (1989).

[93] K.P. Ghatak and B. Nag, *Nanostruct. Mater.*, 5, 769 (1995).

[94] K.P. Ghatak and M. Mondal, *Phys. Stat. Sol.(b)*, 185, K5 (1994).

[95] K.P. Ghatak and B. Mitra, *IlNuovoCimento*, *D*15, 97 (1993).

[96] K.P. Ghatak and S.N. Biswas, *Phys. Stat. Solidi (b)*, 140, K107 (1987).

[97] K.P. Ghatak, *IlNuovoCimentoD*, 13, 1321 (1991).

[98] K.P. Ghatak and A. Ghoshal, *Phys. Stat. Sol. (b)*, 170, K27 (1992).

[99] K.P. Ghatak, B. De, B. Nag, and P.K. Chakraborty, *Molecular Crystals and Liquid Crystals Science and Technology Section B: Nonlinear Optics*, 16, 221 (1996).

[100] K.P. Ghatak, M. Mitra, B. Goswami, and B. Nag, *Molecular Crystals and Liquid Crystals Science and Technology Section B: Nonlinear Optics*, 16, 167 (1996).

[101] K.P. Ghatak, D.K. Basu, D. Basu, and B. Nag, *NuovoCimentodellaSocietaltaliana di Fisica D – Condensed Matter, Atomic, Molecular and Chemical Physics, Biophysics*, 18, 947 (1996).

[102] B. Mitra and K.P. Ghatak, *Phys. Letts. A*, 141, 81 (1989); S.K. Biswas, A.R. Ghatak, A. Neogi, A. Sharma, S. Bhattacharya, and K.P. Ghatak, *Physica E:Low-Dimensional Systems and Nanostructures*, 36, 163 (2007).

[103] M. Mondal, A. Ghoshal, and K.P. Ghatak, *Il NuovoCimento D*, 14, 63 (1992); K.P. Ghatak, and M. Mondal. *Phys. Stat. Sol. (b)*, 135, 819 (1986); L.J. Singh, S. Choudhury, D. Baruah, S.K. Biswas, S. Pahari, and K.P. Ghatak, *Phys. B*, 368, 188 (2005).

[104] K.P. Ghatak, B. De, M. Mondal, and S.N. Biswas, *Materials Research Society Symposium – Proceedings*, 198, 327(1990); K.P. Ghatak, *Proceedings of SPIE – The International Society for Optical Engineering*, 1584, 435 (1992).

[105] K.P. Ghatak, and B. De, *Materials Research Society Symposium – Proceedings*, 234, 55 and 59 (1991).

[106] K.P. Ghatak, B. De, M. Mondal, and S.N. Biswas, *Materials Research Society Symposium – Proceedings*, 184, 261 (1990).

[107] M. Mondal and K.P. Ghatak, *Physica Scripta*, 31, 613 (1985).

[108] K.P. Ghatak and S.N. Biswas, *Materials Research Society Symposium – Proceedings*, 216, 465 (1990).

[109] A.K. Sreedhar and S.C. Gupta, *Phys. Rev. B*, 5, 3160 (1972); R.W. Keyes, IBM. *J. Res. Devlop.* 5, 266 (1961); *Solid State Phys.* 20, 37 (1967).

[110] S. Bhattacharya, S. Chowdhury and K.P. Ghatak, *J. Comput. Theor. Nanosci.*, 3, 423 (2006); K.P. Ghatak, J.Y. Siddiqui and B. Nag, *Phys. Letts. A*, 282, 428 (2001); K.P. Ghatak, J.P. Banerjee and B. Nag, *J. Appl. Phys.*, 83, 1420 (1998); K.P. Ghatak and B. Nag, *Nanostruc. Mater.*, 10, 923 (1998); B. Nag and K.P. Ghatak, *J. Phys. Chem. Sol.*, 58, 427 (1997)

[111] K.P. Ghatak, J.P. Banerjee, B. Goswami, and B. Nag, *Nonlin. Opt. Quant. Opt.*, 16, 241 (1996); K. P. Ghatak, J.P. Banerjee, D. Bhattacharyya, and B. Nag, *Nanotechnology*, 7, 110 (1996); K.P. Ghatak, J.P. Banerjee, M. Mitra, and B. Nag, *Nonlin. Opt.* 17, 193 (1996); K.P. Ghatak, *Inter. J. Electron.*, 71, 239 (1991).

[112] K.P. Ghatak, B. De, S.N. Biswas, and M. Mondal, Mechanical behavior of materials and structures in microelectronics, MRS Symposium Proceedings, Spring Meeting, 2216, 191, (1991); K.P. Ghatak and B. De, MRS Symposium Proceedings, 226, 191 (1991); K.P. Ghatak, B. Nag, and G. Majumdar, MRS, 379, 109, (1995); D. Baruah, S. Choudhury, K.M. Singh, and K.P. Ghatak, *J. Phys.*, Conf. Series, 61, 80, (2007).

[113] P.T. Landsberg, *Eur. J. Phys.*, 2, 213 (1981).

[114] R.B. Dingle, *Philos. Mag.*, 46, 813(1955); D. Redfield, M.A. Afromowitz, *ibid.* 19, 831 (1969); H.C. Casey and F. Stern, *J. Appl. Phys.*, 47, 631 (1976).

[115] S.N. Mohammad, *J. Phys. C*, 13, 2685 (1980).

[116] K.P. Ghatak, S. Bhattacharya, S. Pahari, S.N. Mitra, P.K. Bose, and D. De, *J. Phys. Chem. Sol,.* 70, 122 (2009); S. Chowdhary, L.J. Singh, and K.P. Ghatak, *Physica B: Condensed Matter*, 365, 5 (2005); B. Nag and K.P. Ghatak, *Molecular Crystals and Liquid Crystals Science and Technology Section B: Nonlinear Optics*, 19, 1 (1998); K.P. Ghatak and B. Mitra, *Phys. Scr.*, 46, 182 (1992).

[117] K.P.Ghatak, and S. Bhattacharya, *J. Appl. Phys.*, 102, 073704 (2007); M. Mondal, K.P. Ghatak, *Phys. Lett.*, 102, A, 54 (1984); S. Bhattacharya, N.C Paul, and K.P Ghatak, *Physica B*, 403, 4139 (2008); K.P. Ghatak, S. Bhattacharya, H. Saikia, D. Baruah, A. Saikia, K.M. Singh, A. Ali, S.N. Mitra, P.K. Bose, A. Sinha, *J. Compt. Theor. Nanosci*, 3, 727 (2006); P.K. Chakraborty, G.C. Datta, and K.P. Ghatak, *Phys. Scr.*, 68, 368 (2003). E.O. Kane, *Solid-State Electron.*, 28, 3 (1985); W. Zawadzki, *Handbook on Semiconductors*, edited by W. Paul, North Holland, New York (1982), 1, p. 715.

[118] A.N. Chakravarti and D. Mukherjee, *Phys. Lett.*, 53A, 403 (1975).

[119] B. Mitra, D.K. Basu, B. Nag, and K.P. Ghatak, *Nonlinear Opt.*, 17, 171 (1997).
[120] M. Mondal and K.P. Ghatak, *Phys. Status Solidi B*, 135, 239 (1986).
[121] A.N. Chakravarti, and S. Swaminathan, *Phys. Status Solidi A*, 23, K191 (1974); A.N. Chakravarti, *ibid.* 25, K105 (1974).
[122] A.N. Chakravarti, K.P. Ghatak, K.K. Ghosh, and A. Dhar, *Phys. Status Solidi B*, 103, K55 (1981).
[123] T. Ando, A.H. Fowler, and F. Stern, *Rev. Mod. Phys.*, 54, 437 (1982); P.K. Basu, *Optical Processes in Semiconductors*, Oxford University Press, New York, 2001.
[124] N. Kampf, D. Ben-Yaakov, D. Andelman, S.A. Safran, and J. Klein, *Phys. Rev. Lett.*, 103, 118304 (2009); P. Arnold, and L.G. YafFe, *Phys. Rev. D*, 52, 7208 (1995); G.S. Kulkarni and Z. Zhong, *Nano Letters*, 12, 719 (2012); E. Stern, R. Wagner, F.J. Sigworth, R. Breaker, T.M. Fahmy, M.A. Reed, *Nano Lett.*, 7, 3405 (2007).
[125] A.N. Chakravarti, and D. Mukherji, *Phys. Lett. A.*, 53, 57 (1975); W. Zawadzki, *Adv. Phys.*, 23, 435 (1974).
[126] R.W. Lade, *Proc. IEEE*, 52, 743 (1965).
[127] P.T. Landsberg, *Proc R. Soc. A*, 213, 226 (1952); *Proc. of Phys. Soc.* A 62, 806 (1949)
[128] C.H. Wang, and A. Neugroschel, *IEEE Electron. Dev. Lett.*, ED-11, 576 (1990).
[129] I.Y. Leu, and A. Neugroschel, *IEEE Trans. Electron. Dev*, ED-40, 1872 (1993).
[130] F. Stengel, S.N. Mohammad, and H. Morkoç, *J. Appl. Phys.*, 80, 3031 (1996).
[131] H.J. Pan, W.C. Wang, K.B. Thai, C.C. Cheng, K.H. Yu, K.W. Lin, C.Z. Wu, and W.C. Liu, *Semicond. Sci. Technol*, 15, 1101 (2000)
[132] S.N. Mohammad, *J. Appl. Phys.*, 95, 4856 (2004).
[133] V.K. Arora, *Appl. Phys. Lett.*, 80, 3763 (2002).
[134] S.N. Mohammad, *J. Appl. Phys.*, 95, 7940 (2004).
[135] S.N. Mohammad, *Philos. Mag.*, 84, 2559 (2004).
[136] S.N. Mohammad, *J. Appl. Phys.*, 97, 063703 (2005).
[137] K. Suzue, S.N. Mohammad, Z.F. Fan, W. Kim, O. Aktas, A.E. Botchkarev, and H. Morkoç, *J. Appl. Phys.*, 80, 4467(1996).
[138] S.N. Mohammad, Z.F. Fan, W. Kim, O. Aktas, A.E. Botchkarev, A. Salvador, and H. Morkoç, *Electron. Lett.*, 32, 598 (1996).
[139] Z. Fan, S.N. Mohammad, W. Kim, O. Aktas, A.E. Botchkarev, K. Suzue, and H. Morkoç, *J. Electron. Mater.*, 25, 1703 (1996).
[140] C. Lu, H. Chen, X. Lv, X. Xia, and S.N. Mohammad, *J. Appl. Phys.*, 91, 9216 (2002).
[141] S.G. Dmitriev, and Yu V. Markin, *Semiconductors*, 34, 931 (2000).
[142] M. Tao, D. Park, S.N. Mohammad, D. Li, A.E. Botchkerav, and H. Morkoç, *Philos. Mag. B*, 73, 723 (1996).
[143] D.G. Park, M. Tao, D. Li, A.E. Botchkarev, Z. Fan, S.N. Mohammad, and H. Morkoç, *J. Vac. Sci. Technol. B*, 14, 2674 (1996).
[144] Z. Chen, D.G. Park, S.N. Mohammad, and H. Morkoç, *Appl. Phys. Lett.*, 69, 230 (1996).
[145] A.N. Chakravarti, and D.P. Parui, *Phys. Letts*, 40A, 113 (1972).
[146] A.N. Chakravarti, and D.P. Parui, *Phys. Letts*, 43A, 60(1973); B.R. Nag, A N. Chakravarti, *Solid State Electron.*, 18, 109 (1975), *Phys. Stat. Sol. (a)*, 22, K153 (1974).
[147] B.R. Nag, A.N. Chakravarti, and P.K. Basu, *Phys. Stat. Sol. (a)*, 68, K75(1981).
[148] B.R. Nag, and A.N. Chakravarti, *Phys. Stat. Sol. (a)*, 67, K113 (1981).
[149] A.N. Chakravarti and B.R. Nag, *Phys. Stat. Sol. (a)*, 14, (K55) (1972); *Int. J. Elect.*, 37, 281 (1974); *Phys. Stat. Sol. (a)*, 14, K23 (1972); A.N. Chakravarti and D.P. Parui, *Canad. J. Phys.*, 51, 451 (1973). D. Mukherjee and A.N. Chakravarti, B.R. Nag, *Phys. Stat. Sol (a)*, 26, K27(1974); S. Ghosh and A.N. Chakravarti, *Phys. Stat. Sol. (b)*, 147, 355 (1988); A.N. Chakravarti, K.K. Ghosh, K.P. Ghatak, and H.M. Mukherjee, *Phys. Stat. Sol (b)*, 118, 843 (1983); A.N. Chakravarti, K.P. Ghatak, K.K. Ghosh, and G.B. Rao, *Phys. Stat. Sol (b)*, 111, K61 (1982); A.N. Chakravarti, K.P.

Ghatak, K.K. Ghosh, H.M. Mukherjee, and S. Ghosh, *Phys. Stat. Sol (b)*, 108, 609 (1981); A.N. Chakravarti, K.P. Ghatak, S. Ghosh, and A. Dhar, *Phys. Stat. Sol (b)*, 105, K55(1981); A.N. Chakravarti, A.K. Choudhury, K.P. Ghatak, and D. Roy Choudhury, *Phys. Stat. Sol (b)*, 59, K211 (1980); A.N. Chakravarti, A.K. Chowdhury, K.P. Ghatak, and D.R. Choudhury, *Acta Phys. Polon. A*, 58, 251 (1980); K.P. Ghatak, A.K. Chowdhury, S. Ghosh, and A.N. Chakravarti, *Phys. Stat. Sol (b)*, 99, K55(1980); A.N. Chakravarti, A.K. Chowdhury, K.P. Ghatak, and D.R. Choudhury, *Czech. J. Phys. B*, 30, 1161 (1980); A.N. Chakravarti, K.P. Ghatak, A. Dhar, K.K. Ghosh, and S. Ghosh, *Acta Phys. Polon A*, 60, 151 (1981); A.N. Chakravarti, A.K. Chowdhury, K.P. Ghatak, A. Dhar, and D.R. Choudhury, Czech. *J. Phys. B*, 31, 905 (1981); A.N. Chakravarti, K.P. Ghatak, K.K. Ghosh, S. Ghosh, and H.M. Mukherjee, *Czech. J. Phys. B*, 31, 1138 (1981).

[150] P.T. Landsberg, *J. Appl. Phys.* 56, 1696(1984); P.T. Landsberg, *Phys. Rev. B*, 33, (8321) (1986); Y. Roichman, N. Tessler, *Appl. Phys. Lett.* 80, 1948 (2002); J.M.H. Peters, *Eur. J. Phys.*, 3, 19 (1982); H. Van Cong, S. Brunet, and S. Charar, *Phys. Stat. Solidi B*, 109, K1(1982); H. Van *Cong, Phys. Stat. Solidi A*, 56, 395 (1979); H. Van Cong, *Solid State Electron.*, 24, 495 (1981).

[151] S.N. Mohammad, and S.T.H. Abidi, *J. Appl. Phys.*, 61, 4909 (1987); *Solid State Electron.*, 27, 1153 (1985); *J. Appl. Phys.*, 56, 3341 (1984); M.A. Sobhan, and S.N. Mohammad, *J. Appl. Phys.*, 58, 2634 (1985); S.N. Mohammad and A.V. Bemis, *IEEE Trans. Electron. Dev*, ED-39, 282 (1992); S.N. Mohammad, and R.L. Carter, *Philos. Mag. B*, 72, 13 (1995); S.N. Mohammad, *Solid State Electron.*, 46, 203 (2002); S.N. Mohammad, and J. Chen, J.I. Chyi, H. Morkoç, *Appl. Phys. Lett.*, 56, 937 (1990).

[152] P.T. Landsberg, and S.A. Hope, *Solid State Electr.*, 20, 421 (1977); S.A. Hope, G. Feat, and P.T. Landsberg, *J. Phys. A. Math. Gen.*, 14, 2377 (1981).

[153] W. Elsäber and E.O. Göbel, Electron. Lett. 19, 335(1983); R. Hilfer, A. Blumen, *Phys. Rev.* A 37, (578) (1988); T.G. Castner, *Phys. Rev.* B 55, (4003) (1997); E. Barkai, V.N. Fleurov, *Phys. Rev. E.*, 58, (1296) (1998); T.H. Nguyen, S.K. O'Leary, *Appl. Phys. Lett.* 83, (1998) (2003); T.H. Nguyen, S. K. O'Leary, *J. Appl. Phys.* 98, (076102) (2005); C.G. Rodrigues, Á.R. Vasconcellos, R. Luzzi, *J. Appl. Phys.* 99, (073701) (2006).

[154] R.K. Jain, *Phys. Stat. Sol. (a)*, 42, K221(1977); B.A. Aronzon, E.Z. Meilikhov, *Phys. Stat. Sol. (a)*, 19, 313 (1973).

[155] S. Choudhury, D. De, S. Mukherjee, A. Neogi, A. Sinha, M. Pal, S. K. Biswas, S. Pahari, S. Bhattacharya, K.P. Ghatak, *J. Comp. Theor. Nanosci.*, 5, 375 (2008); S.M. Adhikari, K.P. Ghatak, *Quantum Matter*, 2, (296) (2013); S. Pahari, S. Bhattacharya, D. De, S.M. Adhikari, A. Niyogi, A. Dey, N. Paitya, K.P.Ghatak, *Physica B: Condensed Matter*, 405, 4064 (2010); K.P. Ghatak, S. Bhattacharya, S. Pahari, D. De, R Benedictus, *Superlatt. Microst.*, 46, (387) (2009).

[156] J.P. Bouchaud, A. Georges, *Phys. Rep.*, 195, 127(1996); V. Blickle, T. Speck, C. Lutz, U. Seifert, and C. Bechinger, *Phys. Rev. Lett.*, 98, 210601 (2007);Y. Kang, E. Jean, and C.M. Fortmonn, *Appl. Phys. Lett.*, 88, 112110 (2006); F. Neumann, Y.A. Genenko, and H.V. Seggern, *J. Appl. Phys.*, 99, 013704 (2006); J. van de Lagemaat, *Phys. Rev. B.*, 73, 235319 (2005); Q. Gu, E.A. Schiff, S. Grneber, F. Wang, and R. Schwarz, *Phys. Rev. Lett.*, 76, 3196 (1996); M.Y. Azbel, *Phys. Rev. B.*, 46, 15004 (1992).

[157] A.H. Marshak, *Solid State Electron.*, 30, 1089 (1987); A.H. Marshak, and C.M.V. Vliet, *Proc. IEEE*, 72, 148 (1984); C.M.V. Vliet, A. van der Zeil, *Solid State Electron.*, 20, 931 (1977).

[158] A. Khan, A. Das, *Appl. Phys. A*, 89, 695 (2007).

[159] O. Madelung, *Semiconductors: Data Handbook*, 3rd edn. (Springer, 2004); M. Krieehbaum, P. Kocevar, H. Pascher, G. Bauer, *IEEE QE*, 24, 1727 (1988).

[160] K.P. Ghatak, S.N. Biswas, *J. Appl. Phys.*, 70, 4309 (1991); K.P. Ghatak, B. Mitra, M. Mondal, *Ann. der Physik*, 48, 283 (1991); B. Mitra, and K.P. Ghatak, *Phys. Letts.*, 135A, 397.

[161] K.P. Ghatak, D. Bhattacharyya, *Phys. Letts.*, A184, 366 (1994).

[162] K.P. Ghatak, and D. Bhattacharyya, *Physica Scripta*, 52, 343 (1995); M. Mondal and K.P. Ghatak, *J. Mag. Mag. Mater.*, 62, 115 (1986).
[163] K.P. Ghatak, and B. Nag, D. Bhattacharyya, *J. Low Temp. Phys.*, 14, 1 (1995).
[164] K.P. Ghatak, and M. Mondal, *Thin Solid Films*, 148, 219 (1987).
[165] K.P. Ghatak, and S.N. Biswas, *Nanostruct. Mater.*, 2, 91 (1993).
[166] K.P. Ghatak, *Influence of Band Structure on Some Quantum Processes in Nonlinear optical Semiconductors*, D. Eng. thesis (Jadavpur University, Kolkata, India(1991).
[167] K.P. Ghatak, N. Chattropadhyay, and M. Mondal, *Appl. Phys.* A, 44, 305 (1987).
[168] S.N. Biswas, and K.P. Ghatak, *Proceedings of the Society of Photo-optical and Instrumentation Engineers (SPIE), Quantum Well and Superlattice Physics*, 792, 239 (1987); K.P. Ghatak, M. Mondal, and S. Bhattacharyya, *SPIE*, 1284, 113 (1990); K.P. Ghatak, S. Bhattacharyya, and M. Mondal, *SPIE*, 1307, 588 (1990); K.P. Ghatak, and B. De, *Defect Engineering in Semiconductor Growth, Processing and Device technology Materials Research Society Proceedings, (MRS)* Spring meeting, 262, 911 (1992); S. Bhattacharya, K.P. Ghatak, and S.N. Biswas, *Optoelectronic Materials, Devices, Packaging Interconnects, SPIE*, 836, 72 (1988).
[169] M. Mondal, and K.P. Ghatak, *J. Phys. C, (Sol. State.)*, 20, 1671 (1987); M. Mondal. S.N. Banik, and K.P. Ghatak, Canad. *J. Phys*, 67, 72 (1989); K.P. Ghatak, and M. Mondal, *J. Appl. Phys.*, 70, 1277 (1992).
[170] P.K. Chakraborty, S. Bhattacharya, and K.P. Ghatak, *J. Appl. Phys.*, 98, 053517(2005).
[171] A.S. Filipchenko, I.G. Lang, D.N. Nasledov, S.T. Pavlov, and L.N. Radaikine, *Phys. Stat. Sol. (b)*, 66, 417 (1974).
[172] M. Wegener, *Extreme Nonlinear Optics* (Springer-Verlag, Germany, 2005).
[173 B.S. Wherreff, W. Wolland, C. R. Pidgeon, R.B. Dennis, and S.D. Smith, *Proceedings of the 12th International Conference of the Physics of the Semiconductors*, ed. by M.H. Pilkahn, R.G. Tenbner (Staffgard, 1978), p. 793.
[174] C.H. Peschke, *Phys. Stat. Sol. (b)*, 191, 325 (1995).
[175] P.T Landsberg and H.C Cheng, *Phys. Rev.*, 32B, 8021 (1985); Semiconductors, "*D. van Nostrand Co., Inc.*, New York (1950); H. Reiss, C. S. Fuller, and F. J. Morin, *Bell Syst. Tech. J.*, 35, 535 (1956); P.T. Landsberg and A.G. Guy, *Phys. Rev.* B, 28, 1187 (1983); "Handbook on Semiconductors," W. Paul, Editor, pp. 359–449, North-Holland Publishing Co. (1982)
[176] A. Isihara, Y. Shiwa, *J. Phys. C*, 18, 4703 (1985).
[177] V.G.I. Ravish, B.A. Efimova and I.A. Smirnov, *Semiconducting Lead Chalocogenides* (Plenum Press, New York, p. 163, 1970)
[178] F.J. Blatt, *J. Phys. Chem. Solids*, 17, 177 (1961).
[179] D.K. Roy, *Quantum Mechanical Tunnelling and Its Applications* (World Scientific Publication Co. Ptc. Ltd, p. 309, 1986.
[180] W. Zawadzki, *Physica 1*,27, 388 (1980).
[181] E.V. Rozenfeld, Yu. P. Irkhin and P.G. Guletskii, *Sov. Phys. Solid State*, 29, 1134 (1987).
[182] H. Piller, *Semiconductors and Semimetals*, Ed. by R.K. Willardson and A.C. Beer, 124, 8 (1972).
[183] R.H. Fowler, L. Nordheim, *Proc. Roy. Soc. Lond. Ser-A*, 119,173 (1928); A. Van Der Ziel, *Solid State Physical Electronics* (Prentice-Hall, Inc., Englewood Cliffs U.S.A., 1957), p. 176).
[184] B. Mitra, and K.P. Ghatak, *Phys. Lett. A*, 357,146 (1990); K.P.Ghatak, and M. Mondal, *J. Mag. Mag. Mat.*, 74, 203 (1988); K.P. Ghatak, B. Mitra, *Phys. Lett.*, 156A. 233, (1991); K.P. Ghatak, A. Ghosal, S.N. Biswas, and M. Mondal, *Proc. SPIE, USA*, 1308, 356 (1990).
[185] V.T. Binh and Ch. Adessi, *Phys. Rev. Lett*, 85,864 (2000); R.G. Forbes, *Ultramicroscopy*, 79,11 (1999); J.W. Gadzuk, E.W. Plummer, *Rev. Mod. Phys*, 45,487 (1973); J.M. Beebe, B. Kim, J.W. Gadzuk, C.D. Frisbie, and J.G. Kushmerick, *Phys. Rev. Lett*, 97, 026801 (1999);Y. Feng, J.P. Verboncoeur, *Phys. Plasmas*, 12,103301(2005); W.S. Koh and L.K. Ang, *Nanotechnology*,

19,235402(2008); M. Razavy, *Quantum Theory of Tunneling* (World Scientific Publishing Co. Pte. Ltd, Singapore (2003).

[186] S.I. Baranchuk and N.V. Mileshkina, *Soviet Phys. Solid State*, 23, 1715 (1981); P.G. Borzyak, A.A. dadykin, *Soviet Phys. Docklady*, 27,335 (1982); S. Bono, R.H. Good, *Jr.*, *Surf. Sci.*, 134, 272 (1983); S.M. Lyth, S.R.P. Silva, *Appl. Phys. Letts*. 90, 173124 (2007); C. Xu, X. Sun, *I. J. Nanotech.*, 1, 452 (2004); S.D. Liang, L. Chen, *Phys. Rev. Letts.*, 101, 027602 (2008).

[187] E.C. Heeres, E.P.A.M. Bakkers, A.L. Roest, M. Kaiser, T.H, Oosterkamp, N. de Jonge, *Nano Lett.*, 7,536 (2007); L. Dong, J. Jiao, D. W. Tuggle, J. M. Petty, S. A. Elliff, M. Coulter, *Appl. Phys. Lett.*, 82,1096 (2003); S.Y. Li, P. Lin, C.Y. Lee, and T.Y. Tseng, *J. Appl. Phys.*, 95, 3711 (2004); N.N. Kulkarni, J. Bae, C.K. Shih, S.K. Stanley, S.S. Coffee, and J.G. Ekerdt, *Appl. Phys. Lett.*, 87, 213115 (2005).

[188] K. Senthil, and K. Yong, *Mat. Chem. Phys.*, 112, 88 (2008); R. Zhou, H.C. Chang, V. Protasenko, M. Kuno, A.K. Singh, D. Jena, and H. Xing, *J. Appl. Phys.*, 101, 073704 (2007); K.S. Yeong and J.T. L. Thong, *J. Appl. Phys.*, 100, 114325 (2006); C.H. Oon, S.H. Khong, C.B. Boothroyd, and J.T.L. Thong, *J. Appl. Phys.*, 99, 064309 (2006).

[189] B.H. Kim, M.S. Kim, K.T. Park, J.K. Lee, D.H. Park, J. Joo, S.G. Yu, and S.H. Lee, *Appl. Phys. Lett.*, 83, 539 (2003); Z.S. Wu, S.Z. Deng, N.S. Xu, J. Chen, and J. Zhou, J. Chen, *Appl. Phys. Lett.*, 80, 3829 (2002); Y.W. Zhu, T. Yu, F.C. Cheong, X. J. Xu, C. T. Lim, V. B. C. Tan, J. T. L. Thong, and C. H. Sow, *Nanotechnology*, 16, 88 (2005); Y.W. Zhu, H.Z. Zhang, X.C. Sun, S.Q. Feng, J. Xu, Q. Zhao, B. Xiang, R.M. Wang, and D.P. Yu, *Appl. Phys. Lett.*, 83, 144(2003); S. Bhattacharjee and T. Chowdhury, *Appl. Phys. Lett.*, 95, 061501 (2009); S. Kher, A. Dixit, D.N. Rawat, M.S. Sodha, *Appl. Phys. Lett.*, 96, 044101 (2010).

[190] I. Shigeo, W. Teruo, O. Kazuyoshi, T. Masateru, U. Satoshi, and N. Norio, *J. Vac. Sci. Tech. B*, 13, 487 (2009); C.A. Spindt, I. Brodie, L. Humphrey, and E.R. Westerberg, *J. Appl. Phys.*, 47, 5248 (2009); Q. Fu, A. V. Nurmikko, L.A. Kolodziejski, R.L. Gunshor, and J.W. Wu, *Appl. Phys. Lett.*, 51, 578 (2009).

[191] I. Mertig, E. Mrosan, V.N. Antonov, V.I. Antonov, and P. Ziesche, *Phys. Stat. sol. (b)*, 135, K13 (1986); O. Madelung, *Physics of III-V compounds John Wiley and Sons, Inc. New York*, 1966, p. 80.

[192] L.T. Canham, *Appl. Phys. Letts.*, 57, 1046 (1990).

[193] Z.H. Lu, D.J. Lockwood and J.M. Baribeam, *Nature*, 378,258 (1995).

[194] A.G. Cullis, L.T. Canham, and P.D.O. Calocott, *J. Appl. Phys.*, 82, 909 (1997).

[195] M. Cardona, and L. Ley, *Photoemission in Solids 1 and 2, Topics in Applied Physics*, Vols. 26, 27, (Springer-Verlag, Germany, 1978); S. Hüfner, *Photoelectron Spectroscopy* (Springer, Germany, 2003); S. Hüfner, (Ed.), *Very High Resolution Photoelectron Spectroscopy*, Lecture Notes in Physics, Vol. 715 (Springer-Verlag, Germany, 2007); D.W. Lynch, and C.G. Olson, *Photoemission Studies of High-Temperature Superconductors* (Cambridge University Press, UK, 1999); *Photoemission and the Electronic Properties of Surfaces* ed. B. Feuerbacher, B. Fitton, R. F. Willis (Wiley, New York, 1978); W. Schattke and M.A.V. Hove, *Solid-State Photoemission and Related Methods: Theory and Experiment*, (Wiley, USA, 2003); V. V. Afanas'ev, Internal.

[196] K.P. Ghatak, D. De, and S. Bhattacharya, *Photoemission from Optoelectronic Materials and their Nanostructures*, Springer Series in Nanostructure Science and Technology (Springer, 2009).

[197] R.K. Pathria, *Statistical Mechanics*, 2nd edn. (Butterworth-Heinemann, Oxford, 1996).

[198] D. De, S. Bhattacharya, S.M. Adhikari, A. Kumar, P.K. Bose, and K.P. Ghatak, *Beilstein J. Nanotech.*, 2, 339 (2012); D. De, Kumar, S.M. Adhikari, S. Pahari, N. Islam, P. Banerjee, S.K. Biswas, S. Bhattacharya, and K.P. Ghatak, *Superlatti. Microstruc.*, 47, 377 (2010); K.P. Ghatak, M. Mondal, and S.N. Biswas, *J. Appl. Phys.*, 68, 3032 (1990); M. Mondal, S. Banik, and K.P.

Ghatak, *J. Low Temp. Phys.*, 74, 423 (1989); B. Mitra, A. Ghoshal, and K.P. Ghatak, *Phys. Stat. Sol. (b)*, 150, K67 (1988); K.P. Ghatak, D. Bhattacharyya, B. Nag, and S.N. Biswas, *J. Nonlin. Optics Quant. Optics*, 13, 267 (1995); L. Torres, L. Lopez-Diaz, and J. Iniguez, *Appl. Phys. Lett.*, 73, 3766, (1996); A.Y. Toporov, R.M. Langford, and A.K. Petford-Long, *Appl. Phys. Lett.*, 77, 3063 (2000); L. Torres, L. Lopez-Diaz, O. Alejos, and J. Iniguez, *J. Appl. Phys.*, 85, 6208 (1999); S. Tiwari and S. Tiwari, *Cryst. Res. Technol.*, 41, 78 (2006); R. Houdré, C. Hermann, G. Lampel, P.M. Frijlink, and A.C. Gossard, *Phys. Rev. Lett.*, 55, 734 (1985).

[199] S.M. Adhikari, and K.P. Ghatak, *J. Adv. Phys.*, 2, 130 (2013); D. De, S. Bhattacharaya, S. Ghosh, and K.P. Ghatak, *Adv. Sci., Eng. Med*, 4, 211 (2012); A. Kumar, S. Chowdhury, S.M. Adhikari, S. Ghosh, M. Mitra, D. De, A. Sharma, S. Bhattacharya, A. Dey, and K. P. Ghatak, *J. Comput. Theor. Nanosci.* 7, 115 (2010); R. Houdré, C. Hermann, G. Lampel, and P.M. Frijlink, *Surf. Sci.* 168, 538 (1986); T.C. Chiang, R. Ludeke, and D.E. Eastman, *Phys. Rev. B.*, 25, 6518 (1982); S.P. Svensson, J. Kanski, T.G. Andersson, and P.O. Nilsson, *J. Vacuum Sci. Technol.*, B 2, 235 (1984); S.F. Alvarado, F. Ciccacci, and M. Campagna, *Appl. Phys. Letts.*, 39, 615 (1981); L. Fleming, M.D. Ulrich, K. Efimenko, J. Genzer, A.S.Y. Chan, T.E. Madey, S.J. Oh, O. Zhou, and J.E. Rowe, *Vac. Sci. Technol.*, B 22, 2000 (2004); H. Shimoda, B. Gao, X. –P. Tang, A. Kleinhammes, L. Fleming, Y. Wu, and O. Zhou, *Phys. Rev. Lett.*, 88, 015502 (2002); M. Maillard, P. Monchicourt, and M.P. Pileni, *Chem. Phys. Lett.*, 380, 704 (2003).

[200] B. Mitra, and K.P. Ghatak, *Phys. Scrip.*, 40, 776 (1989); D. De, S. Bhattacharya, and K.P. Ghatak, *International Work Shop in Physics of Semiconductor Devices*, IEEE 897 (2007); S. Bhattacharya, D. De, S. Chowdhury, S. Karmakar, D. K. Basu, S. Pahari, and K.P. Ghatak, *J. Comput. Theor. Nanosci*, 3, 280 (2006).

[201] I. Matsuda, S. Hasegawa, A. Konchenko, Y. Nakayama, Y. Nakamura, and M. Ichikawa, *Phys. Rev. B*, 73, 113311 (2006); V.L. Colvin, A.P. Alivisatos, and J.G. Tobin, *Phys. Rev. Lett.*, 66, 2786 (1991); B. Schroeter, K. Komlev, and W. Richter, *Mat. Sci. Eng.*, B88, 259 (2002); G.F. Bertsch, N. Van Giai, and N. Vinh Mau, *Phys. Rev.*, A 61, 033202 (2000).

[202] K.P. Ghatak, and S.N. Biswas, *Nonlinear Opt.*, 4, 39 (1993).

[203] K.P. Ghatak, and S.N. Biswas, *SPIE, Growth and Characterization of Materials for Infrared Detectors and Nonlinear Optical Switches*, Vol. 1484, p. 136 (USA, 1991).

[204] K.P. Ghatak, and B. De, *Polymeric materials for Integrated Optics and Information Storage, Materials Research Society (MRS) Symposium Proceedings, MRS Spring Meeting*, 1991, Vol. 228, p. 237; L.J. Heyderman, and F. Nolting, Quitmann, *Appl. Phys. Letts.*, 83, 1797 (2003); R.P. Cowburn, A.O. Adeyeye, J.A.C. Bland, *Appl. Phys. Lett.*, 70, 2309 (1997).

[205] K.P. Ghatak, B. Nag, and G. Majumdar, Strained Layer Epitaxy – Materials, Processing, and Device Applications, MRS Symposium Proceedings, MRS Spring Meeting, 1995, Vol. 379, p. 85.

[206] K.P. Ghatak, *SPIE, High Speed Phenomena in Photonic Materials and Optical Bistability*, USA,1990, Vol. 1280, p. 53 .

[207] K.P. Ghatak, Long Wave Length Semiconductor devices, Materials and Processes Symposium Proceedings, *MRS Symposium Proceedings*, MRS Spring Meeting, 1990, Vol. 216, p. 469.

[208] K.P. Ghatak, A. Ghoshal, and S. Bhattacharyya, *SPIE, Nonlinear Optical Materials and Devices for Photonic Switching*, USA, 1990, Vol. 1216, p. 282.

[209] K.P. Ghatak, *SPIE, Nonlinear Optics III*, USA, 1992, Vol. 1626, p. 115.

[210] K.P. Ghatak, A. Ghoshal, and B. De, *SPIE, Optoelectronic Devices and Applications*, USA, 1990, Vol. 1338, p. 111.

[211] K.P. Ghatak, and S.N. Biswas, *Proceedings of the Society of Photo-optical and Instrumentation Engineers (SPIE)*, Nonlinear Optics II, USA, 1991, Vol. 1409, p. 28; K.P. Ghatak, SPIE, Process Module Metrology, USA, 1992, Vol. 1594, p. 110; K.P. Ghatak, SPIE, International Conference on the Application and Theory of Periodic Structures, 1991, Vol. 1545, p. 282.

[212] K.P. Ghatak, and M. Mondal, *Solid State Electron*, 31, 1561 (1988).

[213] K.P. Ghatak, and M. Mondal, *J. Appl. Phys.*, 69, 1666 (1991).

[214] C. Majumdar, A.B. Maity, and A.N. Chakravarti, *Phys. Stat. Sol. (b)*, 140, K7 (1987).

[215] C. Majumdar, A.B. Maity, and A.N. Chakravarti, *Phys. Stat. Sol. (b)*, 141, K35 (1987).

[216] N.R. Das, K.K. Ghosh, and D. Ghoshal, *Phys. Stat. Sol. (b)*, 197, 97 (1996).

[217] C. Majumdar, A.B. Maity, and A.N. Chakravarti, *Phys. Stat. Sol. (b)*, 144, K13 (1987).

[218] N.R. Das, and A.N. Chakravarti, *Phys. Stat. Sol. (b)*, 176, 335 (1993).

[219] S. Sen, N.R. Das and A.N. Chakravarti, *J. Phys Conden. Mat.*, 19, 186205 (2007); N.R. Das, S. Ghosh, A.N. Chakravarti, *Phys. Stat. Sol. (b)*, 174, 45 (1992).

[220] A.B. Maity, C. Majumdar, and A.N. Chakravarti, *Phys. Stat. Sol. (b)*, 144, K93, (1987).

[221] A.B. Maity, C. Majumdar, and A.N. Chakravarti, *Phys. Stat. Sol. (b)*, 149, 565 (1988).

[222] A.V.D. Ziel, *Solid State Physical Electronics* (Prentice Hall, Inc. Eaglewood Cliffs, 1957); A. Modinos, *Field, Thermionic and Secondary Electron Emission Spectroscopy* (Plenum Press, New York, 1984).

[223] L.J. Singh, S. Choudhary, A. Mallik, and K.P. Ghatak, *J. Comput. Theor. Nanosci.*, 2, 287 (2005).

[224] O. Madelung, Physics of III-V compounds (John Wiley and Sons, Inc. New York, 1966).

[225] B. Mitra, A. Ghoshal and K.P. Ghatak, *physica status solidi (b)*, 155, K23(1989).

[226] J.A. Woollam, *Phys. Rev.*, 3, 1148, (1971).

[227] V.V Kaminoskii, N.N. Stepanav, and l.M. Smirnov, *Sov. Phys. Solid State*, 27,1295 (1985).

[228] P.I. Baranskii, V.V. Kolomoets and S.S. Korolyuk, *Phys. Stat Sol. (b)*, 116, K109 (1983).

[229] K.P. Ghatak and S. Biswas, *J. Vac. Sci. Technol.* B7, 104 (1989); D.R. Choudhury, A. K. Chowdhury, and A.N. Chakravarti, *Physica Scripta*, 22, 656 (1981); D.R. Choudhury, A.K. Chowdhury, and A.N. Chakravarti, *Czech. J. Phys.*, B 30 (1980); D.R. Choudhury, A.K. Chowdhury, and A.N. Chakravarti, B.R. Nag, *Phys. Stat. Sol. (a)*, 58, K51 (1980); B. Mitra, A. Ghoshal, and K.P. Ghatak, *Phys. Stat. Sol. (b)*, 153, K209 (1989); M. Mondal, and K.P. Ghatak, *Phys. Stat. Sol. (a)*, 93, 377 (1986); D.R. Choudhury, A.K. Chowdhury, and A.N Chakravarti, *Appl. Phys.*, 22, 145 (1980).

[230] J.L. Shay and J.W. Wernik, *Ternary Chalcoprite Semiconductors: Growth, Electronic Properties and Applications* (Pergamon Press, London, 1975).

[231] E.A. Arushanov, A.A. Kaynzev, A.N. Natepov, and S.I. Radautsan, *Sov. Phys. Semicond.*, 15, 828 (1981).

[232] K.S. Hong, R.F. Speyer and R.A. Condrate, *J. Phys. Chem. Solids*, 51, 969 (1990).

[233] M.H. Cohen, *Phys. Rev.*, 121, 387 (1961).

[234] J.W. McClure and K.H. Choi, *Solid State Commun.*, 21, 1015 (1977).

[235] E. Bangert and P. Kastner, *Phys. Stat. Sol. (b)*, 61, 503, (1974).

[236] G.M.T. Foley and P.N. Langenberg, *Phys. Rev. B*, 15, 4850 (1977).

[237] S. Takaoka, H. Kawamura, K. Murasa, and S. Takano, *Phys. Rev. B*, 13, 1428 (1976).

[238] M. Mondal, S. Banik and K.P. Ghatak, *J. Low. Temp. Phys.*, 74, 423 (1989).

[239] B.R. Nag, *Electron Transport in Compound Semiconductor*, Springer-Verlag, Germany (1980); M. Krieehbaum, P. Kocevar, H. Pascher and G. Bauer, *IEEE QE*, 24, 1727 (1988); J.J. Hopfield, *J. Appl. Phys.*, 32, 2277 (1961).

[240] S. Adachi, *Properties of Group-IV, III-V and II-VI Semiconductors* (John Wiley and Sons, 2005).

[241] O. Madelung, *Semiconductors: Data Handbook*, 3rd edn. (Springer-Verlag, Germany, 2003).

[242] D.J. Newson and A. Kurobe, *Semicond. Sci. Technol.*, 3, 786 (1988).

[243] U. Rossler, *Solid State Commun.*, 49, 943 (1984).

[244] S. Adachi, *J. Appl. Phys.*, 58, R1 (1985).

[245] S. Adachi, *GaAs and Related Materials: Bulk Semiconductors and Superlattice Properties*, (World Scientific, USA, 1994).

[246] G.L. Hansen, J.L. Schmit and T.N. Casselman, *J. Appl. Phys.*, 63, 7079 (1982).

[247] J. Wenus, J. Rutkowski, and A. Rogalski, *IEEE Trans. Elect. Dev.*, 48, 1326, (2001).

[248] S. Adachi, *J. Appl. Phys.*, 53, 8775 (1982).
[249] D.G. Seiler, W.M. Beeker and L.M. Roth, *Phys. Rev.* 1, 764 (1970).
[250] I.V. Skryabinskii and Yu. I. Ukhanov, *Sov. Phys. Solid State*, 14, 2838, (1973); H.I. Zhang, *Phys. Rev. B*, 1, 3450 (1970); P.C. Mathur and S. Jain, *Phys. Rev.*, 19, 1359 (1979).
[251] S. Tiwari and S. Tiwari, *Cryst. Res. Technol.*, 41, 78, (2006).
[252] J.R. Lowney and S.D. Senturia, *J. Appl. Phys.*, 47, 1771 (1976).
[253] W.E. Spicer and G.J. Lapeyre, *Phys. Rev.*, 139, A565 (1965).
[254] J.O. Dimmock, *The Physics of Semimetals and Narrowgap Semiconductors*, ed. by D.L. Carter and R.T. Bates (Pergamon Press, Oxford, 319, 1971).
[255] D.R. Lovett, *Semimetals and Narrow Band Gap Semiconductors* (Pion Limited, London, 1977).
[256] D.G. Seiler, B.D. Bajaj and A.E. Stephens, *Phys. Rev. B*, 16, 2822 (1977); A.V. Germaneko and G. M. Minkov, *Phys. Stat. Sol. (b)*, 184, 9 (1994); G.L. Bir and G.E. Pikus, *Symmetry and Strain – Induced Effects in Semiconductors Nauka*, Russia (1972). (in Russian); M. Mondal and K.P. Ghatak, *Phys. Stat. Sol. (b)*, 135, K21 (1986).
[257] C.C. Wu and C.J. Lin, *J. Low. Temp. Phys.*, 57, 469 (1984).
[258] V.I. Ivanov-Omskii, A. Sh. Mekhtisev, S.A. Rustambekova and E.N. Ukraintsev, *Phys. Stat. Sol. (b)* 119, 159 (1983).
[259] H. Kim, K. Cho, H. Song, B. Min, J. Lee, G. Kim, S. Kim, S.H. Kim and T. Noh, *Appl. Phys. Lett*, 83, 4619 (2003).
[260] P.R. Emtage, *Phys. Rev. A*, 246, 138 (1965).
[261] R.A. Reynolds, M.J. Brau, and R.A. Chapman, *J. Phys. Chem. Solids*, 29, 755, (1968).
[262] J. O'Shaughnessy and C. Smith, *Solid State Commun.*, 8, 481, (1970). XE "PtSb2"
[263] G.J. Rees, *Phys. Compounds*, Proceedings of the 13th International Conference,. Ed. F.G. Fumi, pp. 1166, North Holland Company (1976).
[264] J. Bouat and J.C. Thuillier, *Surface Sci.* 73, 528 (1978).
[265] M.V. Ortenberg and K.J. Button, *Phys. Rev. B*, 16, 2618 (1977).
[266] G. Haeffler, A.E. Klinkmüller, J. Rangell, U. Berzinsh and D. Hanstorp, *Z. Phys. D*, 38, 211 (1996).
[267] N.B. Brandt, V.N. Davydov, V.A. Kulbachinskii and O.M. Nikitina, *Sov. Phys. Sol. Stat.*, 29, 1014, (1987).
[268] L.M. Viculis, J.J. Mack, O.M. Mayer, H.T. Hahn and R.B. Kaner, *J. Mater. Chem.*, 15, 974 (2005).
[269] L.A. Vassilev, *Phys. Stat. Sol. (b)*, 121, 203 (1984); S. Takaoka and K. Murase, *Phys. Rev. B*, 20, 2823 (1979).
[270] D.L. Partin, *Superlattices and Microstructures*, 1, 131, (1985).
[271] Y. Yamada, *J. Phys. Jap.*, 35, 1600 (1973); M. Singh, P. R. Wallace, S. D. Jog and E. Arushanov, *J. Phys. Chem. Solids*, 45, 409, (1984).
[272] W.J. Turner, A.S. Fischler, and W.E. Reese, *Phys. Rev.*, 121, 759 (1961).
[273] G.P. Chuiko, *Sov. Phys. Semi.*, 19, 1381 (1985).
[274] W.E. Swank, and P.G. Le Comber, *Phys. Rev.*, 153, 844 (1967).
[275] M. Stordeur and W. Kuhnberger, *Phys. Stat. Sol. (b)*, 69, 377, (1975); D.R. Lovett, *Semimetals and Narrow-Bandgap Semiconductors, Pion Limited*, 185 (1977); H. Köhler, *Phys. Stat. Sol. (b)*, 74, 591 (1976).
[276] D. Haneman, *J. Phys. Chem. Solids*, 11, 205 (1959).
[277] N. Wei, G. Wu and J. Dong, *Phys. Lett. A*, 325, 403 (2004).
[278] S. Reich, J. Maultzsch, C. Thomsen and P. Ordejo´n, *Phys. Rev. B*, 66, 035412, (2006).
[279] M.S. Lundstrom, and J. Guo, *Nanoscale Transistors: Device Physics, Modeling and Simulation*, (Springer, USA, 2006); J.W. Mintmire and C. T. White, *Phys. Rev. Lett.*, 81, 2506 (1998).
[280] F. Buonocore, F. Trani, D. Ninno, A. Di Matteo, G. Cantele and G. Iadonisi, *Nanotechnology*, 19, 025711 (2008).
[281] J.B. Ketterson, *Phys. Rev.*, 129, 18 (1963).

[282] R.C. Vilão, J.M. Gil, A. Weidinger, H.V. Alberto, J. Piroto Duarte, N.A. de Campos, R.L. Lichti, K. H. Chow, S.P. Cottrell and S.F.J. Cox, *Phys. Rev. B*, 77, 235212 (2008).

[283] I. Kang and F.W. Wise, *Phys. Rev. B*, *J. Opt. Soc. Am. B*, 14, 1632 (1997).

[284] D. Cui, J. Xu, S.-Y. Xu, G. Paradee, B.A. Lewis and M. D. Gerhold, *IEEE*, *Trans. Elect. Dev.*, 5, 362 (2006).

[285] R.W. Cunningham, *Phys. Rev.*, 167, 761 (1968).

[286] A.I. Yekimov, A.A. Onushchenko, A.G. Plyukhin and Al. L. Efros, *J. Exp. Theor. Phys.*, 88, 1490 (1985).

[287] B.J. Roman and A.W. Ewald, *Phys. Rev. B*, 5, 3914 (1972).

[288] E.J. Johnson, and D.H. Dicley, *Phys. Rev.*, 1, 2676 (1970).

[289] V.G. Agafonov, P.M. Valov, B.S. Ryvkin and I.D. Yaroshetskii, *Sov. Phys. Semicond.*, 12, 1182 (1978).

2 Heisenberg's uncertainty principle and Einstein's photoemission from HD optoelectronic nanomaterials in the presence of intense light waves

Time is the ultimate Lord of this World

2.1 Introduction

The importance of Einstein's photoemission (EP) in the whole field of nanoscience and nanotechnology is well known [1–12]. In this chapter, Section 2.2 provides the theoretical background. Section 2.2.1 formulates the EP from heavily doped (HD) III–V, ternary and quaternary materials in the presence of light waves, whose unperturbed electron energy spectrum is described by the three-band model of Kane in the absence of band tailing. In Section 2.2.2, the EP for all the aforementioned cases is studied. Section 2.3 contains the results and discussion.

2.2 Theoretical background

2.2.1 The HUP and EP from HD III–V, ternary and quaternary materials

The velocity along the z-direction and the density of states function in this case for HD optoelectronic Kane-type materials under intense light waves whose conduction electrons in the absence of perturbation obey the three-band model of Kane can, respectively, as follows:

$$v_z(E'_1) = \sqrt{\frac{2}{m_c}}\frac{[T_1(E'_1, \eta_g, \lambda)]^{1/2}}{T'_1(E'_1, \eta_g, \lambda)} \tag{2.1}$$

where $E'_1 = E - E_{01HD}$, $E_{01HD} = \xi_1 + W - h\nu$, and ξ_1 is the root of the equation

$$T_1(\xi_1, \eta_g, \lambda) = 0 \tag{2.2}$$

The EP in this case is given by

https://doi.org/10.1515/9783110610819-002

$$J_{LHD} = \frac{4\pi\alpha_0 e m_c g_v}{h^3} \text{ Real Part of} \int_{E_{01HD}}^{E_{FHDL}} T_1(E'_1, \eta_g, \lambda) dE'_1 \tag{2.3}$$

where E_{FHDL} is the Fermi energy in this case.

Similarly, the EP for perturbed two-band model of Kane and that of parabolic energy bands can, respectively, be expressed as

$$J_{LHD} = \frac{4\pi\alpha_0 e m_c g_v}{h^3} \int_{E_{02HD}}^{E_{FHDL}} T_2(E'_2, \eta_g, \lambda) dE'_2 \tag{2.4}$$

and

$$J_{LHD} = \frac{4\pi\alpha_0 e m_c g_v}{h^3} \int_{E_{03HD}}^{E_{FHDL}} T_3(E'_3, \eta_g, \lambda) dE'_3 \tag{2.5}$$

where $E'_2 = E - E_{02HD}$, $E_{02HD} = \xi_2 + W - h\nu$, and ξ_2 is the root of the equation

$$T_2(\xi_2, \eta_g, \lambda) = 0 \tag{2.6}$$

and $E'_3 = E - E_{03HD}$, $E_{03HD} = \xi_3 + W - h\nu$, and ξ_3 is the root of the equation

$$T_3(\xi_3, \eta_g, \lambda) = 0 \tag{2.7}$$

2.2.2 Results and discussion

Using appropriate equations, the normalized EP from HD n-$Hg_{1-x}Cd_xTe$ has been plotted as functions of normalized I_0 (for a given wavelength and considering red light for which λ is about 640 nm), λ(assuming $I_0 = 10\,\text{nWm}^{-2}$), and the normalized electron degeneracy at $T = 4.2\,\text{K}$ in accordance with the perturbed three- and two-band models of Kane and that of perturbed parabolic energy bands in Figures 2.1–2.3, respectively. Figures 2.4–2.6 exhibit all the aforementioned cases for HD n-$In_{1-x}Ga_xAs_yP_{1-x}$ lattice matched to InP, respectively. It appears that J increases with the increasing electron degeneracy in accordance with all the band models. The combined influence of the energy band constants on the EP from ternary and quaternary materials can easily be assessed from all the figures. It appears that the EP decreases with increasing light intensity for all materials and also decreases as the wavelength shifts from violet to red. The influence of light is immediately apparent from all plots, since the EP depends strongly on the light intensity of all types of perturbed band models, which is

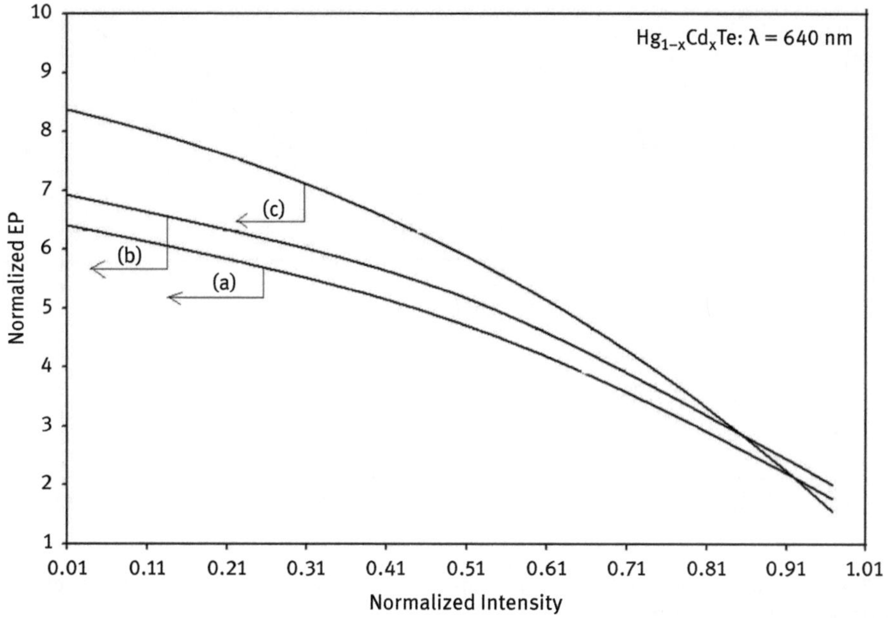

Figure 2.1: Plot of the normalized EP from HD *n*-$Hg_{1-x}Cd_xTe$ as a function of normalized light intensity in which the curves (a), (b), and (c) represent the perturbed three- and two-band models of Kane together with parabolic energy bands, respectively.

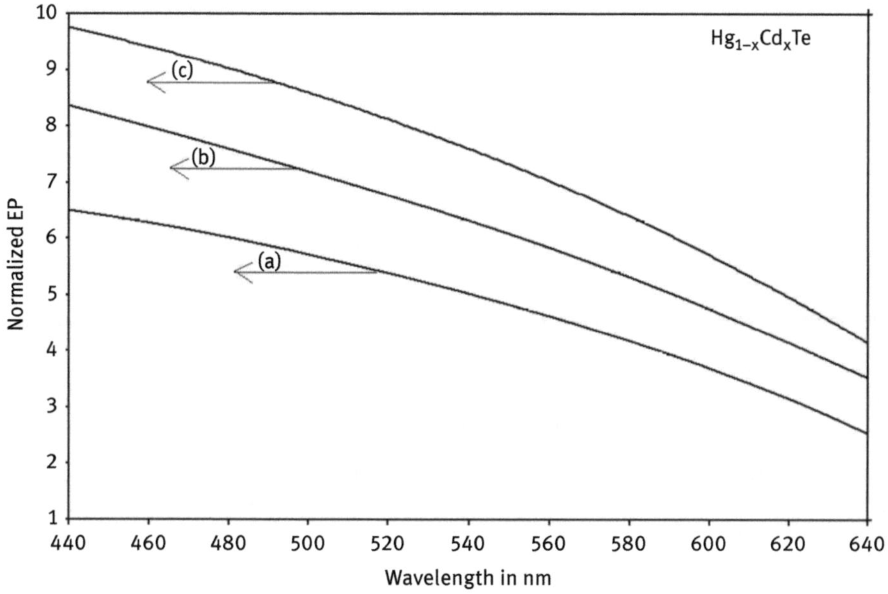

Figure 2.2: Plot of the normalized EP from HD *n*-$Hg_{1-x}Cd_xTe$ as a function of wavelength for all cases of Figure 2.1.

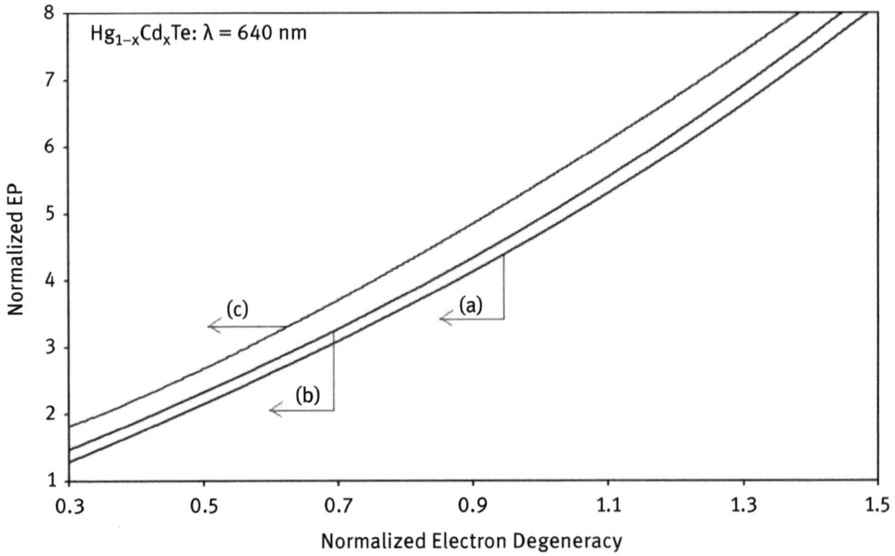

Figure 2.3: Plot of the normalized EP from HD $n\text{-}Hg_{1-x}Cd_xTe$ as a function of normalized electron degeneracy for all cases of Figure 2.1.

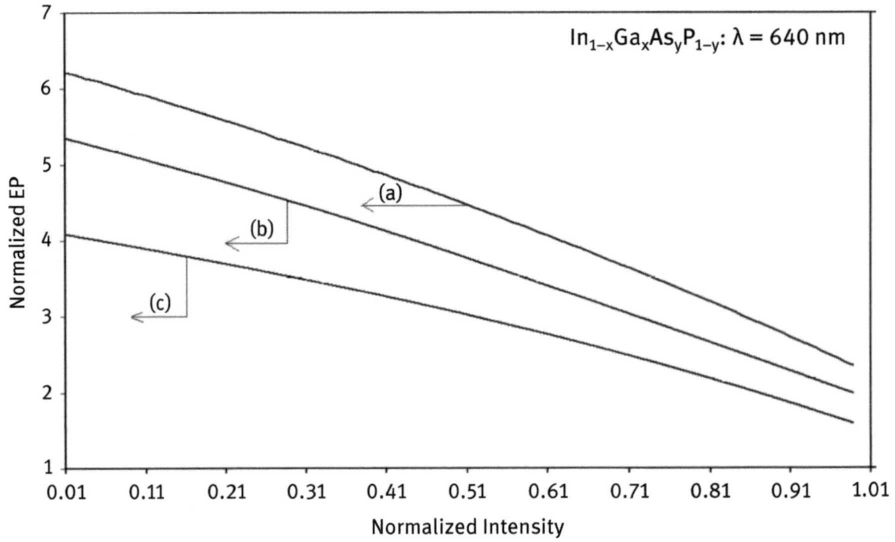

Figure 2.4: Plot of the normalized EP from HD $In_{1-x}Ga_xAs_yP_{1-y}$ lattice matched to InP as a function of normalized light intensity in which curves (a), (b), and (c) represent the perturbed three- and two-band models of Kane together with parabolic energy bands, respectively.

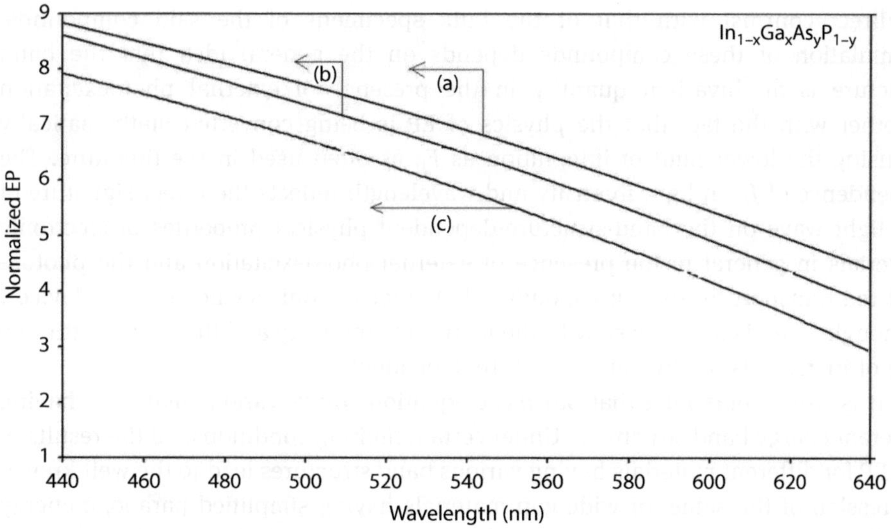

Figure 2.5: Plot of the normalized EP from HD $In_{1-x}Ga_xAs_yP_{1-y}$ lattice matched to InP as a function of wavelength for all cases of Figure 2.4.

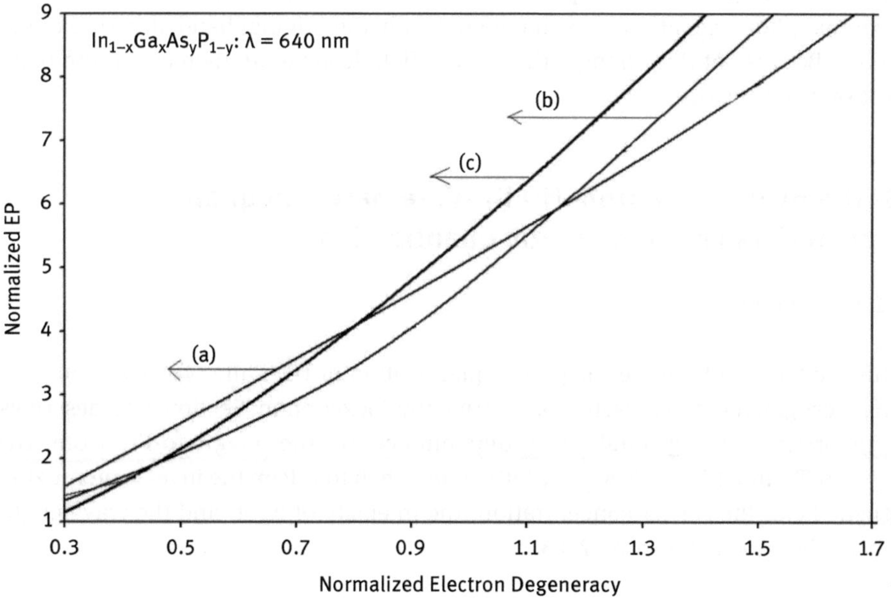

Figure 2.6: Plot of the normalized EP from HD $In_{1-x}Ga_xAs_yP_{1-y}$ lattice matched to InP as a function of normalized electron degeneracy for all cases of Figure 2.4.

in direct contrast with that of the bulk specimens of the said compounds. Formulation of these compounds depends on the general idea that the band structure is an invariant quantity in the presence of external photoexcitation together with the fact that the physics of EP is being converted mathematically by using the lower limit of integration as E_0 as often used in the literature. The dependence of J_L on light intensity and wavelength reflects the direct signature of the light wave on the band-structure-dependent physical properties of electronic materials in general in the presence of external photoexcitation and the photon-assisted transport for the corresponding HD optoelectronic semiconductor devices. Although J_L tends to decrease with the increasing intensity and the wavelength, the rate of increase is totally band structure dependent.

It is worth remarking that our basic equation covers various materials having different energy band structures. Under certain limiting conditions, all the results of the EP for different materials having various band structures lead to the well-known expression of the same for wide-gap materials having simplified parabolic energy bands. This indirect test not only exhibits the mathematical compatibility of the formulation but also shows the fact that the presented simple analysis is a more generalized one, since the well-known result can be obtained under certain limiting conditions of the generalized expressions. It is worth remarking that the influence of an external photoexcitation is to change radically the original band structure of the material. Because of this change, the photon field leads to an increase in the band gap of semiconductors.

2.3 The HUP and EP from HD III–V, ternary and quaternary materials under magnetic quantization

2.3.1 Introduction

In this section, the EP under magnetic quantization in HD Kane-type materials has been investigated in the presence of external photoexcitation. Section 2.3.2 describes the theoretical background. The dependence of the magneto-EP from HD n-$Hg_{1-x}Cd_xTe$ and n-$In_{1-x}Ga_xAs_yP_{1-y}$ lattices matched to InP on the inverse quantizing magnetic field, the carrier concentration, the intensity of light, and the wavelength has been discussed in Section 2.3.3.

2.3.2 Theoretical background

(i) The EP under magnetic quantization in accordance with perturbed three-band model of Kane can be expressed as

$$J = \frac{\alpha_0 e^2 B}{2\pi^2 \hbar^2} \, Real\,Part\,of \sum_{n=0}^{n_{max}} (\eta'_{61HDLB}) \tag{2.8}$$

where $\eta'_{61HDLB} = [E_{FHDLB} - (E_{n_{l1}} + W - h\nu)]$
and $E_{n_{l1}}$ is the Landau sub-band energies in this case and is given by

$$T_1(E_{n_{l1}}, \eta_g, \lambda) = \left(n + \frac{1}{2}\right)\hbar\,\omega_0 \tag{2.9}$$

(ii) The EP under magnetic quantization in accordance with perturbed two-band model of Kane can be expressed as

$$J = \frac{\alpha_0 e^2 B}{2\pi^2 \hbar^2} \sum_{n=0}^{n_{max}} (\eta'_{62HDLB}) \tag{2.10}$$

where $\eta'_{62HDLB} = [E_{FHDLB} - (E_{n12} + W - h\nu)]$
and $E_{n_{l2}}$ is the Landau sub-band energies in this case and is given as

$$T_2(E_{n_{l2}}, \eta_g, \lambda) = \left(n + \frac{1}{2}\right)\hbar\,\omega_0 \tag{2.11}$$

(iii) The EP under magnetic quantization in accordance with perturbed parabolic band model of Kane can be expressed as

$$J = \frac{\alpha_0 e^2 B}{2\pi^2 \hbar^2} \sum_{n=0}^{n_{max}} (\eta'_{63HDLB}) \tag{2.12}$$

where $\eta'_{63HDLB} = [E_{FHDLB} - (E_{n13} + W - h\nu)]$
and $E_{n_{l2}}$ is the Landau sub-band energies in this case and is given as

$$T_3(E_{n_{l2}}, \eta_g, \lambda) = \left(n + \frac{1}{2}\right)\hbar\,\omega_0 \tag{2.13}$$

2.3.3 Results and discussion

Using appropriate equations, we have plotted the normalized magneto-EP from HD n-$Hg_{1-x}Cd_xTe$ versus inverse quantizing magnetic field in accordance with the perturbed three- and two-band models of Kane and that of perturbed parabolic energy bands as shown in Figure 2.7. Figures 2.8–2.10 exhibit the variation of the aforementioned quantity from HD n-$Hg_{1-x}Cd_xTe$ as functions of the normalized electron degeneracy, the normalized intensity of light, and wavelength at $T = 4.2$ K, respectively. Figures 2.11–2.14 represent the said variations of EP under magnetic quantization from HD n-$In_{1-x}Ga_xAs_yP_{1-y}$ lattice matched to InP.

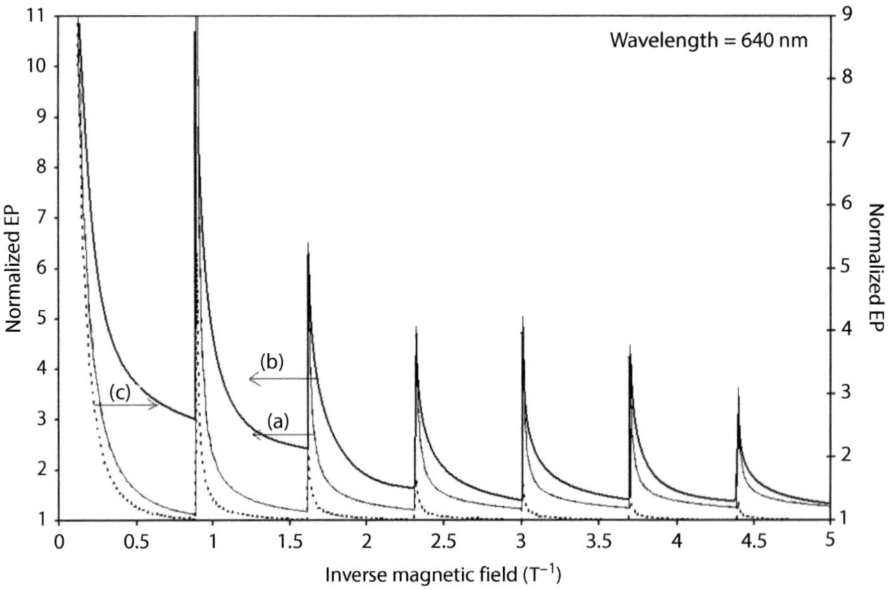

Figure 2.7: Plot of the normalized EP as a function of inverse magnetic field from HD $n\text{-}Hg_{1-x}Cd_xTe$, where curves (a), (b), and (c) represent the perturbed three- and two-band models of Kane together with parabolic energy bands, respectively.

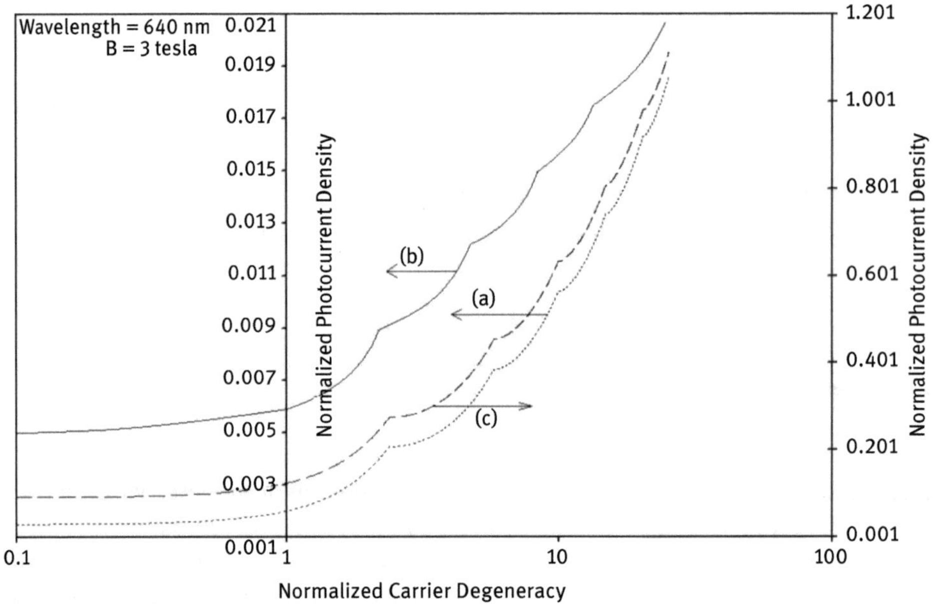

Figure 2.8: Plot of the normalized EP as a function of normalized carrier degeneracy from HD $n\text{-}Hg_{1-x}Cd_xTe$, where curves (a), (b), and (c) represent the perturbed three- and two-band models of Kane together with parabolic energy bands, respectively.

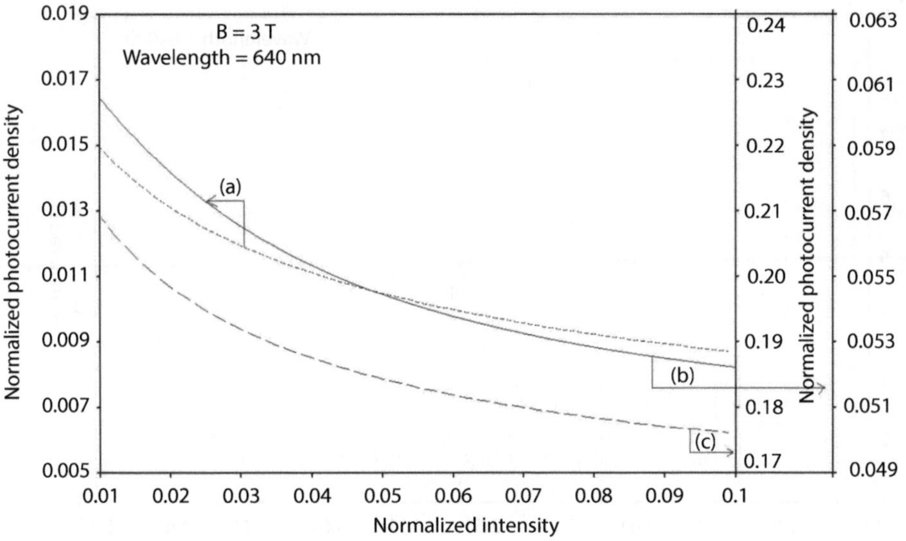

Figure 2.9: Plot of the normalized EP as a function of normalized light intensity from HD *n*-$Hg_{1-x}Cd_xTe$, where curves (a), (b), and (c) represent the perturbed three- and two-band models of Kane together with parabolic energy bands, respectively.

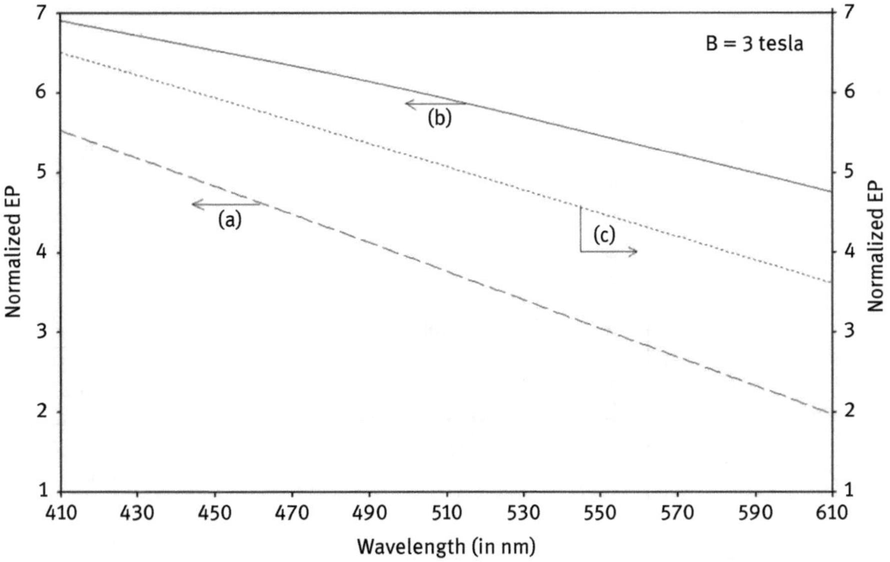

Figure 2.10: Plot of the normalized EP as a function of wavelength from HD *n*-$Hg_{1-x}Cd_xTe$ in which curves (a), (b), and (c) represent the perturbed three- and two-band models of Kane together with parabolic energy bands, respectively.

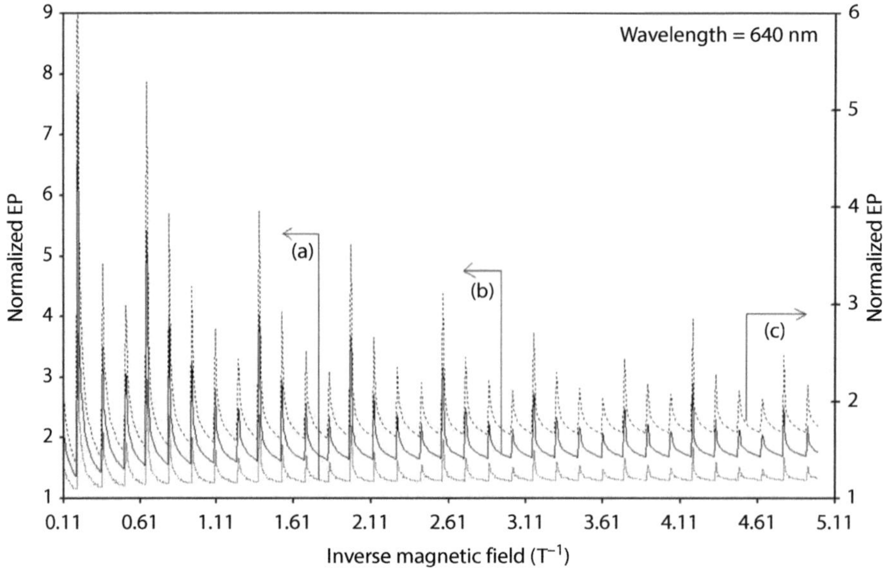

Figure 2.11: Plot of the normalized EP as a function of inverse magnetic field from HD $In_{1-x}Ga_xAs_yP_{1-y}$ lattice matched to InP, where curves (a), (b), and (c) represent the perturbed three- and two-band models of Kane together with parabolic energy bands, respectively.

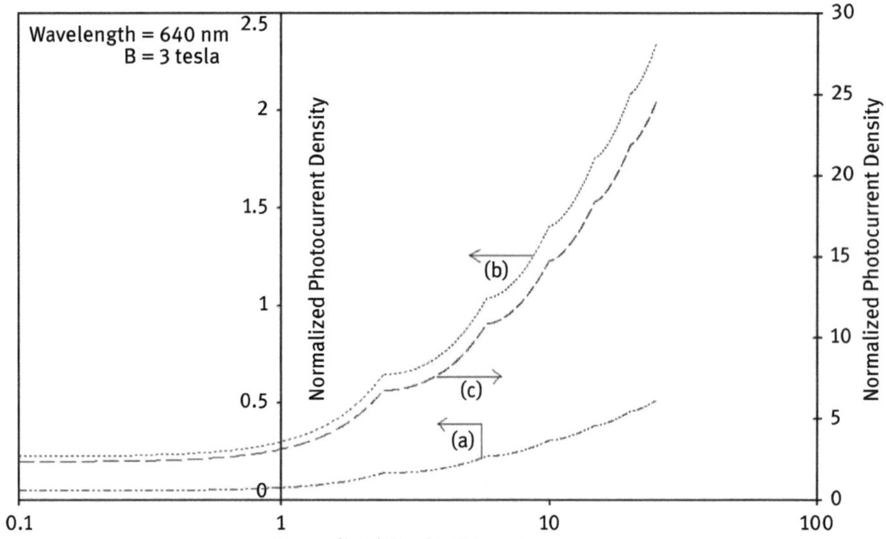

Figure 2.12: Plot of the normalized EP as a function of normalized carrier degeneracy from HD $In_{1-x}Ga_xAs_yP_{1-y}$ lattice matched to InP, where curves (a), (b), and (c) represent the perturbed three- and two-band models of Kane together with parabolic energy bands, respectively.

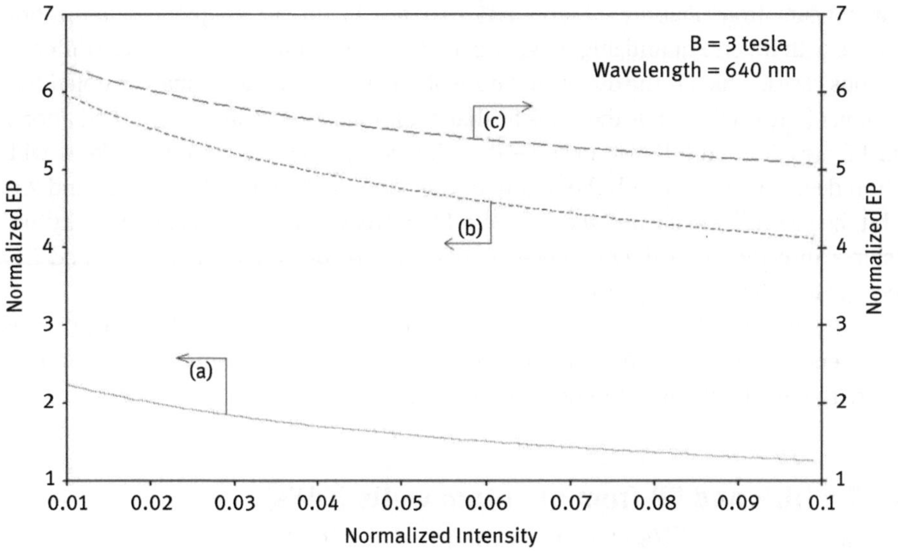

Figure 2.13: Plot of the normalized EP as a function of normalized light intensity from HD $In_{1-x}Ga_xAs_yP_{1-y}$ lattice matched to InP, where curves (a), (b), and (c) represent the perturbed three- and two-band models of Kane together with parabolic energy bands, respectively.

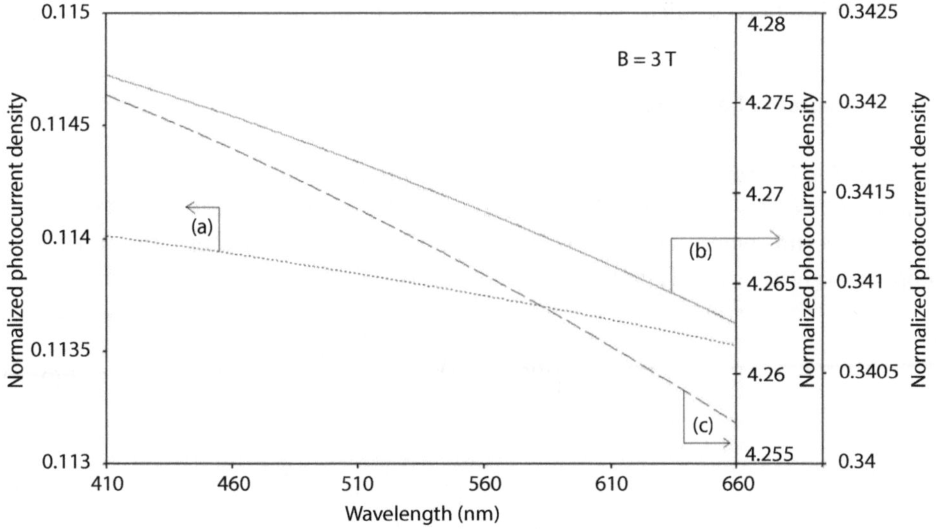

Figure 2.14: Plot of the normalized EP as a function of wave length from HD $In_{1-x}Ga_xAs_yP_{1-y}$ lattice matched to InP, where curves (a), (b), and (c) represent the perturbed three- and two-band models of Kane together with parabolic energy bands, respectively.

It appears from Figures 2.7 and 2.11 that the EP under magnetic quantization oscillates with inverse quantizing magnetic field, and the numerical values are different in various cases, which is the direct signature of the band structure. It may be noted that the origin of the oscillation is the same as that of Shubnikov–de Haas (SdH) oscillations. From Figures 2.8 and 2.12, we observe that the said physical quantity oscillates with electron degeneracy, although the nature of oscillation is different. Figures 2.9 and 2.13 exhibit the fact that the normalized magneto-EP decreases with increasing intensity and the slopes directly reflect the influence of energy band constants. Figures 2.10 and 2.14 reflect the fact that the magneto-EP decreases with increasing wavelength.

Finally, we note that the form of the expression of the said physical quantity in this case is generalized, where the Landau energy and the Fermi energy under magnetic field are the two band-structure-dependent quantities.

2.4 The HUP and EP from quantum wells (QWs), nano wires (NWs), and quantum dots (QDs) of HD III–V, ternary and quaternary materials

2.4.1 Introduction

In this section, the EP from QWs, NWs, and quantm box (QBs) of HD optoelectronic materials has been studied. Section 2.4.3 includes results and discussions.

2.4.2 Theoretical background

2.4.2.1 The EP from HD QWs of optoelectronic materials

The velocity of the electron in the n_{z71th}, n_{z72th}, and n_{z73th} sub-bands for the 2D electron energy spectra, whose bulk dispersion relations in the presence of light and heavy doping as given by eqs. (1.46b), (1.47), and (1.48) can, respectively, be written as

$$\upsilon_z(E_{n_{z71}}) = \left(\frac{m_c}{2}\right)^{-1/2}\left[\frac{\sqrt{T_1(E_{n_{z71}},\eta_g,\lambda)}}{T'_1(E_{n_{z71}},\eta_g,\lambda)}\right] \tag{2.14}$$

$$\upsilon_z(E_{n_{z72}}) = \left(\frac{m_c}{2}\right)^{-1/2}\left[\frac{\sqrt{T_2(E_{n_{z72}},\eta_g,\lambda)}}{T'_2(E_{n_{z72}},\eta_g,\lambda)}\right] \tag{2.15}$$

$$\upsilon_z(E_{n_{z73}}) = \left(\frac{m_c}{2}\right)^{-1/2}\left[\frac{\sqrt{T_3(E_{n_{z73}},\eta_g,\lambda)}}{T'_3(E_{n_{z73}},\eta_g,\lambda)}\right] \tag{2.16}$$

where $n_{z7J}(J=1,2,3)$ is the size quantum number, and the sub-band energies $E_{n_{z71}}$, $E_{n_{z72}}$, and $E_{n_{z73}}$ are, respectively, defined through the following equations:

$$T_1(E_{n_{z71}}, \eta_g, \lambda) = \left(\frac{\hbar^2}{2m_c}\right)\left(\frac{\pi n_{z71}}{d_z}\right)^2 \tag{2.17}$$

$$T_2(E_{n_{z72}}, \eta_g, \lambda) = \left(\frac{\hbar^2}{2m_c}\right)\left(\frac{\pi n_{z72}}{d_z}\right)^2 \tag{2.18}$$

and

$$T_3(E_{n_{z73}}, \eta_g, \lambda) = \left(\frac{\hbar^2}{2m_c}\right)\left(\frac{\pi n_{z73}}{d_z}\right)^2 \tag{2.19}$$

The respective expressions of the photoemission are given by

$$J_{2DL} = \frac{\alpha_0 g_v e}{\pi\hbar^2 d_z}\left(\frac{m_c}{2}\right)^{-1/2} \text{Real Part of} \sum_{n_{z71_{min}}}^{n_{z71_{max}}} \left[\frac{\sqrt{T_1(E_{n_{z71}}, \eta_g, \lambda)}}{T'_1(E_{n_{z71}}, \eta_g, \lambda)}\right] [\phi_{71}(E_{F2DL}, n_{z71})] \tag{2.20}$$

where

$$n_{z71_{min}} \geq \left(\frac{d_z}{\pi}\right)\left(\frac{\sqrt{2m_c}}{\hbar}\right)[T_1(W - h\upsilon, \eta_g, \lambda)]^{1/2}$$

and

$$\phi_{71}(E_{2DF}, n_{z71}) = \left[\frac{2m_c}{\hbar^2} T_1(E_{F2DL}, \eta_g, \lambda) - \left(\frac{\pi n_{z71}}{d_z}\right)^2\right],$$

$$J_{2DL} = \frac{\alpha_0 g_v e}{\pi\hbar^2 d_z}\left(\frac{m_c}{2}\right)^{-1/2} \sum_{n_{z72_{min}}}^{n_{z72_{max}}} \left[\frac{\sqrt{T_2(E_{n_{z72}}, \eta_g, \lambda)}}{T'_2(E_{n_{z72}}, \eta_g, \lambda)}\right] [\phi_{73}(E_{F2DL}, n_{z72})] \tag{2.21}$$

where

$$n_{z72_{min}} \geq \left(\frac{d_z}{\pi}\right)\left(\frac{\sqrt{2m_c}}{\hbar}\right)\sqrt{T_2(W - h\upsilon, \eta_g, \lambda)}$$

and

$$\phi_{73}(E_{F2DL}, n_{z72}) = \left[\frac{2m_c}{\hbar^2} T_2(E_{F2DL}, \eta_g, \lambda) - \left(\frac{\pi n_{z72}}{d_z}\right)^2\right]$$

$$J_{2DL} = \frac{\alpha_0 g_v e}{\pi \hbar^2 d_z}\left(\frac{m_c}{2}\right)^{-1/2} \sum_{n_{z73_{\min}}}^{n_{z73\max}} \left[\frac{\sqrt{T_3(E_{n_{z73}}, \eta_g, \lambda)}}{T'_3(E_{n_{z73}}, \eta_g, \lambda}\right] [\phi_{75}(E_{F2DL}, n_{z73})] \tag{2.22}$$

where

$$n_{z73_{\min}} \geq \left(\frac{d_z}{\pi}\right)\left(\frac{\sqrt{2m_c}}{\hbar}\right)[T_3(W - h\upsilon, \eta_g, \lambda)]^{1/2}$$

and

$$\phi_{75}(E_{F2DL}, n_{z73}) = \left[\frac{2m_c}{\hbar^2} T_3(E_{F2DL}, \eta_g, \lambda) - \left(\frac{\pi n_{z73}}{d_z}\right)^2\right],$$

2.4.2.2 The EP from HD NWs of optoelectronic materials

The generalized expression of photocurrent in this case is given by

$$I_L = \frac{\alpha_0 e g_v}{\pi \hbar} \sum_{n_{x7i}=1}^{n_{x7i\max}} \sum_{n_{z7i}=1}^{n_{z7i\max}} (\eta'_{7i}) \tag{2.23a}$$

where $\eta'_{7i} = E_{F1DL} - (E'_{7i} + W - h\upsilon)$ and E'_{7i} are the sub-band energies in this case and are defined through the following equations:

$$\left.\begin{aligned} T_1(E'_{71}, \eta_g, \lambda) &= G_{71}(n_{x71}, n_{z71}) \\ T_2(E'_{72}, \eta_g, \lambda) &= G_{72}(n_{x72}, n_{z72}) \\ T_3(E'_{73}, \eta_g, \lambda) &= G_{73}(n_{x73}, n_{z73}) \end{aligned}\right\} \tag{2.23b}$$

Real part of eq. (2.22) should be used for computing the EP from NWs of HD optoelectronic materials, whose unperturbed energy band structures are defined by the three-band model of Kane.

2.4.2.3 The EP from QB of HD optoelectronic materials

The dispersion relations of the electrons in QBs of HD optoelectronic materials in the presence of light waves can, respectively, be expressed from eqs. (1.46b), (1.47), and (1.48) as follows:

$$\frac{2m_cT_1(E_{Q1},\eta_g,\lambda)}{\hbar^2}=H_{71}(n_{x71},n_{y71},n_{z71}) \tag{2.24}$$

$$\frac{2m_cT_2(E_{Q2},\eta_g,\lambda)}{\hbar^2}=H_{72}(n_{x72},n_{y72},n_{z72}) \tag{2.25}$$

$$\frac{2m_cT_3(E_{Q3},\eta_g,\lambda)}{\hbar^2}=H_{73}(n_{x73},n_{y73},n_{z73}) \tag{2.26}$$

where E_{Qi} is the totally quantized energy and $H_{7i}(n_{x7i},n_{y7i},n_{z7i})=\left(\frac{\pi n_{x7i}}{d_x}\right)^2+\left(\frac{\pi n_{y7i}}{d_y}\right)^2+\left(\frac{\pi n_{z7i}}{d_z}\right)^2$

The electron concentration can, in general, be written as

$$n_{0DL}=\left(\frac{2g_v}{d_xd_yd_z}\right)\sum_{n_{x7i}=1}^{n_{x7i\max}}\sum_{n_{y7i}=1}^{n_{y7i\max}}\sum_{n_{z7i}=1}^{n_{z7i\max}}F_{-1}(\eta_{7i0D}) \tag{2.27}$$

where $\eta_{7i0D}=\frac{E_{F0DL}-E_{Qi}}{k_BT}$ and E_{F0DL} are the Fermi energies in QBs of HD optoelectronic materials in the presence of light waves as measured from the edge of the conduction band in the vertically upward direction in the absence of any quantization.

Real part of eq. (2.26) should be used for computing the carrier density from QBs of HD optoelectronic materials, whose unperturbed energy band structures are defined by the three-band model of Kane.

The photoemitted current densities in this case are given by the following equations:

$$J_{0DL}=\frac{(\alpha_0 e g_v)}{d_xd_yd_z}\left(\frac{m_c}{2}\right)^{-1/2}\text{Real Part of}\sum_{n_{x71}=1}^{n_{x71\max}}\sum_{n_{y71}=1}^{n_{y71\max}}\sum_{n_{z71\min}}^{n_{z71\max}}\left[\frac{\sqrt{T_1(E_{n_{z71}},\eta_g,\lambda)}}{T'_1(E_{n_{z71}},\eta_g,\lambda)}\right]F_{-1}(\eta_{710D}) \tag{2.28}$$

$$J_{0DL}=\frac{(\alpha_0 e g_v)}{d_xd_yd_z}\left(\frac{m_c}{2}\right)^{-1/2}\sum_{n_{x72}=1}^{n_{x72\max}}\sum_{n_{y72}=1}^{n_{y72\max}}\sum_{n_{z72\min}}^{n_{z72\max}}\left[\frac{\sqrt{T_2(E_{n_{z72}},\eta_g,\lambda)}}{T'_2(E_{n_{z72}},\eta_g,\lambda)}\right]F_{-1}(\eta_{720D}) \tag{2.29}$$

$$J_{0DL}=\frac{(\alpha_0 e g_v)}{d_xd_yd_z}\left(\frac{m_c}{2}\right)^{-1/2}\sum_{n_{x73}=1}^{n_{x73\max}}\sum_{n_{y73}=1}^{n_{y73\max}}\sum_{n_{z73\min}}^{n_{z73\max}}\left[\frac{\sqrt{T_3(E_{n_{z73}},\eta_g,\lambda)}}{T'_3(E_{n_{z73}},\eta_g,\lambda)}\right]F_{-1}(\eta_{730D}) \tag{2.30}$$

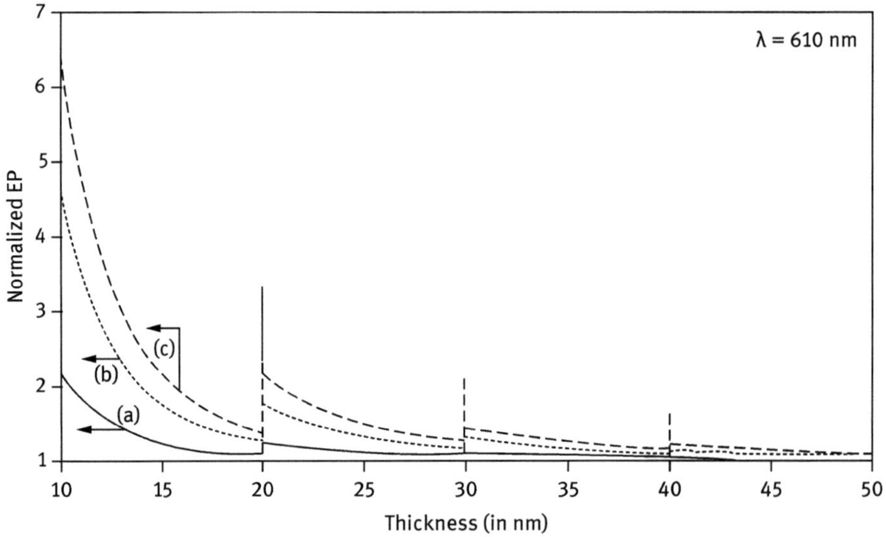

Figure 2.15: Plot of the normalized EP from QWs of HD $n\text{-}Hg_{1-x}Cd_xTe$ as a function of film thickness, where curves (a), (b), and (c) represent the perturbed HD three- and two-band models of Kane together with HD parabolic energy bands, respectively.

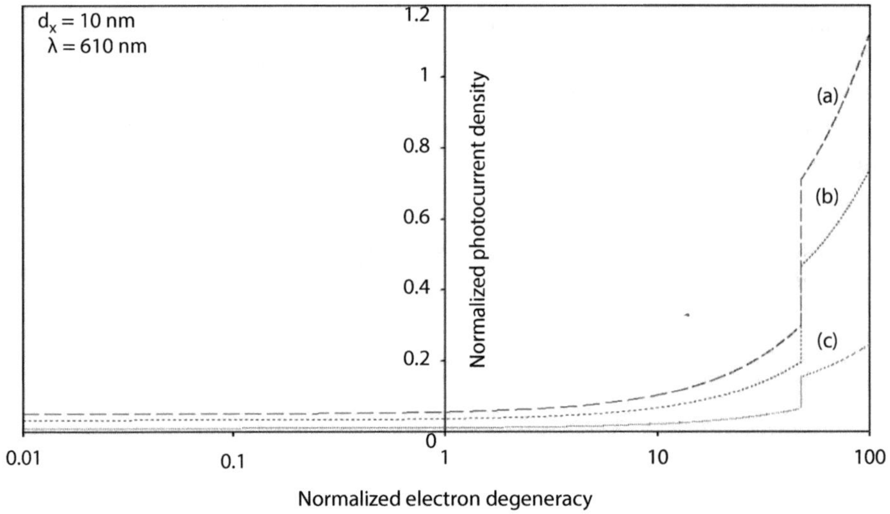

Figure 2.16: Plot of the normalized EP from QWs of HD $n\text{-}Hg_{1-x}Cd_xTe$ as a function of normalized electron degeneracy for all cases of Figure 2.15.

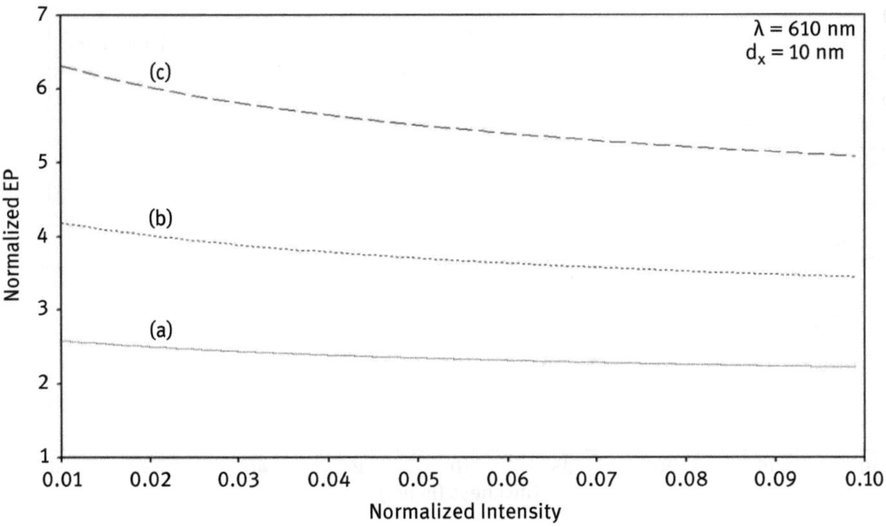

Figure 2.17: Plot of the normalized EP from QWs of HD *n*-$Hg_{1-x}Cd_xTe$ as a function of normalized light intensity for all cases of Figure 2.15.

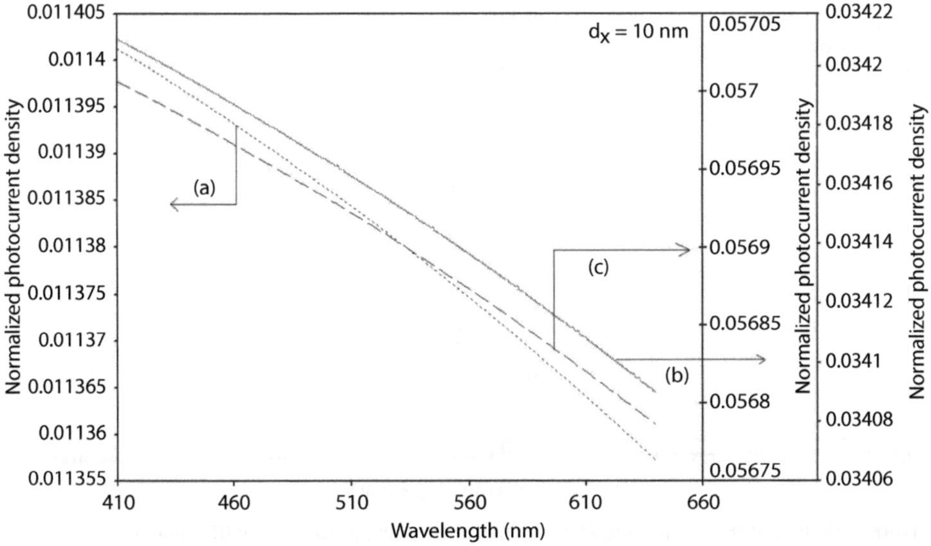

Figure 2.18: Plot of the normalized EP from QWs of HD *n*-$Hg_{1-x}Cd_xTe$ as a function of light wavelength for all cases of Figure 2.15.

2.4.3 Results and discussion

Using numerical values of the energy band constants, the normalized EP has been plotted from QWs of HD *n*-$Hg_{1-x}Cd_xTe$ under external photoexcitation,

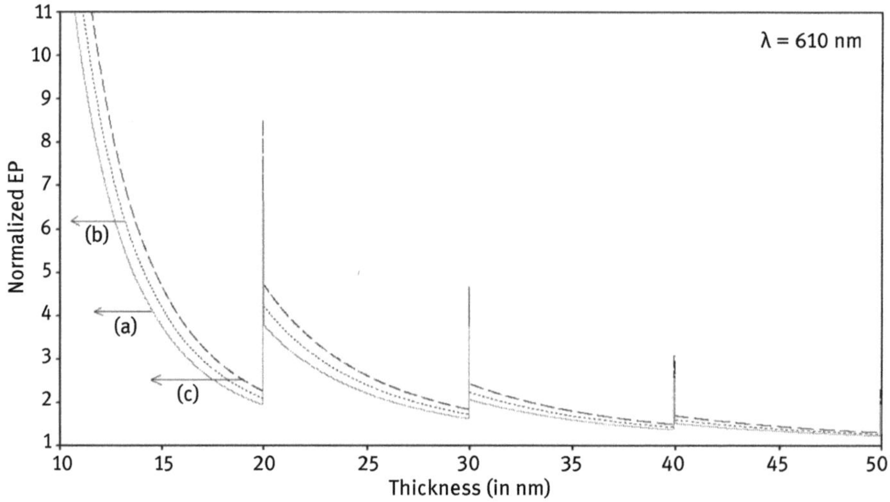

Figure 2.19: Plot of the normalized EP from QWs of HD n-$In_{1-x}Ga_xAs_yP_{1-y}$ lattice matched to InP as a function of film thickness, where curves (a), (b), and (c) represent the perturbed HD three- and two-band models of Kane together with HD parabolic energy bands, respectively.

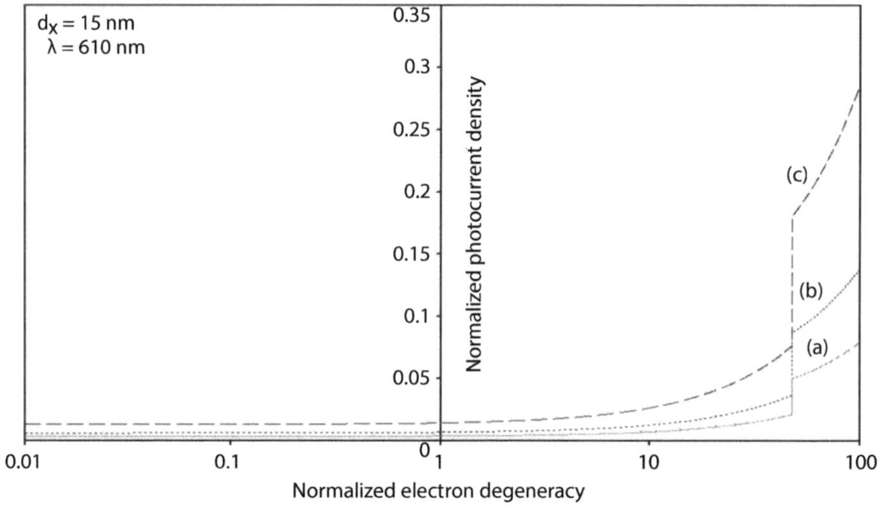

Figure 2.20: Plot of the normalized EP from QWs of HD n-$In_{1-x}Ga_xAs_yP_{1-y}$ lattice matched to InP as a function of normalized electron degeneracy for all cases of Figure 2.19.

whose band structure follows the perturbed HD three- and two-band models of Kane and that of the perturbed HD parabolic energy bands as shown by curves (a), (b), and (c) of Figure 2.15 as functions of film thickness. The plots of Figures 2.16, 2.17, and 2.18 exhibit the dependence of the normalized EP on the normalized electron degeneracy, normalized intensity, and wavelength,

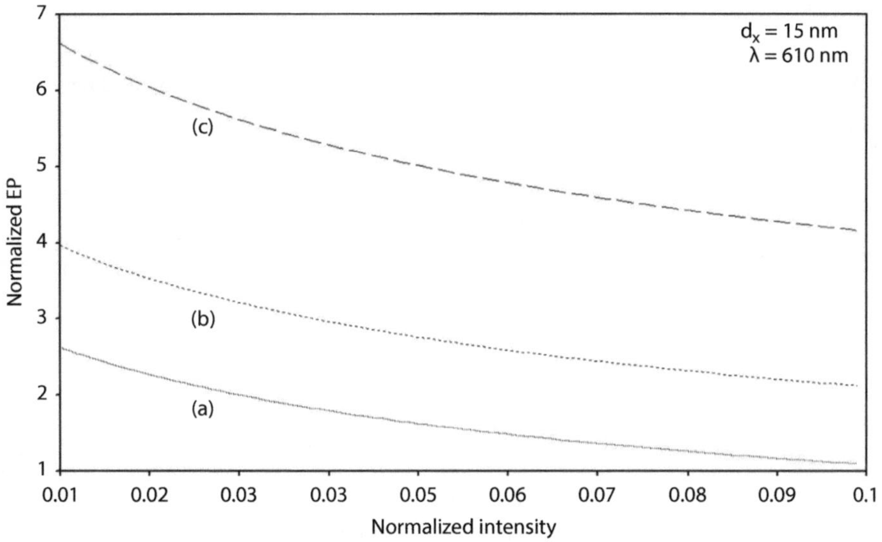

Figure 2.21: Plot of the normalized from QWs of HD $n\text{-}In_{1-x}Ga_xAs_yP_{1-y}$ lattice matched to InP as a function of normalized light intensity for all cases of Figure 2.19.

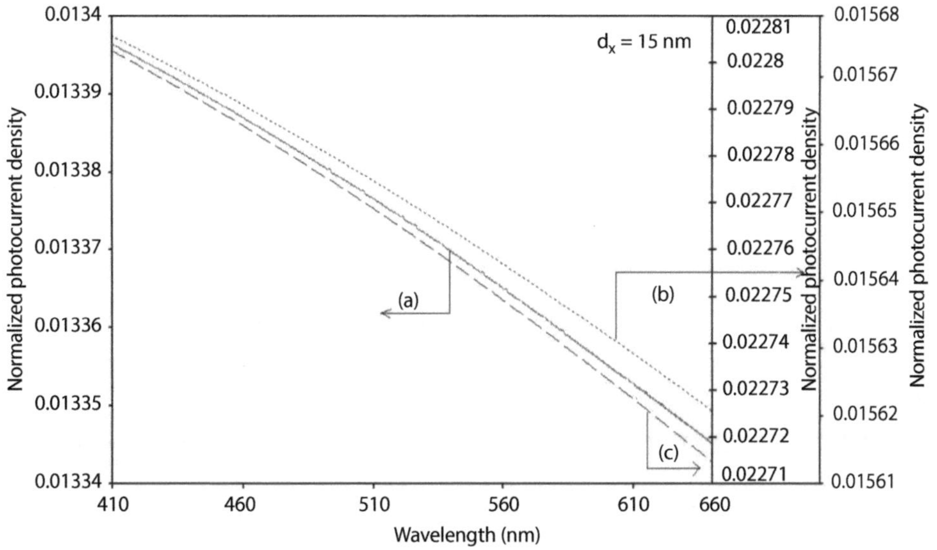

Figure 2.22: Plot of the normalized EP from QWs of HD $n\text{-}In_{1-x}Ga_xAs_yP_{1-y}$ lattice matched to InP as a function of light wavelength for all cases of Figure 2.19.

respectively, for all cases of Figure 2.15. The variations of the normalized EP from QWs of HD $n\text{-}In_{1-x}Ga_xAs_yP_{1-y}$ lattice matched to InP as functions of film thickness, normalized carrier degeneracy, normalized incident light intensity, and wavelength, respectively, have been drawn in Figures 2.19–2.22 for all cases

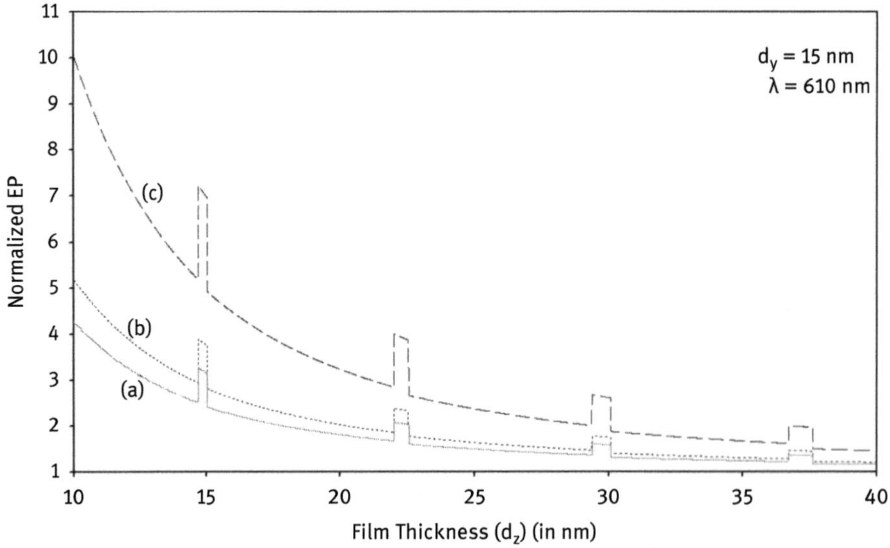

Figure 2.23: Plot of the normalized EP from NWs of HD n-$Hg_{1-x}Cd_xTe$ as a function of film thickness, where curves (a), (b), and (c) represent the perturbed HD three- and two-band models of Kane together with HD parabolic energy bands, respectively.

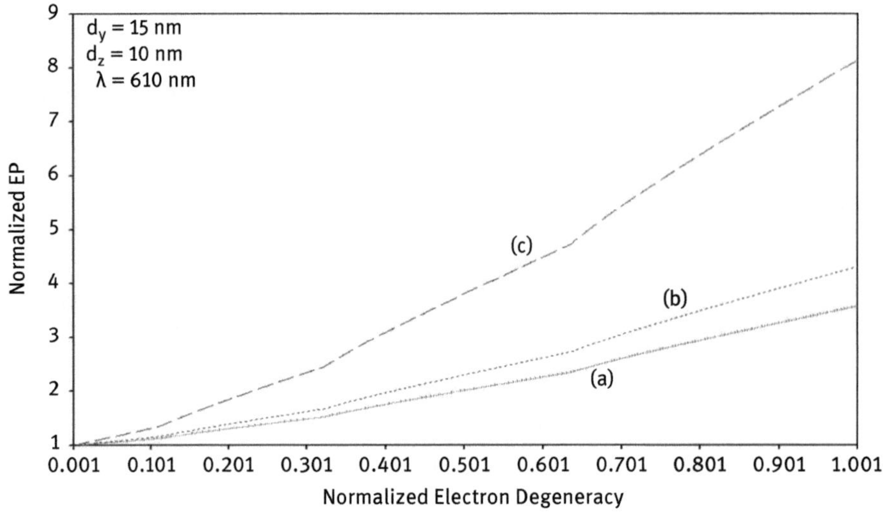

Figure 2.24: Plot of the normalized EP from NWs of HD n-$Hg_{1-x}Cd_xTe$ as a function of normalized electron degeneracy for all cases of Figure 2.23.

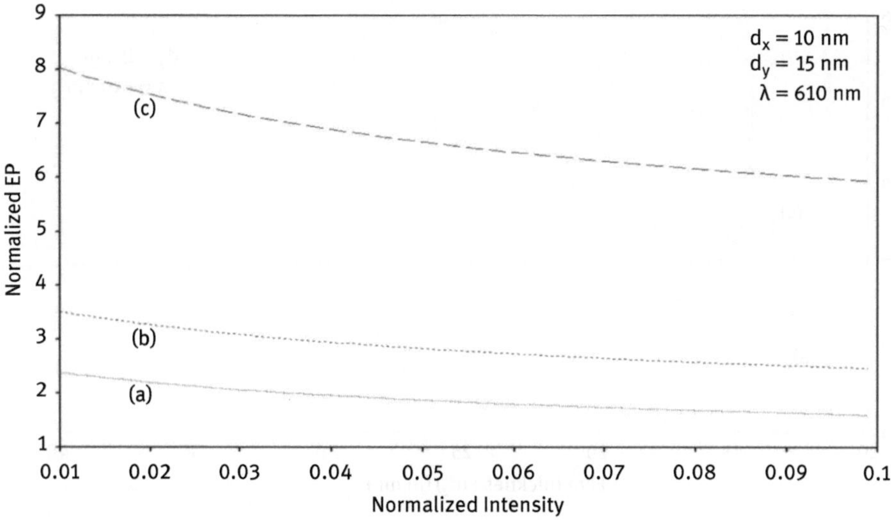

Figure 2.25: Plot of the normalized EP from NWs of HD $n\text{-}Hg_{1-x}Cd_xTe$ as a function of normalized light intensity for all cases of Figure 2.23.

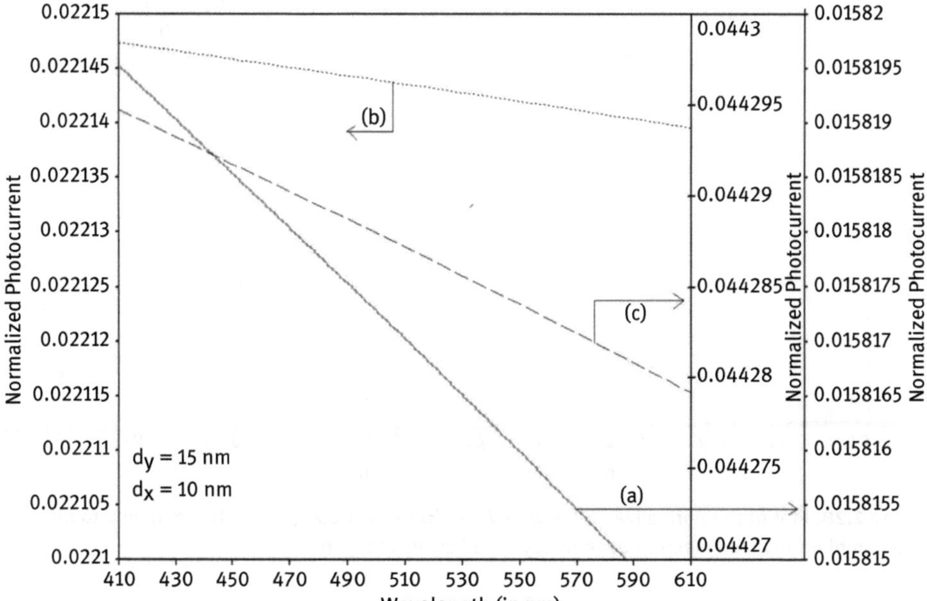

Figure 2.26: Plot of the normalized EP from NWs of HD $n\text{-}Hg_{1-x}Cd_xTe$ as a function of light wavelength for all cases of Figure 2.23.

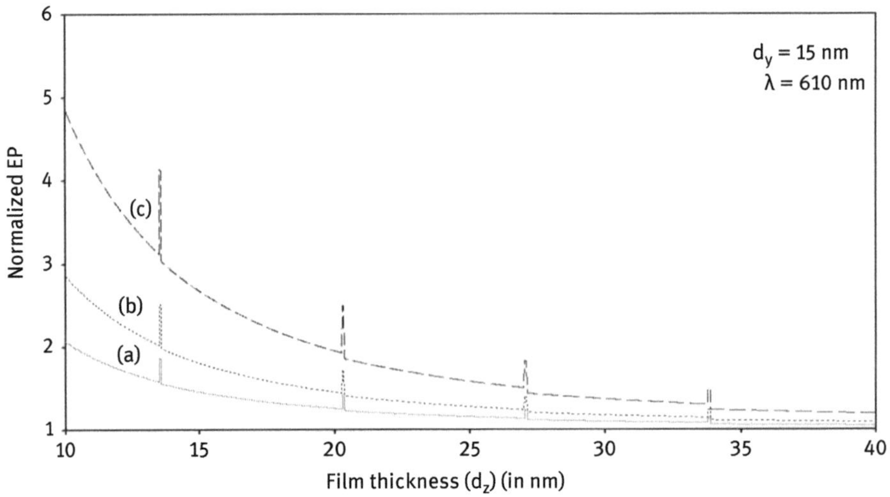

Figure 2.27: Plot of the normalized EP from NWs of HD $n\text{-}In_{1-x}Ga_xAs_yP_{1-y}$ lattice matched to InP as a function of film thickness, where curves (a), (b), and (c) represent the perturbed HD three- and two-band models of Kane together with HD parabolic energy bands, respectively.

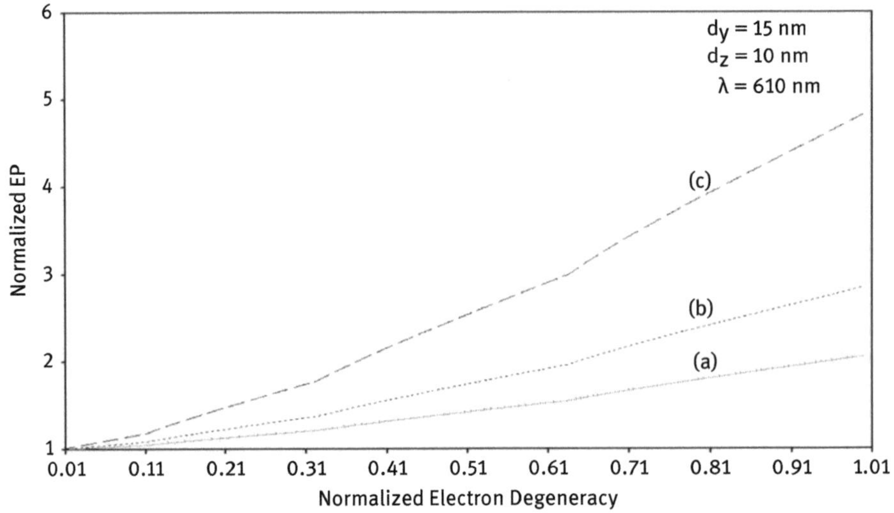

Figure 2.28: Plot of the normalized EP from NWs of HD $n\text{-}In_{1-x}Ga_xAs_yP_{1-y}$ lattice matched to InP as a function of normalized electron degeneracy for all cases of Figure 2.27.

of Figure 2.15. The dependences of the normalized EP from NWs of HD $n\text{-}Hg_{1-x}Cd_xTe$ with respect to film thickness, normalized carrier degeneracy, normalized light intensity, and wavelength have been drawn in Figures 2.23–2.26 in accordance with perturbed HD three-band (using eqs. (2.22) and (2.19)) and HD two-band (using eqs. (2.22) and (2.20)) models of Kane together with HD parabolic (using eqs. (2.22) and (2.21))

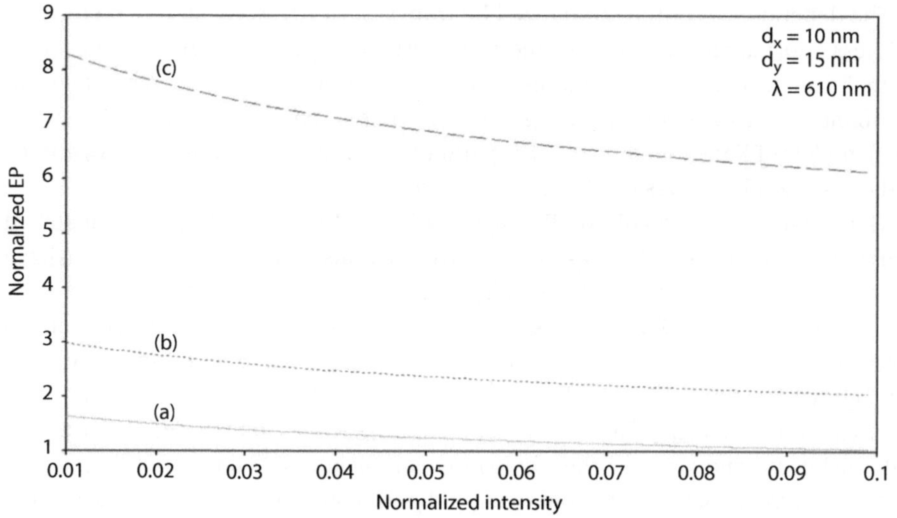

Figure 2.29: Plot of the normalized EP from NWs of HD n-$In_{1-x}Ga_xAs_yP_{1-y}$ lattice matched to InP as a function of normalized light intensity for all cases of Figure 2.27.

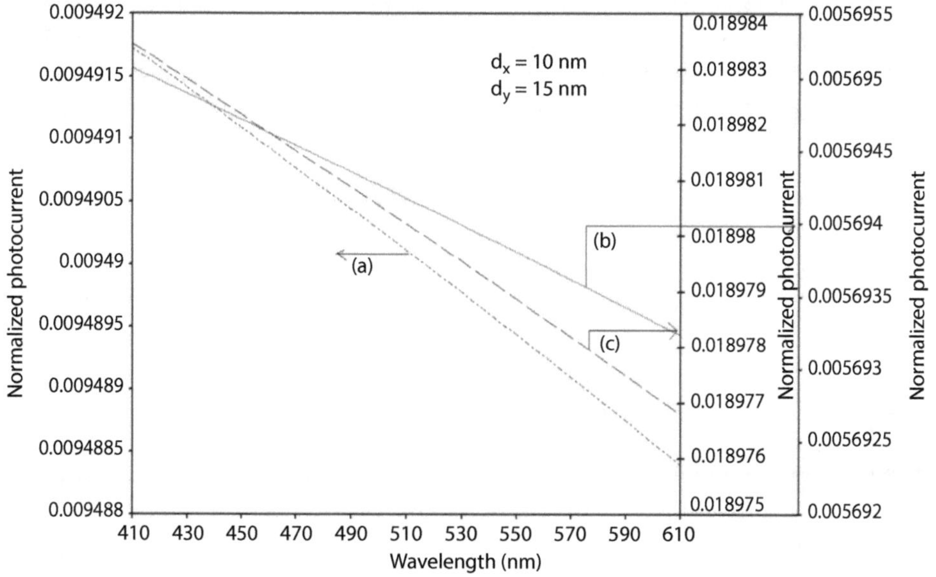

Figure 2.30: Plot of the normalized EP from NWs of HD n-$In_{1-x}Ga_xAs_yP_{1-y}$ lattice matched to InP as a function of light wavelength for all cases of Figure 2.27.

energy bands as shown by curves (a), (b), and (c), respectively. The variations of normalized EP from NWs of HD n-$In_{1-x}Ga_xAs_yP_{1-y}$ lattice matched to InP have been drawn in Figures 2.27–2.30 as functions of film thickness, normalized carrier degeneracy, normalized incident light intensity, and wavelengths, respectively.

The dependences of the normalized EP from QBs of HD $n\text{-}Hg_{1-x}Cd_xTe$ on the film thickness, normalized carrier degeneracy, normalized light intensity, and wavelength have been drawn in Figures 2.31–2.34 in accordance with perturbed HD three-band (using eqs. (2.27) and (2.26) and HD two-band (using eqs. (2.28) and (2.26)) models of Kane together with HD parabolic (using eqs. (2.29) and (2.26)) energy bands as shown by curves (a), (b), and (c), respectively.

The variations of normalized EP from QBs of HD $n\text{-}In_{1-x}Ga_xAs_yP_{1-y}$ lattice matched to InP have been drawn in Figures 2.35–2.38 as functions of film thickness, normalized carrier degeneracy, normalized incident light intensity, and wavelengths, respectively, for all cases of Figure 2.31. Figures 2.15 and 2.19 show that EP from QWs of HD optoelectronic materials decreases with increasing film thickness in oscillatory manners. Figures 2.23 and 2.27 show that the EP from NWs of HD optoelectronic materials increases with decreasing film thickness, exhibiting trapezoidal variation for a very small thickness bandwidth for the whole range of thicknesses considered. The widths of the trapezoids depend on the energy band constants of $n\text{-}Hg_{1-x}Cd_xTe$ and $n\text{-}In_{1-x}Ga_xAs_yP_{1-y}$ lattice matched to InP, respectively.

Figures 2.31 and 2.35 show that the EP from HD QBs of optoelectronic materials decreases with increasing film thickness, exhibiting prominent trapezoidal variation for relatively large thickness bandwidth. These three types of variations are special signatures of 1D confinement in HD QWs, 2D confinement in HD NWs, and 3D confinement in HD QBs of optoelectronic materials, respectively, in the presence of light. Figures 2.16 and 2.20 show that the normalized EP from HD QWs increases with

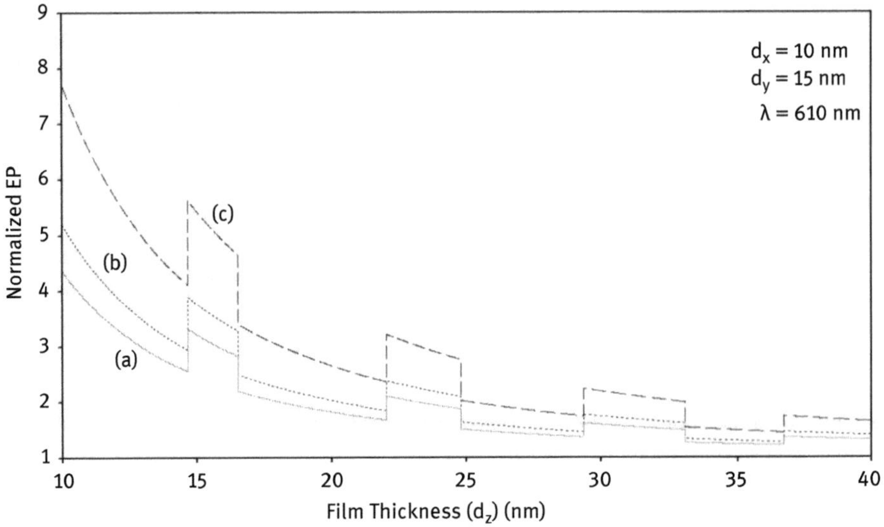

Figure 2.31: Plot of the normalized EP from HD QBs of $n\text{-}Hg_{1-x}Cd_xTe$ as a function of film thickness, where curves (a), (b), and (c) represent the perturbed HD three- and two-band models of Kane together with HD parabolic energy bands, respectively.

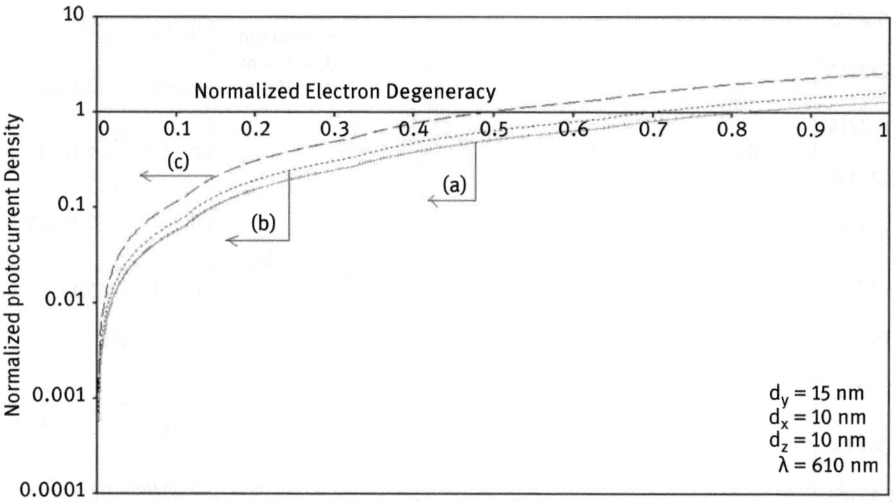

Figure 2.32: Plot of the normalized EP from HD QBs of $n\text{-}Hg_{1-x}Cd_xTe$ as a function of normalized electron degeneracy for all cases of Figure 2.31.

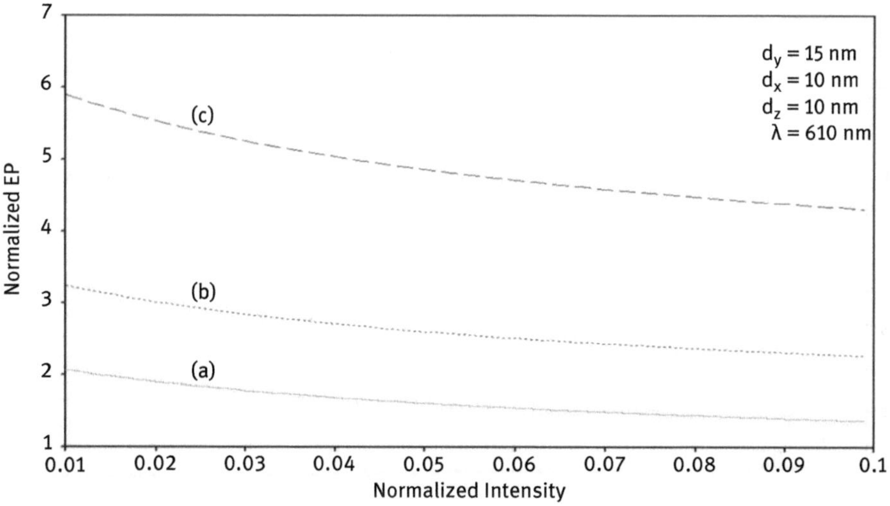

Figure 2.33: Plot of the normalized EP from HD QBs of $n\text{-}Hg_{1-x}Cd_xTe$ as a function of normalized light intensity for all cases of Figure 2.31.

increasing carrier degeneracy, and for relatively large values of the same variable, it exhibits quantum jumps for all types of band models when the size quantum number changes from one fixed value to another. Figures 2.24 and 2.28 show, respectively, that the normalized EP in HD NWs of optoelectronic materials increases with increasing normalized electron degeneracy.

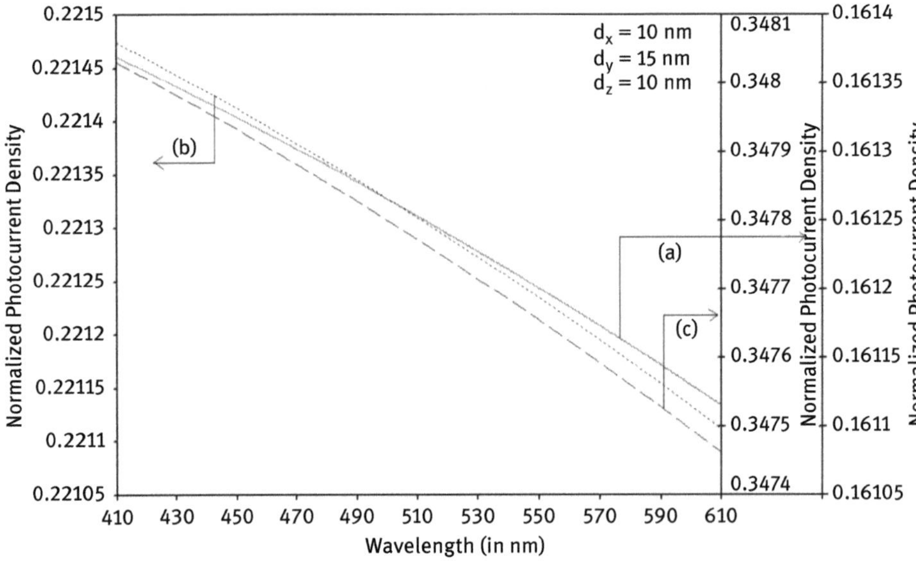

Figure 2.34: Plot of the normalized EP from HD QBs of $n\text{-}Hg_{1-x}Cd_xTe$ as a function of light wavelength for all cases of Figure 2.31.

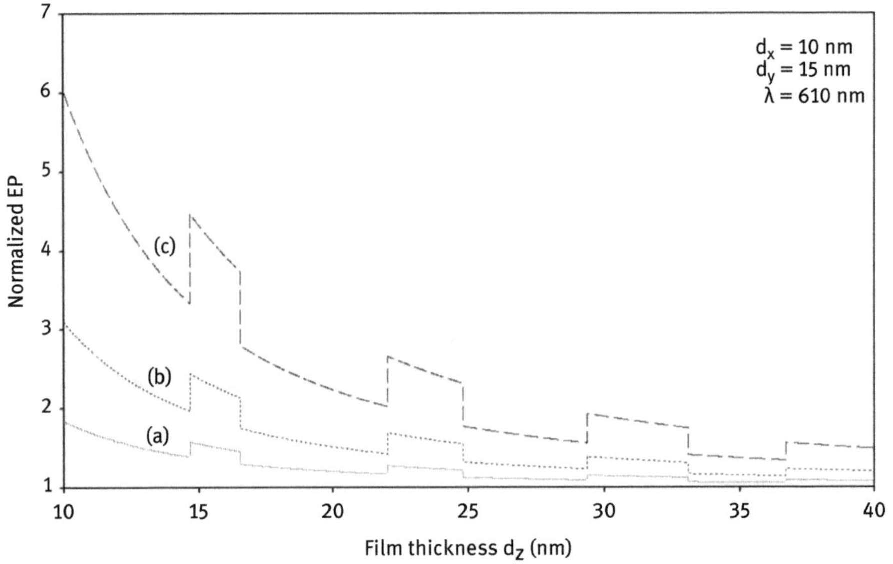

Figure 2.35: Plot of the normalized EP from HD QBs of $n\text{-}In_{1-x}Ga_xAs_yP_{1-y}$ lattice matched to InP as a function of film thickness, where curves (a), (b), and (c) represent the perturbed HD three- and two-band models of Kane together with HD parabolic energy bands, respectively.

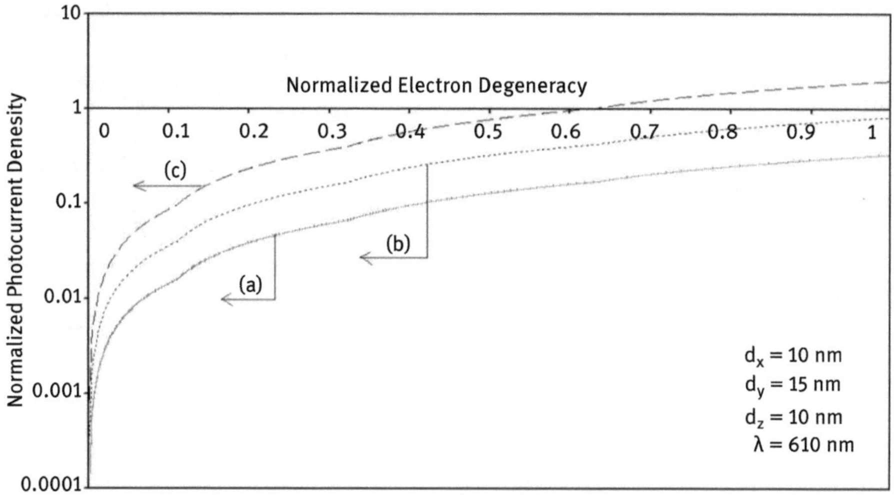

Figure 2.36: Plot of the normalized EP from HD QBs of $n\text{-}In_{1-x}Ga_xAs_yP_{1-y}$ lattice matched to InP as a function of normalized electron degeneracy for all cases of Figure 2.35.

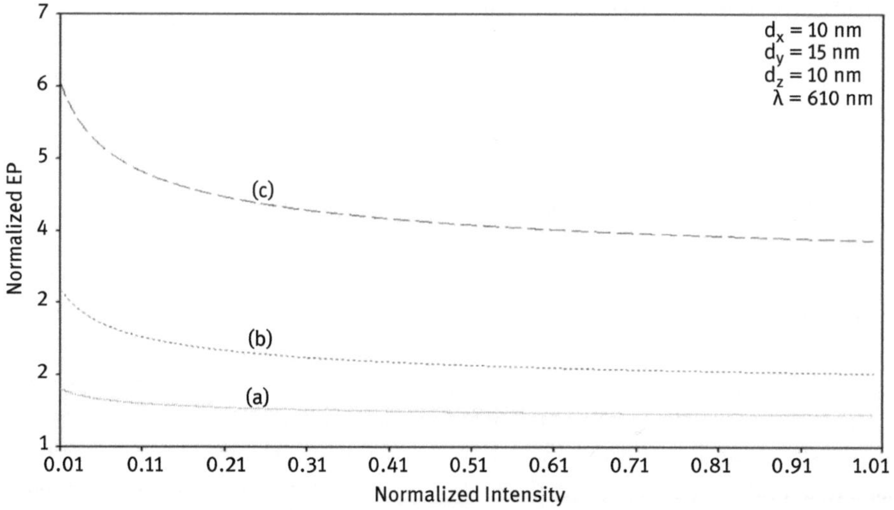

Figure 2.37: Plot of the normalized EP from HD QBs of $n\text{-}In_{1-x}Ga_xAs_yP_{1-y}$ lattice matched to InP as a function of normalized light intensity for all cases of Figure 2.35.

Figures 2.32 and 2.36 demonstrate that the EP from HD QBs of optoelectronic materials increases with increasing electron degeneracy again in a different oscillatory manner. From Figures 2.17, 2.21, 2.25, 2.29, 2.33, and 2.37, it appears that the EP increases with decreasing intensity for all types of quantum confinement. From Figures 2.18, 2.22, 2.26, 2.30, 2.34, and 2.38, we can conclude that the normalized EP decreases with increasing wavelength for HD QWs, NWs, and QBs of optoelectronic materials. Finally, it is apparent from all figures that the EP from quantum-confined

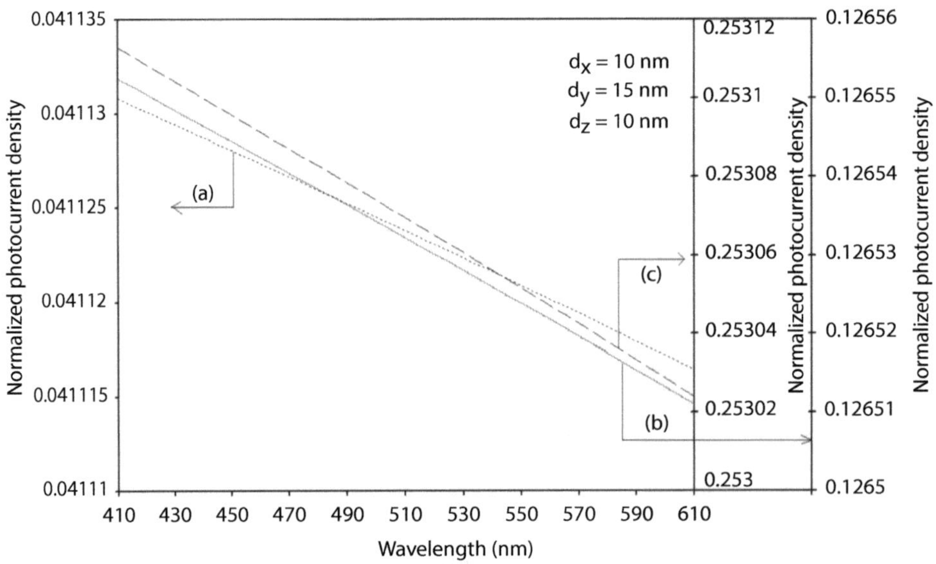

Figure 2.38: Plot of the normalized EP from HD QBs of $n\text{-}In_{1-x}Ga_xAs_yP_{1-y}$ lattice matched to InP as a function of light wavelength for all cases of Figure 2.35.

HD ternary materials is larger as compared with the quantum-confined HD quaternary compounds for all types of quantum confinement.

2.5 The EP from HD effective mass superlattices of optoelectronic materials

2.5.1 Introduction

In Section 2.5.2, an attempt is made to study the magneto-EP from HD effective mass quantum well super lattices (QWSL) of optoelectronic materials. In Section 2.5.2.1, the photoemission from HD effective mass NW SLs of optoelectronic materials has been investigated, and in Section 2.5.2.2, the EP from HD effective mass QB SLs of optoelectronic materials has been studied. Section 2.5.2.3 explores the magneto-EP from HD effective mass SLs of optoelectronic materials. Section 2.5.3 contains results and discussions.

2.5.2 Theoretical background

2.5.2.1 The magneto-EP from HD QWs effective mass superlattices

The electron dispersion law in III–V effective mass superlattices can be written as

$$k_x^2 = \left[\frac{1}{L_0^2}[Cos^{-1}\left\{f_{HD1}(E,\lambda,\eta_g,k_y,k_z)\right\}\right]^2 - k_\perp^2] \tag{2.31}$$

where

$$f_{HD1}(E,\lambda,\eta_g,k_y,k_z) = [[\bar{a}_{1HD}Cos[a_0\bar{C}_{1HD}(E,\eta_{g1},\lambda,k_\perp) + b_0\bar{D}_{1HD}(E,\eta_{g2},\lambda,k_\perp)]]$$
$$- [\bar{a}_{2HD}Cos[a_0\bar{C}_{1HD}(E,\eta_{g1},\lambda,k_\perp) - b_0\bar{D}_{1HD}(E,\eta_{g2},\lambda,k_\perp)]]],$$

$$\bar{a}_{1HD} = \left[\sqrt{\frac{m_{c2}T'_1(0,\lambda,\eta_{g2})}{m_{c1}T'_1(0,\lambda,\eta_{g1})}}+1\right]^2\cdot\left[4\sqrt{\frac{m_{c2}T'_1(0,\lambda,\eta_{g2})}{m_{c1}T'_1(0,\lambda,\eta_{g1})}}\right]^{-1}$$

$$a_{2HD} = \left[\sqrt{\frac{m_{c2}T'_1(0,\lambda,\eta_{g2})}{m_{c1}T'_1(0,\lambda,\eta_{g1})}}-1\right]^2\cdot\left[4\sqrt{\frac{m_{c2}T'_1(0,\lambda,\eta_{g2})}{m_{c1}T'_1(0,\lambda,\eta_{g1})}}\right]^{-1}$$

$$\bar{C}_{1HD}(E,\eta_{g1},\lambda,k_\perp) = \left[\frac{2m_{c1}}{\hbar^2}T_1(E,\eta_{g1},\lambda) - k_\perp^2\right]^{1/2}$$

and

$$\bar{D}_{1HD}(E,\eta_{g1},\lambda,k_\perp) = \left[\frac{2m_{c2}}{\hbar^2}T_1(E,\eta_{g2},\lambda) - k_\perp^2\right]^{1/2}$$

In the presence of a quantizing magnetic field B along k_x direction, the magnetoelectron energy spectrum can be written as

$$k_x^2 = \bar{\omega}_{HD}(E,\eta_g,\lambda,n) \tag{2.32}$$

where

$$\bar{\omega}_{HD}(E,\eta_g,\lambda,n) = \left[\frac{1}{L_0^2}[Cos^{-1}\{f_{HD1}(E,\lambda,\eta_g,n)\}]^2 - \frac{2eB}{\hbar}\left(n+\frac{1}{2}\right)\right]$$

$$f_{HD1}(E,\lambda,\eta_g,n) = [[\bar{a}_{1HD}Cos[a_0\bar{C}_{1HD}(E,\eta_{g1},\lambda,n) + b_0\bar{D}_{1HD}(E,\eta_{g2},\lambda,n)]]$$
$$- [\bar{a}_{2HD}Cos[a_0\bar{C}_{1HD}(E,\eta_{g1},\lambda,n) - b_0\bar{D}_{1HD}(E,\eta_{g2},\lambda,n)]]],$$

$$\bar{C}_{1HD}(E,\eta_{g1},\lambda,n) = \left[\frac{2m_{c1}}{\hbar^2}T_1(E,\eta_{g1},\lambda) - \frac{2eB}{\hbar}\left(n+\frac{1}{2}\right)\right]^{1/2}$$

and

$$\bar{D}_{1HD}(E,\eta_{g1},\lambda,n) = \left[\frac{2m_{c2}}{\hbar^2}T_1(E,\eta_{g2},\lambda) - \frac{2eB}{\hbar}\left(n+\frac{1}{2}\right)\right]^{1/2}$$

The total energy e_{TQ1} in this case can be expressed as

The Z part of the energy E_{ZQ1} in this case can be written as

$$\left(\frac{n_x\pi}{d_x}\right)^2 = \bar{\omega}_{HD}(e_{TQ1}, \eta_g, \lambda, n) \tag{2.33}$$

$$\left(\frac{n_x\pi}{d_x}\right)L_0 = Cos^{-1}[f_{HD1}(E_{ZQ1}, \lambda, \eta_g, 0)] \tag{2.34}$$

where

$$[f_{HD1}(E_{ZQ1}, \lambda, \eta_g, 0)] = [[\bar{a}_{1HD}Cos[a_0\bar{C}_{1HD}(E_{ZQ1}, \eta_{g1}, \lambda, 0) + b_0\bar{D}_{1HD}(E_{ZQ1}, \eta_{g2}, \lambda, 0)]]$$
$$- [\bar{a}_{2HD}Cos[a_0\bar{C}_{1HD}(E_{ZQ1}, \eta_{g1}, \lambda, 0) - b_0\bar{D}_{1HD}(E_{ZQ1}, \eta_{g2}, \lambda, 0)]]],$$

$$\bar{C}_{1HD}(E_{ZQ1}, \eta_{g1}, \lambda, 0) = \left[\frac{2m_{c1}}{\hbar^2}T_1(E_{ZQ1}, \eta_{g1}, \lambda)\right]^{1/2}$$

and

$$\bar{D}_{1HD}(E_{ZQ1}, \eta_{g1}, \lambda, 0) = \left[\frac{2m_{c2}}{\hbar^2}T_1(E_{ZQ1}, \eta_{g2}, \lambda)\right]^{1/2}$$

The electron concentration is given by

$$n_0 = \frac{g_v eB}{\pi\hbar}\ \text{Real Part of} \sum_{n=0}^{n_{\max}}\sum_{n=1}^{n_{\max}} F_{-1}(\eta_{8SL1}) \tag{2.35}$$

where $\eta_{8SL1} = (k_B T)^{-1}[E_{FF} - E_{TQ1}]$ and E_{FF} is the Fermi energy.

EP can be written as

$$J = \frac{g_v e^2 B\alpha_0}{hd_x}\ \text{Real Part of} \sum_{n=0}^{n_{\max}}\sum_{n_{x_{\min}}}^{n_{x_{\max}}} F_{-1}(\eta_{8SL1})v_z(E_{ZQ1}) \tag{2.36}$$

where

$$v_z(E_{ZQ1}) = \frac{L_0\sqrt{1 - f_{HD1}^2(E_{ZQ1}, \lambda, \eta_g, 0)}}{\hbar f'_{HD1}(E_{ZQ1}, \lambda, \eta_g, 0)}$$

2.5.2.2 The EP from HD NW effective mass superlattices

The dispersion relation in this case is given by

$$k_x^2 = \left[\frac{1}{L_0^2}\left[Cos^{-1}\left\{f_{HD1}(E,\lambda,\eta_g,n_y,n_z)\right\}\right]^2 - G_{881}\right] \tag{2.37}$$

where

$$f_{HD1}(E,\lambda,\eta_g,n_y,n_z) = [[\bar{a}_{1HD}\,Cos[a_0\bar{C}_{1HD}(E,\eta_{g1},\lambda,n_y,n_z) + b_0\bar{D}_{1HD}(E,\eta_{g2},\lambda,n_y,n_z)]]$$
$$- [\bar{a}_{2HD}Cos[a_0\bar{C}_{1HD}(E,\eta_{g1},\lambda,n_y,n_z) - b_0\bar{D}_{1HD}(E,\eta_{g2},\lambda,n_y,n_z)]]],$$

$$\bar{C}_{1HD}(E,\eta_{g1},\lambda,n_y,n_z) = \left[\frac{2m_{c1}}{\hbar^2}T_1(E,\eta_{g1},\lambda) - G_{881}\right]^{1/2},$$

$$G_{881} = \left[\left(\frac{n_y\pi}{d_y}\right)^2 + \left(\frac{n_z\pi}{d_z}\right)^2\right]$$

and

$$\bar{D}_{1HD}(E,\eta_{g1},\lambda,n_y,n_z) = \left[\frac{2m_{c2}}{\hbar^2}T_1(E,\eta_{g2},\lambda) - G_{881}\right]^{1/2}$$

The sub-band energy E_{831} is given by

$$0 = \left[\frac{1}{L_0^2}\left[Cos^{-1}\left\{f_{HD1}(E_{831},\lambda,\eta_g,n_y,n_z)\right\}\right]^2 - G_{881}\right] \tag{2.38}$$

Equation (2.37) can be written as

$$k_x = [\Delta(E,\lambda,\eta_g,n_y,n_z)] \tag{2.39}$$

where $\Delta(E,\lambda,\eta_g,n_y,n_z) = \left[\frac{1}{L_0^2}\left[Cos^{-1}\left\{f_{HD1}(E,\lambda,\eta_g,n_y,n_z)\right\}\right]^2 - G_{881}\right]^{1/2}$

The electron concentration is given by

$$n_0 = \frac{2g_v}{\pi}\ \text{Real Part of} \sum_{n_y=1}^{n_{y\max}}\sum_{n_z=1}^{n_{z\max}}[\Delta(E_{F81},\lambda,\eta_g,n_y,n_z)]$$

where E_{F81} is the Fermi energy.

The EP can be written as

$$I_{1LHD} = \left(\frac{\alpha_0 g_v e}{\pi\hbar}\right)\ \text{Real Part of} \sum_{n_y=1}^{n_{y\max}}\sum_{n_z=1}^{n_{z\max}}(\Upsilon_{SLHD1}) \tag{2.41}$$

where $\Upsilon_{SLHD1} = [E_{F81} - (E_{831} + W - h\nu)]$

2.5.2.3 The EP from HD QB effective mass superlattices

The totally quantized energy E_{TQSL88} in this case is given by

$$\left(\frac{n_x\pi}{d_x}\right)^2 = \left[\frac{1}{L_0^2}\left[Cos^{-1}\left\{f_{HD1}(E_{TQSL88}, \lambda, \eta_g, n_y, n_z)\right\}\right]^2 - G_{881}\right] \tag{2.42}$$

The electron concentration in this case is given by

$$N_{0L} = \frac{2g_v}{d_x d_y d_z} \text{ Real Part of } \sum_{n_x=1}^{n_{x\max}} \sum_{n_y=1}^{n_{y\max}} \sum_{n_z=1}^{n_{z\max}} F_{-1}(\eta_{32HD}) \tag{2.43}$$

where $\eta_{32HD} = (k_B T)^{-1}[E_{FLEMHD} - E_{TQSL88}]$ and $E_{FQDSLEMHD}$ is the Fermi energy in this case.

EP in this case can be expressed as

$$J = \frac{\alpha_0 e g_v}{d_x d_y d_z} \text{ Real Part of } \sum_{n_{x\min}}^{n_{x\max}} \sum_{n_y=1}^{n_{y\max}} \sum_{n_z=1}^{n_{z\max}} F_{-1}(\eta_{32HD}) v_z(E_{ZQ1}) \tag{2.44}$$

2.5.2.4 The magneto-EP from HD effective mass superlattices

Equation (2.32) can be written as

$$\Delta_3(E, \eta_g, \lambda, n) = k_x \tag{2.45}$$

where $\Delta_3(E, \eta_g, \lambda, n) = [\bar{\omega}_{HD}(E, \eta_g, \lambda, n)]^{1/2}$

The Landau sub-band energy E_{33HD} in this case can be expressed as

$$\Delta_3(E_{33HD}, \eta_g, \lambda, n) = 0 \tag{2.46}$$

The electron concentration is given by

$$n_0 = \frac{g_v eB}{\pi^2 \hbar} \text{ Real Part of } \sum_{n=0}^{n_{y\max}} [\Delta_3(E_{FBSLEMHD}, \eta_g, \lambda, n)] \tag{2.47}$$

where $E_{FBSLEMHD}$ is the Fermi energy in this case.

The EP assumes the form

$$J_{ML} = \left(\frac{\alpha_0 e^2 B}{2\pi^2 \hbar^2}\right) \text{ Real Part of } \sum_{n=0}^{n_{\max}} (\eta_{701HD}) \tag{2.48}$$

where $\eta_{701HD} = [E_{FBSLEMHD} - (E_{33HD} + W - h\nu)]$.

2.5.3 Results and discussion

Using appropriate equations, the normalized EP from QW $HgTe/Hg_{1-x}Cd_xTe$ HD effective mass SL has been plotted as a function of inverse quantizing magnetic field as shown in Figure 2.39(a) whose constituent materials obey the perturbed HD three-band model of Kane in the presence of external photoexcitation. Curves (b) and (c) of the same figure have been drawn for perturbed HD two-band model of Kane and that of perturbed HD parabolic energy bands, respectively. Curves (d), (e), and (f) in the same figure exhibit the corresponding plots of QW $In_xGa_{1-x}As/InP$ effective mass HD SL. Figures 2.40–2.43 show the variations of the normalized EP from the said HD SLs as functions of normalized electron degeneracy, normalized intensity, wavelength, and thickness, respectively, for all cases of Figure 2.39. Using appropriate equations, the normalized EP from NW HD effective mass $HgTe/Hg_{1-x}Cd_xTe$ SL as a function of film thickness has been depicted in Figure 2.44(a), whose constituent materials obey the perturbed HD three-band model of Kane in the presence of external light waves.

Curves (b) and (c) of the same figure have been drawn for perturbed HD two-band model of Kane and perturbed HD parabolic energy bands, respectively. Curves (d), (e),

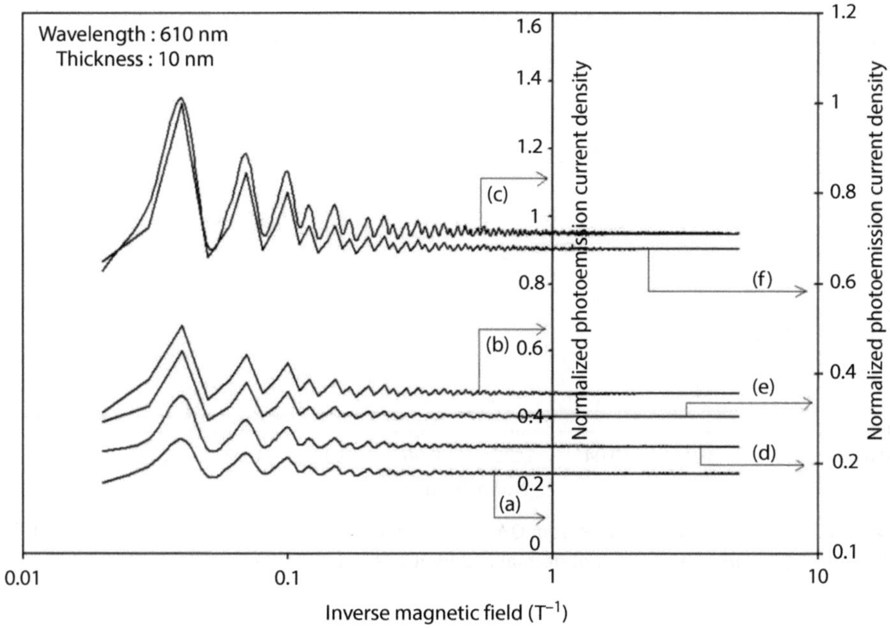

Figure 2.39: Plot of the normalized EP from HD QW effective mass superlattices of $HgTe/Hg_{1-x}Cd_xTe$ as a function of inverse magnetic field, where curves (a), (b), and (c) represent the perturbed HD three- and two-band models of Kane together with HD parabolic energy bands, respectively. Curves (d), (e), and (f) exhibit the corresponding plots of HD $In_xGa_{1-x}As/InP$ SL.

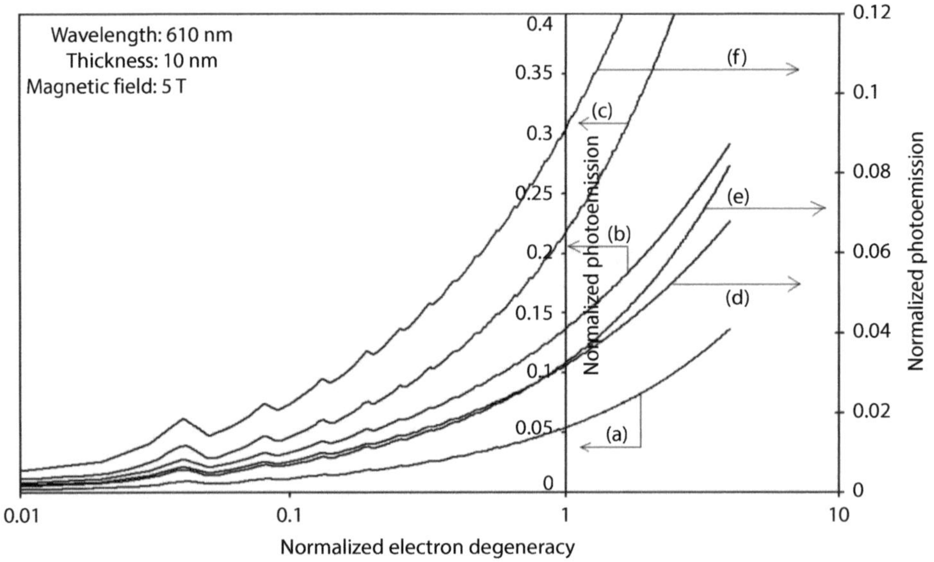

Figure 2.40: Plot of the normalized EP from QW HD effective mass superlattices of $HgTe/Hg_{1-x}Cd_xTe$ and $In_xGa_{1-x}As/InP$ as a function of normalized electron degeneracy for all cases of Figure 2.39.

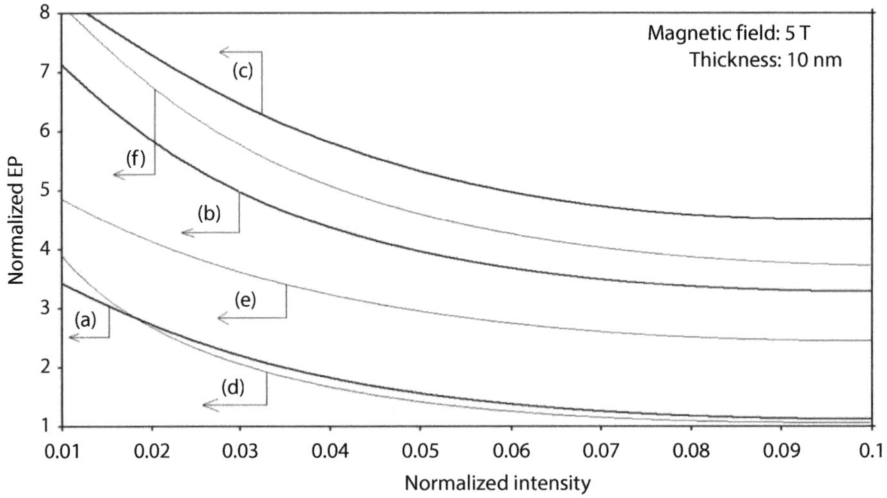

Figure 2.41: Plot of the normalized EP from QW HD effective mass superlattices of $HgTe/Hg_{1-x}Cd_xTe$ *and HD* $In_xGa_{1-x}As/InP$ as a function of normalized light intensity for all cases of Figure 2.39.

and (f) in the same figure exhibit the corresponding plots of $In_xGa_{1-x}As/InP$ NW HD effective mass SL. Figures 2.45–2.48 exhibit the plots of the normalized EP as functions of normalized carrier concentration, normalized intensity, wavelength, and normalized incident photon energy, respectively, for all cases of Figure 2.44.

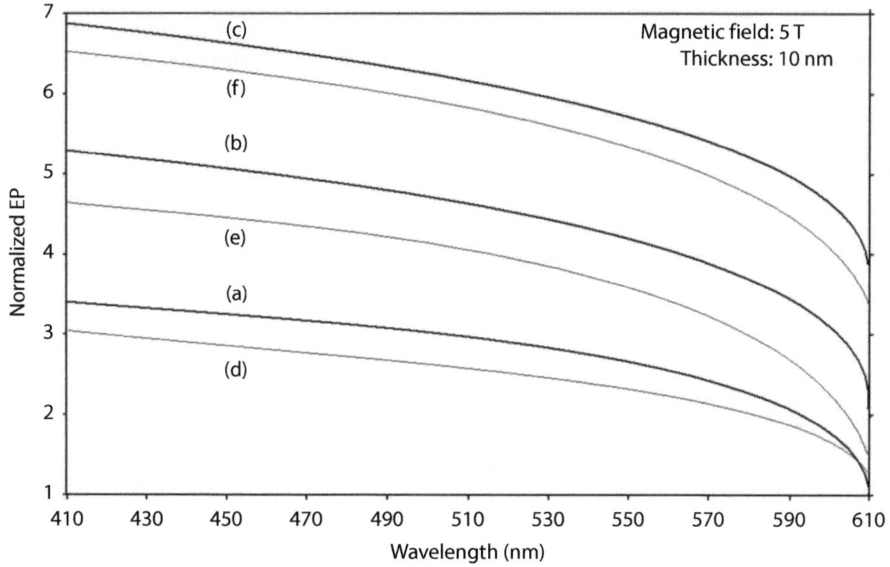

Figure 2.42: Plot of the normalized EP from QW HD effective mass superlattices of $HgTe/Hg_{1-x}Cd_xTe$ and $In_xGa_{1-x}As/InP$ as a function of light wavelength for all cases of Figure 2.39.

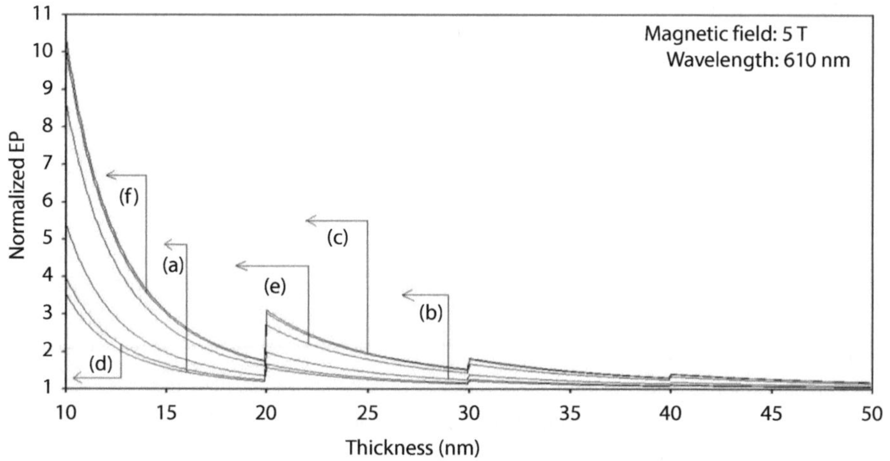

Figure 2.43: Plot of the normalized EP from QW HD effective mass superlattices of $HgTe/Hg_{1-x}Cd_xTe$ as a function of film thickness, where curves (a), (b), and (c) represent the perturbed three- and two-band models of Kane together with HD parabolic energy bands, respectively. Curves (d), (e), and (f) exhibit the corresponding plots of HD $In_xGa_{1-x}As/InP$ HDSL.

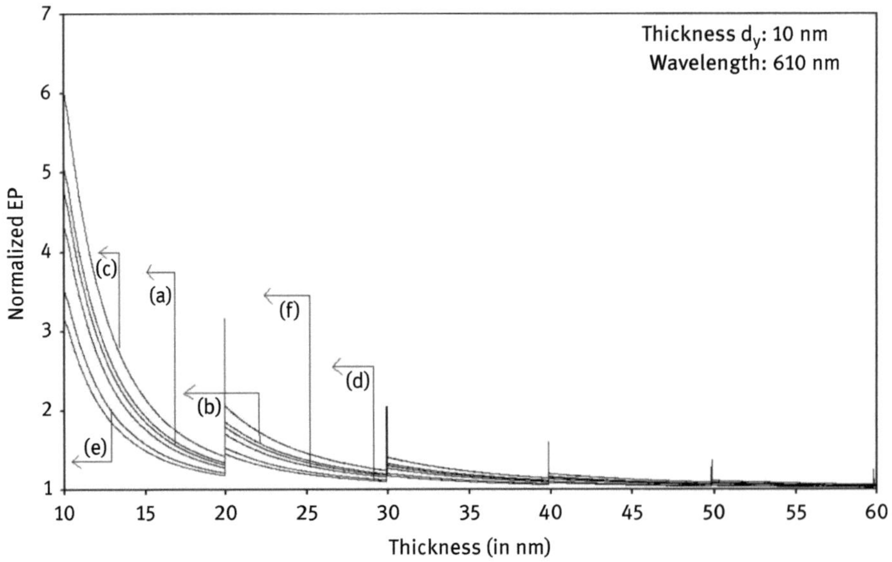

Figure 2.44: Plot of the normalized EP from NW HD effective mass superlattices of $HgTe/Hg_{1-x}Cd_xTe$ and $In_xGa_{1-x}As/InP$ as a function of film thickness for all cases of Figure 2.43.

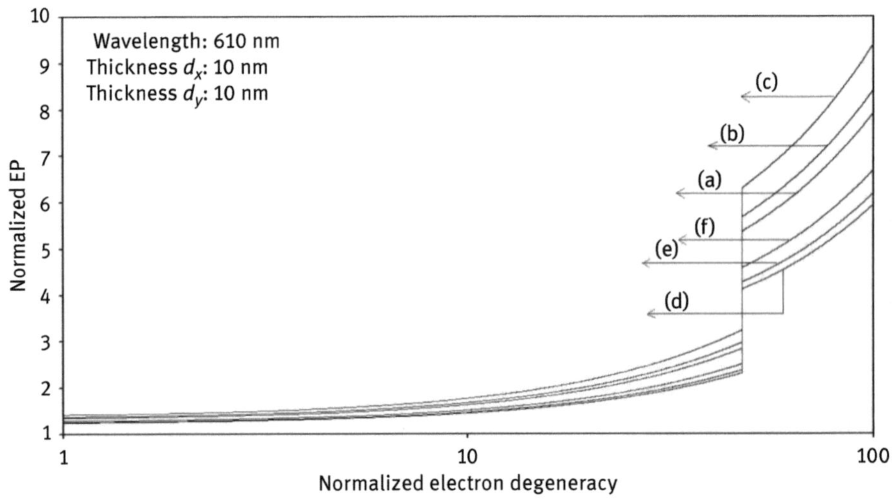

Figure 2.45: Plot of the normalized EP from NW HD effective mass superlattices of $HgTe/Hg_{1-x}Cd_xTe$ and HD $In_xGa_{1-x}As/InP$ as a function of normalized electron degeneracy for all cases of Figure 2.43.

Using appropriate equations, the normalized EP from $HgTe/Hg_{1-x}Cd_xTe$ and $In_xGa_{1-x}As/InP$ effective mass QB HD SLs, respectively, has been plotted for all types of band models as a function of film thickness as shown in Figure 2.49. Figures 2.50–2.53 exhibit the plots of normalized EP from the said QB HD SLs as

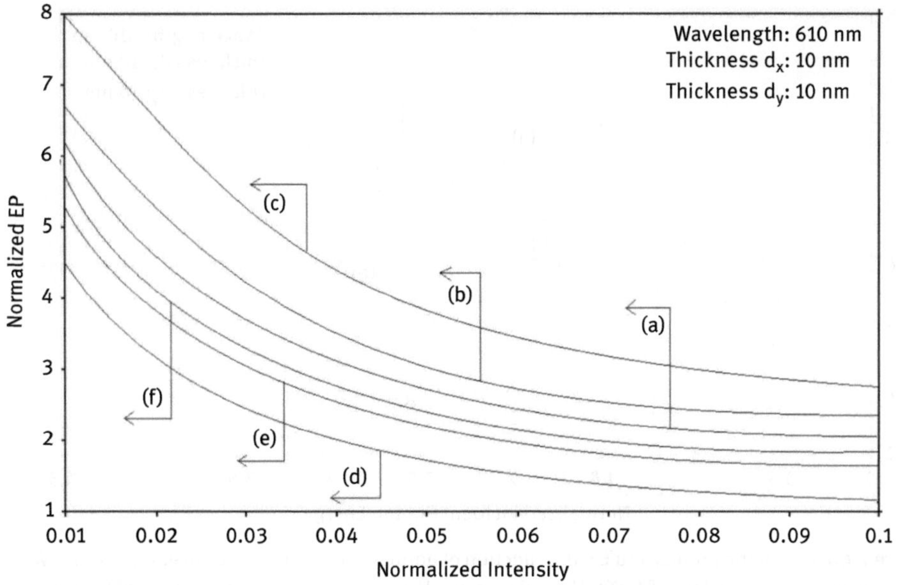

Figure 2.46: Plot of the normalized EP from NW HD effective mass superlattices of $HgTe/Hg_{1-x}Cd_xTe$ and HD $In_xGa_{1-x}As/InP$ as a function of normalized light intensity for all cases of Figure 2.43.

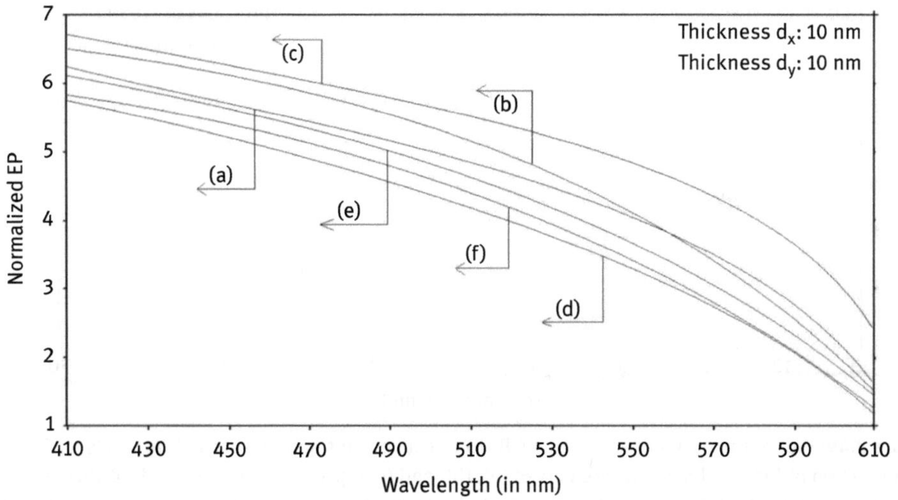

Figure 2.47: Plot of the normalized EP from NW HD effective mass superlattices of $HgTe/Hg_{1-x}Cd_xTe$ and HD $In_xGa_{1-x}As/InP$ as a function of light wavelength for all cases of Figure 2.43.

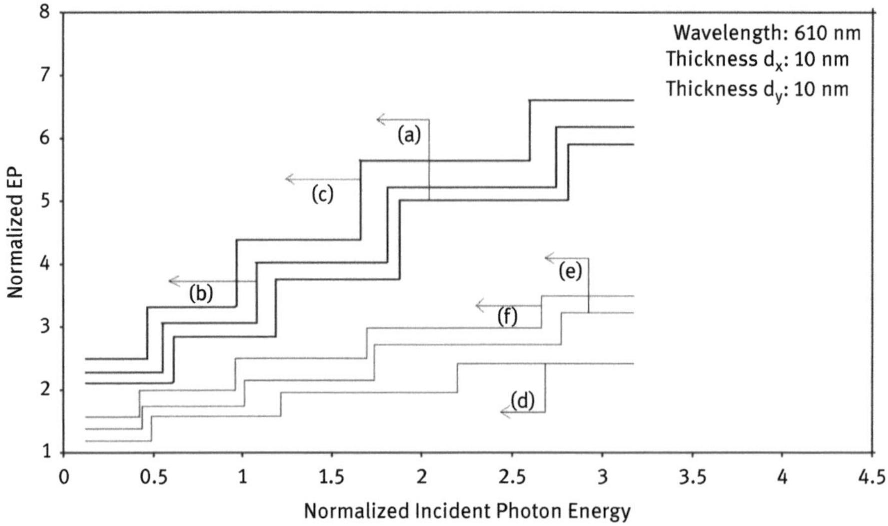

Figure 2.48: Plot of the normalized EP as a function of normalized incident photon energy from NW HD effective mass superlattices of $HgTe/Hg_{1-x}Cd_xTe$ and HD $In_xGa_{1-x}As/InP$ for all cases of Figure 2.43.

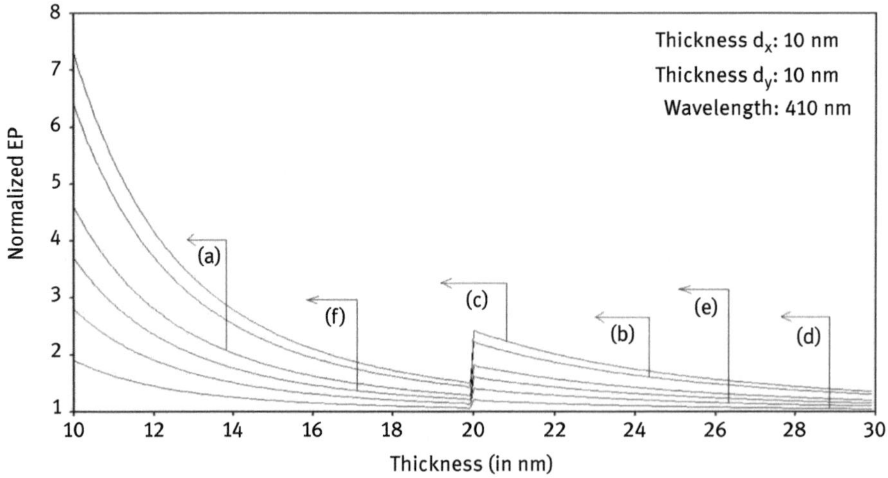

Figure 2.49: Plot of the normalized EP from QB HD effective mass superlattices of $HgTe/Hg_{1-x}Cd_xTe$ as a function of film thickness, where curves (a), (b), and (c) represent the perturbed HD three- and two-band models of Kane together with HD parabolic energy bands, respectively. Curves (d), (e), and (f) exhibit the corresponding plots of HD $In_xGa_{1-x}As/InP$.

functions of normalized electron degeneracy, normalized intensity, wavelength, and normalized incident photon energy, respectively, for all cases of Figure 2.49. Using appropriate equations, the normalized EP from effective mass $HgTe/Hg_{1-x}Cd_xTe$ HD SL under magnetic quantization has been plotted as a function of quantizing inverse magnetic field as shown in Figure 2.54(a), whose constituent HD materials obey the perturbed HD three-band model of Kane in the presence of external photoexcitation. Curves (b) and (c) of the same figure have been drawn for perturbed HD two-band model of Kane and perturbed HD parabolic energy bands, respectively. Curves (d), (e), and (f) in the same figure exhibit the corresponding plots of $In_xGa_{1-x}As/InP$ HD

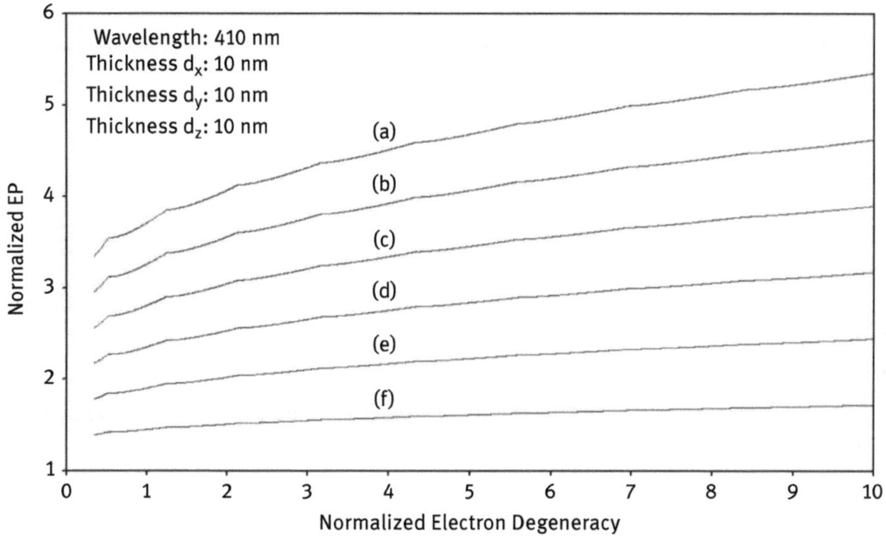

Figure 2.50: Plot of the normalized EP from QB HD effective mass superlattices of $HgTe/Hg_{1-x}Cd_xTe$ and HD $In_xGa_{1-x}As/InP$ as a function of normalized electron degeneracy for all cases of Figure 2.49.

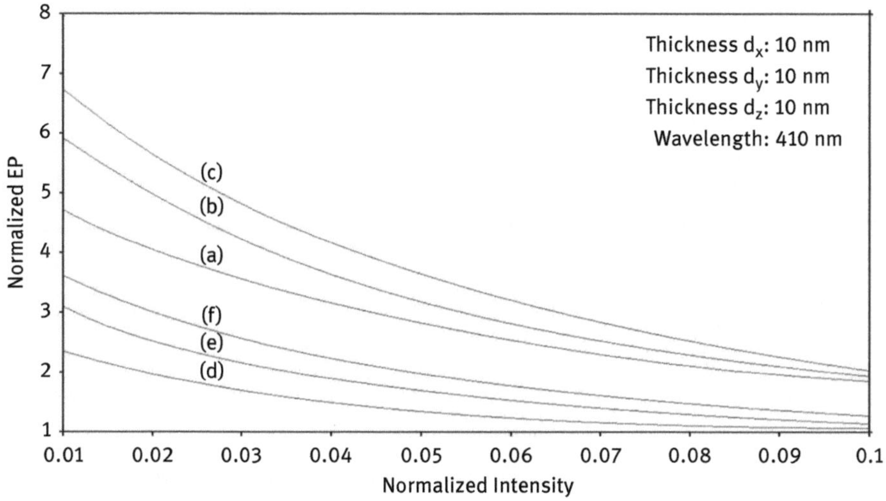

Figure 2.51: Plot of the normalized EP from QB HD effective mass superlattices of $HgTe/Hg_{1-x}Cd_xTe$ and HD $In_xGa_{1-x}As/InP$ as a function of normalized light intensity for all cases of Figure 2.49.

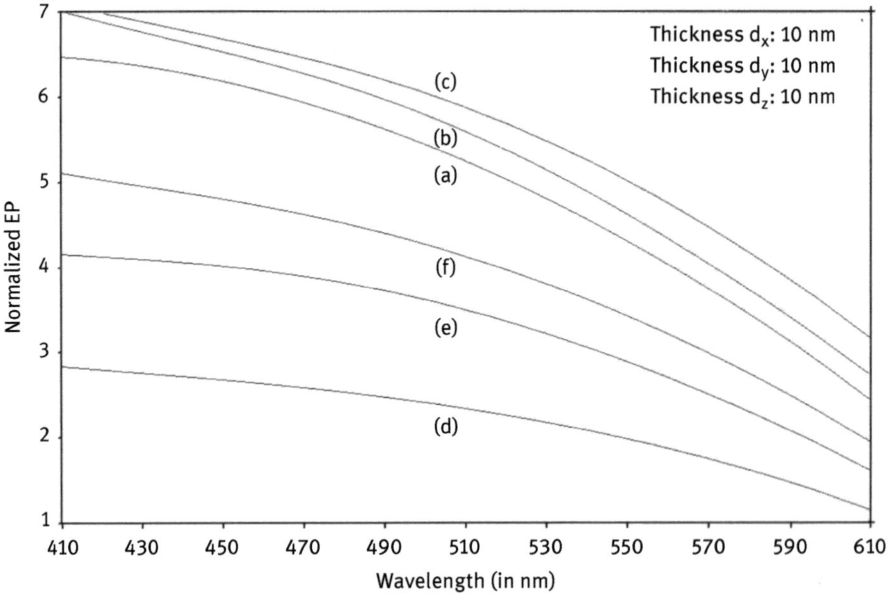

Figure 2.52: Plot of the normalized EP from QB HD effective mass superlattices of $HgTe/Hg_{1-x}Cd_xTe$ and HD $In_xGa_{1-x}As/InP$ as a function of light wavelength for all cases of Figure 2.49.

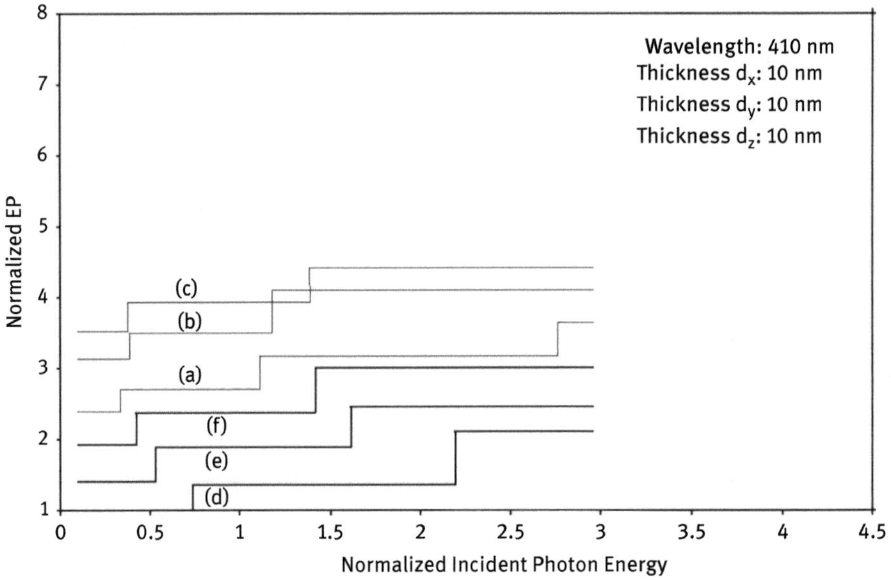

Figure 2.53: Plot of the normalized EP from QB HD effective mass superlattices of $HgTe/Hg_{1-x}Cd_xTe$ and HD $In_xGa_{1-x}As/InP$ as a function of normalized incident photon energy for all cases of Figure 2.49.

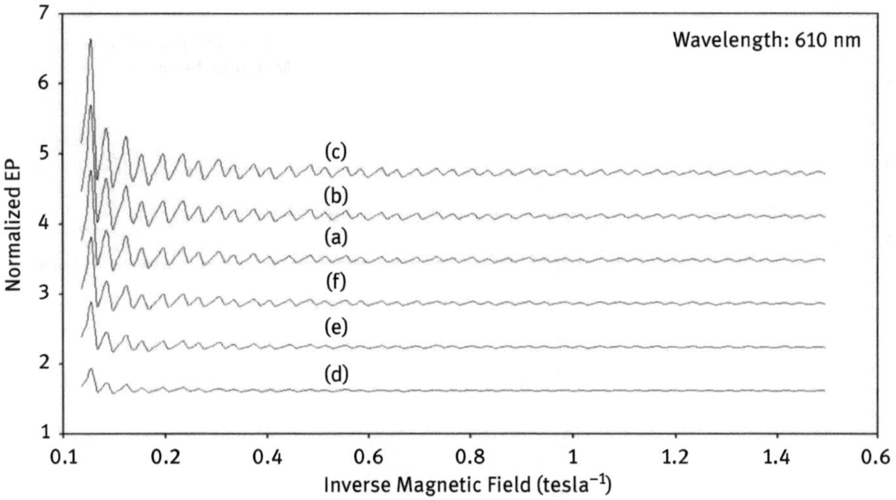

Figure 2.54: Plot of the normalized EP from HD effective mass superlattices of $HgTe/Hg_{1-x}Cd_xTe$ as a function of inverse magnetic field, where curves (a), (b), and (c) represent the perturbed three- and two-band models of Kane together with parabolic energy bands, respectively. Curves (d), (e), and (f) exhibit the corresponding plots of HD $In_xGa_{1-x}As/InP$.

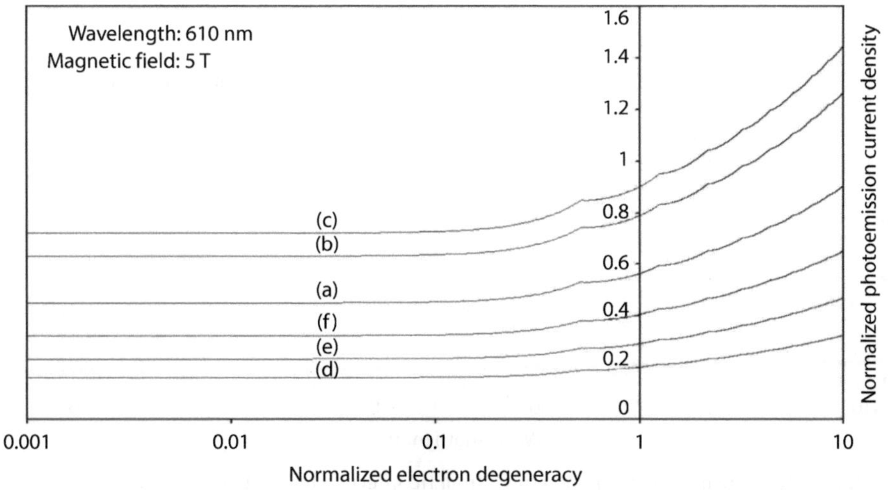

Figure 2.55: Plot of the normalized magneto-EP from HD effective mass superlattices of $HgTe/Hg_{1-x}Cd_xTe$ and HD $In_xGa_{1-x}As/InP$ as a function of normalized electron degeneracy for all cases of Figure 2.54.

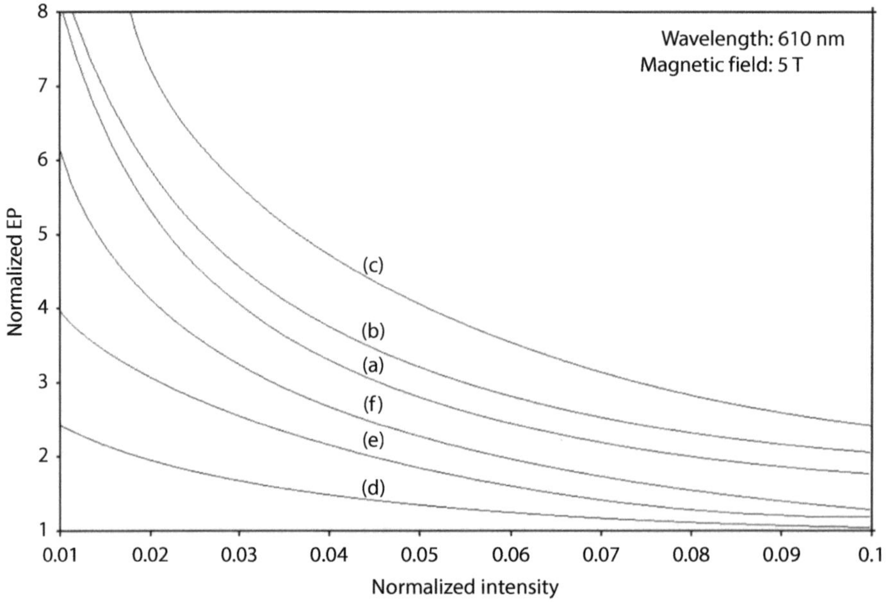

Figure 2.56: Plot of the normalized magneto-EP from HD effective mass superlattices of $HgTe/Hg_{1-x}Cd_xTe$ and HD $In_xGa_{1-x}As/InP$ as a function of normalized light intensity for all cases of Figure 2.54.

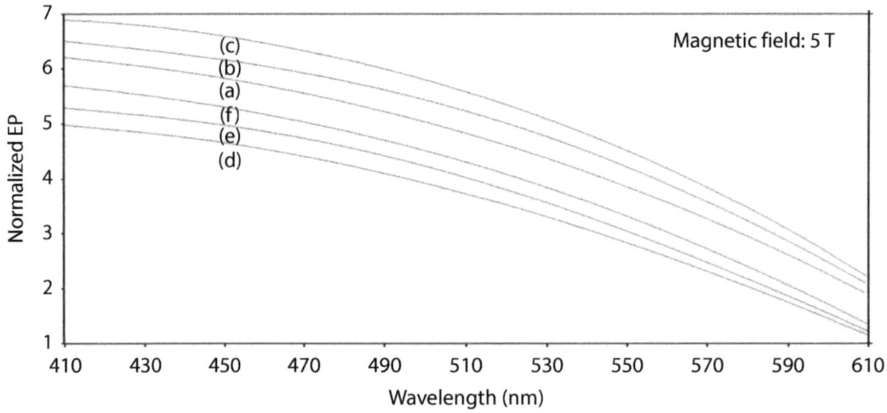

Figure 2.57: Plot of the normalized magneto-EP from HD effective mass superlattices of $HgTe/Hg_{1-x}Cd_xTe$ and HD $In_xGa_{1-x}As/InP$ as a function of light wavelength for all cases of Figure 2.54.

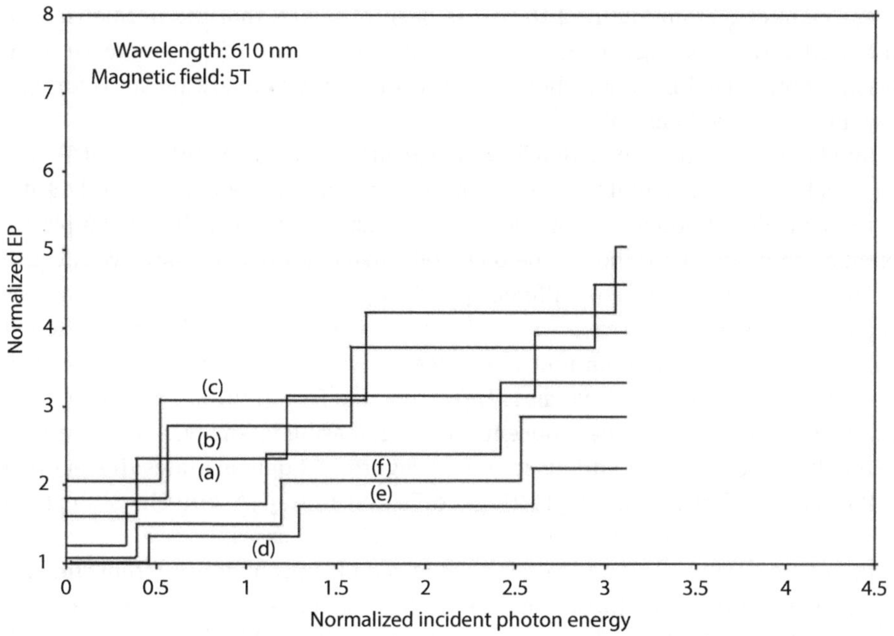

Figure 2.58: Plot of the normalized magneto-EP from HD effective mass superlattices of $HgTe/Hg_{1-x}Cd_xTe$ and HD $In_xGa_{1-x}As/InP$ as a function of normalized incident photon energy for all cases of Figure 2.54.

SL. Figures 2.55–2.58 exhibit the said variation in this case as functions of normalized electron degeneracy, normalized intensity, wavelength, and normalized incident photon energy, respectively, for all cases of Figure 2.54.

It appears from Figure 2.9 that the normalized EP from QW effective mass HgTe/$Hg_{1-x}Cd_xTe$ and $In_xGa_{1-x}As/InP$ HD SLs oscillate with the inverse quantizing magnetic field due to SdH effect, where the oscillatory amplitudes and the numerical values are determined by the respective energy band constants. Figure 2.40 show that the EP increases with increasing carrier concentration in an oscillatory way. Figures 2.41 and 2.42 show that the EP decreases with increasing intensity and wavelength in different manners. Figure 2.43 shows that the normalized EP from QW effective mass $HgTe/Hg_{1-x}Cd_xTe$ and $In_xGa_{1-x}As/InP$ HD SLs decreases with increasing film thickness in an oscillatory manner with different numerical values as specified by the energy band constants of the aforementioned HD SLs. Figure 2.44 shows that the normalized EP from NW effective mass $HgTe/Hg_{1-x}Cd_xTe$ and $In_xGa_{1-x}As/InP$ HD SLs increases with decreasing thickness and exhibit large oscillations. From Figure 2.45, it appears that normalized EP for the said system increases with increasing carrier concentration, exhibiting a quantum jump for a particular value of the said variable for all the models of both the HD SLs. From Figures 2.46 and 2.47, it can be inferred that the normalized EP in this case increases with decreasing intensity and wavelength in different manners. From Figure 2.48, it has been observed that the normalized EP from NW effective mass $HgTe/Hg_{1-x}Cd_xTe$ and $In_xGa_{1-x}As/InP$ HD SLs increases with increasing normalized incident photon energy and exhibits quantum steps for specific values of the said variable.

From Figure 2.49, it appears that EP from QB effective mass $HgTe/Hg_{1-x}Cd_xTe$ and $In_xGa_{1-x}As/InP$ HD SLs exhibit the same type of variations as given in Figures 2.43 and 2.44, respectively, although the physics of QB effective mass HD SLs is completely different as compared with the magneto-QW effective mass HD SLs and NW effective mass HD SLs, respectively. The different physical phenomena in the former one as compared with the latter two cases yield different numerical values of EP and different thicknesses for exhibiting quantum jump, respectively. From Figures 2.50–2.52, it appears that EP from QB effective mass $HgTe/Hg_{1-x}Cd_xTe$ and $In_xGa_{1-x}As/InP$ HD SLs increases with increasing carrier concentration, decreasing intensity, and decreasing wavelength, respectively, in various manners. Figure 2.53 demonstrates the fact that the EP from QB effective mass $HgTe/Hg_{1-x}Cd_xTe$ and $In_xGa_{1-x}As/InP$ HD SLs exhibit quantum steps with increasing photon energy for both the cases.

Figure 2.54 exhibits the fact that the normalized EP current density from effective mass $HgTe/Hg_{1-x}Cd_xTe$ and $In_xGa_{1-x}As/InP$ HD SLs oscillates with inverse quantizing magnetic field. Figure 2.55 exhibits the fact that the EP in this case increases with increasing carrier concentration. Figures 2.56 and 2.57 demonstrate that EP decreases with increasing intensity and wavelength in different manners. Finally, from Figure 2.58, it can be inferred that EP exhibits step functional dependence with increasing photon energy for both the HD SLs with different numerical magnitudes.

2.5.4 Open research problems

(R2.1) Investigate the EP in the presence of intense external light waves for all HD materials, whose respective dispersion relations of the carriers in the absence of any field are given in R 1.1 of Chapter 1.

(R2.2) Investigate the EP for HD semiconductors in the presences of Gaussian, exponential, Kane, Halperin, Lax, and Bonch-Burevich types of band tails for all systems whose unperturbed carrier energy spectra are defined in (R 1.1) in the presence of external light waves.

(R2.3) Investigate the EP in the presence of external light waves for bulk specimens of HD negative refractive index, organic, magnetic, and other advanced optical materials in the presence of an arbitrarily oriented alternating electric field.

(R2.4) Investigate all the appropriate HD problems of this chapter for a Dirac electron.

(R2.5) Investigate all appropriate problems of this chapter by including the many body, broadening, and hot carrier effects, respectively.

(R2.6) Investigate all appropriate problems of this chapter by removing all mathematical approximations and establishing the respective appropriate uniqueness conditions.

(R2.7) Investigate the multiphoton EP from all HD materials whose unperturbed dispersion relations are given in (R1.1) of Chapter 1 in the presence of arbitrarily oriented photoexcitation and quantizing magnetic field, respectively.

(R2.8) Investigate the multiphoton EP from all HD materials whose unperturbed dispersion relations are given in (R1.1) of Chapter 1 in the presence of an arbitrarily oriented nonquantizing nonuniform electric field, photoexcitation, and quantizing magnetic field, respectively.

(R2.9) Investigate the multiphoton EP from all HD materials whose unperturbed dispersion relations are given in (R1.1) of Chapter 1 in the presence of an arbitrarily oriented nonquantizing alternating electric field, photoexcitation, and quantizing magnetic field, respectively.

(R2.10) Investigate the multiphoton EP from all the HD materials whose unperturbed dispersion relations are given in (R1.1) of Chapter 1 in the presence of an arbitrarily oriented nonquantizing alternating electric field, photoexcitation, and quantizing alternating magnetic field, respectively.

(R2.11) Investigate the multiphoton EP from all HD materials whose unperturbed dispersion relations are given in (R1.1) of Chapter 1 in the presence of an arbitrarily oriented photoexcitation and crossed electric and quantizing magnetic fields, respectively.

(R2.12) Investigate the multiphoton EP for arbitrarily oriented photoexcitation and quantizing magnetic field from HD materials in the presence of Gaussian, exponential, Kane, Halperin, Lax, and Bonch-Bruevich types of band for all materials whose unperturbed carrier energy spectra are defined in Chapter 1.

(R2.13) Investigate the multiphoton EP for arbitrarily oriented photoexcitation and quantizing alternating magnetic field for all cases of R2.12.

(R2.14) Investigate the multiphoton EP for arbitrarily oriented photoexcitation and nonquantizing alternating electric field and quantizing magnetic field for all cases of R2.12.

(R2.15) Investigate the multiphoton EP for arbitrarily oriented photoexcitation and nonuniform alternating electric field and quantizing magnetic field for all cases of R2.12.

(R2.16) Investigate the multiphoton EP for arbitrarily oriented photoexcitation and crossed electric and quantizing magnetic fields for all cases of R2.12.

(R2.17) Investigate the multiphoton EP from HD negative refractive index, organic, magnetic, HD, disordered, and other advanced optical materials in the presence of arbitrary oriented photoexcitation and quantizing magnetic field.

(R2.18) Investigate the multiphoton EP in the presence of arbitrary oriented photoexcitation, quantizing magnetic field, and alternating nonquantizing electric field for all problems of R2.17.

(R2.19) Investigate the multiphoton EP in the presence of arbitrary oriented photoexcitation, quantizing magnetic field, and nonquantizing nonuniform electric field for all problems of R2.17.

(R2.20) Investigate the multiphoton EP in the presence of arbitrary oriented photoexcitation, alternating quantizing magnetic field, and crossed alternating nonquantizing electric field for all problems of R2.17.

(R2.21) Investigate all problems from R2.7 to R2.20 by removing all the mathematical approximations and establishing the respective appropriate uniqueness conditions.

(R2.22) Investigate the multiphoton EP from all the quantum-confined HD materials (i.e., HD multiple QWs, NWs, and QBs) whose unperturbed carrier energy spectra are defined in (R1.1) of Chapter 1 in the presence of arbitrary oriented photoexcitation and quantizing magnetic field, respectively.

(R2.23) Investigate the multiphoton EP in the presence of arbitrary oriented photoexcitation and alternating quantizing magnetic field, respectively, for all problems of R2.22.

(R2.24) Investigate the multiphoton EP in the presence of arbitrary oriented photoexcitation, alternating quantizing magnetic field, and an additional arbitrary oriented nonquantizing nonuniform electric field, respectively, for all problems of R2.22.

(R2.25) Investigate the multiphoton EP in the presence of arbitrary oriented photoexcitation, alternating quantizing magnetic field, and additional arbitrary oriented nonquantizing alternating electric field, respectively, for all problems of R2.22.

(R2.26) Investigate the multiphoton EP in the presence of arbitrary oriented photoexcitation, and crossed quantizing magnetic and electric fields, respectively, for all problems of R2.22.

(R2.27) Investigate the multiphoton EP for arbitrarily oriented photoexcitation and quantizing magnetic field from the entire quantum-confined HD materials in the presence of exponential, Kane, Halperin, Lax, and Bonch-Bruevich types of band tails for all materials whose unperturbed carrier energy spectra are defined in (R1.1) of Chapter 1.

(R2.28) Investigate the multiphoton EP for arbitrarily oriented photoexcitation and alternating quantizing magnetic field for all cases of R2.27.

(R2.29) Investigate the multiphoton EP in the presence of arbitrarily oriented photoexcitation, alternating quantizing magnetic field and an additional arbitrarily oriented nonquantizing nonuniform electric field for all cases of R2.27.

(R2.30) Investigate the multiphoton EP in the presence of arbitrary oriented photoexcitation, alternating quantizing magnetic field, and additional arbitrary oriented nonquantizing alternating electric field, respectively, for all cases of R2.27.

(R2.31) Investigate the multiphoton EP in the presence of arbitrary oriented photoexcitation, and crossed quantizing magnetic and electric fields, respectively, for all cases of R2.27.

(R2.32) Investigate the multiphoton EP for all the appropriate problems from R2.22 to R2.31 in the presence of finite potential wells.

(R2.33) Investigate the multiphoton EP for all the appropriate HD problems from R2.22 to R2.31 in the presence of parabolic potential wells.

(R2.34) Investigate the multiphoton EP for all the above appropriate HD problems for quantum rings.

(R2.35) Investigate the multiphoton EP for all the above appropriate HD problems in the presence of elliptical Hill and quantum square rings, respectively.

(R2.36) Investigate the multiphoton EP from HD nanotubes in the presence of arbitrary photoexcitation.

(R2.37) Investigate the multiphoton EP from HD nanotubes in the presence of arbitrary photoexcitation and nonquantizing alternating electric field.

(R2.38) Investigate the multiphoton EP from HD nanotubes in the presence of arbitrary photoexcitation and nonquantizing alternating magnetic field.

(R2.39) Investigate the multiphoton EP from HD nanotubes in the presence of arbitrary photoexcitation and crossed electric and quantizing magnetic fields.

(R2.40) Investigate the multiphoton EP from HD semiconductor nanotubes in the presence of arbitrary photoexcitation for all materials whose unperturbed carrier dispersion laws are defined in (R1.1) of Chapter 1.

(R2.41) Investigate the multiphoton EP from HD semiconductor nanotubes in the presence of nonquantizing alternating electric field and arbitrary photoexcitation for all materials whose unperturbed carrier dispersion laws are defined in (R1.1) of Chapter 1.

(R2.42) Investigate the multiphoton EP from HD semiconductor nanotubes in the presence of nonquantizing alternating magnetic field and arbitrary photoexcitation for all materials whose unperturbed carrier dispersion laws are defined in (R1.1) of Chapter 1.

(R2.43) Investigate the multiphoton EP from HD semiconductor nanotubes in the presence of arbitrary photoexcitation and nonuniform electric field for all materials whose unperturbed carrier dispersion laws are defined in (R1.1) of Chapter 1.

(R2.44) Investigate the multiphoton EP from HD semiconductor nanotubes in the presence of arbitrary photoexcitation and alternating quantizing magnetic fields for all materials whose unperturbed carrier dispersion laws are defined in (R1.1) of Chapter 1.

(R2.45) Investigate the multiphoton EP from HD semiconductor nanotubes in the presence of arbitrary photoexcitation and crossed electric and quantizing magnetic fields for all materials whose unperturbed carrier dispersion laws are defined in (R1.1) of Chapter 1.

(R2.46) Investigate the multiphoton EP in the presence of arbitrary photoexcitation for all the appropriate HD nipi structures of the materials whose unperturbed carrier energy spectra are defined in (R1.1) of Chapter 1.

(R2.47) Investigate the multiphoton EP in the presence of arbitrary photoexcitation for all the appropriate HD nipi structures of materials whose unperturbed carrier energy spectra are defined in (R1.1) of Chapter 1 in the presence of an arbitrarily oriented nonquantizing nonuniform additional electric field.

(R2.48) Investigate the multiphoton EP for all the appropriate HD nipi structures of materials whose unperturbed carrier energy spectra are defined in (R1.1) of Chapter 1 in the presence of an arbitrarily oriented photoexcitation and nonquantizing alternating additional magnetic field.

(R2.49) Investigate the multiphoton EP for all the appropriate HD nipi structures of materials whose unperturbed carrier energy spectra are defined in (R1.1) of Chapter 1 in the presence of an arbitrarily oriented photoexcitation and quantizing alternating additional magnetic field.

(R2.50) Investigate the multi-photon EP for all the appropriate HD nipi structures of materials whose unperturbed carrier energy spectra are defined in (R1.1) of Chapter 1 in the presence of an arbitrarily oriented photoexcitation and crossed electric and quantizing magnetic fields.

(R2.51) Investigate the multiphoton EP from HD nipi structures for all appropriate cases of all the above problems.

(R2.52) Investigate the multiphoton EP in the presence of arbitrary photoexcitation for the appropriate accumulation layers of all materials whose unperturbed carrier energy spectra are defined in (R1.1) of Chapter 1.

(R2.53) Investigate the multi-photon EP in the presence of arbitrary photoexcitation for the appropriate accumulation layers of all materials whose unperturbed carrier energy spectra are defined in (R1.1) of Chapter 1 in the presence of an arbitrarily oriented nonquantizing nonuniform additional electric field.

(R2.54) Investigate the multiphoton EP for the appropriate accumulation layers of all materials whose unperturbed carrier energy spectra are defined in (R1.1) of Chapter 1 in the presence of an arbitrarily oriented photoexcitation and nonquantizing alternating additional magnetic field.

(R2.55) Investigate the multiphoton EP for the appropriate accumulation layers of all materials whose unperturbed carrier energy spectra are defined in (R1.1) of Chapter 1 in the presence of an arbitrarily oriented photoexcitation and quantizing alternating additional magnetic field.

(R2.56) Investigate the multiphoton EP for the appropriate accumulation layers of all materials whose unperturbed carrier energy spectra are defined in (R1.1) of Chapter 1 in the presence of an arbitrarily oriented photoexcitation and crossed electric and quantizing magnetic fields by considering electron spin and broadening of Landau levels.

(R2.57) Investigate the multiphoton EP in the presence of arbitrary photoexcitation from wedge-shaped and cylindrical HD QBs of all materials whose unperturbed carrier energy spectra are defined in (R1.1) of Chapter 1.

(R2.58) Investigate the multiphoton EP in the presence of arbitrary photoexcitation from wedge-shaped and cylindrical HD QBs of all materials whose unperturbed carrier energy spectra are defined in (R1.1) of Chapter 1 in the presence of an arbitrarily oriented nonquantizing nonuniform additional electric field.

(R2.59) Investigate the multiphoton EP from wedge-shaped and cylindrical HD QBs of all materials whose unperturbed carrier energy spectra are defined in (R1.1) of Chapter 1 in the presence of an arbitrarily oriented photoexcitation and nonquantizing alternating additional magnetic field.

(R2.60) Investigate the multiphoton EP from wedge-shaped and cylindrical HD QBs of all materials whose unperturbed carrier energy spectra are defined in (R1.1) of Chapter 1 in the presence of an arbitrarily oriented photoexcitation and quantizing alternating additional magnetic field.

(R2.61) Investigate the multiphoton EP from wedge-shaped and cylindrical HD QBs of all materials whose unperturbed carrier energy spectra are defined in (R1.1) of Chapter 1 in the presence of an arbitrarily oriented photoexcitation and crossed electric and quantizing magnetic fields.

(R2.62) Investigate the multiphoton EP from wedge-shaped and cylindrical HD QBs for all appropriate cases of the above problems.

(R2.63) Investigate all problems from R2.22 to R2.62 by removing all mathematical approximations and establishing the respective appropriate uniqueness conditions.

(R2.64) Investigate the EP from quantum-confined HD III–V, II–VI, IV–VI, HgTe/CdTe effective mass superlattices together with short period, strained layer, random, Fibonacci, polytype, and sawtooth superlattices in the presence of arbitrarily oriented photoexcitation and strain.

(R2.65) Investigate the multiphoton EP in the presence of arbitrarily oriented photoexcitation and quantizing magnetic field for all cases of R2.64.

(R2.66) Investigate the multiphoton EP in the presence of arbitrarily oriented photoexcitation and nonquantizing nonuniform electric field for all cases of R2.64.

(R2.67) Investigate the multiphoton EP in the presence of arbitrarily oriented photoexcitation and nonquantizing alternating electric field for all cases of R2.64.

(R2.68) Investigate the multiphoton EP in the presence of arbitrarily oriented photoexcitation and crossed electric and quantizing magnetic fields for all cases of R2.64.

(R2.69) Investigate the multiphoton EP from HD quantum-confined superlattices for all problems of R2.64.

(R2.70) Investigate the multiphoton EP in the presence of arbitrarily oriented photoexcitation and quantizing magnetic field for all cases of R2.64.

(R2.71) Investigate the multiphoton EP in the presence of arbitrarily oriented photoexcitation and nonquantizing nonuniform electric field for all cases of R2.64.

(R2.72) Investigate the multiphoton EP in the presence of arbitrarily oriented photoexcitation and nonquantizing alternating electric field for all cases of R2.64.

(R2.73) Investigate the multiphoton EP in the presence of arbitrarily oriented photoexcitation and crossed electric and quantizing magnetic fields for all cases of R2.64.

(R2.74) Investigate the EP from quantum-confined HD III–V, II–VI, IV–VI, HgTe/CdTe superlattices with graded interfaces together with short period, strained layer, random, Fibonacci, polytype, and sawtooth superlattices in this context in the presence of arbitrarily oriented photoexcitation.

(R2.75) Investigate the multiphoton EP from HD quantum-confined superlattices for all problems of R2.74 in the presence nonuniform strain.

(R2.76) Investigate the multiphoton EP in the presence of arbitrarily oriented photoexcitation and quantizing magnetic field for all cases of R2.74.

(R2.77) Investigate the multiphoton EP in the presence of arbitrarily oriented photoexcitation and nonquantizing nonuniform electric field for all cases of R2.74.

(R2.78) Investigate the multiphoton EP in the presence of arbitrarily oriented photoexcitation and nonquantizing alternating electric field for all cases of R2.74.

(R2.79) Investigate the multiphoton EP in the presence of arbitrarily oriented photoexcitation and crossed electric and quantizing magnetic fields, respectively, for all cases of R2.74.

(R2.80) Investigate all problems from R2.64 to R2.79 by removing all mathematical approximations and establishing the respective appropriate uniqueness conditions.

References

[1] P.K. Basu, *Theory of Optical Process in Semiconductors, Bulk and Microstructures* (Oxford University Press, Oxford, England, 1997).

[2] K.P. Ghatak, S. Bhattacharya, S. Bhowmik, R. Benedictus, S. Chowdhury, J. Appl. Phys. **103**, 094314 (2008).

[3] K.P. Ghatak, S. Bhattacharya, J. Appl. Phys. 102, 073704 (2007).

[4] K.P. Ghatak, S. Bhattacharya, S.K. Biswas, A. De, A.K. Dasgupta, Phys. Scr., 75, 820 (2007).

[5] P.K. Bose, N. Paitya, S. Bhattacharya, D. De, S. Saha, K.M. Chatterjee, S. Pahari, K.P. Ghatak, Quantum Matter, **1**, 89 (2012)

[6] K. P. Ghatak, S. Bhattacharya, A. Mondal, S. Debbarma, P. Ghorai, and A. Bhattacharjee, Quantum Matter, **2**, 25 (2013).

[7] S. Bhattacharya, D. De, S. Ghosh, J. P. Bannerje, M. Mitra, B. Nag, S. Saha, S. K. Bishwas, M. Paul, Jour. Comp. Theor. Nanosci., **7**, 1066 (2010).

[8] K.P. Ghatak, S. Bhattacharya, S. Pahari, S.N. Mitra, P.K. Bose, J. Phys. Chem. Solids, **70**, 122 (2009).

[9] S. Bhattacharya, D. De, R. Sarkar, S. Pahari, A. De, A.K. Dasgupta, S. N. Biswas, K.P. Ghatak, J. Comp. Theor. Nanosci., **5**, 1345 (2008).

[10] S. Mukherjee, D. De, D. Mukherjee, S. Bhattacharya, A. Sinha, K.P. Ghatak, Physica B., **393**, 347 (2007)

[11] K. Seeger, *Semiconductor Physics*, 7th edn. (Springer-Verlag, Germany, 2006).

[12] B. R. Nag, *Physics of Quantum Well Devices* (Kluwer Academic Publishers, The Netherlands, 2000).

3 The Heisenberg's uncertainty principle and the diffusivity to mobility ratio from HD optoelectronic nanomaterials in the presence of intense light waves

I hear, I know, I see,
I remember, I do,
and I understand.

3.1 Introduction

In Section 3.2.1, we have formulated the diffusivity to mobility ratio (DMR) in III–V, ternary and quaternary heavily doped (HD) materials in the presence of intense light waves by formulating the electron statistics using Heisenberg's uncertainty principle (HUP). Section 3.2.1.1 consists of suggestions for the experimental determination of DMR. The DMR has been numerically investigated by taking HD n-InAs and n-InSb as examples of III–V compounds, HD n-$Hg_{1-x}Cd_xTe$ as an example of ternary compounds and HD n-$In_{1-x}Ga_xAs_yP_{1-y}$ lattice matched to InP as an example of quaternary compounds in accordance with the said band models for the purpose of relative assessment. Section 3.2.1.2 consists of results and discussion.

3.2 Theoretical background

3.2.1 The DMR in the presence of light waves in HD III–V, ternary and quaternary semiconductors

The DMR in HD materials can be written as follows:

$$\frac{D}{\mu} = Real\ Part\ of \left[\frac{n_0}{e}\left[\frac{\partial n_0}{\partial(E_{FHD} - E_{0HD})}\right]^{-1}\right] \tag{3.1}$$

Using eqs. (1.51e), (1.51f), (1.51g), and (3.1), we can study the DMR in this case.

For inversion layers and nipi structures, under the condition of electric quantum limit, the DMR assumes the following form:

$$\frac{D}{\mu} = Real\ Part\ of \left[\frac{n_{02D}}{e}\left[\frac{\partial n_{02D}}{\partial(E_{F2D} - E_{02D})}\right]^{-1}\right] \tag{3.2}$$

In the absence of band tails, we can write the following equation:

https://doi.org/10.1515/9783110610819-003

$$\frac{D}{\mu} = \frac{n_0}{e}\left[\frac{\partial n_0}{\partial E_F}\right]^{-1} \tag{3.3}$$

Using eqs. (1.52a), (1.52b), (1.52c), and (3.3), the DMR in this case can be expressed as where the primes denote the differentiation with respect to Fermi energy.

In the absence of band tails and photon energy, the DMR for three- and two-band models of Kane under the condition of extreme degeneracy can be expressed as follows:

$$\frac{D}{\mu} = \frac{n_0}{eg_v}\left(\frac{2m_c}{h^2}\right)^{-3/2}[[I_{11}(E_F)]^{3/2\prime}]^{-1} \tag{3.4}$$

$$\frac{D}{\mu} = \frac{n_0}{eg_v}\left(\frac{2m_c}{h^2}\right)^{-3/2}[[E_F(1+\alpha E_F)]^{3/2\prime}]^{-1} \tag{3.5}$$

At finite temperature, the DMR in accordance with two-band Kane model under the condition $\alpha E \ll 1$ can be expressed as follows:

$$\frac{D}{\mu} = \frac{k_BT}{e}\left[F_{1/2}(\eta) + \left(\frac{15ak_BT}{4}\right)F_{3/2}(\eta)\right]\left[F_{-1/2}(\eta) + \left(\frac{15ak_BT}{4}\right)F_{1/2}(\eta)\right]^{-1} \tag{3.6}$$

For relatively wide gap materials $E_g \to \infty$, we get the following equation:

$$\frac{D}{\mu} = \left(\frac{k_BT}{|e|}\right)\left[\frac{F_{1/2}(\eta)}{F_{-1/2}(\eta)}\right] \tag{3.7}$$

Eq. (3.19) was derived for the first time by Landsberg [1]

Combining eqs. (3.18) and (3.19) and using the formula $\frac{d}{d\eta}[F_j(\eta)] = F_{j-1}(\eta)$ [2–3] as easily derived from the eqs. (3.16) and (3.17) together with the fact that under the condition of extreme carrier degeneracy

$$F_{1/2}(\eta) = \left[\frac{4}{3\sqrt{\pi}}\right](\eta)^{3/2} \tag{3.8}$$

we can write

$$\frac{D}{\mu} = \frac{1}{|e|}\left(\frac{2}{3}\right)E_F\frac{(1+\alpha E_F)}{(1+2\alpha E_F)} \tag{3.9}$$

For $\alpha \to 0$, we get the following equation:

$$\frac{D}{\mu} = \frac{2E_F}{3|e|} \tag{3.10}$$

Under the condition of nondegenerate electron concentration, $\eta \ll 0$ and $F_j(\eta) \cong \exp(\eta)$ for all j [2.24]. Therefore the eqs. (2.18) and (2.19) assume the well-known forms as follows [4]

$$\frac{D}{\mu} = \frac{k_B T}{|e|} \tag{3.11}$$

3.2.1.1 Suggestion for the experimental determination of DMR

Using eqs. (1.391) and (3.1), (1.390) and (3.2), and (1.389) we can write

$$\frac{D}{\mu} = \frac{\pi^2 k_B^2 T}{3e^2 G} \tag{3.12}$$

Thus, the DMR for degenerate materials can be determined by knowing the experimental values of the thermoelectric power under strong magnetic field (G).

The suggestion for the experimental determination of the DMR for degenerate semiconductors having arbitrary dispersion laws as given in eq. (3.14) does not contain any energy band constants. For a fixed temperature, the DMR varies inversely as G. Only the experimental values of G for any material as a function of electron concentration will generate experimental values of the DMR for that range of n_0 for that system. Since G decreases with increasing n_0, from eq. (3.14), one can infer that the DMR will increase with increase in n_0. This statement is the compatibility test so far as the suggestion for the experimental determination of DMR for degenerate materials is concerned. Equation (3.14) is valid for inversion layers and nipi structures. For quantum wires and heterostructures with small charge densities, the same eq. (3.14) is valid.

Equation (3.14) is also valid under magnetic quantization and also for cross-field configuration. Thus, eq. (3.14) is independent of the dimensions of quantum confinement. We should note that the present analysis is not valid for totally k-space quantized systems such as quantum dots, magneto-inversion and accumulation layers, magneto-size quantization, magneto nipis, quantum dot superlattices, and quantum well superlattices under magnetic quantization. Under the said conditions, the electron motion is possible in the broadened levels. The experimental results of G for degenerate materials will provide an experimental check on the DMR and also a technique for probing the band structure of degenerate compounds having arbitrary dispersion laws.

In accordance with Nag and Chakravarti [5]

$$\frac{D}{\mu} = P_n |e| b \tag{3.13}$$

where P_n is the available noise power in the band width b. We wish to remark that eq. (3.13) is valid only for semiconductors having nondegenerate electron concentration, whereas the compound small gap semiconductors are degenerate in general.

3.2.1.2 Results and discussion

Using the appropriate equations, the plot of the DMR as a function of electron concentration at T = 4.2 K has been shown in Figures 3.1–3.4 by taking HD n-InAs, n-InSb, $Hg_{1-x}Cd_xTe$, and n-$In_{1-x}Ga_xAs_yP_{1-y}$ lattice matched to InP in the presence of

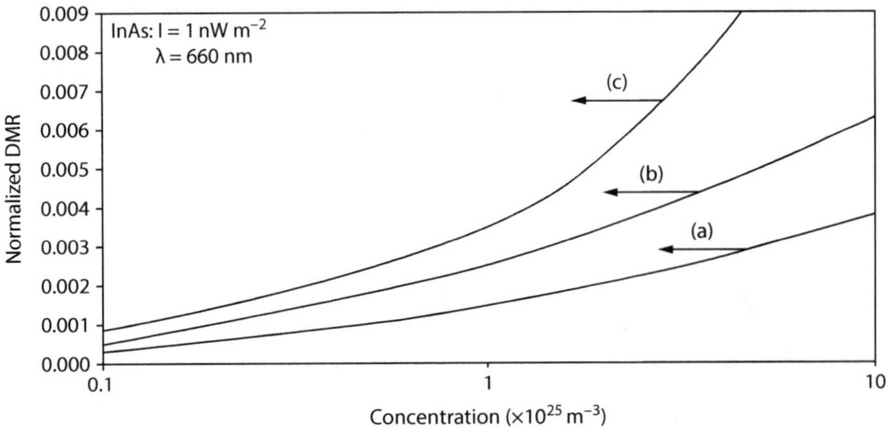

Figure 3.1: Plot of the DMR as a function of electron concentration for bulk HD n-InAs in the presence of light waves, where the curves (a), (b), and (c) represent the perturbed three- and two-band models of Kane and that of the parabolic energy bands respectively.

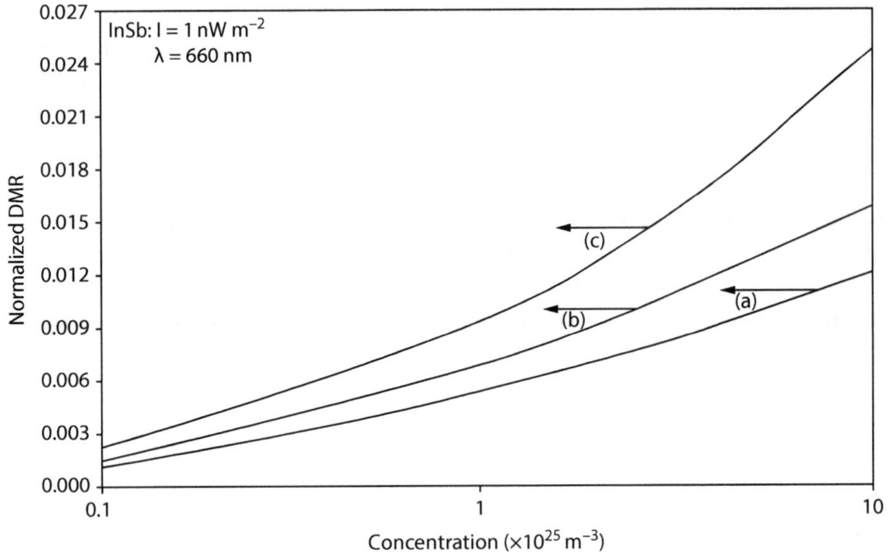

Figure 3.2: Plot of the DMR as a function of electron concentration for bulk HD n-InSb in the presence of light waves, where the curves (a), (b), and (c) represent the perturbed three- and two-band models of Kane and that of the parabolic energy bands respectively.

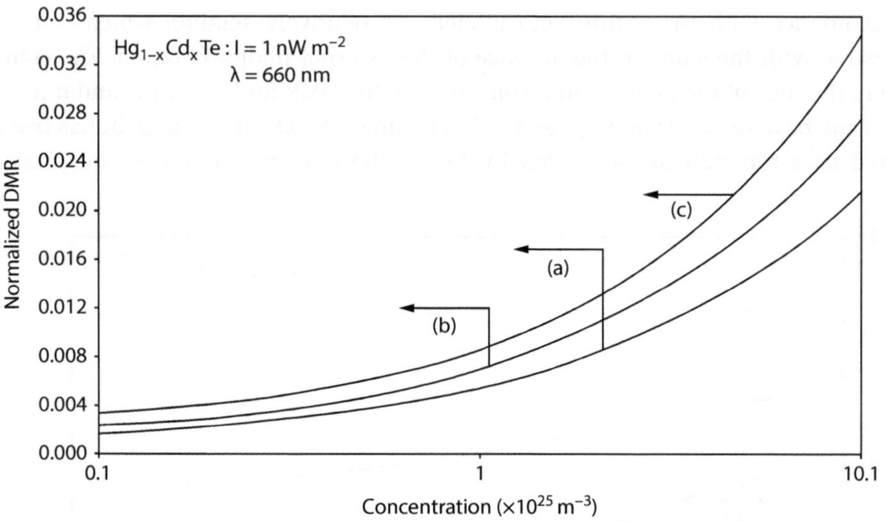

Figure 3.3: Plot of the DMR as a function of electron concentration for bulk HD n-$Hg_{1-x}Cd_xTe$ in the presence of light waves, where the curves (a), (b), and (c) represent the perturbed three- and two-band models of Kane and that of the parabolic energy bands respectively.

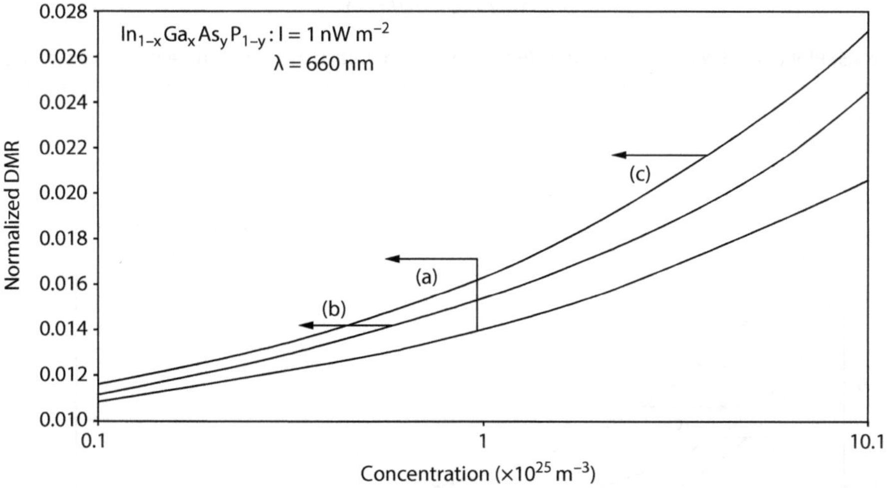

Figure 3.4: Plot of the DMR as a function of electron concentration for bulk HD n-$In_{1-x}Ga_xAs_yP_{1-y}$ lattice matched to InP in the presence of light waves, where the curves (a), (b), and (c) represent the perturbed three- and two-band models of Kane and that of the parabolic energy bands respectively.

light waves, whose unperturbed conduction electrons obey the HD three: the two-band models of Kane and that of parabolic energy bands respectively. Figures 3.1–3.4 show that the DMR increases with the increasing electron concentration and the numerical values of the DMR in the presence of light waves for all the HD materials in accordance with the entire band models are relatively smaller, which can be compared with the same in the absence of the external photoexcitation. The combined influence of the energy band constants on the DMR for HD n-InAs and n-InSb can easily be assessed from Figures 3.1–3.2. Figures 3.5–3.8 show the DMR has been plotted as a function of light intensity for all the aforementioned HD materials.

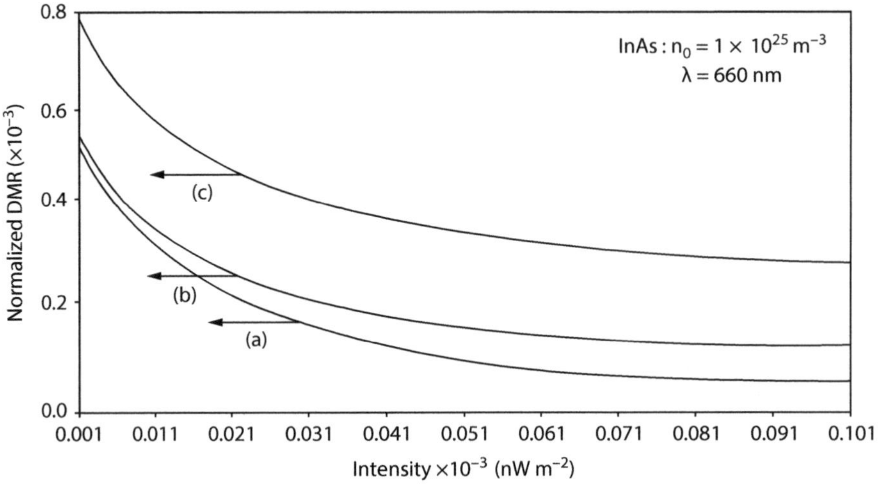

Figure 3.5: Plot of the DMR as a function of light intensity for bulk HD n-InAs, where the curves (a), (b), and (c) represent the perturbed three- and two-band models of Kane and that of parabolic energy bands respectively.

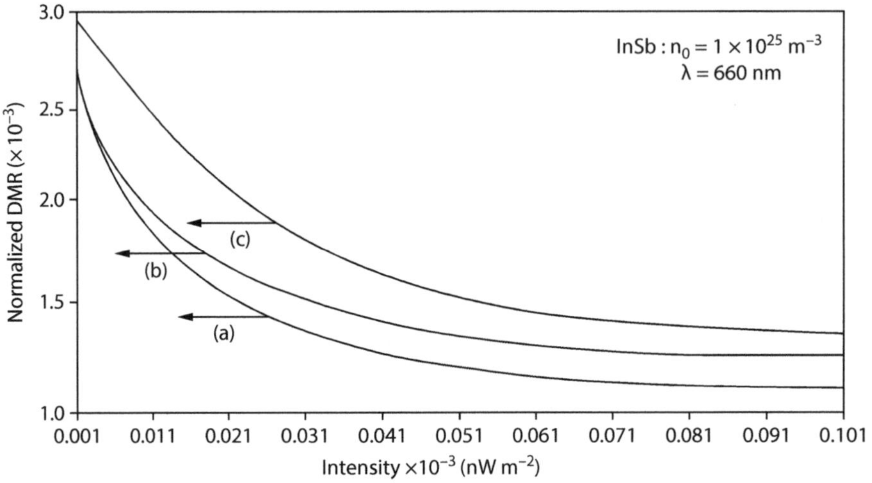

Figure 3.6: Plot of the DMR as a function of light intensity for bulk HD n-InSb, where the curves (a), (b), and (c) represent the perturbed three- and two-band models of Kane and that of parabolic energy bands respectively.

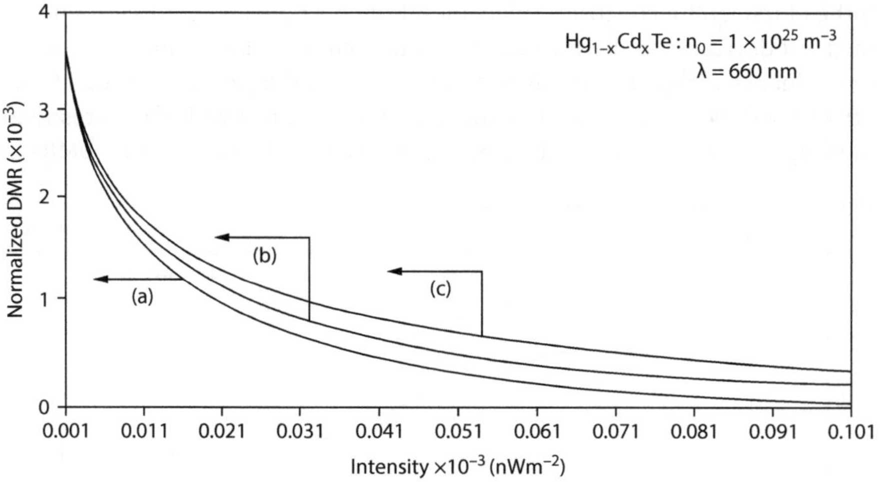

Figure 3.7: Plot of the DMR as a function of light intensity for bulk HD n-$Hg_{1-x}Cd_xTe$, where the curves (a), (b), and (c) represent the perturbed three- and two-band models of Kane and that of parabolic energy bands respectively.

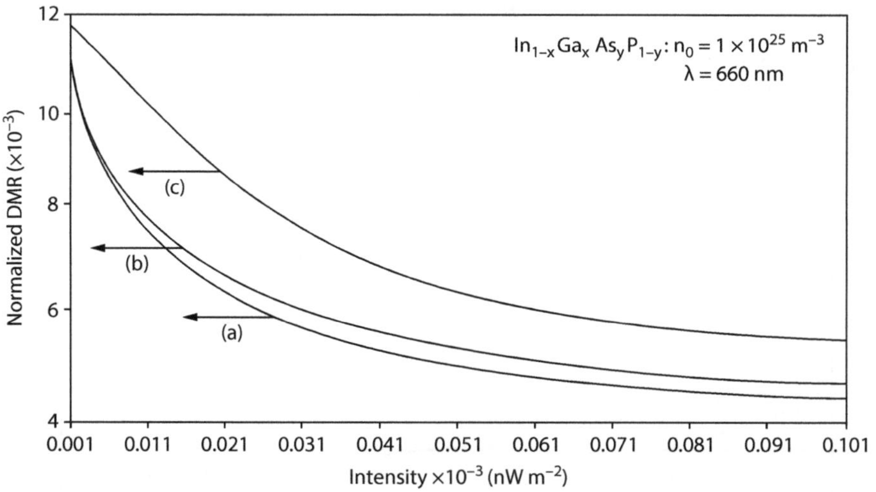

Figure 3.8: Plot of the DMR as a function of light intensity for bulk HD n-$In_{1-x}Ga_xAs_yP_{1-y}$ lattice matched to InP, where the curves (a), (b), and (c) represent the perturbed three- and two-band models of Kane and that of parabolic energy bands respectively.

Figures 3.5–3.8 show that the DMR decreases with increasing light intensity for all the materials. It should be noted from the aforementioned equations that in the absence of external photoexcitation, the DMR is independent of light intensity. Figures 3.9–3.12 show the DMR has been plotted as a function of wavelengths in the visible region for all the aforementioned HD materials for all the energy band models. It appears that the DMR decreases as the wavelength shifts from red color to violet.

The influence of light is immediately apparent from the plots in Figures 3.5–3.12 since the DMR depends strongly on I and λ in direct contrast with the corresponding bulk specimens of the said compounds. The variations of the DMRs in

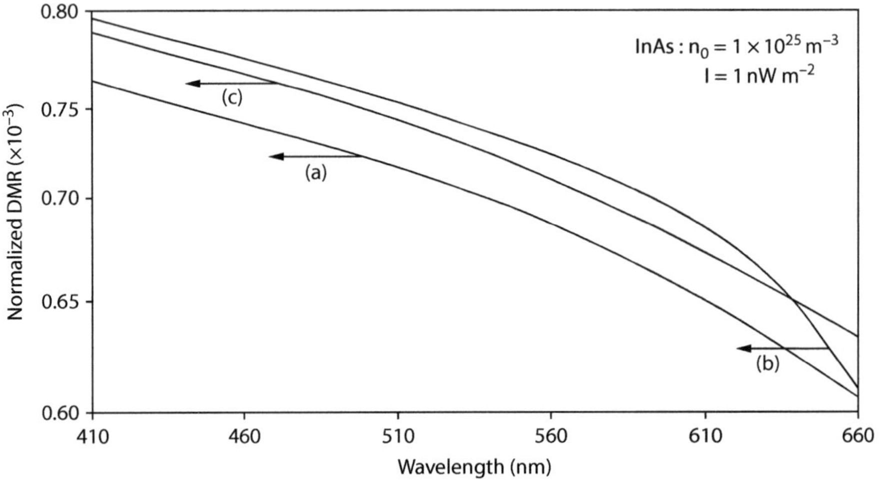

Figure 3.9: Plot of the DMR as a function of wavelength for bulk HD *n*-InAs, where the curves (a), (b), and (c) represent the perturbed three- and two-band models of Kane and that of parabolic energy bands respectively.

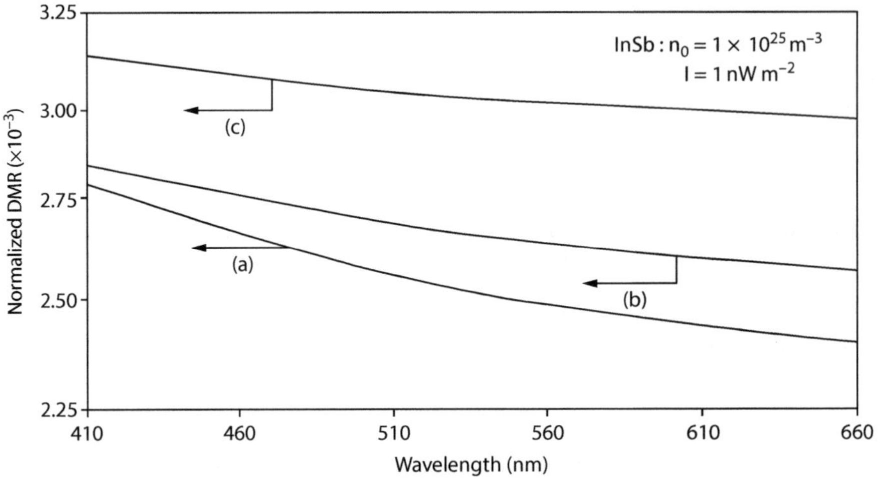

Figure 3.10: Plot of the DMR as a function of wavelength for bulk HD *n*-InSb, where the curves (a), (b), and (c) represent the perturbed three- and two-band models of Kane and that of parabolic energy bands respectively.

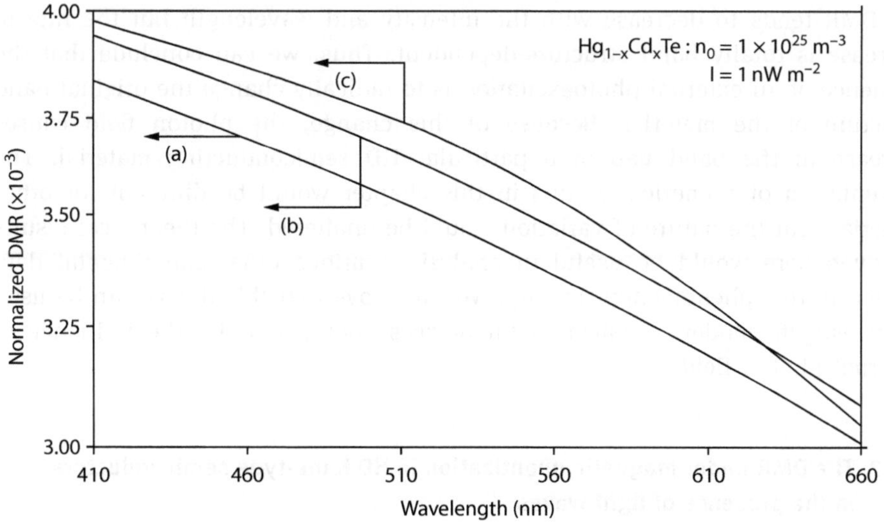

Figure 3.11: Plot of the DMR as a function of wavelength for bulk HD n-$Hg_{1-x}Cd_xTe$, where the curves (a), (b), and (c) represent the perturbed three- and two-band models of Kane and that of parabolic energy bands respectively.

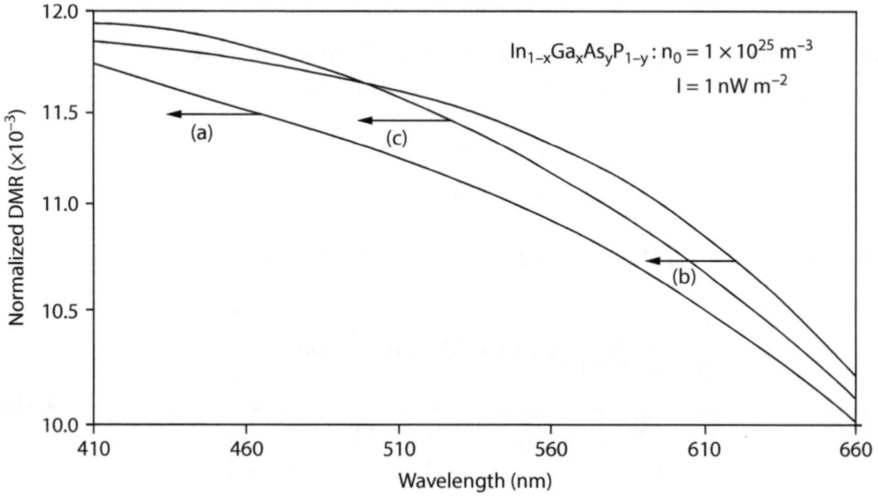

Figure 3.12: Plot of the DMR as a function of wavelength for bulk HD n-$In_{1-x}Ga_xAs_yP_{1-y}$ lattice matched to InP, where the curves (a), (b), and (c) represent the perturbed three- and two-band models of Kane and that of parabolic energy bands respectively.

Figures 3.5–3.12 reflect the direct signature of the light wave on the band structure-dependent physical properties of HD materials in the presence of light waves and the photon-assisted transport for the corresponding photonic devices, although the DMR tends to decrease with the intensity and wavelength but the rate of decrease is totally band structure-dependent. Thus, we can conclude that the influence of an external photoexcitation is to radically change the original band structure of the material. Because of this change, the photon field causes increase in the band gap of a particular HD semiconducting material. The presentation of numerical results in this chapter would be different for other materials but the nature of variation would be unaltered. The theoretical results as given here would be useful in analyzing various other experimental data related to this phenomenon. Finally, we can prove that this theory can be used to investigate modern semiconductor devices operated under the influence of external photon field.

3.2.2 The DMR under magnetic quantization in HD Kane-type semiconductors in the presence of light waves

(a) The DMR for HD materials can be written as follows:

$$\frac{D}{\mu} = \text{Real Part of} \left[\frac{n_0}{e} \left[\frac{\partial n_0}{\partial (E_{FHDLB} - E_{0HDB})} \right]^{-1} \right] \tag{3.14}$$

Using eq. (1.66a) and eq. (3.14), we can study the DMR in this case.

In the absence of band tails, we can write as follows:

$$\frac{D}{\mu} = \frac{n_0}{e} \left[\frac{\partial n_0}{\partial E_{FLB}} \right]^{-1} \tag{3.15}$$

Using eq. (3.15) and eq. (1.68), we derive the following equation:

$$\frac{D}{\mu} = \frac{1}{e} \left[\sum_{n=0}^{n_{max}} \left[\left\{ \beta_0(E_{FLB}, \lambda) - \left(n + \frac{1}{2} \right) \hbar\omega_0 \right\}^{1/2} \right] \right] \left[\sum_{n=0}^{n_{max}} \left[\left\{ \beta_0(E_{FLB}, \lambda) - \left(n + \frac{1}{2} \right) \hbar\omega_0 \right\}^{1/2} \right]' \right]^{-1} \tag{3.16}$$

In the absence of light waves and heavy doping, the DMR can be written as follows:

$$\frac{D}{\mu} = \frac{n_0}{e} \left[\frac{\partial n_0}{\partial E_{FB}} \right]^{-1} \tag{3.17}$$

Using eq. (1.73) and eq. (3.17), we get the following equation:

$$\frac{D}{\mu}=\frac{1}{e}\left[\left[\sum_{n=0}^{n_{\max}}\left[\left\{I_{11}(E_{FB})-\left(n+\frac{1}{2}\right)\hbar\omega_0\right\}^{1/2}\right]\right]^{-1}\right.$$
$$\left.\left[\left[\sum_{n=0}^{n_{\max}}\left[\left\{I_{11}(E_{FB})-\left(n+\frac{1}{2}\right)\hbar\omega_0\right\}^{1/2}\right]'\right]^{-1}\right] \tag{3.18}$$

(b) Using eqs. (3.14) and (3.19), the magneto-DMR in the absence of spin for HD III–V, ternary and quaternary semiconductors and in the presence of photoexcitation in which unperturbed conduction electrons obey the two-band Kane model can be investigated in this case.

Using eqs. (1.81) and (3.14), the magneto-DMR in the absence of spin and band tails for III–V, ternary and quaternary semiconductors and in the presence of photoexcitation, in which unperturbed conduction electrons obey the two-band model of Kane is given by the following equation:

$$\frac{D}{\mu}=\frac{1}{e}\left[\sum_{n=0}^{n_{\max}}\left[\left\{\tau_0(E_{FLB},\lambda)-\left(n+\frac{1}{2}\right)\hbar\omega_0\right\}^{1/2}\right]\right]$$
$$\left[\sum_{n=0}^{n_{\max}}\left[\left\{\tau_0(E_{FLB},\lambda)-\left(n+\frac{1}{2}\right)\hbar\omega_0\right\}^{1/2}\right]'\right]^{-1} \tag{3.19}$$

In the absence of light waves and band tails, the DMR for two-band model of Kane in the presence of magnetic quantization can be written using eqs. (1.84) and (3.17) as follows:

$$\frac{D}{\mu}=\frac{1}{e}\left[\sum_{n=0}^{n_{\max}}\left[\left\{E_{FB}(1+\alpha E_{FB})-\left(n+\frac{1}{2}\right)\hbar\omega_0\right\}^{1/2}\right]\right]$$
$$\left[\sum_{n=0}^{n_{\max}}\left[\left\{E_{FB}(1+\alpha E_{FB})-\left(n+\frac{1}{2}\right)\hbar\omega_0\right\}^{1/2}\right]'\right]^{-1} \tag{3.20}$$

(c) Using eq. (1.88) and (3.14), the magneto-DMR in the absence of spin for HD III–V, ternary and quaternary semiconductors and in the presence of photoexcitation in which unperturbed conduction electrons obey the parabolic energy bands can be investigated.

Using eqs. (3.15) and (1.90), the magneto-DMR in the absence of spin and band tails for III–V, ternary and quaternary semiconductors and in the presence of photoexcitation in which unperturbed conduction electrons obey the parabolic energy bands is given by the following equation:

$$\frac{D}{\mu}=\frac{1}{e}\left[\sum_{n=0}^{n_{\max}}\left[\left\{\rho_0(E_{FLB},\lambda)-\left(n+\frac{1}{2}\right)\hbar\omega_0\right\}^{1/2}\right]\right]\left[\sum_{n=0}^{n_{\max}}\left[\left\{\rho_0(E_{FLB},\lambda)-\left(n+\frac{1}{2}\right)\hbar\omega_0\right\}^{1/2}\right]'\right]^{-1} \tag{3.21}$$

In the absence of light waves and band tails, the DMR for isotropic parabolic energy bands can be written under magnetic quantization as follows:

$$\frac{D}{\mu}=\frac{1}{e}\left[\sum_{n=0}^{n_{\max}}\left[E_{FB}-\left(n+\frac{1}{2}\right)\hbar\omega_0\right]^{\frac{1}{2}}\right]\left[\left[\sum_{n=0}^{n_{\max}}\left[E_{FB}-\left(n+\frac{1}{2}\right)\hbar\omega_0\right]^{\frac{1}{2}}\right]'\right]^{-1} \tag{3.22}$$

Equation (3.22) is well known in the literature.

Under the condition $\alpha E_{FB} \ll 1$, the DMR at a finite temperature in this case can be expressed as follows:

$$\frac{D}{\mu}=\frac{k_BT}{e}\left[\sum_{n=0}^{n_{\max}}\frac{1}{\sqrt{a_{01}}}\left[\left(1+\frac{3}{2}\alpha b_{01}\right)F_{\frac{-1}{2}}(\bar{\eta}_{B_1})+\frac{3}{4}\alpha k_BTF_{\frac{1}{2}}(\bar{\eta}_{B1})\right]\right]\left[\sum_{n=0}^{n_{\max}}\frac{1}{\sqrt{a_{01}}}\left[\left(1+\frac{3}{2}\alpha b_{01}\right)F_{\frac{-3}{2}}(\bar{\eta}_{B_1})+\frac{3}{4}\alpha k_BTF_{\frac{-1}{2}}(\bar{\eta}_{B1})\right]\right]^{-1} \tag{3.23}$$

3.2.2.1 Results and discussion

Using the appropriate equations, the plot of the DMR as a function of inverse magnetic field in the presence of light waves at T = 4.2 K is shown in Figures 3.13–3.16 by taking n-InAs, n-InSb, $Hg_{1-x}Cd_xTe$, and n-$In_{1-x}Ga_xAs_yP_{1-y}$ lattice matched to InP respectively. Figures 3.17–3.20 exhibit the variation of the DMR as a function of electron concentration under quantizing magnetic field in the presence of light waves for the aforementioned materials. The DMR again shows the oscillatory dependence with different numerical magnitude emphasizing the influence of the energy band constants. The origin of the oscillation is same as that of Shubnikov -de Haas (SdH) oscillations and all discussions of the relevant portions of Section 3.3 are applicable in this case. Figures 3.21–3.24 show the variation of the DMR as a function of light intensity in the presence of quantizing magnetic field, whereas Figures 3.25–3.28 exhibit the same as a function of wavelength, where the variations of the wavelengths are in the zone of visible region. One can observe that the DMR decreases with increase in the light intensity and wavelengths in different ways, as appears from Figures 3.21–3.28. The nature of variations in all the cases depends strongly on the energy spectrum constants of the respective materials and the external physical conditions.

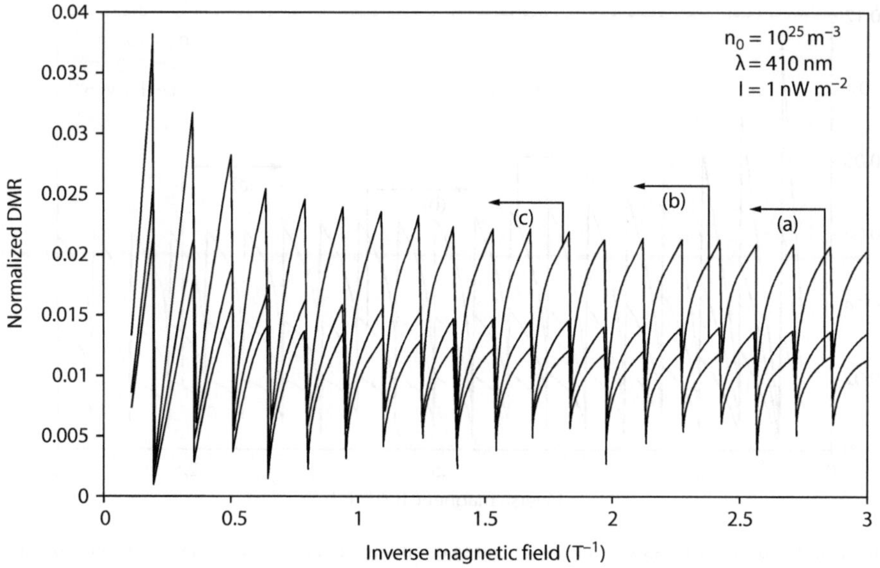

Figure 3.13: Plot of the DMR as a function of inverse quantizing magnetic field in the presence of light waves for HD *n*-InAs, where the curves (a), (b), and (c) represent the perturbed three- and two-band models of Kane and that of parabolic energy bands respectively.

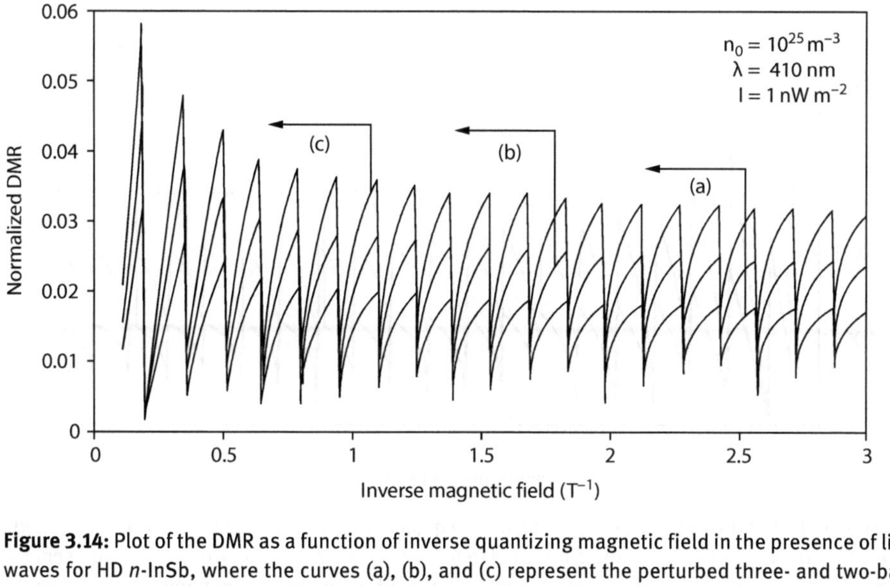

Figure 3.14: Plot of the DMR as a function of inverse quantizing magnetic field in the presence of light waves for HD *n*-InSb, where the curves (a), (b), and (c) represent the perturbed three- and two-band models of Kane and that of parabolic energy bands respectively.

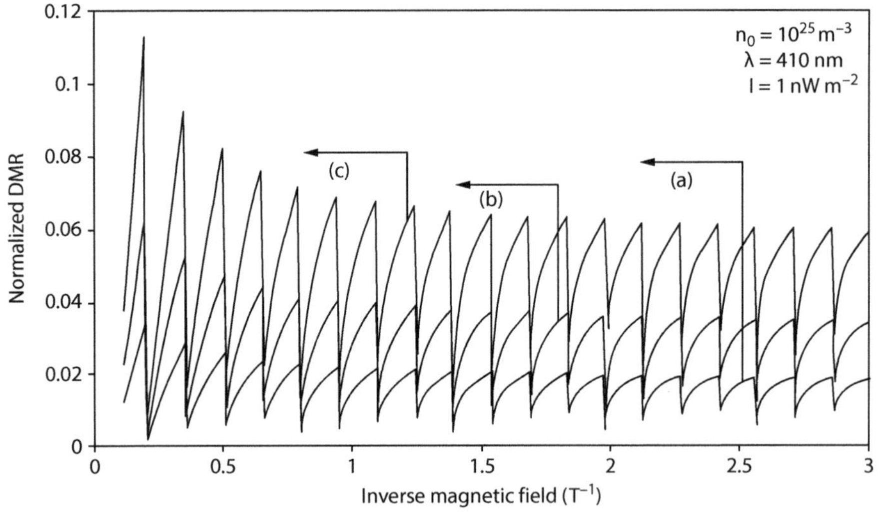

Figure 3.15: Plot of the DMR as a function of inverse quantizing magnetic field in the presence of light waves for HD n-$Hg_{1-x}Cd_xTe$, where the curves (a), (b), and (c) represent the perturbed three- and two-band models of Kane and that of parabolic energy bands respectively.

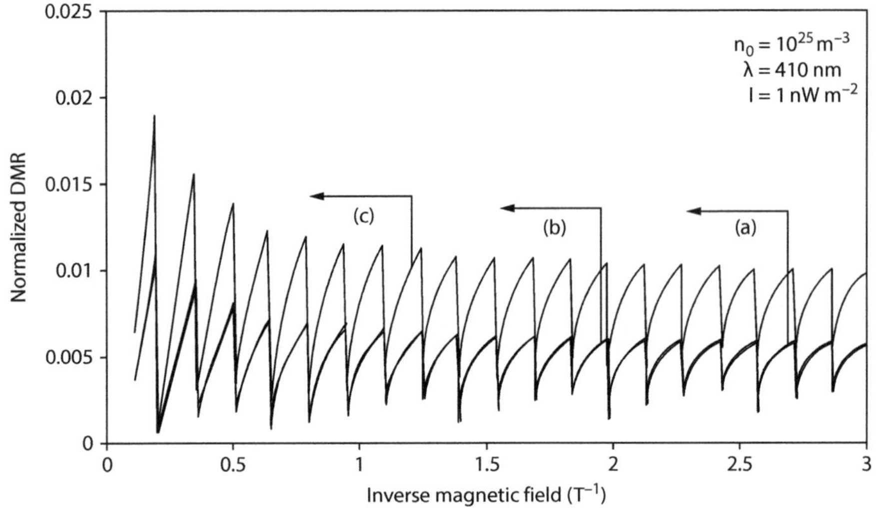

Figure 3.16: Plot of the DMR as a function of inverse quantizing magnetic field in the presence of light waves for HD n-$In_{1-x}Ga_xAs_yP_{1-y}$ lattice matched to InP, where the curves (a), (b), and (c) represent the perturbed three- and two-band models of Kane and that of parabolic energy bands respectively.

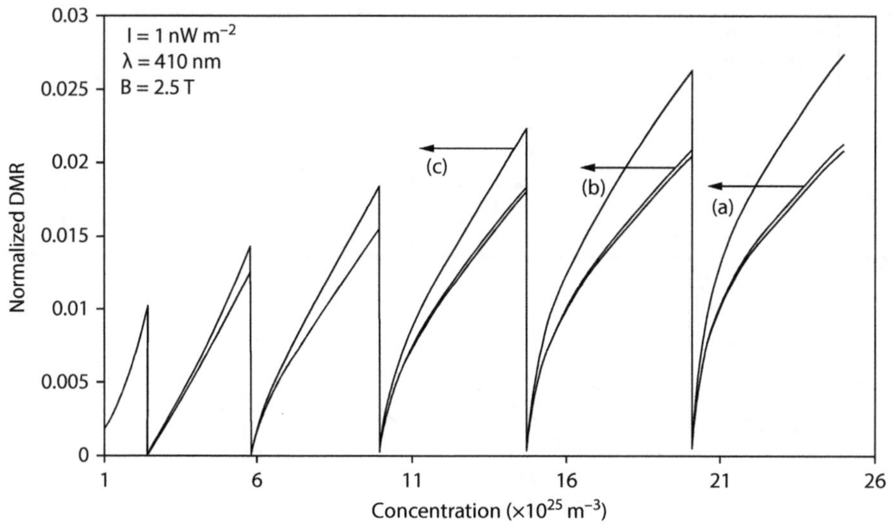

Figure 3.17: Plot of the DMR as a function of electron concentration under quantizing magnetic field in the presence of light waves for HD *n*-InAs, where the curves (a), (b), and (c) represent the perturbed three- and two-band models of Kane and that of parabolic energy bands respectively.

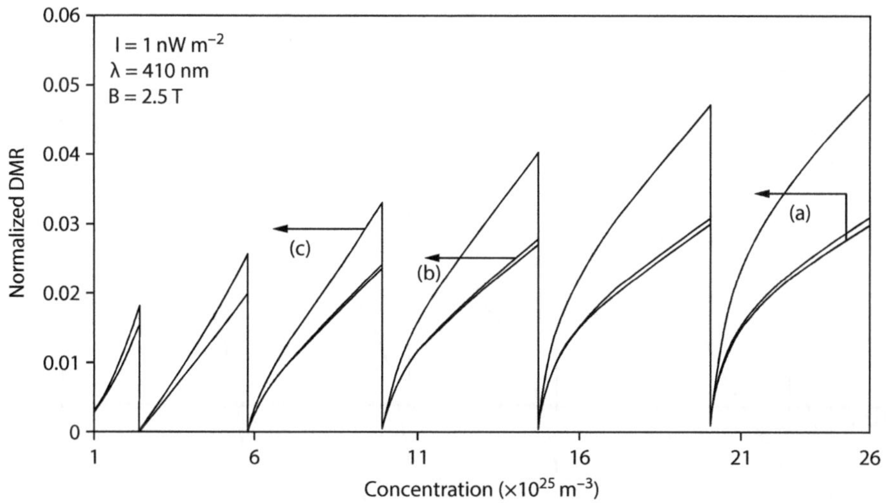

Figure 3.18: Plot of the DMR as a function of electron concentration under quantizing magnetic field in the presence of light waves for HD *n*-InSb, where the curves (a), (b), and (c) represent the perturbed three- and two-band models of Kane and that of parabolic energy bands respectively.

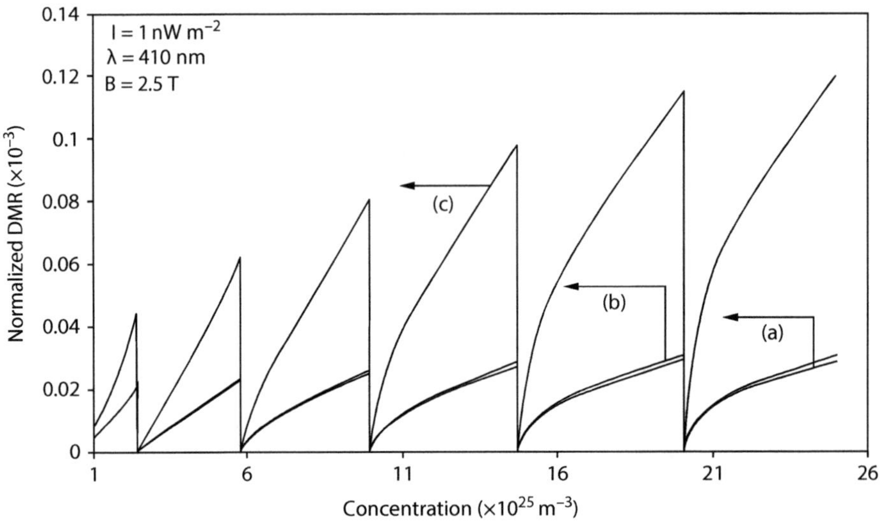

Figure 3.19: Plot of the DMR as a function of electron concentration under quantizing magnetic field in the presence of light waves for HD n-$Hg_{1-x}Cd_xTe$, where the curves (a), (b), and (c) represent the perturbed three- and two-band models of Kane and that of parabolic energy bands respectively.

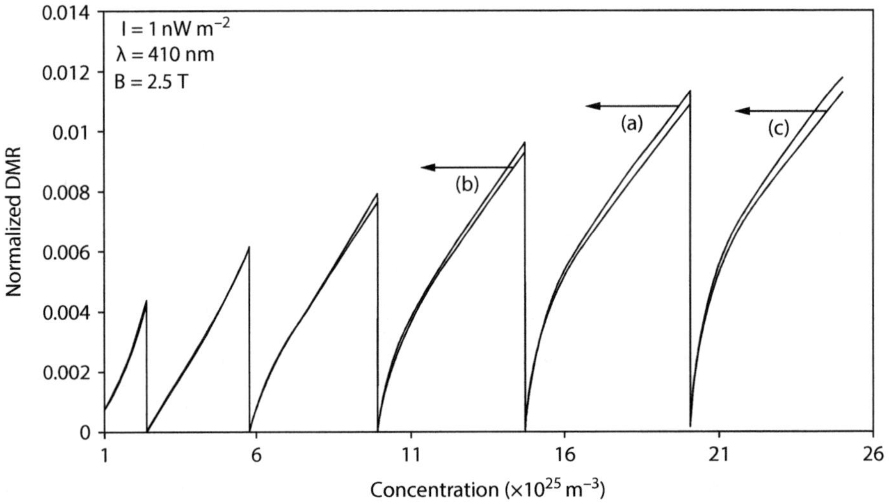

Figure 3.20: Plot of the DMR as a function of electron concentration under quantizing magnetic field in the presence of light waves for HD n-$In_{1-x}Ga_xAs_yP_{1-y}$ lattice matched to InP, where the curves (a), (b), and (c) represent the perturbed three- and two-band models of Kane and that of parabolic energy bands respectively.

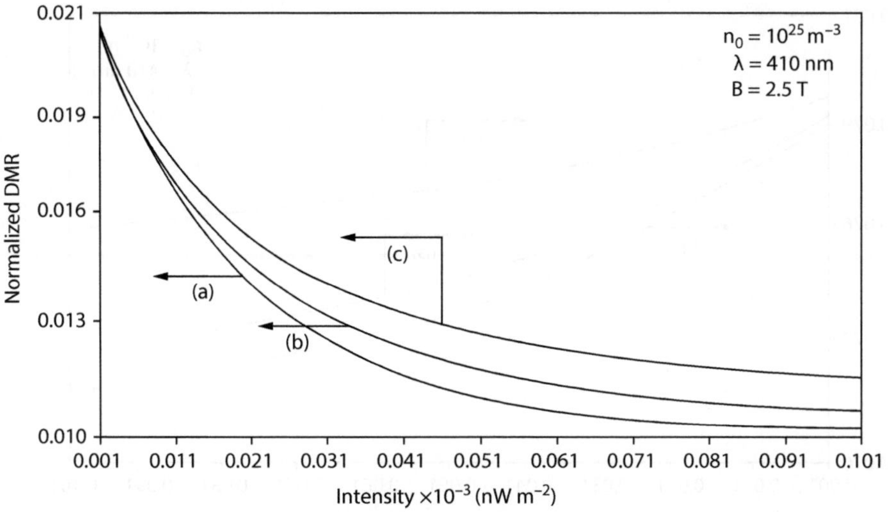

Figure 3.21: Plot of the DMR as a function of light intensity under quantizing magnetic field for HD *n*-InAs, where the curves (a), (b), and (c) represent the perturbed three- and two-band models of Kane and that of parabolic energy bands respectively.

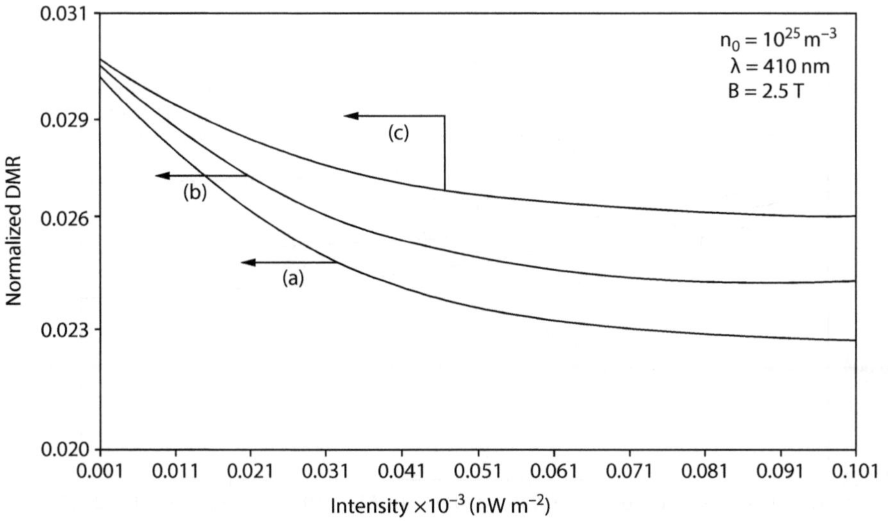

Figure 3.22: Plot of the DMR as a function of light intensity under quantizing magnetic field for HD *n*-InSb, where the curves (a), (b), and (c) represent the perturbed three- and two-band models of Kane and that of parabolic energy bands respectively.

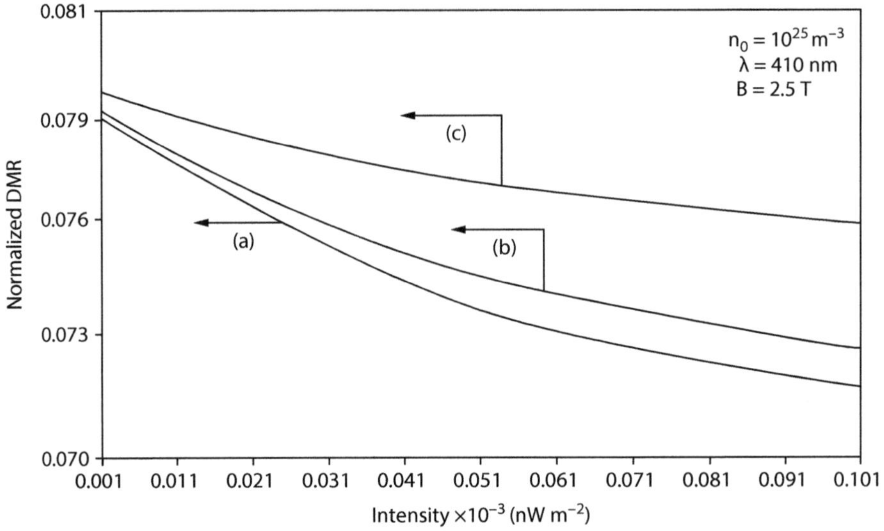

Figure 3.23: Plot of the DMR as a function of light intensity under quantizing magnetic field for HD n-$Hg_{1-x}Cd_xTe$, where the curves (a), (b), and (c) represent the perturbed three- and two-band models of Kane and that of parabolic energy bands respectively.

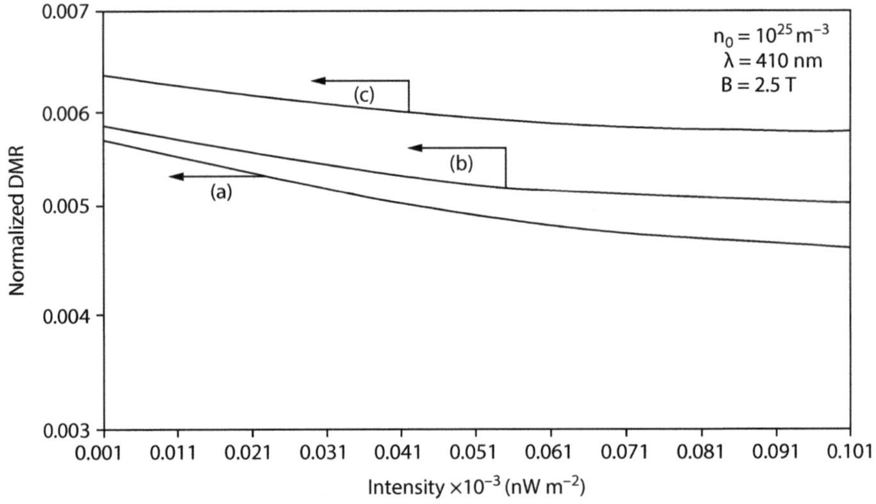

Figure 3.24: Plot of the DMR as a function of light intensity under quantizing magnetic field for HD n-$In_{1-x}Ga_xAs_yP_{1-y}$ lattice matched to InP, where the curves (a), (b), and (c) represent the perturbed three- and two-band models of Kane and that of parabolic energy bands respectively.

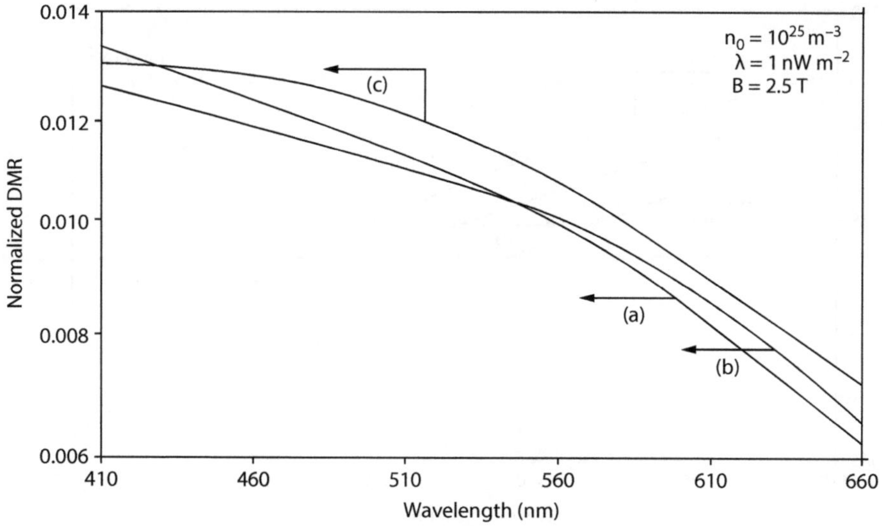

Figure 3.25: Plot of the DMR as a function of wavelength under quantizing magnetic field for HD *n*-InAs, where the curves (a), (b), and (c) represent the perturbed three- and two-band models of Kane and that of parabolic energy bands respectively.

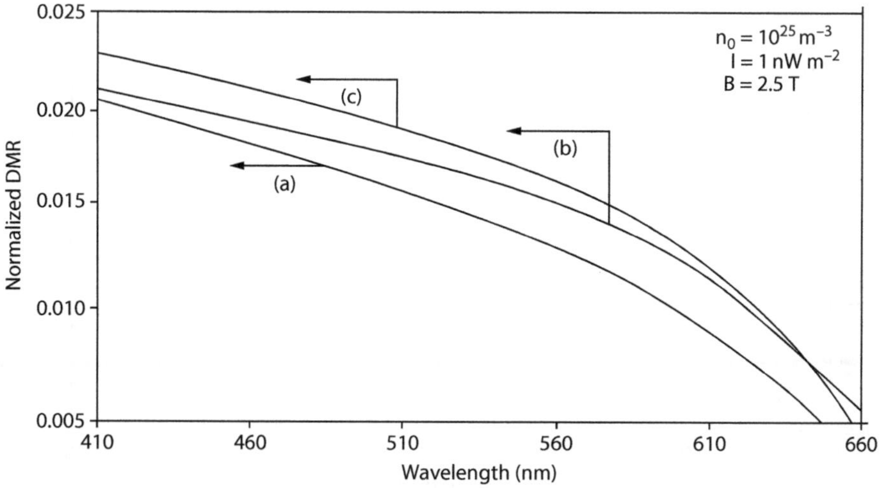

Figure 3.26: Plot of the DMR as a function of wavelength under quantizing magnetic field for HD *n*-InSb, where the curves (a), (b), and (c) represent the perturbed three- and two-band models of Kane and that of parabolic energy bands respectively.

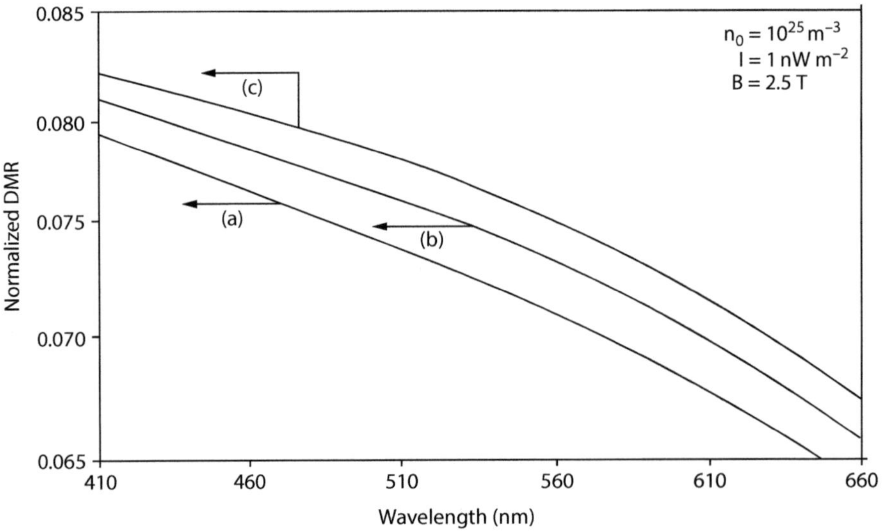

Figure 3.27: Plot of the DMR as a function of wavelength under quantizing magnetic field for HD n-$Hg_{1-x}Cd_xTe$, where the curves (a), (b), and (c) represent the perturbed three- and two-band models of Kane and that of parabolic energy bands respectively.

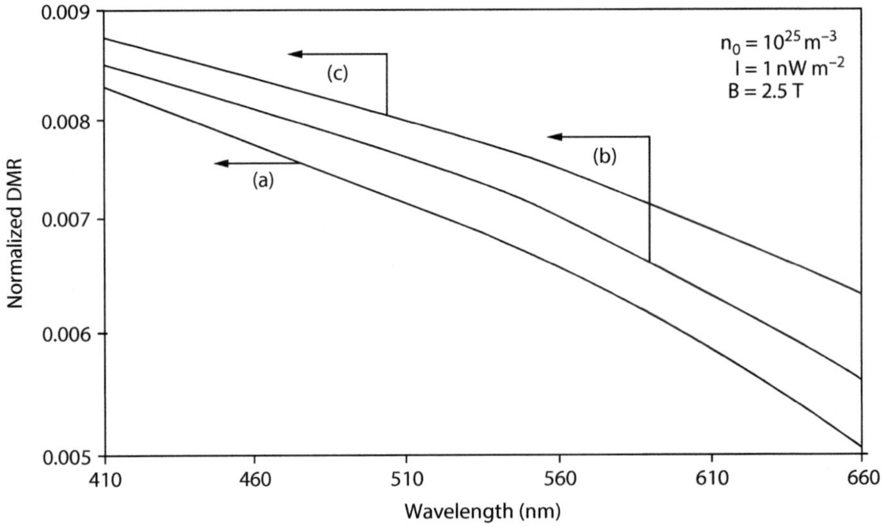

Figure 3.28: Plot of the DMR as a function of wavelength under quantizing magnetic field for HD n-$In_{1-x}Ga_xAs_yP_{1-y}$ lattice matched to InP, where the curves (a), (b), and (c) represent the perturbed three- and two-band models of Kane and that of parabolic energy bands respectively.

3.2.3 The DMR under crossed electric and quantizing magnetic fields in HD Kane-type semiconductors in the presence of light waves

(a) The DMR in this case can be written as follows:

$$\frac{D}{\mu} = \text{Real Part of } \left[\frac{n_0}{e}\left[\frac{\partial n_0}{\partial(E_{F_{BLHDC}} - E_{0HDB1})}\right]^{-1}\right] \tag{3.24}$$

where E_{0HDB1} is obtained by putting $k_z(E) = 0$ and $k_y = 0$in the corresponding dispersion relation (DR) under cross-field configuration.

Using eqs. (3.24) and (1.103), we can study the DMR in this case.

The DMR in the absence of band tails in the present case can be written as follows;

$$\frac{D}{\mu} = \frac{n_0}{e}\left[\frac{\partial n_0}{\partial E_{F_{BLC}}}\right]^{-1} \tag{3.25}$$

Using eqs. (3.25) and (1.106a) in this case, we get the following equation:

$$\frac{D}{\mu} = \frac{1}{e}\left[\sum_{n=0}^{n_{max}}\left[M_{1612}(E_{F_{BLHDC}}, n, E_0, B, \lambda)\right]\right]\left[\sum_{n=0}^{n_{max}}[M_{1612}(E_{F_{BLHDC}}, n, E_0, B, \lambda)]'\right]^{-1} \tag{3.26}$$

(b) The DMR in Kane-type materials in the presence of light waves whose energy band structure in the present case is given by eq. (1.109) can be studied by using eqs. (3.24) and (1.110) respectively.

The DMR in the present case in the absence of band tails is given by the following equation:

$$\frac{D}{\mu} = \frac{1}{e}\left[\sum_{n=0}^{n_{max}}\left[M_{1614}(E_{F_{BLC}}, n, E_0, B, \lambda)\right]\right]\left[\sum_{n=0}^{n_{max}}[M_{1614}(E_{F_{BLC}}, n, E_0, B, \lambda)]'\right]^{-1} \tag{3.27}$$

(c) The DMR in Kane-type materials in the presence of light waves whose energy band structure in the present case is given by eq. (1.116) can be studied by using eqs. (3.24) and (1.117) respectively

The electron dispersion law in the present case in the absence of band tails is given by the following equation:

$$\rho_0(E,\lambda) = \left(n + \frac{1}{2}\right)\hbar\omega_0 + \frac{[\hbar k_z(E)]^2}{2m_c} - \frac{E_0}{B}\hbar k_y\{\rho_0(E,\lambda)\}' - \left\{\frac{m_c E_0^2[\{\rho_0(E,\lambda)\}']^2}{2B^2}\right\} \tag{3.28}$$

The DMR in the present case in the absence of band tails is given by the following equation:

$$\frac{D}{\mu}=\frac{1}{e}\left[\sum_{n=0}^{n_{\max}}\left[M_{1615}(E_{F_{BLC}},n,E_0,B,\lambda)\right]\right]\left[\sum_{n=0}^{n_{\max}}\left[M_{1615}(E_{F_{BLC}},n,E_0,B,\lambda)\right]'\right]^{-1} \tag{3.29}$$

(d) In the absence of light waves and heavy doping the DMR in III–V semiconductors whose energy band structures are defined by the three-band model of Kane can be written using eqs. (1.124) and (3.24)in the presence of cross-field configuration as the following:

$$\frac{D}{\mu}=\frac{1}{e}\left[\sum_{n=0}^{n_{\max}}\left[T_{45}(n,\bar{E}_{FB})\right]\right]\left[\sum_{n=0}^{n_{\max}}\left[T_{45}(n,\bar{E}_{FB})\right]'\right]^{-1} \tag{3.30}$$

where $\bar{E}_{FB}$ is the Fermi energy in this case.

Under the condition $\Delta \gg E_g$two band Kane model the DMR in the present case can be expressed as follows:

$$\frac{D}{\mu}=\frac{1}{e}\left[\sum_{n=0}^{n_{\max}}\left[T_{45}(n,\bar{E}_{FB})\right]\right]\left[\sum_{n=0}^{n_{\max}}\left[T_{45}(n,\bar{E}_{FB})\right]'\right]^{-1} \tag{3.31}$$

For parabolic energy band$\alpha \to 0$and the DMR in this case can be expressed at a finite temperature as follows:

$$\frac{D}{\mu}=\frac{k_BT}{|e|}\left[\sum_{n=0}^{n_{\max}}\left[F_{\frac{1}{2}}(\eta_1)-F_{\frac{1}{2}}(\eta_2)\right]\right]\left[\sum_{n=0}^{n_{\max}}\left[F_{\frac{-1}{2}}(\eta_1)-F_{\frac{-1}{2}}(\eta_2)\right]\right]^{-1} \tag{3.32}$$

3.2.3.1 Results and discussion

XUsing the appropriate equations the plot of the DMR as a function of inverse magnetic field under cross-field configurations in the presence of external photo-excitation at T = 4.2 K is shown in Figures 3.29–3.32 by taking HD *n*-InAs, *n*-InSb, $Hg_{1-x}Cd_xTe$, and HD *n*-$In_{1-x}Ga_xAs_yP_{1-y}$ lattice matched to InP respectively. It appears that the DMR oscillates with the inverse quantizing magnetic field with different numerical magnitudes for all the cases. Figures 3.33–3.36, 3.37–3.40, and 3.41–3.45 exhibit the variation of the DMR in this case as functions of electron concentration, light intensity, and wavelength respectively. Figures 3.46–3.48 show the variation of the same function of electric field for all the respective cases as mentioned earlier. It appears from Figures 3.37–3.40 and 3.41–3.45 that the DMR decreases with the increase in light intensity and the wavelength which is in the visible region.

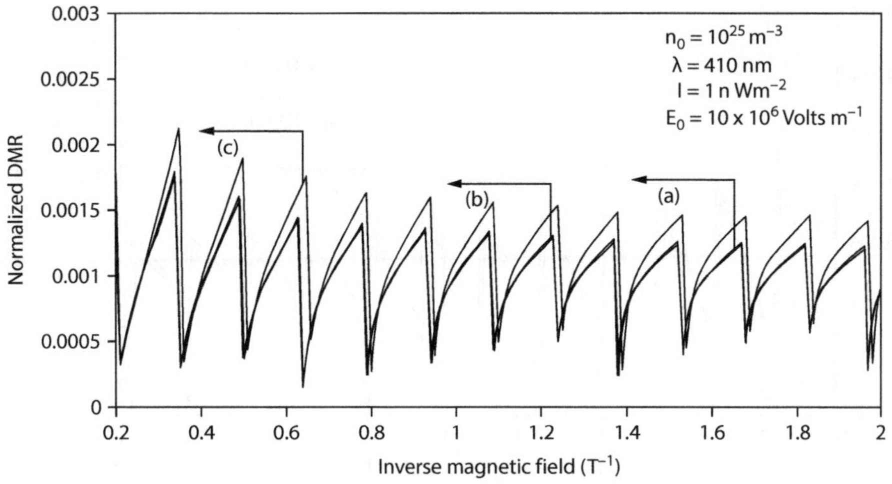

Figure 3.29: Plot of the DMR as a function of inverse quantizing magnetic field under cross-field configuration in external photoexcitation for HD *n*-InAs, where the curves (a), (b), and (c) represent the perturbed three- and two-band models of Kane and that of parabolic energy bands respectively.

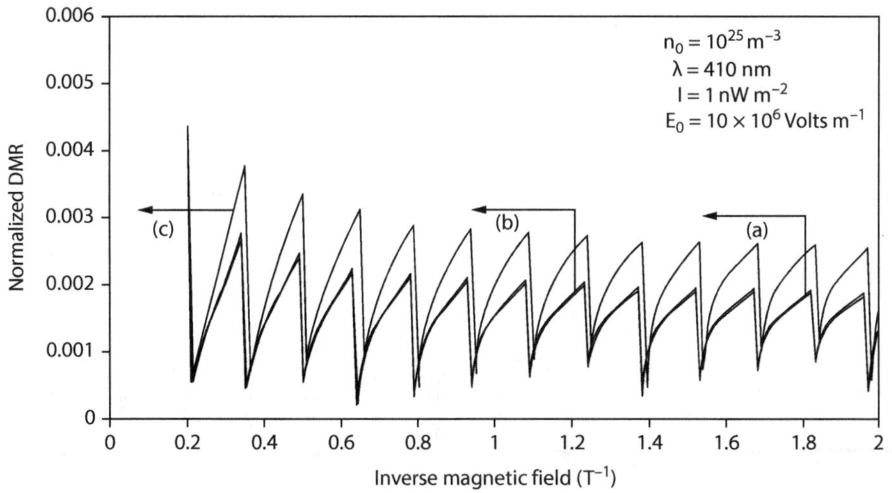

Figure 3.30: Plot of the DMR as a function of inverse quantizing magnetic field under cross-field configuration in external photoexcitation for HD *n*-InSb, where the curves (a), (b), and (c) represent the perturbed three- and two-band models of Kane and that of parabolic energy bands respectively.

From Figures 3.46–3.48, it appears that the DMR increases with the increase in the electric field. It should be noted that the rate of change of the DMR in the respective cases are totally energy spectrum dependent. The rest of the discussion of Section 4.4 is also applicable here in this case.

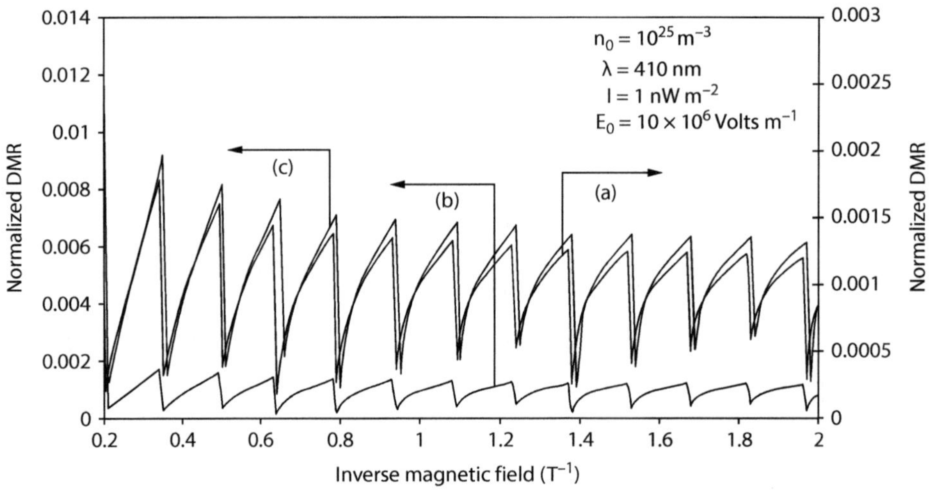

Figure 3.31: Plot of the DMR as a function of inverse quantizing magnetic field under cross-field configuration in external photoexcitation for HD n-$Hg_{1-x}Cd_xTe$, where the curves (a), (b), and (c) represent the perturbed three- and two-band models of Kane and that of parabolic energy bands respectively.

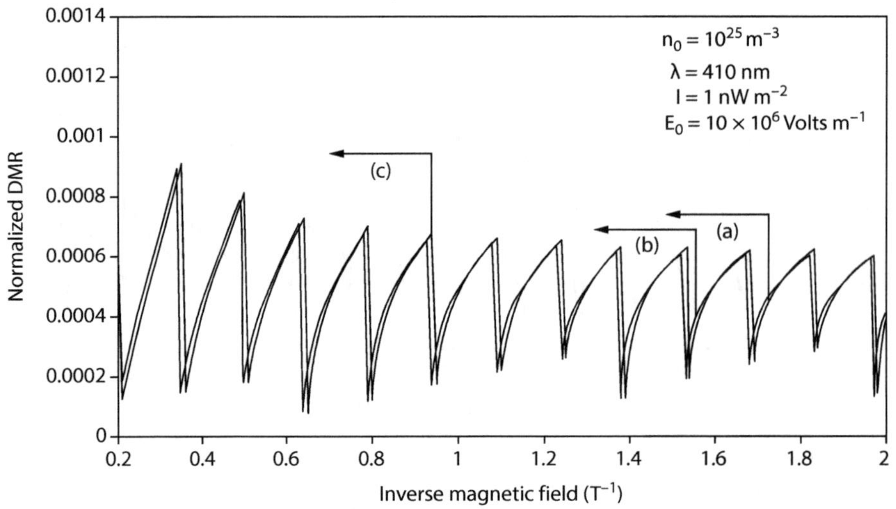

Figure 3.32: Plot of the DMR as a function of inverse quantizing magnetic field under cross-field configuration in external photoexcitation for HD n-$In_{1-x}Ga_xAs_yP_{1-y}$ lattice matched to InP, where the curves (a), (b), and (c) represent the perturbed three- and two-band models of Kane and that of parabolic energy bands respectively.

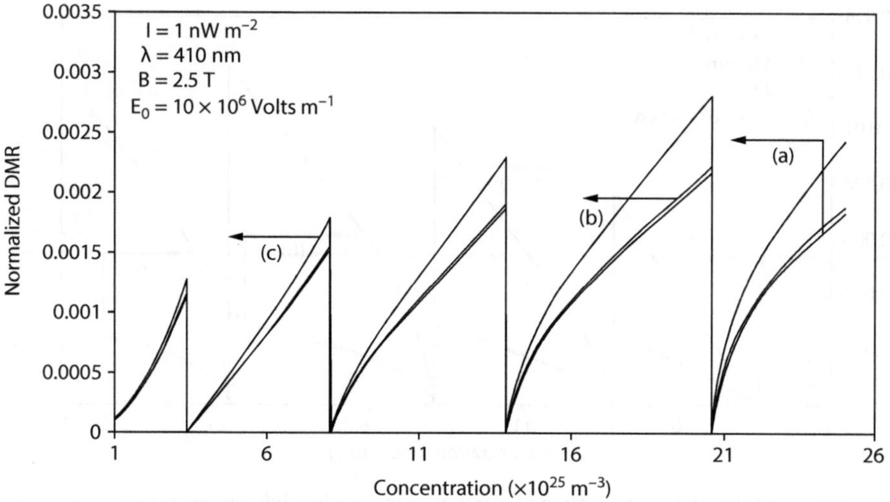

Figure 3.33: Plot of the DMR as a function of electron concentration field under cross-field configuration in external photoexcitation for HD *n*-InAs, where the curves (a), (b), and (c) represent the perturbed three- and two-band models of Kane and that of parabolic energy bands respectively.

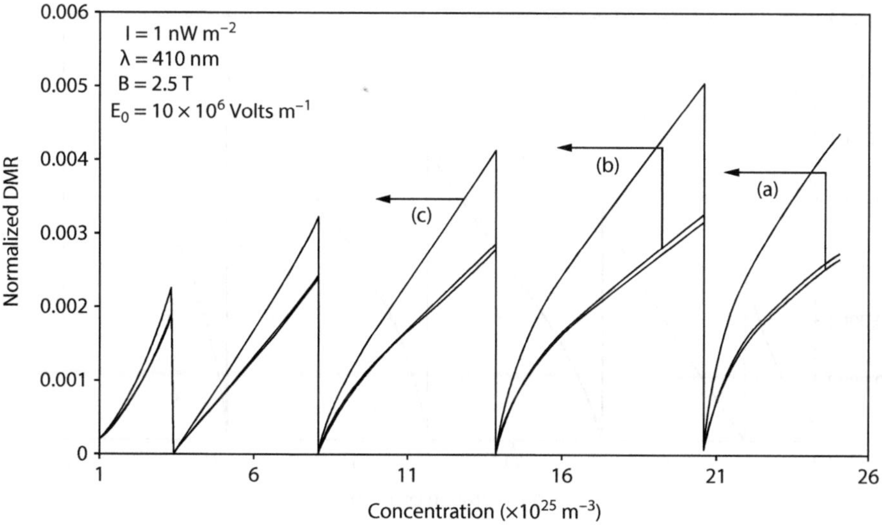

Figure 3.34: Plot of the DMR as a function of electron concentration field under cross-field configuration in external photoexcitation for HD *n*-InSb, where the curves (a), (b), and (c) represent the perturbed three- and two-band models of Kane and that of parabolic energy bands respectively.

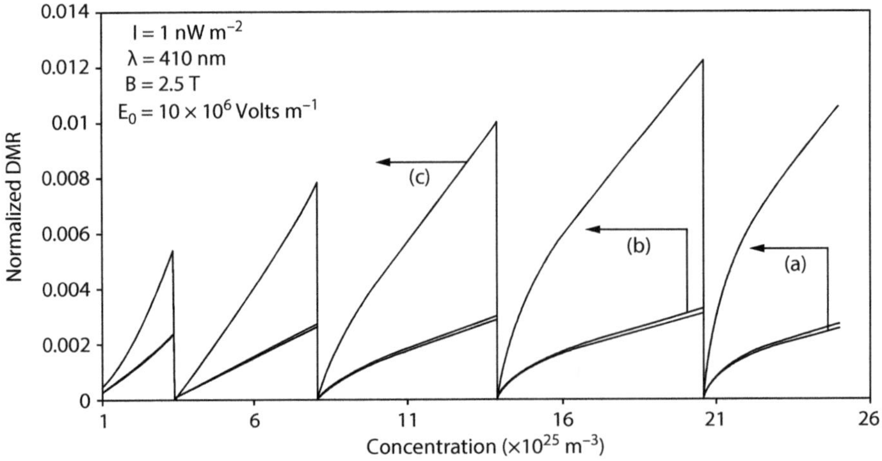

Figure 3.35: Plot of the DMR as a function of electron concentration field under cross-field configuration in external photoexcitation for HD n-$Hg_{1-x}Cd_xTe$, where the curves (a), (b), and (c) represent the perturbed three- and two-band models of Kane and that of parabolic energy bands respectively.

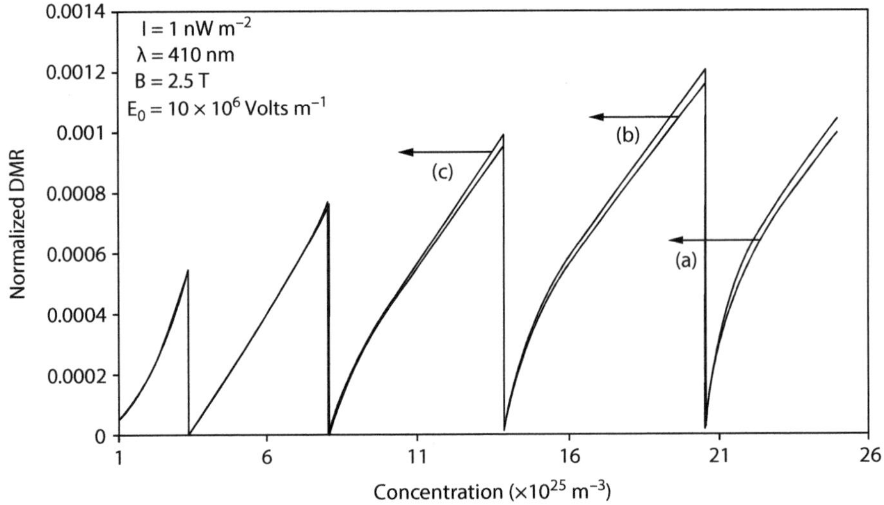

Figure 3.36: Plot of the DMR as a function of electron concentration field under cross-field configuration in external photoexcitation for HD n-$In_{1-x}Ga_xAs_yP_{1-y}$ lattice matched to InP, where the curves (a), (b), and (c) represent the perturbed three- and two-band models of Kane and that of parabolic energy bands respectively.

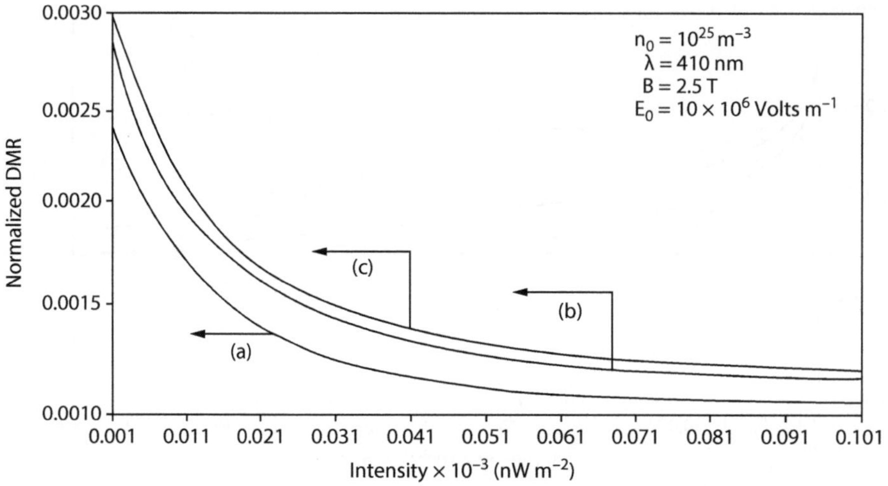

Figure 3.37: Plot of the DMR as a function of light intensity under cross-field configuration in external photoexcitation for HD *n*-InAs, where the curves (a), (b), and (c) represent the perturbed three- and two-band models of Kane and that of parabolic energy bands respectively.

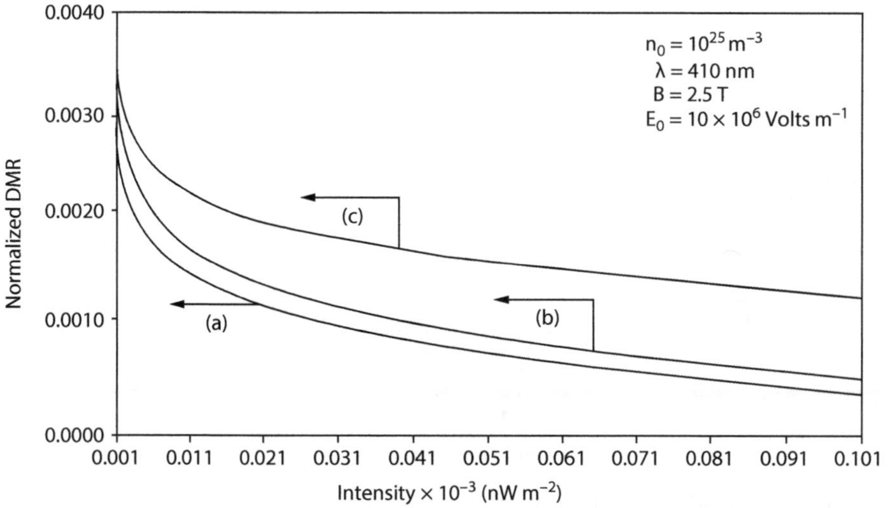

Figure 3.38: Plot of the DMR as a function of light intensity under cross-field configuration in external photoexcitation for HD *n*-InSb, where the curves (a), (b), and (c) represent the perturbed three- and two-band models of Kane and that of parabolic energy bands respectively.

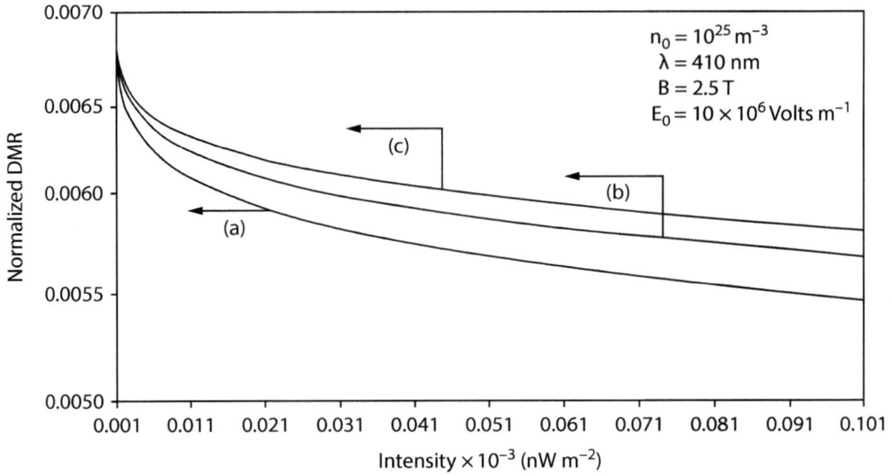

Figure 3.39: Plot of the DMR as a function of light intensity under cross-field configuration in external photoexcitation for HD n-$Hg_{1-x}Cd_xTe$, where the curves (a), (b), and (c) represent the perturbed three- and two-band models of Kane and that of parabolic energy bands respectively.

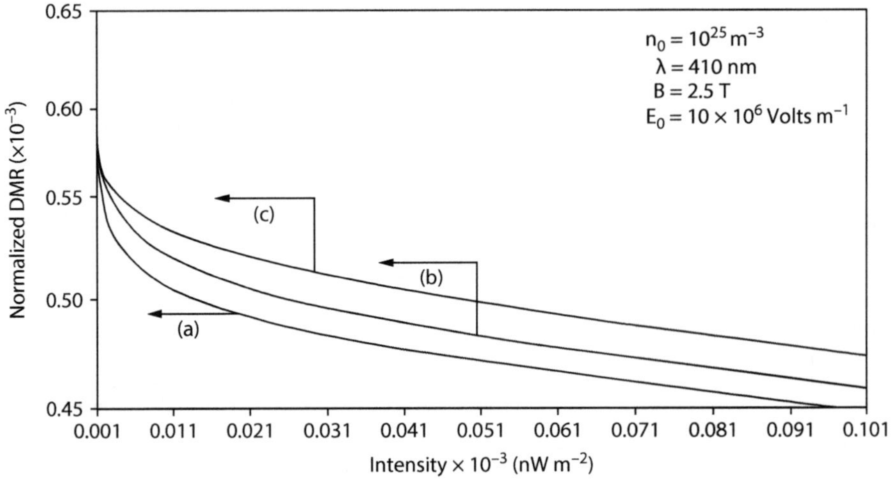

Figure 3.40: Plot of the DMR as a function of light intensity under cross-field configuration in external photoexcitation for HD n-$In_{1-x}Ga_xAs_yP_{1-y}$ lattice matched to InP, where the curves (a), (b), and (c) represent the perturbed three- and two-band models of Kane and that of parabolic energy bands respectively.

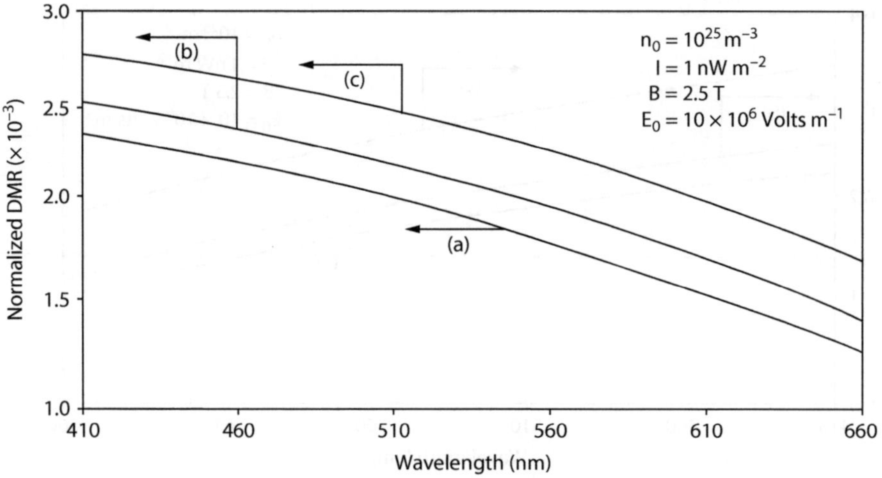

Figure 3.41: Plot of the DMR as a function of wavelength under cross-field configuration in external photoexcitation for HD *n*-InAs, where the curves (a), (b), and (c) represent the perturbed three- and two-band models of Kane and that of parabolic energy bands respectively.

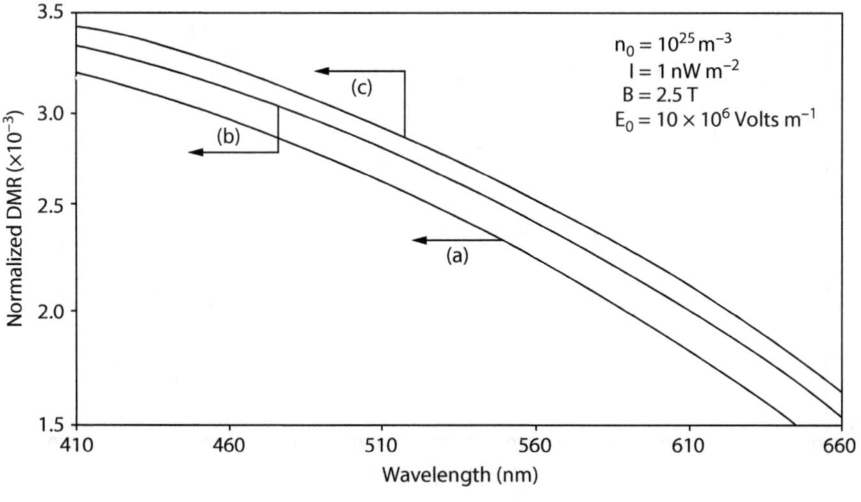

Figure 3.42: Plot of the DMR as a function of wavelength under cross-field configuration in external photoexcitation for HD *n*-InSb, where the curves (a), (b), and (c) represent the perturbed three- and two-band models of Kane and that of parabolic energy bands respectively.

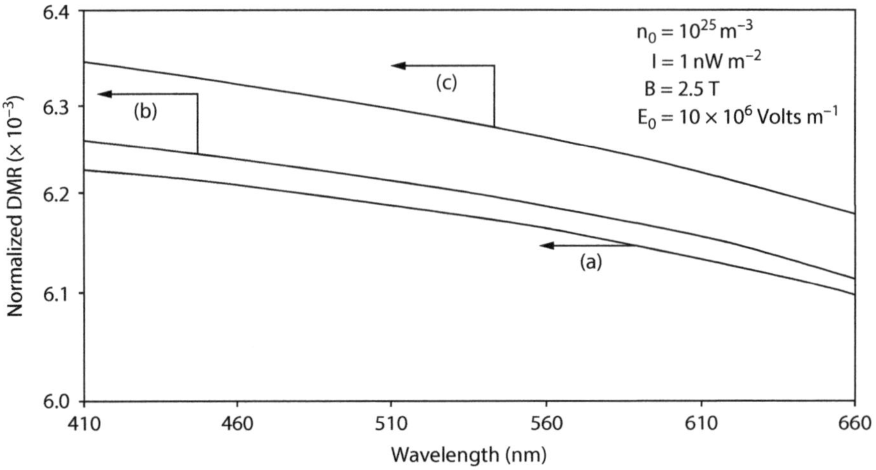

Figure 3.43: Plot of the DMR as a function of wavelength under cross-field configuration in external photoexcitation for HD n-$Hg_{1-x}Cd_xTe$, where the curves (a), (b), and (c) represent the perturbed three- and two-band models of Kane and that of parabolic energy bands respectively.

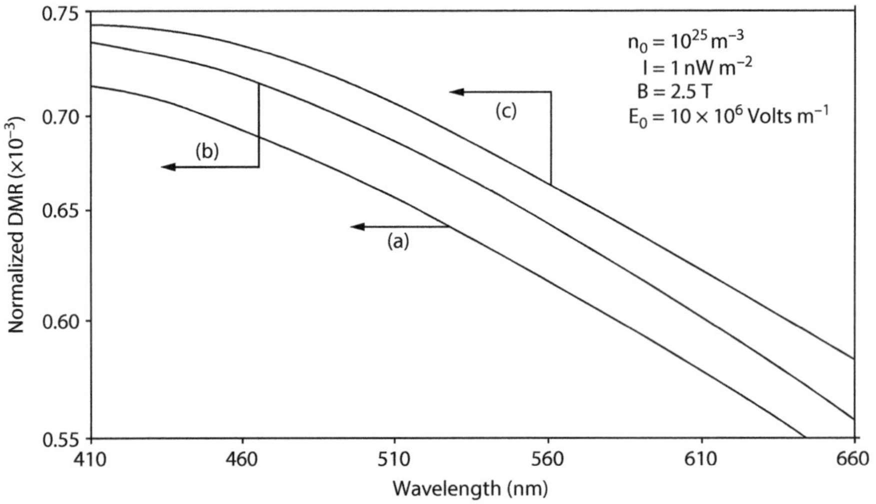

Figure 3.44: Plot of the DMR as a function of wavelength under cross-field configuration in external photoexcitation for HD n-$In_{1-x}Ga_xAs_yP_{1-y}$ lattice matched to InP, where the curves (a), (b), and (c) represent the perturbed three- and two-band models of Kane and that of parabolic energy bands respectively.

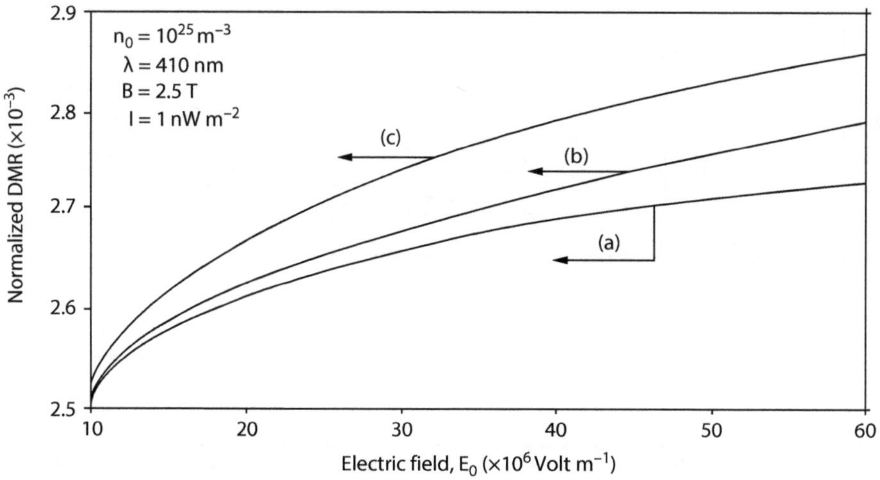

Figure 3.45: Plot of the DMR as a function of electric field under cross-field configuration in external photoexcitation for HD *n*-InAs, where the curves (a), (b), and (c) represent the perturbed three- and two-band models of Kane and that of parabolic energy bands respectively.

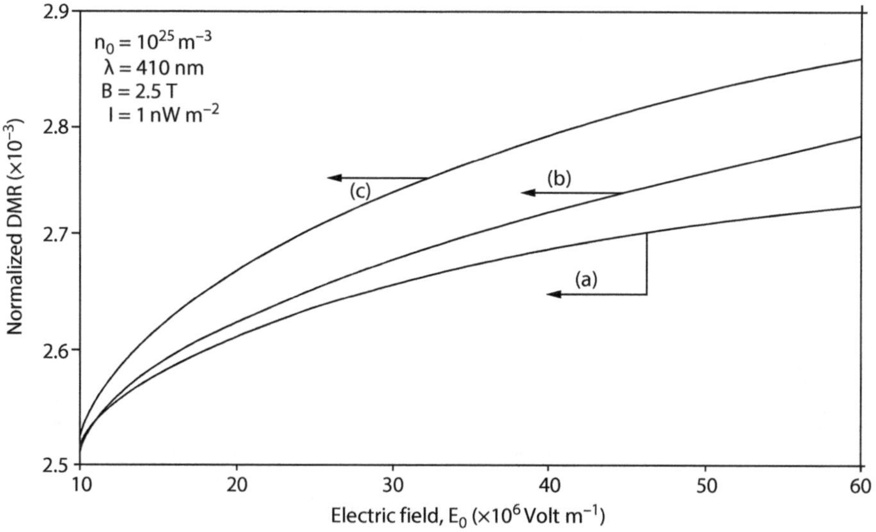

Figure 3.46: Plot of the DMR as a function of electric field under cross-field configuration in external photoexcitation for HD *n*-InSb, where the curves (a), (b), and (c) represent the perturbed three- and two-band models of Kane and that of parabolic energy bands respectively.

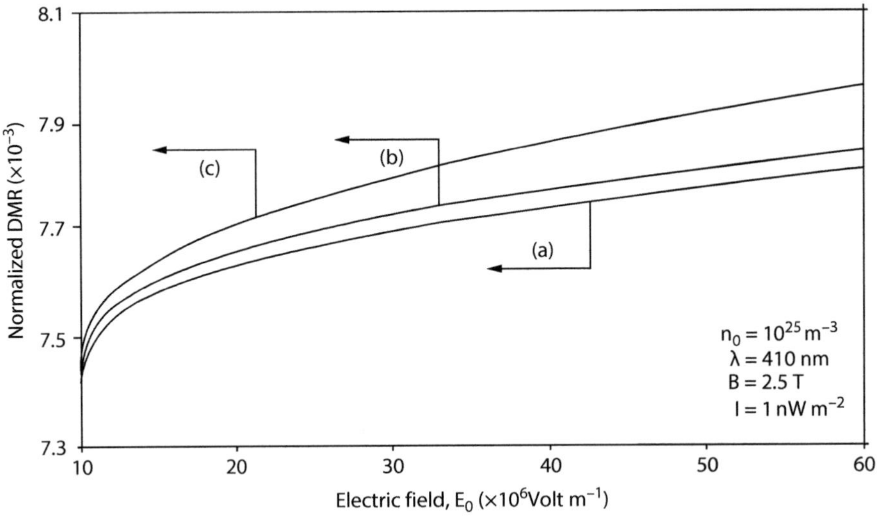

Figure 3.47: Plot of the DMR as a function of electric field under cross-field configuration in external photoexcitation for HD n-$Hg_{1-x}Cd_xTe$, where the curves (a), (b), and (c) represent the perturbed three- and two-band models of Kane and that of parabolic energy bands respectively.

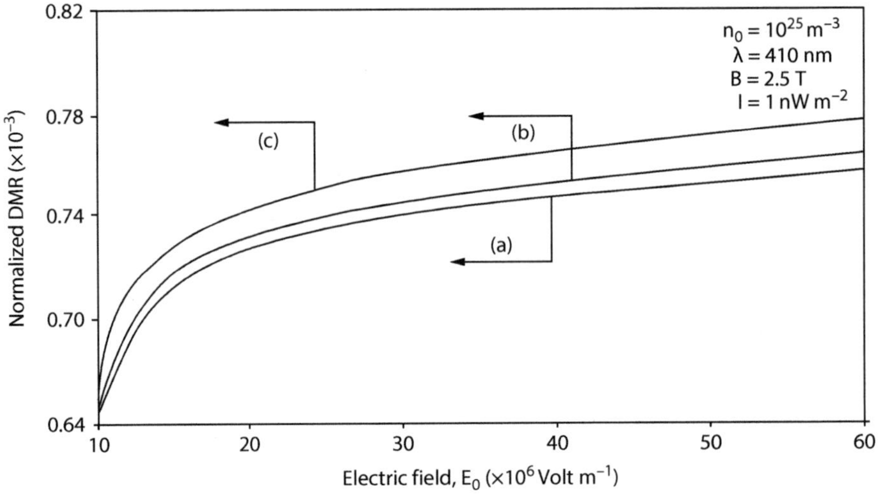

Figure 3.48: Plot of the DMR as a function of electric field under cross-field configuration in external photoexcitation for HD n-$In_{1-x}Ga_xAs_yP_{1-y}$ lattice matched to InP, where the curves (a), (b), and (c) represent the perturbed three- and two-band models of Kane and that of parabolic energy bands respectively.

3.2.4 The DMR in 2D systems of HD Kane-type semiconductors in the presence of light waves

3.2.4.1 Introduction

In this section, we shall study the DMR in quantum wells (QWs), Doping superlattices and inversion and accumulation layers of HD Kane-type semiconductors in the presence of light waves in Sections 3.2.4.2.1, 3.2.4.2.2, and 3.2.4.2.3 respectively.

3.2.4.2 Theoretical background

3.2.4.2.1 The DMR in QWs of HD Kane-type semiconductors in the presence of light waves

(a) The DMR in this case for HD materials can be written as follows:

$$\frac{D}{\mu} = \text{Real Part of}\left[\frac{n_0}{e}\left[\frac{\partial n_0}{\partial(E_{F2DLHD} - E_{OH2D})}\right]^{-1}\right] \tag{3.33}$$

where E_{OH2D} is the sub-band energy in this case.

Using eqs. (1.136) and (3.33), we can study the DMR in this case

The 2D DMR in QWs of III–V, ternary and quaternary materials in the absence of band tails, whose unperturbed band structure is defined by the three-band model of Kane, in the presence of light waves, can be expressed as follows:

$$\frac{D}{\mu} = \frac{n_0}{e}\left[\frac{\partial n_0}{\partial E_{F2DL}}\right]^{-1} \tag{3.34}$$

Using eqs. (1.140) and (3.34) we can write the following equation:

$$\frac{D}{\mu} = \frac{1}{e}\left[\sum_{n_z=1}^{n_{z\max}}\left[\beta_0(E_{F2DL}, n_z, \lambda) - \frac{\hbar^2}{2m_c}\left(\frac{\pi n_z}{d_z}\right)^2\right]\right]\left[\sum_{n_z=1}^{n_{z\max}}[\beta'_0(E_{F2DL}, n_z, \lambda)]\right]^{-1} \tag{3.35}$$

In the absence of band tails and light waves and for isotropic three-band model of Kane, the 2D DMR in this case can be written as follows:

$$\frac{D}{\mu} = \frac{1}{e}\left[\sum_{n_z=1}^{n_{z\max}}[T_{53}(E_{Fs}, n_z)]\right]\left[\sum_{n_z=1}^{n_{z\max}}[T_{53}(E_{Fs}, n_z)]'\right]^{-1} \tag{3.36}$$

(b) The 2D DMR in QWs of HD III–V, ternary and quaternary materials, whose unperturbed band structure is defined by the two-band model of Kane, in the presence of light waves, can be investigated by using eqs. (3.33) and (1.152) respectively

The 2D DMR in QWs of III–V, ternary and quaternary materials in the absence of band tails, whose unperturbed band structure is defined by the two-band model of Kane, in the presence of light waves, can be expressed as follows:

$$\frac{D}{\mu}=\frac{1}{e}\left[\sum_{n_z=1}^{n_{z\max}}\left[\tau_0(E_{F2DL},n_z,\lambda)-\frac{\hbar^2}{2m_c}\left(\frac{\pi n_z}{d_z}\right)^2\right]\right]\left[\sum_{n_z=1}^{n_{z\max}}[\tau_0(E_{F2DL},n_z,\lambda)]'\right]^{-1} \tag{3.37}$$

In the absence of light waves and heavy doping, the 2D DMR for isotropic two-band model of Kane can be written as follows:

$$\frac{D}{\mu}=\frac{1}{e}\left[\sum_{n_z=1}^{n_{z\max}}\left[E_{Fs}(1+\alpha E_{Fs})-\frac{\hbar^2}{2m_c}\left(\frac{n_z\pi}{d_z}\right)^2\right]\right]\left[\sum_{n_z=1}^{n_{z\max}}[(1+2\alpha E_{Fs})]\right]^{-1} \tag{3.38}$$

The DMR at a finite temperature can be written as follows:

$$\frac{D}{\mu}=\frac{k_BT}{e}\left[\sum_{n_z=1}^{n_{z\max}}[(1+2\alpha E_{n_{z_3}})F_0(\eta_{n_1})+2\alpha k_BTF_1(\eta_{n_1})]\right]\left[\sum_{n_z=1}^{n_{z\max}}[(1+2\alpha E_{n_{z_3}})F_{-1}(\eta_{n_1})+2\alpha k_BTF_0(\eta_{n_1})]\right]^{-1} \tag{3.39}$$

The eq. (3.39) is well known in the literature.

(c) The 2D DMR in QWs of HD III–V, ternary and quaternary materials, whose unperturbed band structure is defined by the parabolic energy bands in the presence of light waves, can be studied by using eqs. (1.164) and (3.33) respectively

The 2D DMR in QWs of III–V, ternary and quaternary materials in the absence of band tails, whose unperturbed band structure is defined by the parabolic energy band in the presence of light waves, can be expressed as follows:

$$\frac{D}{\mu}=\frac{1}{e}\left[\sum_{n_z=1}^{n_{z\max}}[\rho_0(E_{F2DL},n_z,\lambda)-\frac{\hbar^2}{2m_c}\left(\frac{\pi n_z}{d_z}\right)^2]\right]\left[\sum_{n_z=1}^{n_{z\max}}[\rho_0(E_{F2DL},n_z,\lambda)]'\right]^{-1} \tag{3.40}$$

In the absence of light waves and heavy doping for isotropic parabolic energy band the 2D DMR can be written as follows:

$$\frac{D}{\mu}=\frac{1}{e}\left[\sum_{n_z=1}^{n_{z\max}}[(E_{Fs}-E_{n_{z33}})]\right]\left[\sum_{n_z=1}^{n_{z\max}}[1]\right]^{-1} \tag{3.41}$$

The DMR at a finite temperature can be written as follows:

$$\frac{D}{\mu} = \frac{k_B T}{e}\left[\sum_{n_z=1}^{n_{z\max}} [F_0(\eta_{n_{11}})]\right]\left[\sum_{n_z=1}^{n_{z\max}} [F_{-1}(\eta_{n_{11}})]\right]^{-1} \tag{3.42}$$

3.2.4.2.2 The DMR in doping superlattices of HD Kane-type semiconductors in the presence of light waves

(a) The sub-band energy (E_{3251}) in doping superlattices of HD III–V, ternary and quaternary materials in the presence of external photoexcitation whose DR is given by eq. (1.173) can be expressed as follows:

$$T_1(E_{3251}, \eta_g, \lambda) = \left(n_i + \frac{1}{2}\right)\hbar\omega_{91HD}(E_{3251}, \eta_g, \lambda) \tag{3.43}$$

The DMR in this case can be written as follows:

$$\frac{D}{\mu} = \frac{n_0}{e}\left[\frac{\partial n_0}{\partial(E_{F2DLHDD} - E_{SE})}\right]^{-1} \tag{3.44}$$

where E_{SE} is the sub-band energy in this case.

The DMR in this case can be studied by using eqs. (1.174) and (3.44) respectively.

In the absence of band tails, sub-band energy in doping superlattices of III–V, ternary and quaternary materials in the presence of external photoexcitation whose DR is given by eq. (1.175) can be expressed as follows:

$$\beta_0(E_{3251UP}, \lambda) = \left(n_i + \frac{1}{2}\right)\hbar\omega_{911}(E_{3251UP}, \lambda) \tag{3.45}$$

The DMR in this case for HD materials can be written as follows:

$$\frac{D}{\mu} = \text{Real Part of}\left[\frac{n_0}{e}\left[\frac{\partial n_0}{\partial(E_{F2DLD} - E_{3251UP})}\right]^{-1}\right] \tag{3.46}$$

where E_{3251UP} is the sub-band energy in this case.

Using eqs. (1.116), (3.45), and (3.46), we can find the DMRin this case.

The DMR in nipi structures of III–V, ternary and quaternary materials can be expressed in the absence of both band tails and light waves whose DR is given by eq. (1.179) as follows:

$$\frac{D}{\mu} = \frac{n_0}{e}\left[\frac{\partial n_0}{\partial(\bar{E}_{Fn_i} - E_{2ni})}\right]^{-1} \tag{3.47}$$

where $\bar{E}_{00}$ is the sub-band energy in this case.

The sub-band energy E_{2ni} in nipi structures of III–V, ternary and quaternary materials can be expressed in the absence of both band tails and light waves whose DR is given by eq. (1.179) as follows:

$$I_{11}(E_{2ni}) = \left(n_i + \frac{1}{2}\right)\hbar\omega_9(E_{2ni}) \tag{3.48}$$

Thus using eqs. (3.47), (3.48), and (1.180), we can study the DMR in this case.(b) The sub-band energy (E_{3252}) in doping superlattices of HD III–V, ternary and quaternary materials in the presence of external photoexcitation whose electrons are defined by eq. (1.183) can be expressed as follows:

$$T_2(E_{3252}, \eta_g, \lambda) = \left(n_i + \frac{1}{2}\right)\hbar\omega_{92HD}(E_{3252}, \eta_g, \lambda) \tag{3.49}$$

Using eqs. (3.44), (1.184), and (3.49), we can investigate the DMR in this case.

In the absence of band tails, sub-band energy in doping superlattices of III–V, ternary and quaternary materials in the presence of external photoexcitation whose unperturbed electrons are defined by eq. (1.185) can be expressed as follows:

$$\tau_0(E_{3252UP}, \lambda) = \left(n_i + \frac{1}{2}\right)\hbar\omega_{912}(E_{3252UP}, \lambda) \tag{3.50}$$

The DMR in this case for HD materials can be written as

$$\frac{D}{\mu} = \frac{n_0}{e}\left[\frac{\partial n_0}{\partial(E_{F2DLD} - E_{3252UP})}\right]^{-1} \tag{3.51}$$

where E_{3252UP} is the sub-band energy in this case.

Using eqs. (3.50), (3.51), and (1.185), we can study the DMR in this case.

In the absence of band tails, sub-band energy(E_{3253UP}) in doping superlattices of III–V, ternary and quaternary materials in the presence of external photoexcitation whose unperturbed electrons are defined by the two-band model of Kane can be expressed as follows:

$$\tau_0(E_{3253UP}, \lambda) = \left(n_i + \frac{1}{2}\right)\hbar\omega_{912}(E_{3253UP}, \lambda) \tag{3.52}$$

The DMR in this case for HD materials can be written as follows:

$$\frac{D}{\mu} = \frac{n_0}{e}\left[\frac{\partial n_0}{\partial(E_{F2DLD} - E_{3253UP})}\right]^{-1} \tag{3.53}$$

Using eqs. (1.187), (3.52), and (3.53), we can find the DMR in this case.

The sub-band energy(E_{188}) in nipi structures of III–V, ternary and quaternary materials can be expressed in the absence of both band tails and light waves and whose unperturbed DR is given by the two-band model of Kane as follows:

$$E_{188}(1+\alpha E_{188}) = \left(n_i + \frac{1}{2}\right)\hbar\omega_{10}(E_{188}) \tag{3.54}$$

The DMR in this case for HD materials can be written as follows:

$$\frac{D}{\mu} = \frac{n_0}{e}\left[\frac{\partial n_0}{\partial(\bar{E}_{Fn_i} - E_{188})}\right]^{-1} \tag{3.55}$$

Using eqs. (1.190), (3.54), and (3.55), we can find the DMR in this case.

(c) The sub-band energy (E_{3253}) in doping superlattices of HD III–V, ternary and quaternary materials in the presence of external photoexcitation whose electrons are defined by eq. (1.183) can be expressed as follows:

$$T_3(E_{3255}, \eta_g, \lambda) = \left(n_i + \frac{1}{2}\right)\hbar\omega_{93HD}(E_{3255}, \eta_g, \lambda) \tag{3.56}$$

Using eqs. (3.44), (1.192), and (3.55), we can investigate the DMR in this case.

In the absence of band tails, sub-band energy in doping superlattices of III–V, ternary and quaternary materials in the presence of external photoexcitation whose unperturbed electrons are defined by eq. (1.193) can be expressed as follows:

$$\rho_0(E_{3255UP}, \lambda) = \left(n_i + \frac{1}{2}\right)\hbar\omega_{913}(E_{3255UP}, \lambda) \tag{3.57}$$

The DMR in this case for HD materials can be written as follows:

$$\frac{D}{\mu} = \frac{n_0}{e}\left[\frac{\partial n_0}{\partial(E_{F2DLD} - E_{3255UP})}\right]^{-1} \tag{3.58}$$

Using eqs. (3.57), (3.58), and (1.195), we can study the DMR in this case.

In the absence of band tails, sub-band energy(E_{3256UP}) in doping superlattices of III–V, ternary and quaternary materials in the presence of external photoexcitation whose unperturbed electrons are defined by the two-band model of Kane can be expressed as follows:

$$\rho_0(E_{3256UP}, \lambda) = \left(n_i + \frac{1}{2}\right)\hbar\omega_{912}(E_{3256UP}, \lambda) \tag{3.59}$$

The DMR in this case can be written as follows:

$$\frac{D}{\mu} = \frac{n_0}{e}\left[\frac{\partial n_0}{\partial(E_{F2DLD} - E_{3256UP})}\right]^{-1} \tag{3.60}$$

Using eqs. (1.194), (3.59), and(3.60), we can study the DMR in this case.

The sub-band energy(E_{1881}) in nipi structures of III–V, ternary and quaternary materials can be expressed in the absence of both band tails and light waves and whose unperturbed DR is given by the two-band model of Kane as follows:

$$E_{1881} = \left(n_i + \frac{1}{2}\right)\hbar\omega_{11} \tag{3.61}$$

The DMR in this case can be written as follows:

$$\frac{D}{\mu} = \frac{n_0}{e}\left[\frac{\partial n_0}{\partial(\bar{E}_{Fn_i} - E_{1881})}\right]^{-1} \tag{3.62}$$

Using eqs. (1.196), (3.61), and(3.62), we can study the DMR in this case.

3.2.4.2.3 The DMR in accumulation and inversion layers of Kane-type semiconductors in the presence of light waves

(a) Using eq. (1.255) under the weak electric field limit, the sub-band energy (E_{321}) in accumulation layers of HD III–V, ternary and quaternary materials, whose unperturbed band structure is defined by the three-band model of Kane, in the presence of light waves can be expressed as follows:

$$T_1(E_{321}, \eta_g, \lambda) = S_i\left[\frac{\hbar|e|F_s[T_1(E_{321}, \eta_g, \lambda)]'}{\sqrt{2m_c}}\right]^{2/3} \tag{3.63}$$

The DMR for HD materials in this case can be written as follows:

$$\frac{D}{\mu} = \frac{n_0}{e}\,Real\,Part\,of\left[\frac{\partial n_0}{\partial(E'_{fL} - E_{321})}\right]^{-1} \tag{3.64}$$

Thus, using eqs. (1.256), (3.63), and (3.64), we can find the DMR in this case.

Under the weak electric field limit, the sub-band energy (E_{322})in accumulation layers of HD III–V, ternary and quaternary materials, whose unperturbed band structure is defined by the three-band model of Kane, in the absence of light waves, can be expressed from eq. (1.260) as follows:

$$\bar{T}_1(E_{322}, \eta_g) = S_i \left[\frac{\hbar|e|F_s[\bar{T}_1(E_{322}, \eta_g)]'}{\sqrt{2m_c}} \right]^{2/3} \tag{3.65a}$$

The DMR for HD materials in this case can be written as

$$\frac{D}{\mu} = Real\ Part\ of \left[\frac{n_0}{e} \left[\frac{\partial n_0}{\partial(E'_{f1} - E_{322})} \right]^{-1} \right] \tag{3.65b}$$

Thus Using (1.261), (3.65a) and (3.65b) we can find the DMR in this case.

In the absence of band tails and light waves, the sub-band energy (E_{323}) in III-V, ternary and quaternary materials whose bulk electrons obey the three band model of Kane under the condition of weak electric field limit, assumes the form

$$[I_{11}(E_{323}) = S_i \left[\frac{\hbar|e|F_s[I_{11}(E_{323})]'}{\sqrt{2m_c}} \right]^{2/3} \tag{3.66}$$

The DMR for HD materials in this case can be written as follows:

$$\frac{D}{\mu} = \frac{n_0}{e} \left[\frac{\partial n_0}{\partial(E_{Fiw} - E_{323})} \right]^{-1} \tag{3.67}$$

Thus, using eqs. (1.266), (3.66), and (3.67), we can find the DMR in this case.

(b) Using eq. (1.269) under the weak electric field limit, the sub-band energy (E_{3212}) in accumulation layers of HD III–V, ternary and quaternary materials, whose unperturbed band structure is defined by the two-band model of Kane, in the presence of light waves can be expressed as follows:

$$T_2(E_{3212}, \eta_g, \lambda) = S_i \left[\frac{\hbar|e|F_s[T_2(E_{3212}, \eta_g, \lambda)]'}{\sqrt{2m_c}} \right]^{2/3} \tag{3.68}$$

The DMR for HD materials in this case can be written as follows:

$$\frac{D}{\mu} = \frac{n_0}{e} \left[\frac{\partial n_0}{\partial(E'_{fL} - E_{3212})} \right]^{-1} \tag{3.69}$$

Thus, using eqs. (1.270), (3.68), and (3.69), we can find the DMR in this case.

Under the weak electric field limit, the sub-band energy (E_{3222})in accumulation layers of HD III–V, ternary and quaternary materials, whose unperturbed band structure is defined by the two-band model of Kane, in the absence of light waves, can be expressed from eq. (1.274) as follows:

$$\bar{T}_2(E_{3222},\eta_g)=S_i\left[\frac{\hbar|e|F_s[\bar{T}_2(E_{3222},\eta_g)]'}{\sqrt{2m_c}}\right]^{2/3} \tag{3.70}$$

The DMR for HD materials in this case can be written as follows:

$$\frac{D}{\mu}=\frac{n_0}{e}\left[\frac{\partial n_0}{\partial(E'_{f1}-E_{3222})}\right]^{-1} \tag{3.71}$$

Thus, using eqs. (1.275), (3.70), and (3.71), we can find the DMR in this case.

In the absence of band tails and light waves, the sub-band energy (E_{3232}) in III–V, ternary and quaternary materials whose bulk electrons obey the two-band model of Kane under the condition of weak electric field limit, assumes the following form:

$$E_{3232}(1+\alpha E_{3232})=S_i\left[\frac{\hbar|e|F_s(1+2\alpha E_{3232})}{\sqrt{2m_c}}\right]^{2/3} \tag{3.72}$$

The DMR for HD materials in this case can be written as follows:

$$\frac{D}{\mu}=\frac{n_0}{e}\left[\frac{\partial n_0}{\partial(E_{Fiw}-E_{3232})}\right]^{-1} \tag{3.73}$$

Thus, using eqs. (1.280), (3.72), and (3.73), we can find the DMR in this case.

(c) Using eq. (1.283) under the weak electric field limit, the sub-band energy (E_{3213}) in accumulation layers of HD III–V, ternary and quaternary materials, whose unperturbed band structure is defined by the parabolic band model, in the presence of light waves can be expressed as follows:

$$T_3(E_{3213},\eta_g,\lambda)=S_i\left[\frac{\hbar|e|F_s[T_3(E_{3213},\eta_g,\lambda)]'}{\sqrt{2m_c}}\right]^{2/3} \tag{3.74}$$

The DMR for HD materials in this case can be written as follows:

$$\frac{D}{\mu}=\frac{n_0}{e}\left[\frac{\partial n_0}{\partial(E'_{fL}-E_{3213})}\right]^{-1} \tag{3.75}$$

Thus, using eqs. (1.284), (3.74), and (3.75), we can find the DMR in this case.

Under weak electric field limit, the sub-band energy (E_{3223})in accumulation layers of HD III–V, ternary and quaternary materials, whose unperturbed band structure is defined by the parabolic band model in the absence of light waves, can be expressed from eq. (1.288) as follows:

$$\bar{T}_3(E_{3223}, \eta_g) = S_i \left[\frac{\hbar |e| F_s [\bar{T}_3(E_{3223}, \eta_g)]'}{\sqrt{2m_c}} \right]^{2/3} \tag{3.76}$$

The DMR for HD materials in this case can be written as follows:

$$\frac{D}{\mu} = \frac{n_0}{e} \left[\frac{\partial n_0}{\partial (E'_{f1} - E_{3223})} \right]^{-1} \tag{3.77}$$

Thus, using eqs. (1.289), (3.76), and (3.77), we can find the DMR in this case.

In the absence of band tails and light waves, the sub-band energy (E_{3233}) in III–V, ternary and quaternary materials whose bulk electrons obey the parabolic band model under the conditions of weak electric field limit, assumes the following form:

$$E_{3233} = S_i \left[\frac{\hbar |e| F_s}{\sqrt{2m_c}} \right]^{2/3} \tag{3.78}$$

The DMR for HD materials in this case can be written as follows:

$$\frac{D}{\mu} = \frac{n_0}{e} \left[\frac{\partial n_0}{\partial (E_{Fiw} - E_{3233})} \right]^{-1} \tag{3.79}$$

Thus, using eqs. (1.294), (3.78), and (3.79), we can find the DMR in this case.

3.2.4.3 Results and discussion

Using the appropriate equations in Figures 3.49–3.52, DMR as function of the film thickness in the presence of external photoemission was plotted for QWs of *n*-InAs, *n*-InSb, *n*-$Hg_{1-x}Cd_xTe$, and *n*-$In_{1-x}Ga_xAs_yP_{1-y}$, where the curves (a), (b), and (c) represent the respective DMR for the perturbed three- and the two-band models of Kane and the parabolic energy bands. It has been observed that the DMR in this case also decreases with the increase in film thickness in a stepwise manner. Figures 3.53–3.56 exhibit the plot of the DMR as a function of electron concentration in the presence of light waves for the QWs of the aforementioned materials. The variations of the DMR against light intensity for the QWs of aforementioned materials have been plotted in Figures 3.57–3.60. The DMR decreases with the increase in the light intensity. Figures 3.61–3.64 exhibit variation of the DMR as functions of wavelength in the visible region. The DMR in this case decreases as the wavelength shifts from red to violet. Figures 3.65 and 3.66 show that the DMR decreases with increasing alloy composition for ternary and quaternary materials.

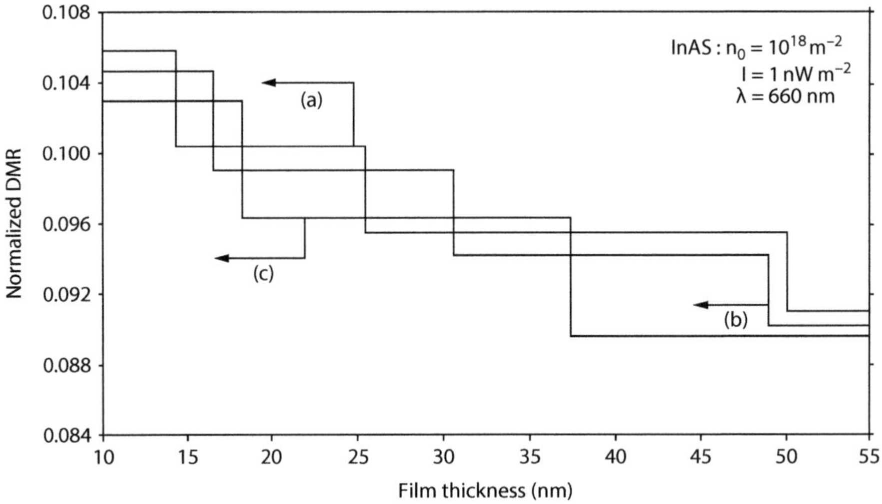

Figure 3.49: Plot of the normalized 2D DMR as a function of film thickness for ultrathin films of *HD* *n*-InAs in the presence of light waves, where the curves (a), (b), and (c) represent the perturbed three- and two-band models of Kane and that of the parabolic energy bands respectively.

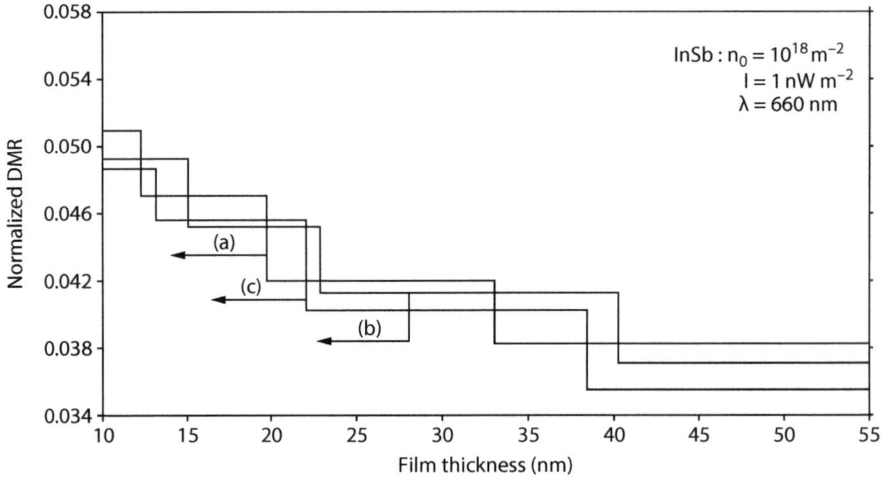

Figure 3.50: Plot of the normalized 2D DMR as a function of film thickness for ultrathin films of *HD* *n*-InSb in the presence of light waves, where the curves (a), (b), and (c) represent the perturbed three- and two-band models of Kane and that of the parabolic energy bands respectively.

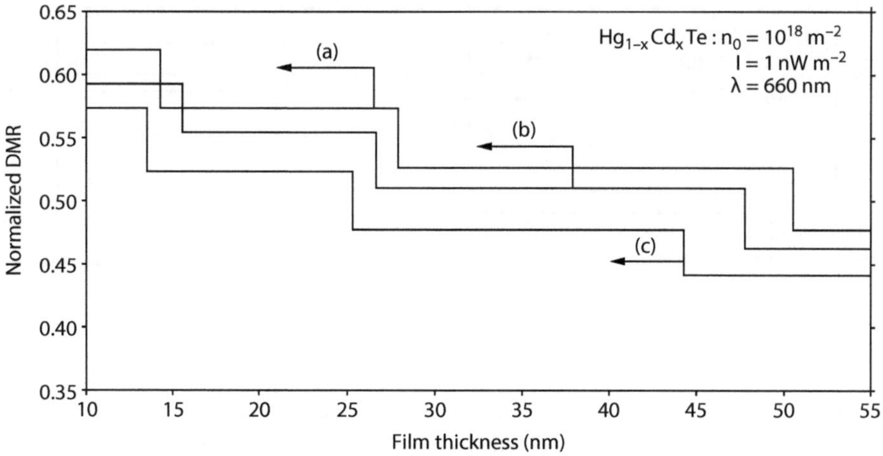

Figure 3.51: Plot of the normalized 2D DMR as a function of film thickness for ultrathin films of *HD* n-$Hg_{1-x}Cd_xTe$ in the presence of light waves, where the curves (a), (b), and (c) represent the perturbed three- and two-band models of Kane and that of the parabolic energy bands respectively.

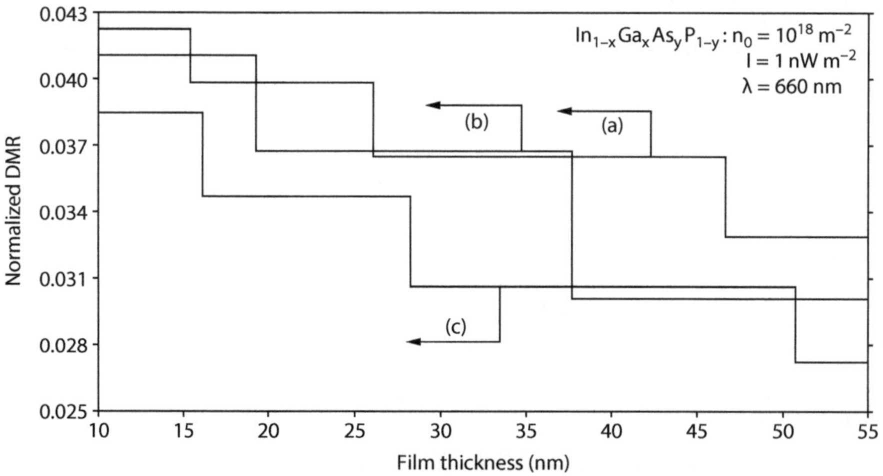

Figure 3.52: Plot of the normalized 2D DMR as a function of film thickness for ultrathin films of HD n-$In_{1-x}Ga_xAs_yP_{1-y}$ lattice matched to InP in the presence of light waves, where the curves (a), (b), and (c) represent the perturbed three- and two-band models of Kane and that of the parabolic energy bands respectively.

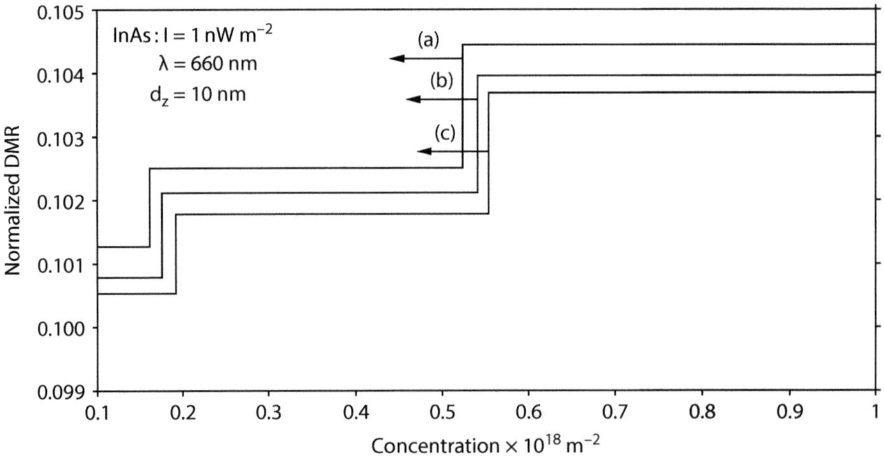

Figure 3.53: Plot of the normalized 2D DMR as a function of electron concentration for ultrathin films of HD *n*-InAs in the presence of light waves, where the curves (a), (b), and (c) represent the perturbed three- and two-band models of Kane and that of the parabolic energy bands respectively.

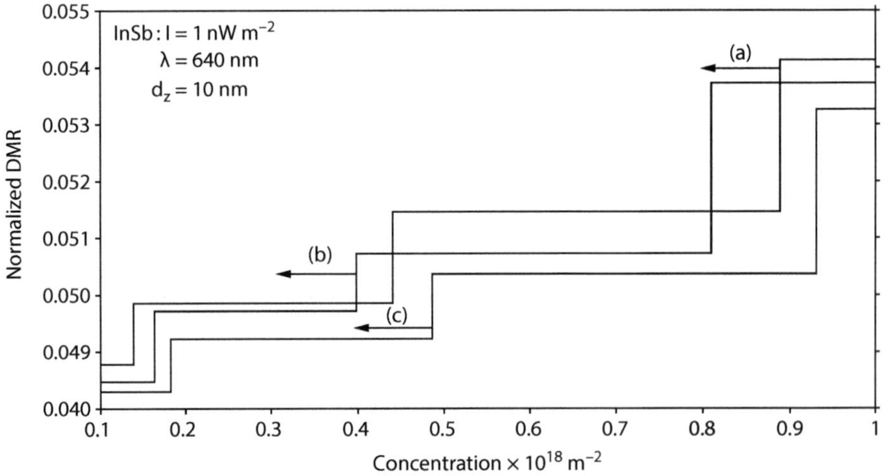

Figure 3.54: Plot of the normalized 2D DMR as a function of electron concentration for ultrathin films of HD *n*-InSb in the presence of light waves, where the curves (a), (b), and (c) represent the perturbed three- and two-band models of Kane and that of the parabolic energy bands respectively.

3.2.5 The DMR in nano wire (NW) of HD Kane-type semiconductors in the presence of light waves

(a) The sub-band energy $(E'_{3HDNWL1})$ in NWs of HD III–V, ternary and quaternary materials, whose unperturbed band structure is defined by the three-band model of

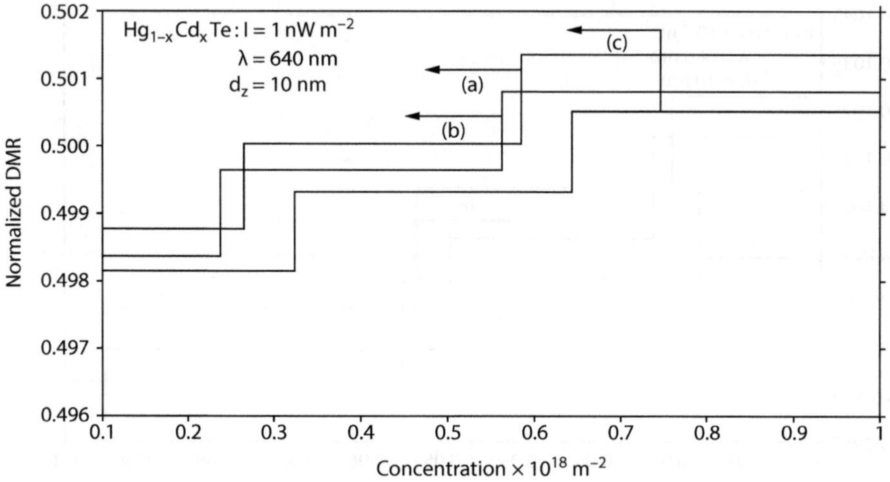

Figure 3.55: Plot of the normalized 2D DMR as a function of electron concentration for ultrathin films of HD n-$Hg_{1-x}Cd_xTe$ in the presence of light waves, where the curves (a), (b), and (c) represent the perturbed three- and two-band models of Kane and that of the parabolic energy bands respectively.

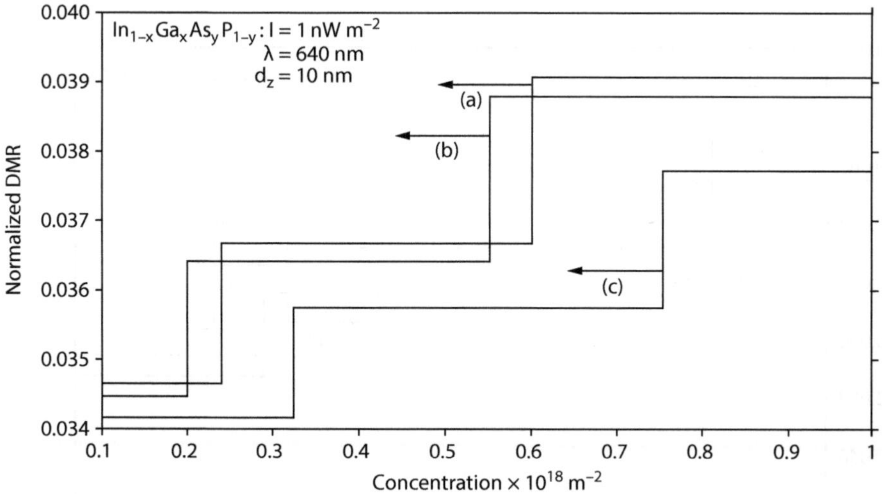

Figure 3.56: Plot of the normalized 2D DMR as a function of electron concentration for ultrathin films of HD n-$In_{1-x}Ga_xAs_yP_{1-y}$ lattice matched to InP in the presence of light waves, where the curves (a), (b), and (c) represent the perturbed three- and two-band models of Kane and that of the parabolic energy bands respectively.

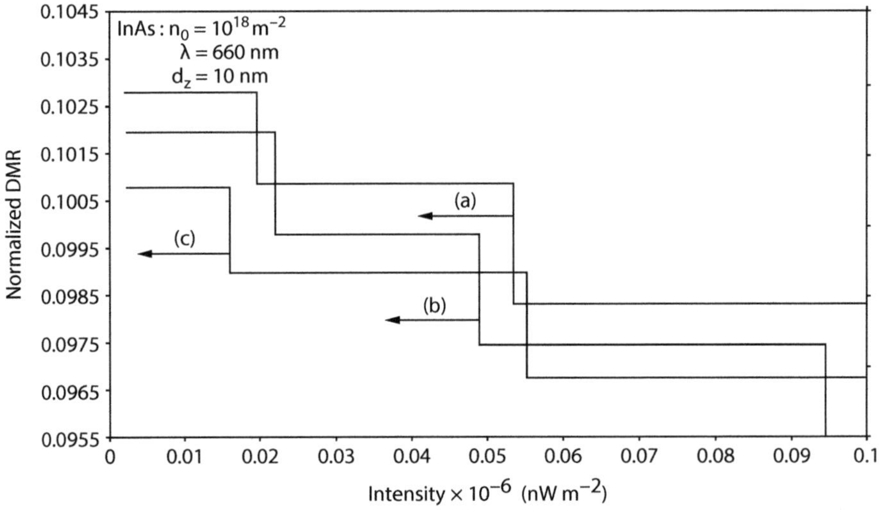

Figure 3.57: Plot of the normalized 2D DMR as a function of light intensity for ultrathin films of HD *n*-InAs in the presence of light waves, where the curves (a), (b), and (c) represent the perturbed three- and two-band models of Kane and that of the parabolic energy bands respectively.

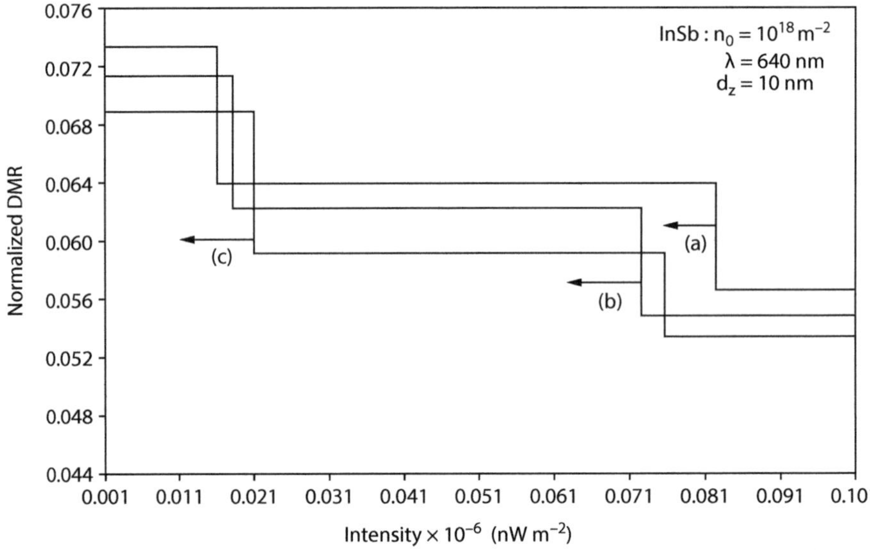

Figure 3.58: Plot of the normalized 2D DMR as a function of light intensity for ultrathin films of HD *n*-InSb in the presence of light waves, where the curves (a), (b), and (c) represent the perturbed three- and two-band models of Kane and that of the parabolic energy bands respectively.

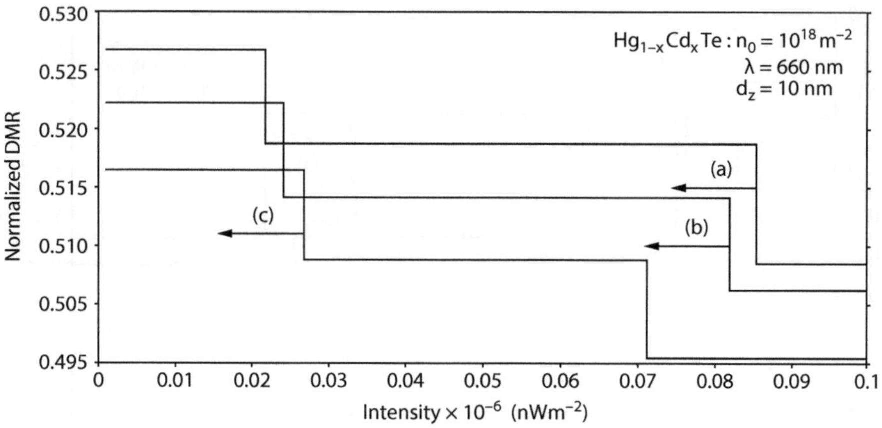

Figure 3.59: Plot of the normalized 2D DMR as a function of light intensity for ultrathin films of HD n-$Hg_{1-x}Cd_xTe$ in the presence of light waves, where the curves (a), (b), and (c) represent the perturbed three- and two-band models of Kane and that of the parabolic energy bands respectively.

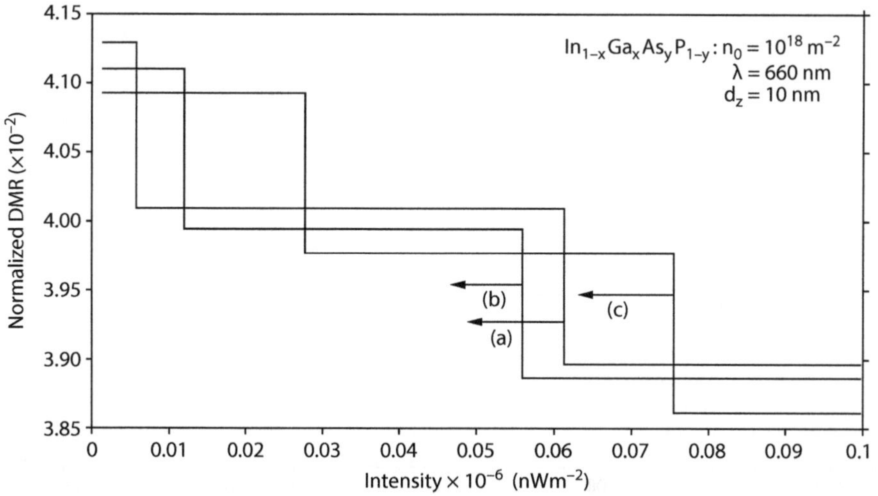

Figure 3.60: Plot of the normalized 2D DMR as a function of light intensity for ultrathin films of HD n-$In_{1-x}Ga_xAs_yP_{1-y}$ lattice matched to InP in the presence of light waves, where the curves (a), (b), and (c) represent the perturbed three- and two-band models of Kane and that of the parabolic energy bands respectively.

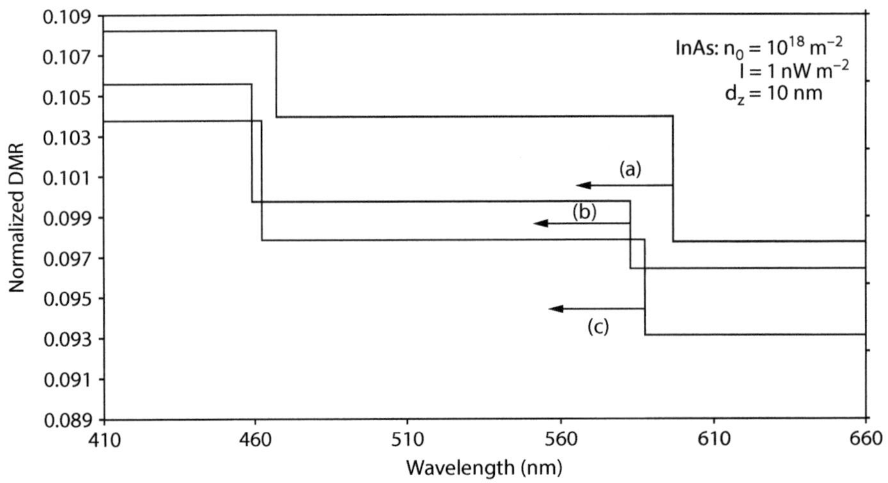

Figure 3.61: Plot of the normalized 2D DMR as a function of wavelength for ultrathin films of HD *n*-InAs in the presence of light waves, where the curves (a), (b), and (c) represent the perturbed three- and two-band models of Kane and that of the parabolic energy bands respectively.

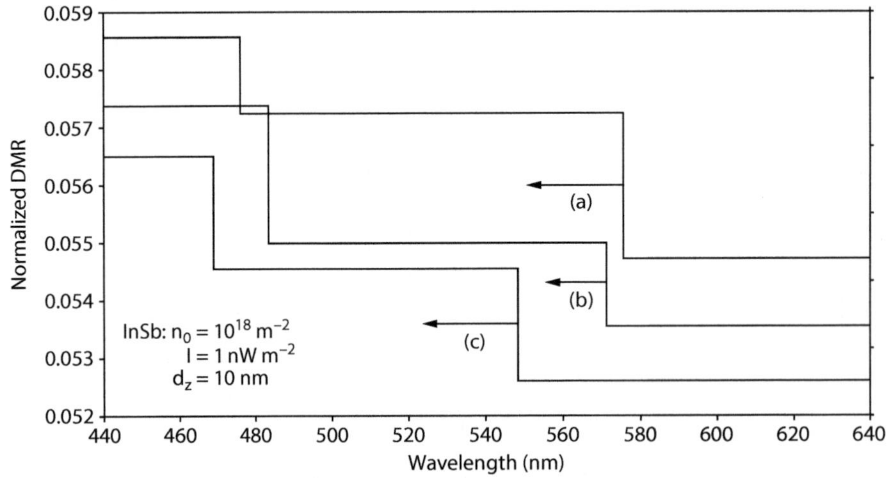

Figure 3.62: Plot of the normalized 2D DMR as a function of wavelength for ultrathin films of HD *n*-InSb in the presence of light waves, where the curves (a), (b), and (c) represent the perturbed three- and two-band models of Kane and that of the parabolic energy bands respectively.

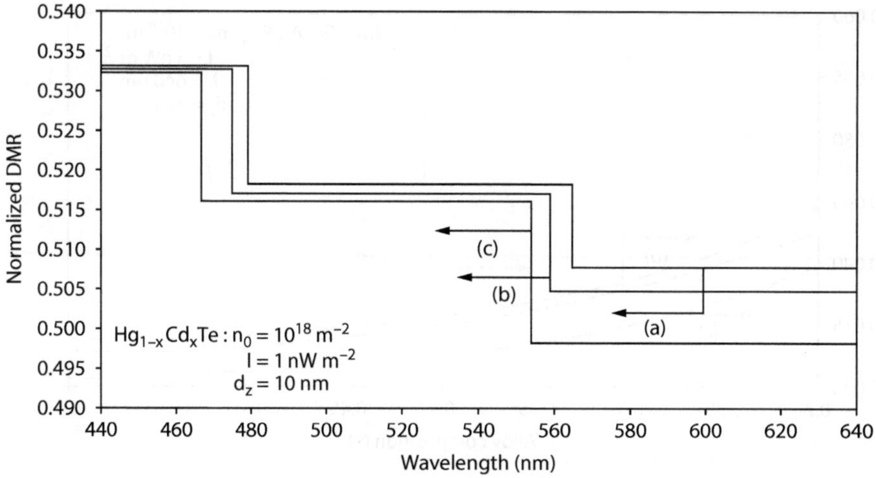

Figure 3.63: Plot of the normalized 2D DMR as a function of wavelength for ultrathin films of HD n-$Hg_{1-x}Cd_xTe$ in the presence of light waves, where the curves (a), (b), and (c) represent the perturbed three- and two-band models of Kane and that of the parabolic energy bands respectively.

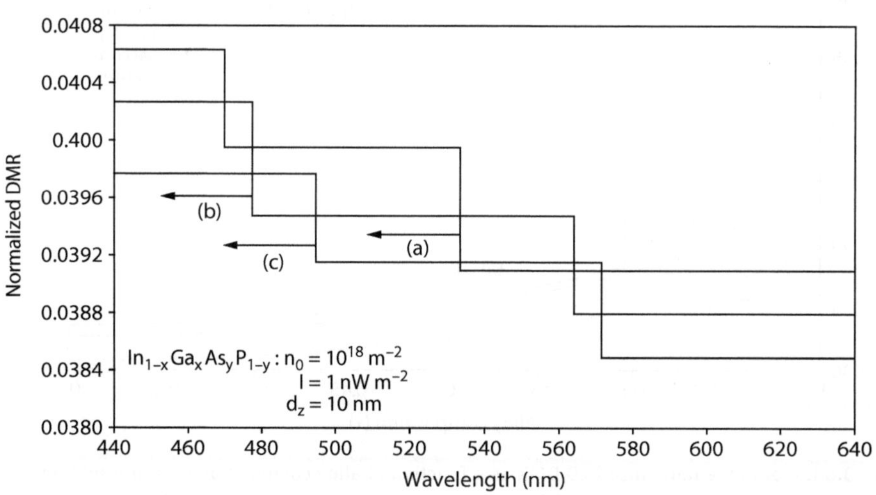

Figure 3.64: Plot of the normalized 2D DMR as a function of wavelength for ultrathin films of HD n-$In_{1-x}Ga_xAs_yP_{1-y}$ in the presence of light waves, where the curves (a), (b), and (c) represent the perturbed three- and two-band models of Kane and that of the parabolic energy bands respectively.

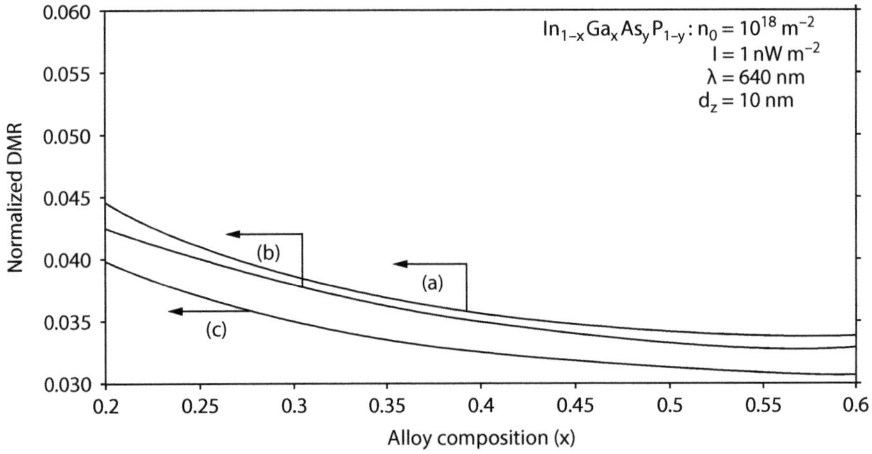

Figure 3.65: Plot of the normalized 2D DMR as a function of alloy composition for ultrathin films of HD n-$Hg_{1-x}Cd_xTe$ in the presence of light waves, where the curves (a), (b), and (c) represent the perturbed three- and two-band models of Kane and that of the parabolic energy bands respectively.

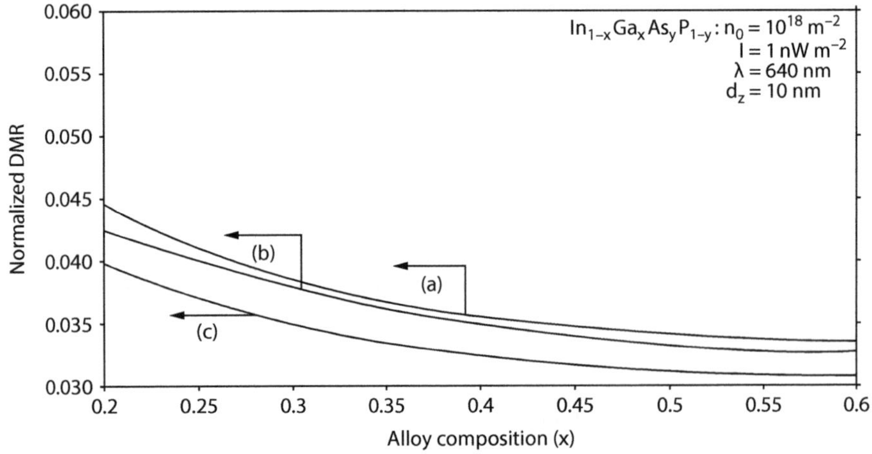

Figure 3.66: Plot of the normalized 2D DMR as a function of alloy composition for ultrathin films of *HD* n-$In_{1-x}Ga_xAs_yP_{1-y}$ in the presence of light waves, where the curves (a), (b), and (c) represent the perturbed three- and two-band models of Kane and that of the parabolic energy bands respectively.

Kane in the absence of any field and in the presence of light waves can be expressed from eq. (1.296) as follows:

$$\frac{\hbar^2(n_z\pi/d_z)^2}{2m_c}+\frac{\hbar^2(n_y\pi/d_y)^2}{2m_c}=T_1(E'_{3HDNWL1},\eta_g,\lambda) \tag{3.80}$$

The 1D DMR for HD materials in this case can be written as follows:

$$\frac{D}{\mu}=\text{Real Part of}\left[\frac{n_0}{e}\left[\frac{\partial n_0}{\partial(E_{F1HDNWL1}-E'_{3HDNWL1})}\right]^{-1}\right] \tag{3.81}$$

Thus, using eqs. (1.296b), (3.80), and (3.81), we can find the DMR in this case.

The sub-band energy $(E_{327})$for NWs of III–V materials whose energy band structures are defined by the three-band model of Kane in the absence of band tailing can be written using eq. (1.298) as follows:

$$\frac{\hbar^2(n_z\pi/d_z)^2}{2m_c}+\frac{\hbar^2(n_y\pi/d_y)^2}{2m_c}=\beta_0(E_{327},\lambda) \tag{3.82}$$

The DMR in this case can be written as follows:

$$\frac{D}{\mu}=\frac{n_0}{e}\left[\frac{\partial n_0}{\partial(E_{F1NWL2})}\right]^{-1} \tag{3.83}$$

Thus, using eqs. (1.299) and (3.83), we can find the DMR in this case.

(b) The sub-band energy $(E'_{3HDNWL2})$ in NWs of HD III–V, ternary and quaternary materials, whose unperturbed band structure is defined by the two-band model of Kane in the absence of any field and in the presence of light waves can be expressed from eq. (1.301) as follows:

$$\frac{\hbar^2(n_z\pi/d_z)^2}{2m_c}+\frac{\hbar^2(n_y\pi/d_y)^2}{2m_c}=T_2(E'_{3HDNWL2},\eta_g,\lambda) \tag{3.84}$$

The DMR for HD materials in this case can be written as follows:

$$\frac{D}{\mu}=\frac{n_0}{e}\left[\frac{\partial n_0}{\partial(E_{F1HDNWL2}-E'_{3HDNWL2})}\right]^{-1} \tag{3.85}$$

Thus, using eqs. (1.302) and (3.85), we can find the DMR in this case.

The sub-band energy (E_{328}) for NWs of III–V materials whose energy band structures are defined by the two-band model of Kane in the absence of band tailing assumes the following form:

$$\frac{\hbar^2(n_z\pi/d_z)^2}{2m_c}+\frac{\hbar^2(n_y\pi/d_y)^2}{2m_c}=\tau_0(E_{328},\lambda) \tag{3.86}$$

The DMR in this case can be written as follows:

$$\frac{D}{\mu}=\frac{n_0}{e}\left[\frac{\partial n_0}{\partial(E_{F1NWL21})}\right]^{-1} \tag{3.87}$$

Thus, using eqs. (1.305) and (3.87), we can find the DMR in this case.

(c) The sub-band energy ($E'_{3HDNWL3}$) in NWs of HD III–V, ternary and quaternary materials, whose unperturbed band structure is defined by the parabolic energy bands in the absence of any field and in the presence of light waves can be expressed from eq. (1.307) as follows:

$$\frac{\hbar^2(n_z\pi/d_z)^2}{2m_c}+\frac{\hbar^2(n_y\pi/d_y)^2}{2m_c}=T_3(E'_{3HDNWL3},\eta_g,\lambda) \tag{3.88}$$

The DMR for HD materials in this case can be written as follows:

$$\frac{D}{\mu}=\frac{n_0}{e}\left[\frac{\partial n_0}{\partial(E_{F1HDNWL3}-E'_{3HDNWL3})}\right]^{-1} \tag{3.89}$$

Thus, using eqs. (1.308), (3.88), and (3.89), we can find the DMR in this case.

The sub-band energy (E_{329}) for NWs of III–V materials whose energy band structures are defined by the parabolic energy bands in the absence of band tailing can be written using eq. (1.310) as follows:

$$\frac{\hbar^2(n_z\pi/d_z)^2}{2m_c}+\frac{\hbar^2(n_y\pi/d_y)^2}{2m_c}=\rho_0(E_{329},\lambda) \tag{3.90}$$

The DMR in this case can be written as follows:

$$\frac{D}{\mu}=\frac{n_0}{e}\left[\frac{\partial n_0}{\partial(E_{F1NWL3})}\right]^{-1} \tag{3.91}$$

Thus, using eqs. (1.311) and (3.91), we can find the DMR in this case.

3.2.5.1 Results and discussion

Using the appropriate equations in Figures 3.67–3.70, the DMR has been plotted as a function of film thickness in QWs of n-InAs, n-InSb, n-$Hg_{1-x}Cd_xTe$ and n-$In_{1-x}Ga_xAs_yP_{1-y}$ lattice matched to InP in the presence of external photoexcitation respectively. The curves (a), (b), and (c) chronologically exhibit the DMR in QWs of the aforementioned

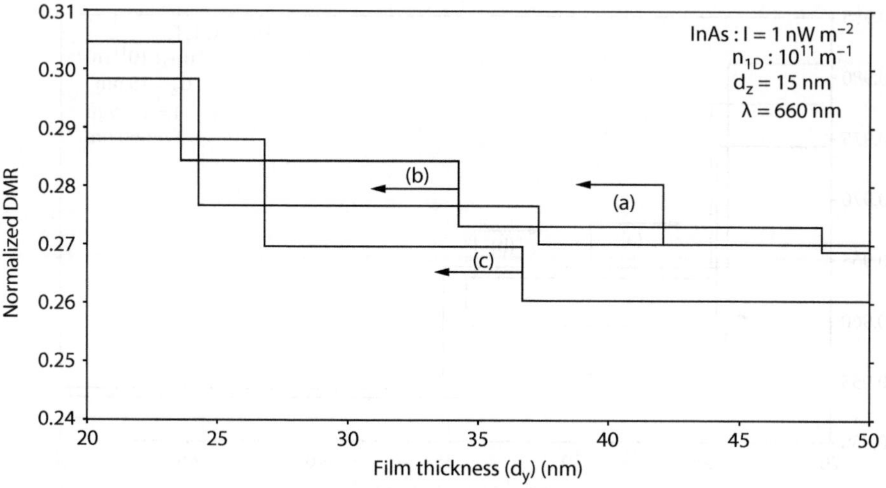

Figure 3.67: Plot of the normalized 1D DMR as a function of film thickness for QWs of HD *n*-InAs in the presence of light waves, where the curves (a), (b), and (c) represent the three- and two-band models of Kane and that of the parabolic energy bands respectively.

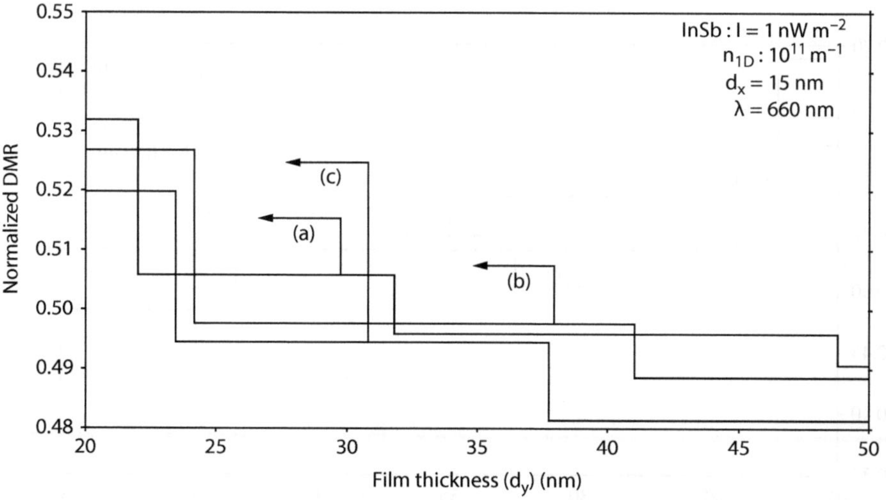

Figure 3.68: Plot of the normalized 1D DMR as a function of film thickness for QWs of HD *n*-InSb in the presence of light waves, where the curves (a), (b), and (c) represent the perturbed three- and two-band models of Kane and that of the parabolic energy bands respectively.

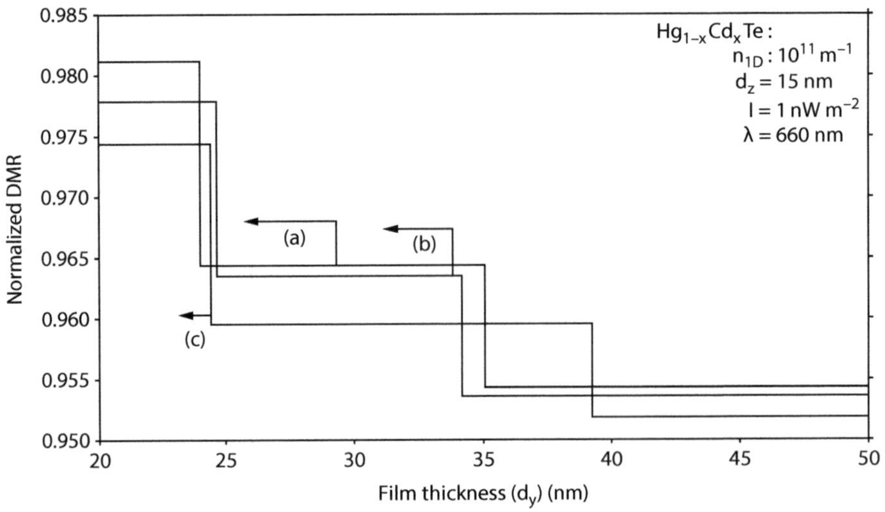

Figure 3.69: Plot of the normalized 1D DMR as a function of film thickness for QWs of HD n-$Hg_{1-x}Cd_xTe$ in the presence of light waves, where the curves (a), (b), and (c) represent the perturbed three- and two-band models of Kane and that of the parabolic energy bands respectively.

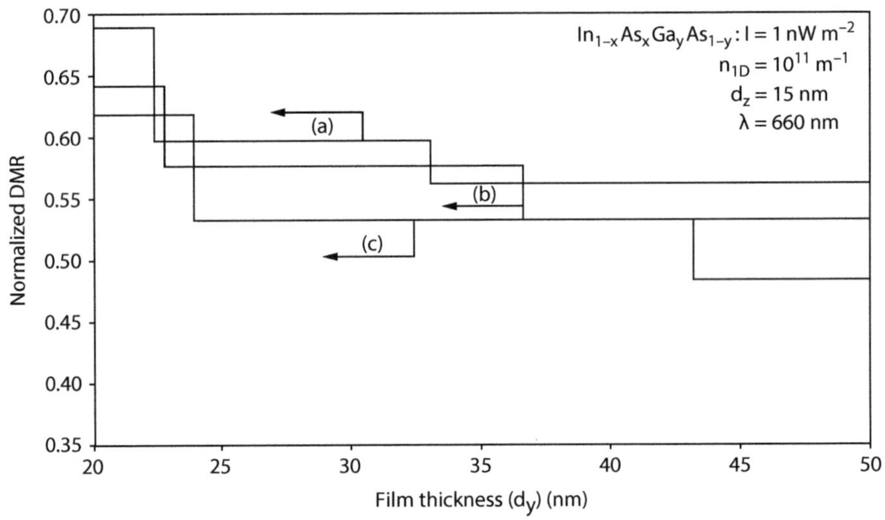

Figure 3.70: Plot of the normalized 1D DMR as a function of film thickness for QWs of HD n-$In_{1-x}Ga_xAs_yP_{1-y}$ in the presence of light waves, where the curves (a), (b), and (c) represent the perturbed three- and two-band models of Kane and that of the parabolic energy bands respectively.

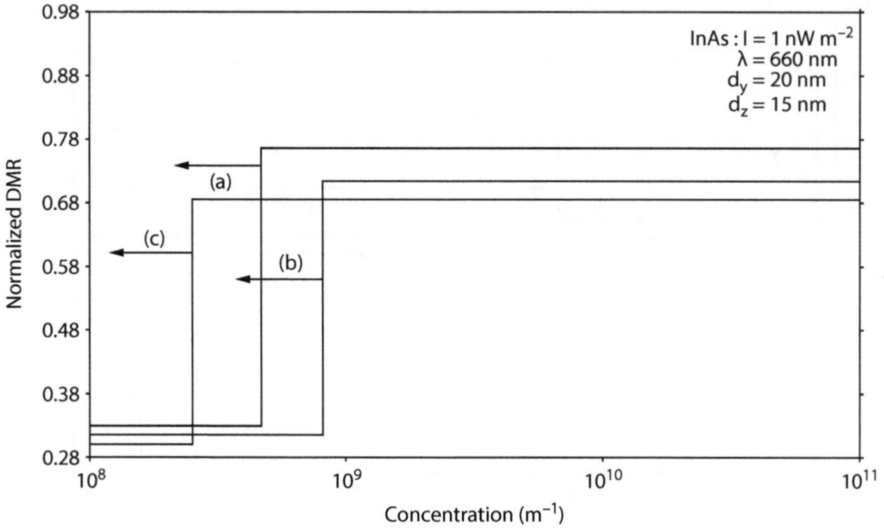

Figure 3.71: Plot of the normalized 1D DMR as a function of electron concentration per unit length for QWs of HD *n*-InAs in the presence of light waves, where the curves (a), (b), and (c) represent the perturbed three- and two-band models of Kane and that of the parabolic energy bands respectively.

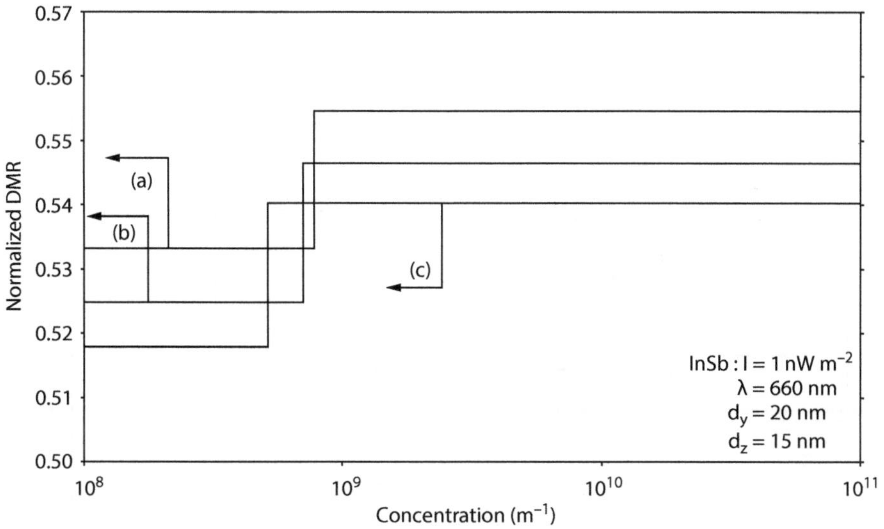

Figure 3.72: Plot of the normalized 1D DMR as a function of electron concentration per unit length for QWs of HD *n*-InSb in the presence of light waves, where the curves (a), (b), and (c) represent the perturbed three- and two-band models of Kane and that of the parabolic energy bands respectively.

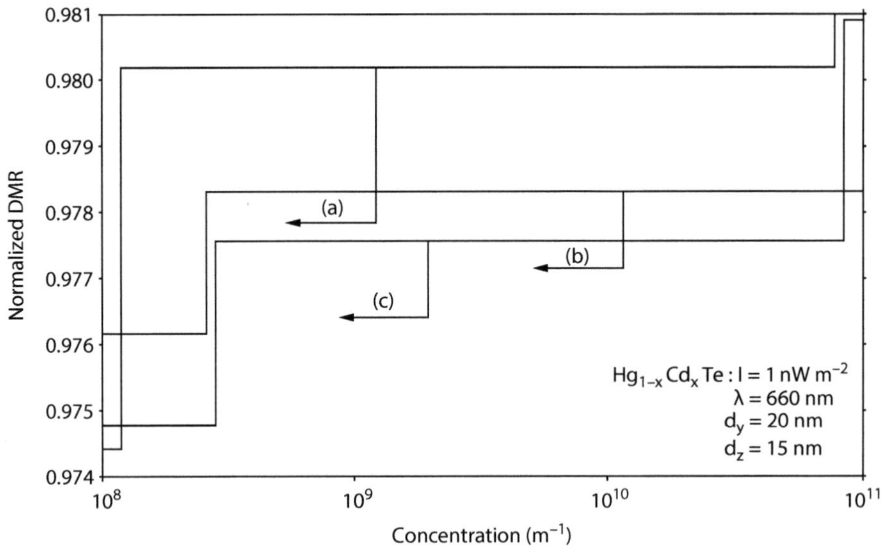

Figure 3.73: Plot of the normalized 1D DMR as a function of electron concentration per unit length for QWs of HD n-$Hg_{1-x}Cd_xTe$ in the presence of light waves, where the curves (a), (b), and (c) represent the perturbed three- and two-band models of Kane and that of the parabolic energy bands respectively.

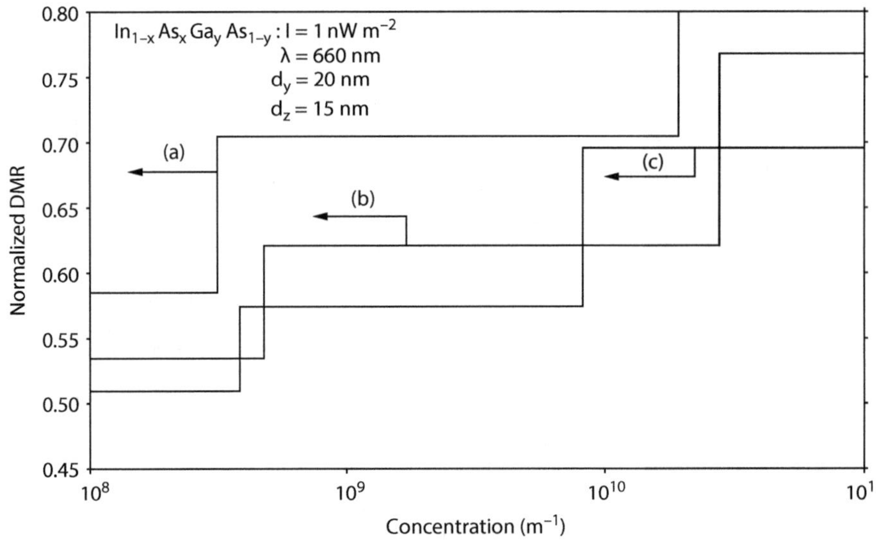

Figure 3.74: Plot of the normalized 1D DMR as a function of electron concentration per unit length for QWs of HD n-$In_{1-x}Ga_xAs_yP_{1-y}$ lattice matched to InP in the presence of light waves, where the curves (a), (b), and (c) represent the perturbed three- and two-band models of Kane and that of the parabolic energy bands respectively.

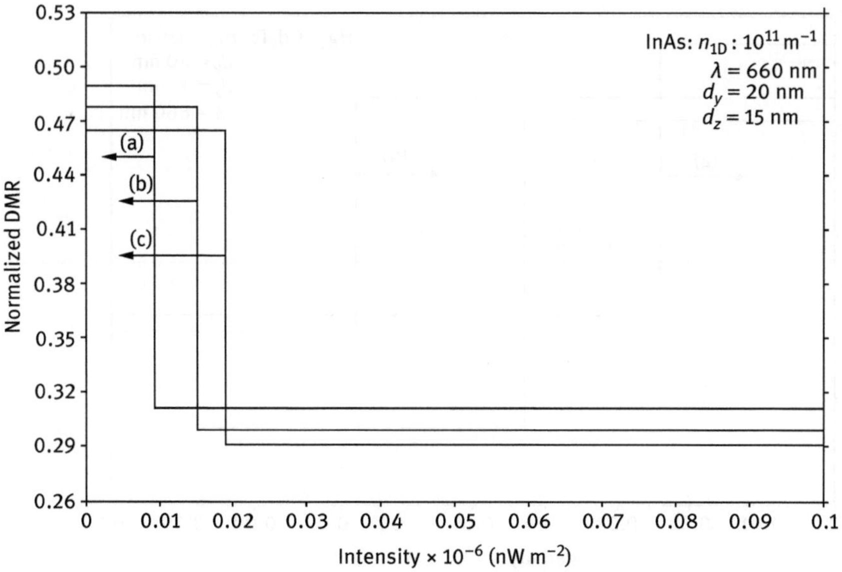

Figure 3.75: Plot of the normalized 1D DMR as a function of light intensity for QWs of HD *n*-InAs in the presence of light waves, where the curves (a), (b), and (c) represent the perturbed three- and two-band models of Kane and that of the parabolic energy bands respectively.

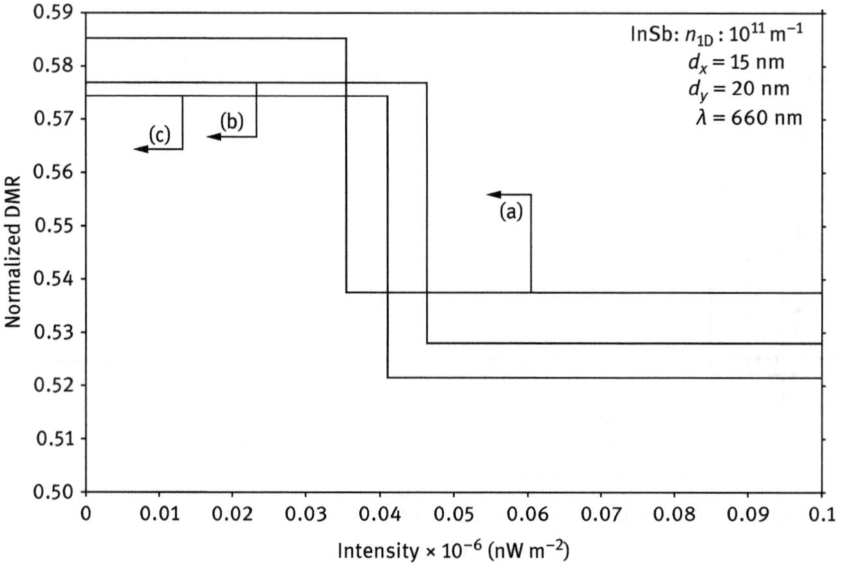

Figure 3.76: Plot of the normalized 1D DMR as a function of light intensity for QWs of HD *n*-InSb in the presence of light waves, where the curves (a), (b), and (c) represent the perturbed three- and two-band models of Kane and that of the parabolic energy bands respectively.

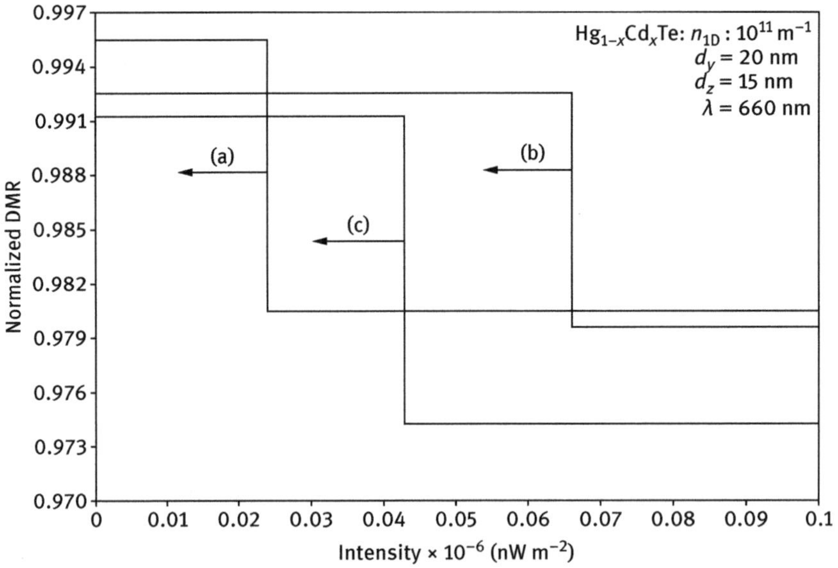

Figure 3.77: Plot of the normalized 1D DMR as a function of light intensity for QWs of HD n-$Hg_{1-x}Cd_xTe$ in the presence of light waves, where the curves (a), (b), and (c) represent the perturbed three- and two-band models of Kane and that of the parabolic energy bands respectively.

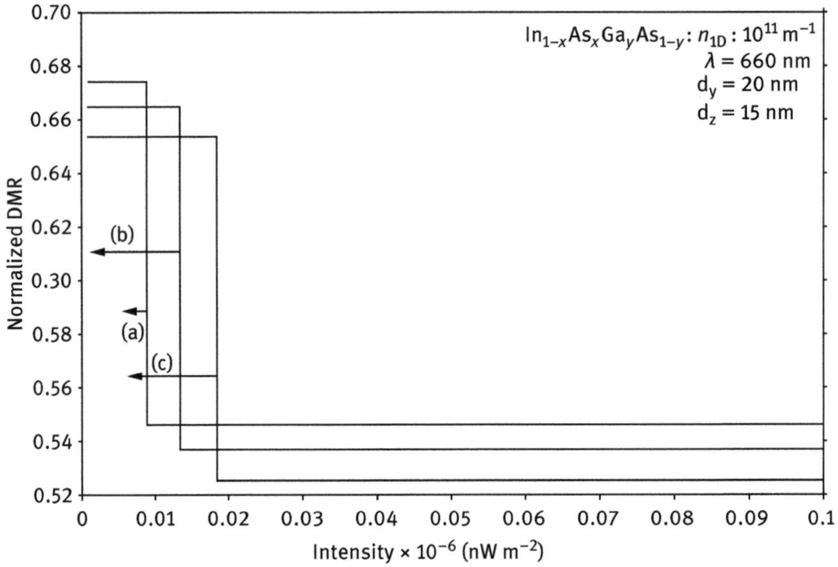

Figure 3.78: Plot of the normalized 1D DMR as a function of light intensity for QWs of HD n-$In_{1-x}Ga_xAs_yP_{1-y}$ lattice matched to InP in the presence of light waves, where the curves (a), (b), and (c) represent the perturbed three- and two-band models of Kane and that of the parabolic energy bands respectively.

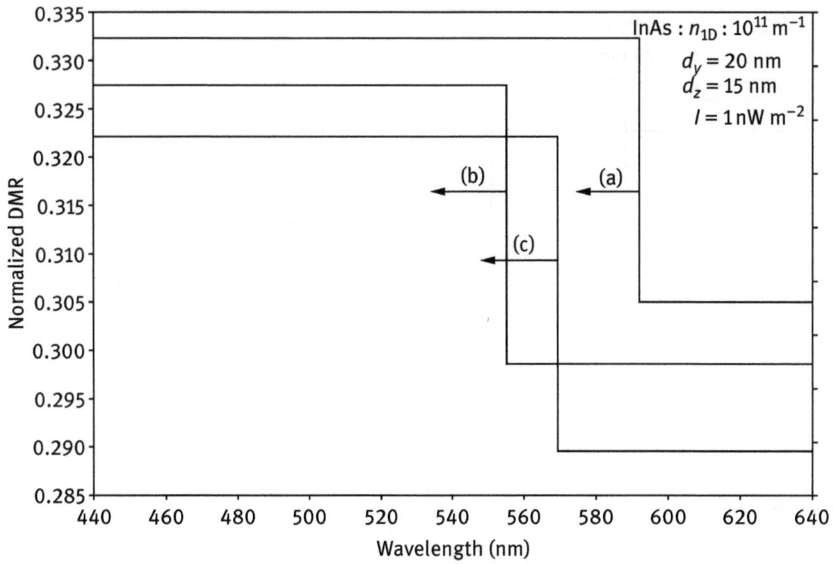

Figure 3.79: Plot of the normalized 1D DMR as a function of wavelength for QWs of HD *n*-InAs in the presence of light waves, where the curves (a), (b), and (c) represent the perturbed three- and two-band models of Kane and that of the parabolic energy bands respectively.

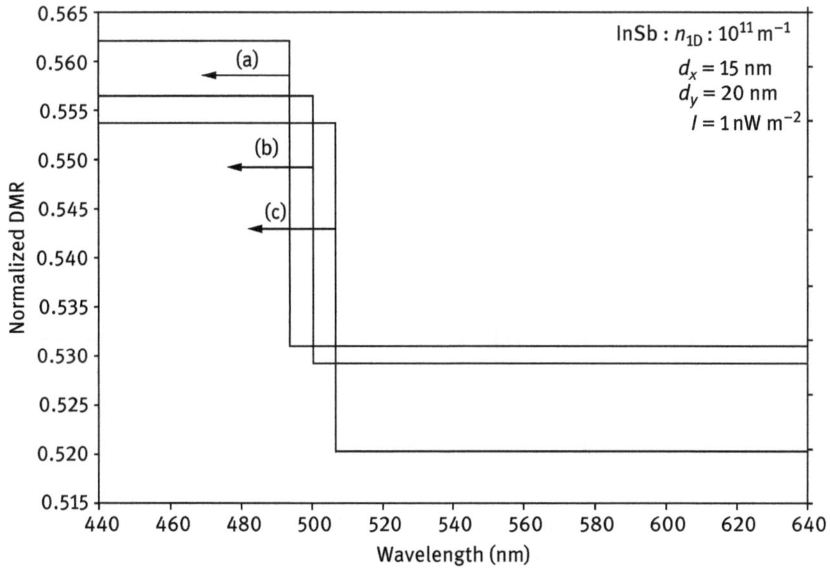

Figure 3.80: Plot of the normalized 1D DMR as a function of wavelength for QWs of HD *n*-InSb in the presence of light waves, where the curves (a), (b), and (c) represent the perturbed three- and two-band models of Kane and that of the parabolic energy bands respectively.

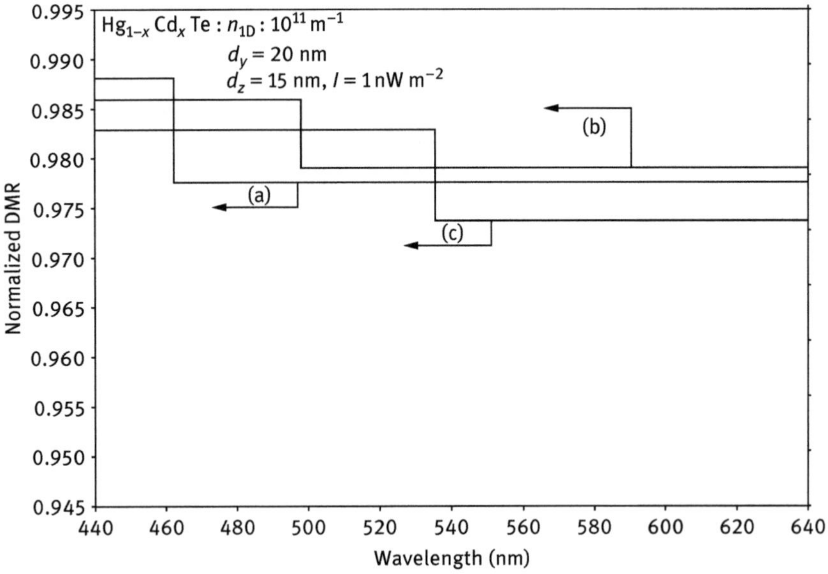

Figure 3.81: Plot of the normalized 1D DMR as a function of wavelength for QWs of HD n-$Hg_{1-x}Cd_xTe$ in the presence of light waves, where the curves (a), (b), and (c) represent the perturbed three- and two-band models of Kane and that of the parabolic energy bands respectively.

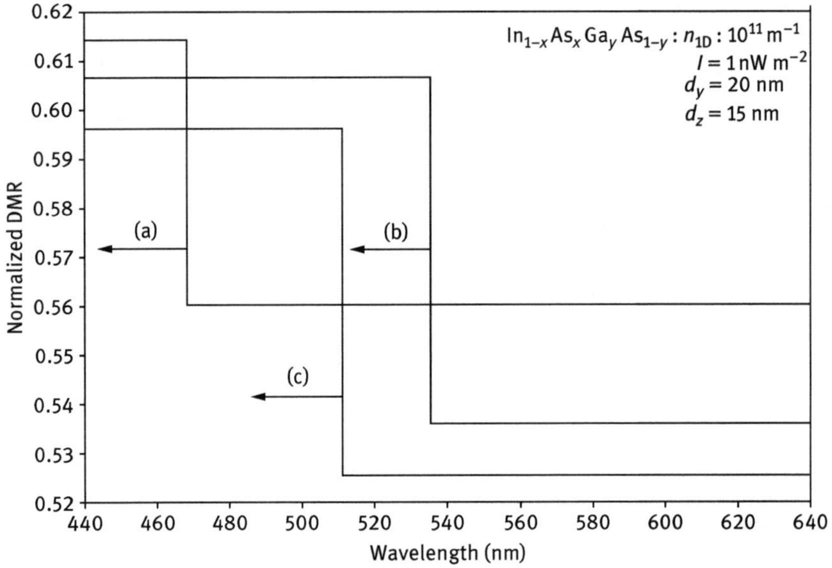

Figure 3.82: Plot of the normalized 1D DMR as a function of wavelength for QWs of HD n-$In_{1-x}Ga_xAs_yP_{1-y}$ lattice matched to InP in the presence of light waves, where the curves (a), (b), and (c) represent the perturbed three- and two-band models of Kane and that of the parabolic energy bands respectively.

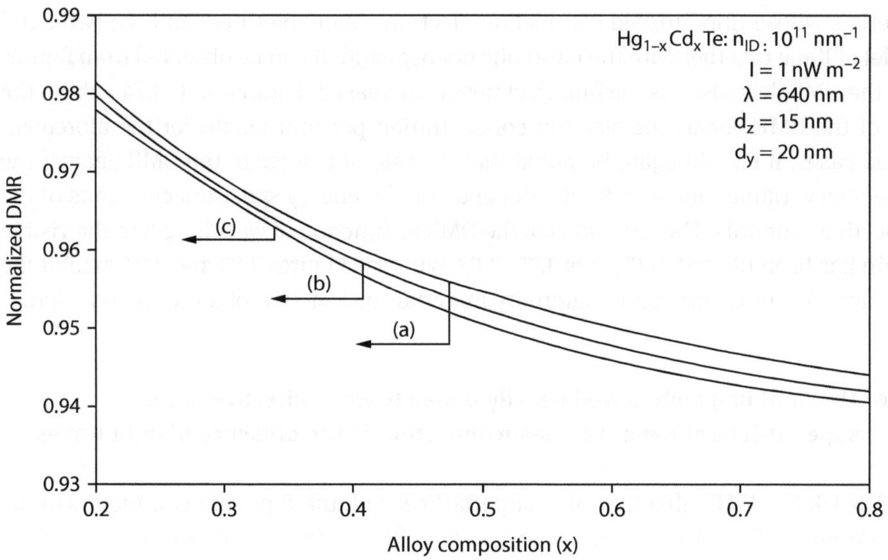

Figure 3.83: Plot of the normalized 1D DMR as a function of alloy composition for QWs of HD *n*-$Hg_{1-x}Cd_xTe$ in the presence of light waves, where the curves (a), (b), and (c) represent the perturbed three- and two-band models of Kane and that of the parabolic energy bands respectively.

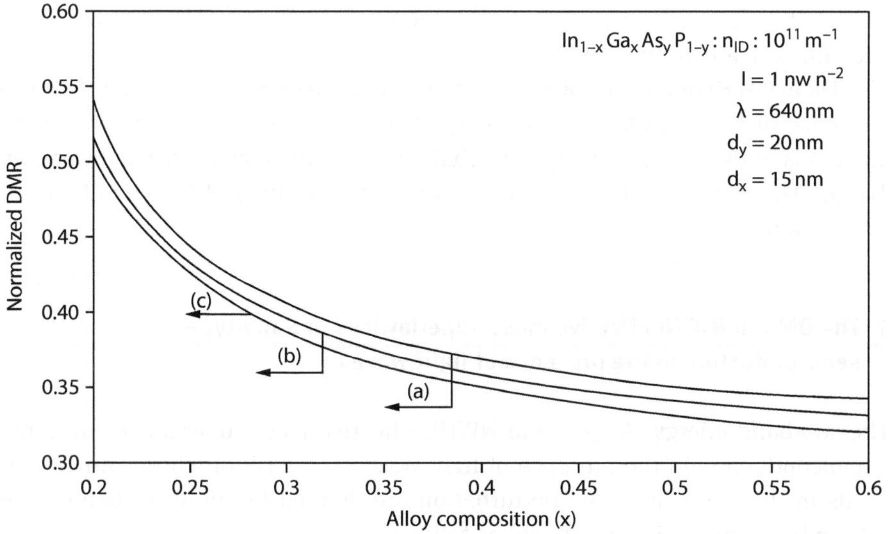

Figure 3.84: Plot of the normalized 1D DMR as a function of alloy composition for QWs of *HD* *n*-$In_{1-x}Ga_xAs_yP_{1-y}$ lattice matched to InP in the presence of light waves, where the curves (a), (b), and (c) represent the perturbed three- and two-band models of Kane and that of the parabolic energy bands respectively.

materials whose unperturbed conduction electrons obey the three- and the two-band model of Kane together with the parabolic energy band. It can be observed from figures that the DMR decreases as the film thickness is increased. Figures 3.71–3.74 exhibit the plot of the DMR versus the electron concentration per unit length for the aforementioned cases. It should again be noted that the rate of change of the DMR against the respective variable functions totally depends on the energy spectrum constants of the respective materials. The variations of the DMR as function of wavelength in the visible region has been plotted in Figures 3.75–3.82, whereas Figures 3.83 and 3.84 exhibit the same for QWs of ternary and quaternary materials as function of alloy composition.

3.2.6 The DMR in quantum well heavily doped (QWHD) effective mass superlattices of Kane-type semiconductors in the presence of light waves

(a) The DR in QWHD effective mass superlattices of Kane-type semiconductors in the presence of light waves is given in eq. (1.341) – the DR of whose constituent materials in the absence of any perturbation are defined by the three-band Kane model. The electron concentration per unit area and the DMR have to be evaluated numerically in this case.

(b) The DR in QWHD effective mass superlattices of Kane-type semiconductors in the presence of light waves is given in eq. (1.343) – the DR of whose constituent materials in the absence of any perturbation are defined by the two-band Kane model.The electron concentration per unit area and the DMR have to be evaluated numerically in this case.

(c) The DR in QWHD effective mass superlattices of Kane-type semiconductors in the presence of light waves is given in eq. (1.345) – the DR of whose constituent materials in the absence of any perturbation are defined by the parabolic energy bands. The electron concentration per unit area and the DMR have to be evaluated numerically in this case

3.2.7 The DMR in NWHD effective mass superlattices of Kane-type semiconductors in the presence of light waves

(a) The sub-band energy ($E_{nSL5HD4}$) in NWHD effective mass superlattices of Kane-type semiconductors in the presence of light waves – the DR of whose constituent materials in the absence of any perturbation are defined by the three-band Kane model can be written using eq. (1.346) as follows:

$$\left[\frac{1}{L_0^2}\left\{\cos^{-1}\left(f_{HD1}\left(E_{nSL5HD4},\frac{n_y\pi}{d_y},\frac{n_z\pi}{d_z},\lambda\right)\right)\right\}^2-\left[\left(\frac{n_y\pi}{d_y}\right)^2+\left(\frac{n_z\pi}{d_z}\right)^2\right]\right]=0 \quad (3.92)$$

The DMR for HD materials in this case can be written as follows:

$$\frac{D}{\mu} = \text{Real Part of}\left[\frac{n_0}{e}\left[\frac{\partial n_0}{\partial(E_{F1.2.13} - E_{nSL5HD4})}\right]^{-1}\right] \tag{3.93}$$

Thus, using eqs. (1.347), (3.92), and (3.93), we can find the DMR in this case.

(b) The sub-band energy $E_{nSL5HD5}$ in NWHD effective mass superlattices of Kane-type semiconductors in the presence of light waves, the DR of whose constituent materials in the absence of any perturbation are defined by the two-band Kane model can be written by following eq. (1.349) as follows:

$$\left[\frac{1}{L_0^2}\left\{\cos^{-1}\left(f_{HD2}\left(E_{nSL5HD5}, \frac{n_y\pi}{d_y}, \frac{n_z\pi}{d_z}, \lambda\right)\right)\right\}^2 - \left[\left(\frac{n_y\pi}{d_y}\right)^2 + \left(\frac{n_z\pi}{d_z}\right)^2\right]\right] = 0] \tag{3.94}$$

The DMR in this case can be written as follows:

$$\frac{D}{\mu} = \frac{n_0}{e}\left[\frac{\partial n_0}{\partial(E_{F1.2.131} - E_{nSL5HD5})}\right]^{-1} \tag{3.95}$$

Thus, using eqs. (1.350), (3.94), and (3.95), we can find the DMR in this case.

(c) The sub-band energy $(E_{nSL5HD6})$in NWHD effective mass superlattices of Kane-type semiconductors in the presence of light waves – the DR of whose constituent materials in the absence of any perturbation are defined by the parabolic energy bands can be written as follows using eq. (1.352) :

$$\left[\frac{1}{L_0^2}\left\{\cos^{-1}\left(f_{HD3}\left(E_{nSL5HD6}, \frac{n_y\pi}{d_y}, \frac{n_z\pi}{d_z}, \lambda\right)\right)\right\}^2 - \left[\left(\frac{n_y\pi}{d_y}\right)^2 + \left(\frac{n_z\pi}{d_z}\right)^2\right]\right] = 0 \tag{3.96}$$

The DMR in this case can be written as follows:

$$\frac{D}{\mu} = \frac{n_0}{e}\left[\frac{\partial n_0}{\partial(E_{F1.2.132} - E_{nSL5HD6})}\right]^{-1} \tag{3.97}$$

Thus, using eqs. (1.353), (3.96), and (3.97), we can find the DMR in this case.

3.2.8 The DMR in QWHD superlattices of Kane-type semiconductors with graded interfaces in the presence of light waves

The simplified DR of HDQWs of III–V superlattices with graded interfaces is given by eq. (1.375). The electron concentration and the DMR have to be evaluated numerically.

3.2.9 The DMR in NWHD superlattices of Kane-type semiconductors with graded interfaces in the presence of light waves

The DMR for HD materials in this case can be written as follows:

$$\frac{D}{\mu} = \text{Real Part of}\left[\frac{n_0}{e}\left[\frac{\partial n_0}{\partial(E_{F8.17.51} - E_{8.17.52})}\right]^{-1}\right] \tag{3.98}$$

The $(E_{8.17.52})$may be formulated from the following equation:

$$G_{8,17,50} + iH_{8,17,50}]|_{E = E_{F8.17,52}}] = 0 \tag{3.99}$$

Thus, using eqs. (1.377a), (3.98b), and (3.99), we can study the DMR in this case.

3.2.10 The magneto DMR in HD super lattices of Kane-type semiconductors with graded interfaces in the presence of light waves

The DMR for HD materials in this case can be written as follows:

$$\frac{D}{\mu} = \text{Real Part of}\left[\frac{n_0}{e}\left[\frac{\partial n_0}{\partial(E_{F8.17.54} - E_{8.17.55})}\right]^{-1}\right] \tag{3.100}$$

The $(E_{8.17.55})$may be formulated from the following equation:

$$\left[G_{8,17,54} + iH_{8,17,54}\right]|_{E = E_{F8.17,55}}] = 0 \tag{3.101}$$

Thus, using eqs. (1.383), (3.100), and (3.101), we can study the DMR in this case.

3.3 Open research problems

(R.3.1) Investigate the DMR in the presence of intense external light waves for all the HD materials whose respective DR of the carriers in the absence of any field are given in R 1.1.

(R.3.2) Investigate the DMR for heavily doped semiconductors in the presence of Gaussian, exponential, Kane, Halperian, Lax, and Bonch–Burevich types of band tails for all systems whose unperturbed carrier energy spectra are defined in the presence of external light waves in (R 1.1).

(R.3.3) Investigate the DMR in the presence of external light waves for bulk specimens of the negative refractive index, organic, magnetic, and other advanced optical materials in the presence of an arbitrarily oriented alternating electric field.

(R.3.4) Investigate all the appropriate problems of this chapter for a Dirac electron.

(R.3.5) Investigate all the appropriate problems of this chapter by including the many body, broadening, and hot carrier effects respectively.

(R.3.6) Investigate all the appropriate problems of this chapter by removing all the mathematical approximations and establishing the respective appropriate uniqueness conditions.

References

[1] P.T. Landsberg, Eur. J. Phys, 2, 213 (1981).

[2] J.S. Blakemore, *Semiconductor Statistics* (Dover, New York, 1987).

[3] K.P. Ghatak, S. Bhattacharya, S. K. Biswas, A. Dey and A. K. Dasgupta, Phys. Scr., 75, 820 (2007).

[4] A. Einstein, Ann. der Physik, 17, 549 (1905).

[5] B.R. Nag and A. N. Chakravarti, Phys. Stal Sol. (a), 67, K113 (1981).

4 Heisenberg's uncertainty principle and the screening length in heavily doped optoelectronic nano materials in the presence of intense light waves

Blessed are those who can give without remembering but take without forgetting.

4.1 Introduction

In Section 4.2, we formulate the screening length (SL) in III–V, ternary, and quaternary heavily-doped (HD) materials in the presence of intense light waves by formulating the electron statistics using Heisenberg's Uncertainty Principle. SL has been investigated numerically by taking HD n-InAs and n-InSb as examples of III-V compounds, HD n-$Hg_{1-x}Cd_xTe$ as an example of ternary compounds and HD n-$In_{1-x}Ga_xAs_yP_{1-y}$ lattice matched to InP as an example of quaternary compounds in accordance with the said band models for the purpose of relative assessment. Section 4.3 contains the result and discussion.

4.2 Theoretical background

4.2.1 The SL in the presence of light waves in HD III-V, ternary, and quaternary semiconductors

SL in HD materials can be written as [1–4]

$$\frac{1}{L_D^2} = \frac{e^2}{\varepsilon_{sc}} \operatorname{Re} al\, Part\, of \left[\frac{\partial n_0}{\partial (E_{FHD} - E_{OHD})} \right] \tag{4.1a}$$

Using eqs. (1.51e), (1.51f), (1.51g), and (4.1), we can study SL in this case.

For inversion layers and NIPI structures, under the condition of electric quantum limit, SL assumes the form

$$\frac{1}{L_D} = \frac{e^2}{2\varepsilon_{sc}} \operatorname{Re} al\, Part\, of \left[\frac{\partial n_{02D}}{\partial (E_{F2D} - E_{02D})} \right] \tag{4.1b}$$

In the absence of band-tails, we can write

https://doi.org/10.1515/9783110610819-004

$$\frac{1}{L_D^2} = \frac{e^2}{\varepsilon_{sc}} \left[\frac{\partial n_0}{\partial E_F} \right] \tag{4.2}$$

Using eqs. (1.52a), (1.52b), (1.52c), and (4.2), SL in this case can be expressed as where the primes denote the differention with respect to Fermi energy.

In the absence of band tails and photon energy SL for three- and two-band models of Kane under the condition of extreme degeneracy can be expressed as:

$$\frac{1}{L_D^2} = \frac{e^2}{\varepsilon_{sc}} \frac{g_v}{3\pi^2} \left(\frac{2m_c}{h^2} \right)^{3/2} \frac{3}{2} [I_{11}(E_F)]^{1/2} [I_{11}(E_F)]' \tag{4.3}$$

$$\frac{1}{L_D^2} = \frac{e^2}{\varepsilon_{sc}} \frac{g_v}{3\pi^2} \left(\frac{2m_c}{h^2} \right)^{3/2} \frac{3}{2} [E_F(1+\alpha E_F)]^{1/2} [(1+2\alpha E_F)] \tag{4.4}$$

At finite temperature, SL in accordance with two-band Kane model under the condition $\alpha E = 1$, can be expressed as

$$\frac{1}{L_D^2} = \left(\frac{e^2 N_c}{\varepsilon_{sc} k_B T} \right) \left[F_{-1/2}(\eta) + \left(\frac{15 a k_B T}{4} \right) F_{1/2}(\eta) \right] \tag{4.5}$$

For relatively wide gap materials $E_g \to \infty$, we get

$$\frac{1}{L_D^2} = \left(\frac{e^2 N_c}{\varepsilon_{sc} k_B T} \right) [F_{-1/2}(\eta)] \tag{4.6}$$

Under the condition of non-degeneracy, eq. (4.6) gets transformed into the well-known classical value of SL, which is equal to $[\varepsilon_{sc} k_B T/(e^2 n_0)]^{1/2}$, valid for both the carriers. In this conventional form, SL decreases with increasing carrier concentration at a constant temperature. It is interesting to note that under the condition of extreme degeneracy the expression of SL for materials having parabolic energy bands can be written as $L_D = (\pi^{2/3} \hbar \sqrt{\varepsilon_{sc}})(e g_v^{1/3} 3^{1/6} n_0^{1/6} \sqrt{m_c})^{-1}$. Thus we observe that the result is independent of temperature but depends on n_0,g_v, and m_c. Besides, the indices of inverse electron variation changes from half in the former case to one-sixth in the latter case.

4.2.2 Suggestion for the experimental determination of SL

Using the appropriate equations, SL in 3D (L_{3D}) in bulk materials assumes the form

$$L_{3D} = \left[\frac{\pi^2 k_B^2 T \varepsilon_{sc}}{3 e^3 n_0 G} \right]^{\frac{1}{2}} \tag{4.7}$$

For inversion layers, NIPI structures and quantum wells, SL in 2D (L_{2D}) can be written as:

$$L_{2D} = \frac{2\pi^2 k_B^2 T \varepsilon_{sc}}{3e^3 n_{02D} G} \tag{4.8}$$

From the experimental determination of SL in 3D for degenerate materials having arbitrary dispersion laws as given by eq. (9.4), we observe that for a constant T, SL varies inversely with the square root of Gn_0. Only the experimental values of G for any material as a function of electron concentration will generate the experimental values of SL in 3D for that range of $n_0$for that material. Since $(Gn_0)^{-1/2}$ decreases with increasing n_0 for a constant T, from eq. (4.7) we can conclude that SL in 3D will decrease with increasingn_0. For SL in 2D, L_D at a constant temperature varies inversely with $n_{02D}G$ as it appears from eq. (4.8). Since $n_{02D}G$ increases with decreasing surface concentration, from eq. (4.8) we can infer that 2D SL will increase with decreasing n_{02D} for the appropriate 2D systems. This statement provides a compatibility test of our theoretical analysis. Thus, eqs. (4.8) and (4.9) provide experimental checks of both 3D and 2D SLs and also a technique for probing the band structures of the materials having arbitrary band structures.

4.2.3 Results and discussion

Using the appropriate equations, we have plotted SL as a function of electron concentration at T = 4.2 K in Figures 4.1–4.4 for HD n-InAs, n-InSb, n-$Hg_{1-x}Cd_xTe$, and HD n-$In_{1-x}Ga_xAs_yP_{1-y}$ lattice matched to InP as examples of III-V, ternary, and quaternary materials, respectively, which are used for the purpose of numerical computations in accordance with three- and two-band models of Kane together with the parabolic energy bands. In Figures 4.1–4.4, we plot the classical SL equation for the purpose of fixing the reference. In Figures 4.1–4.4, we observe that SL decreases from the light off case to the light on case, since the value of the Fermi energy in the presence of light waves increases due to the increase in the carrier concentration as compared with the same in the absence of photoexcitation. Therefore, the numerical magnitude of SL in the presence of light is smaller as compared with the same in the light off case in the whole range of the concentration considered, although SL decreases with increase in carrier degeneracy. The combined influence of the energy band constants on SL for HD n-InAs and HD n-InSb can easily be assessed from Figures 4.1 and 4.2. For the purpose of relative assessment, all the plots in the absence of light waves have further been drawn. In Figures 4.5–4.8, we have plotted SL as a function of light intensity and observe that SL decreases with increasing light intensity for the materials, whereas in the absence of external photoexcitation, SL is independent of intensity. In Figures 4.9–4.12, we have plotted SL as a function of wavelengths in the visible region for HD n-InAs, n-$Hg_{1-x}Cd_xTe$, and HD n-$In_{1-x}Ga_xAs_yP_{1-y}$ lattice matched to InP for all types of

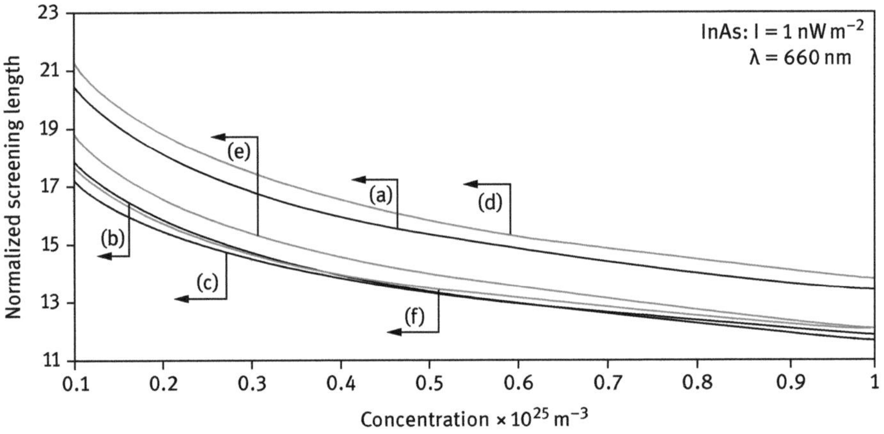

Figure 4.1: Plot of SL as a function of electron concentration for HD n-InAs in the presence of light waves in which the curves (a), (b), and (c) represent the three- and two-band models of Kane and that of the parabolic energy bands, respectively. The curves (d), (e), and (f) represent the same in the absence of external photoexcitation. The plot (g) indicates the classical SL equation.

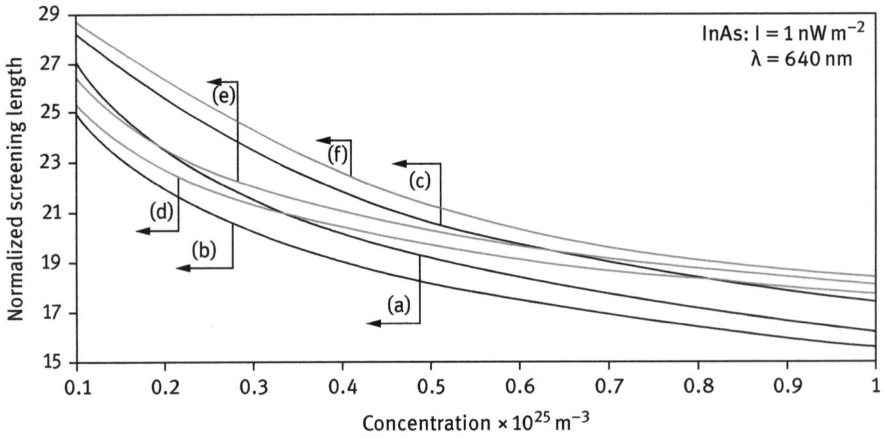

Figure 4.2: Plot of SL as a function of electron concentration for HD n-InSb for all cases of Figure 4.1.

energy band models. SL decreases as the wavelength shifts from red color to violet. In Figures 4.13 and 4.14, SL has been plotted as a function of alloy composition for HD n-$Hg_{1-x}Cd_xTe$ and HD n-$In_{1-x}Ga_xAs_yP_{1-y}$ lattice matched to InP in which all the cases of Figure 4.1 have further been plotted for the purpose of relative comparison. The plots of Figure 4.9 are valid for $x > 0.17$, since for $x < 0.17$, the band gap becomes negative in n-$Hg_{1-x}Cd_xTe$ leading to semi-metallic state. The plots of Figure 4.14 exhibit the variation of SL with the alloy composition for HD n-$In_{1-x}Ga_xAs_yP_{1-y}$ lattice

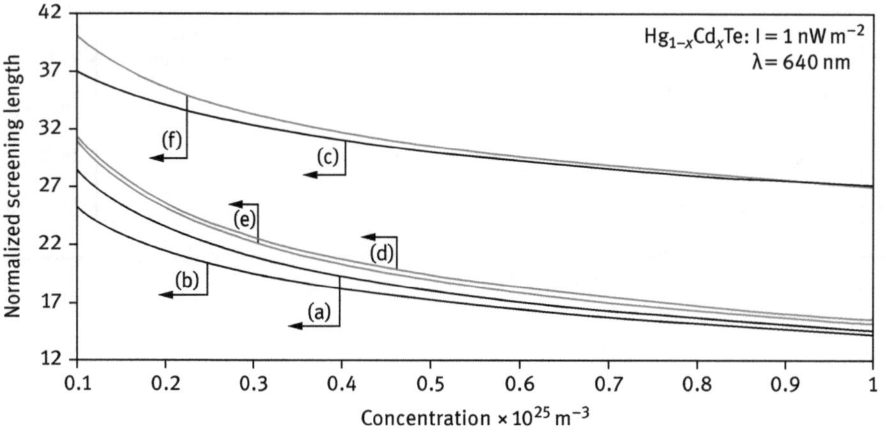

Figure 4.3: Plot of SL as a function of electron concentration for HD n-$Hg_{1-x}Cd_xTe$ for all cases of Figure 4.1.

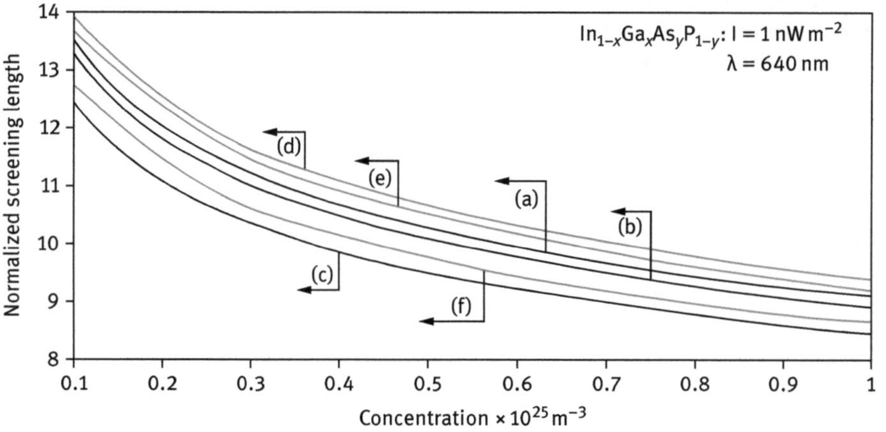

Figure 4.4: Plot of SL as a function of electron concentration for HD n-$In_{1-x}Ga_xAs_yP_{1-y}$ lattice matched to InP for all cases of Figure 4.1.

matched to InP. In Figures 4.5–4.14, we have further included the classical SL equation for the purpose of fixation of the reference.

The influence of light is immediately apparent from the plots in Figures 4.9–4.16 since SL depends strongly on I and, which is in direct contrast to the corresponding cases for the bulk specimens of the said compounds in the absence of external photoexcitation. The variations of SLs in Figures 4.9–4.16 reflect the direct signature of the light wave on the electronic, optic, and the other band structure-dependent properties of semiconducting materials in the presence of light waves and

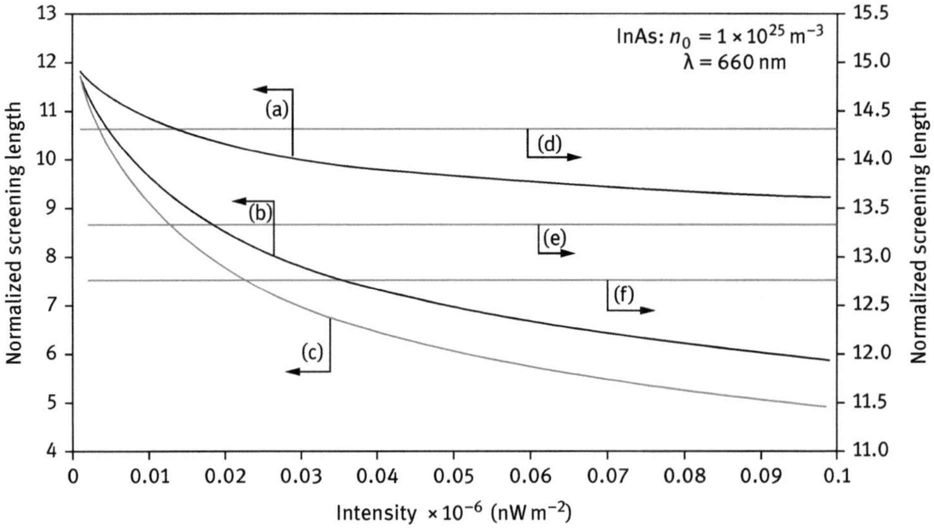

Figure 4.5: Plot of SL as a function of light intensity for HD n-InAs in which the curves (a), (b), and (c) represent the three- and two-band models of Kane and that of parabolic energy bands, respectively. The curves (d), (e), and (f) represent the same in the absence of external photoexcitation. The plot (g) indicates the classical SL equation.

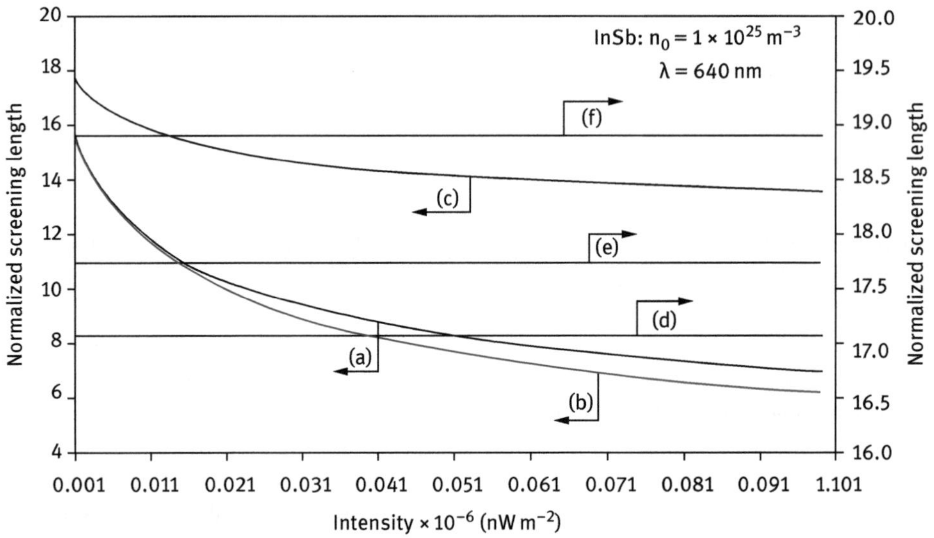

Figure 4.6: Plot of SL as a function of light intensity for HD n-InSb for all cases of Figure 4.5.

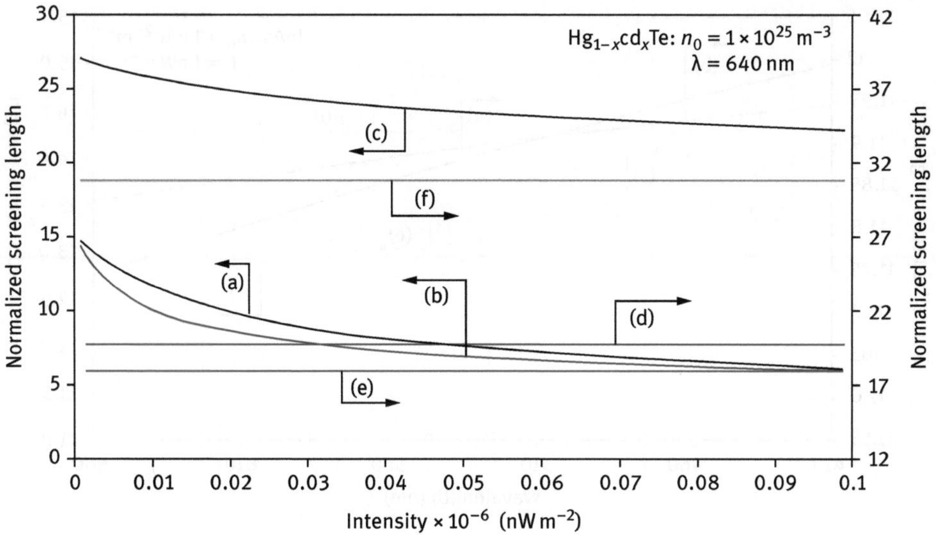

Figure 4.7: Plot of SL as a function of light intensity for HD n-$Hg_{1-x}Cd_xTe$ for all cases of Figure 4.5

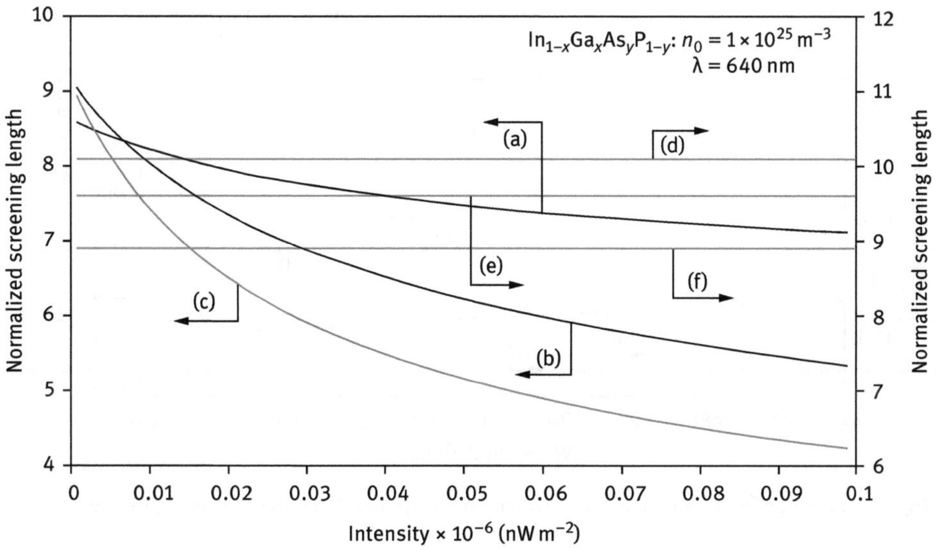

Figure 4.8: Plot of SL as a function of light intensity for HD n-$In_{1-x}Ga_xAs_yP_{1-y}$ lattice matched to InP for all cases of Figure 4.5.

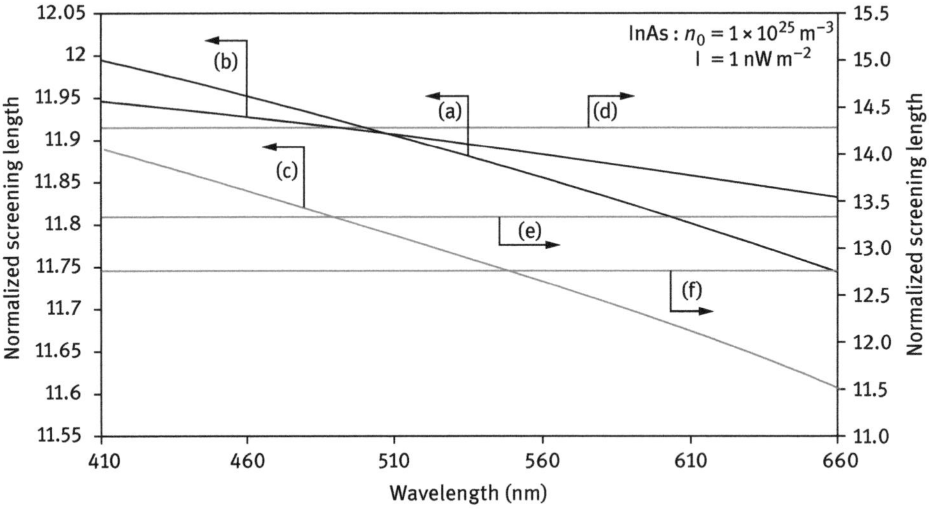

Figure 4.9: Plot of SL as a function of wavelength for HD n-InAs in which the curves (a), (b), and (c) represent the three- and two-band models of Kane and that of parabolic energy bands, respectively. The curves (d), (e), and (f) represent the same in the absence of external photoexcitation. The plot (g) indicates the classical SL equation.

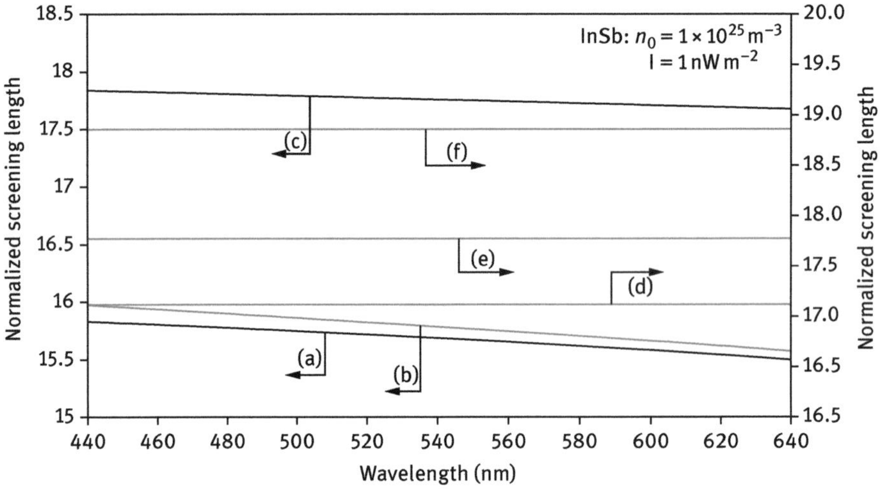

Figure 4.10: Plot of SL as a function of wavelength for HD n-InSb for all cases of Figure 4.9

the photon-assisted transport for the corresponding semiconductor devices. Although SL tends to decrease with the intensity and the wavelength, the rate of decrease is totally dependent on band structure. The numerical values of SL are greatest for ternary materials and least for quaternary compounds. We note that our basic

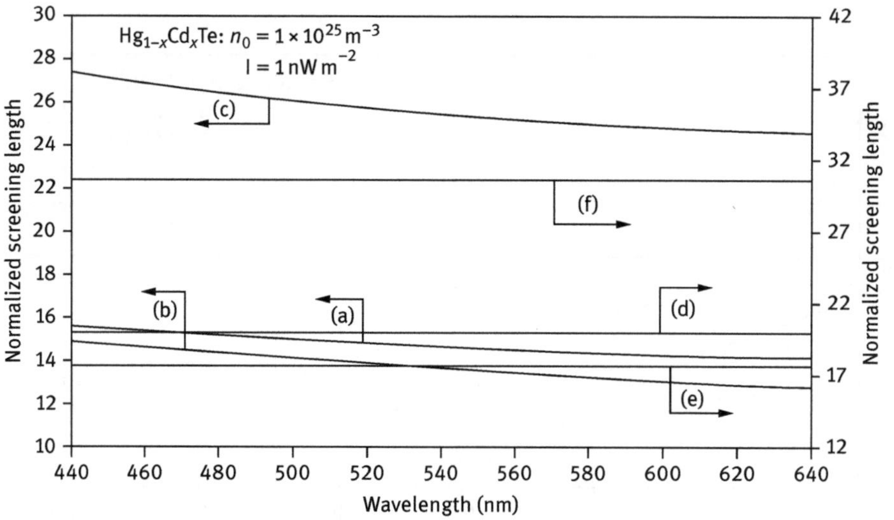

Figure 4.11: Plot of SL as a function of wavelength for HD n-$Hg_{1-x}Cd_xTe$ for all cases of Figure 4.9.

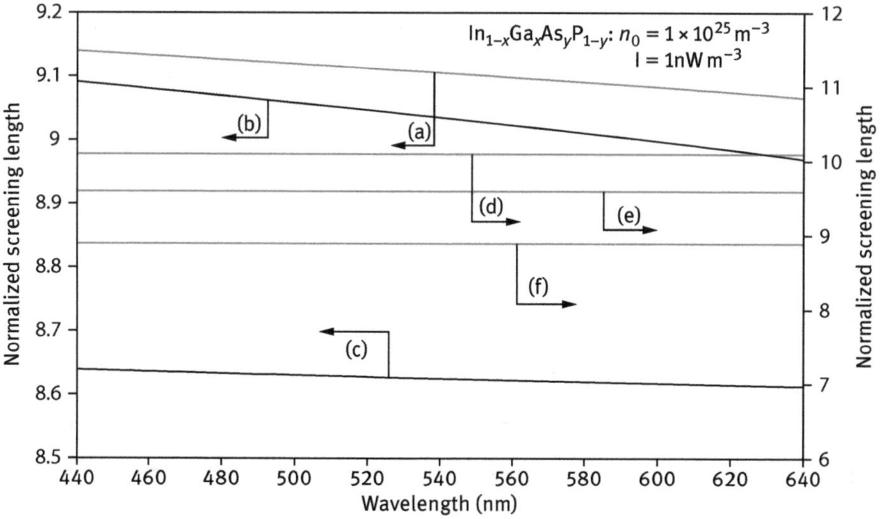

Figure 4.12: Plot of SL as a function of wavelength for HD n-$In_{1-x}Ga_xAs_yP_{1-y}$ lattice matched to InP for all cases of Figure 4.9.

eq. (4.41) covers various materials having different energy band structures. In this chapter, the concentration, the alloy composition, the light intensity and the wavelength dependences of SL for HD n-InAs, n-InSb, n-$Hg_{1-x}Cd_xTe$, and HD n-$In_{1-x}Ga_xAs_yP_{1-y}$ lattice matched to InP have been studied. Thus, we have covered a wide class of optoelectronic and allied compounds whose energy band structures are defined by the

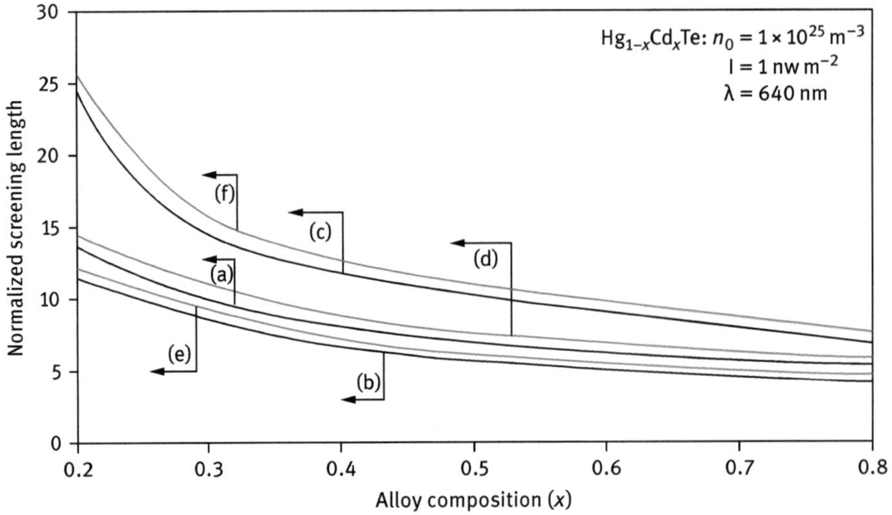

Figure 4.13: Plot of SL as a function of alloy composition for HD n- $Hg_{1-x}Cd_xTe$ in presence of light waves in which the curves (a), (b), and (c) represent the three- and two-band models of Kane and that of parabolic energy band, respectively. The curves (d), (e), and (f) represent the same in the absence of external photoexcitation. The plot (g) indicates the classical SL equation.

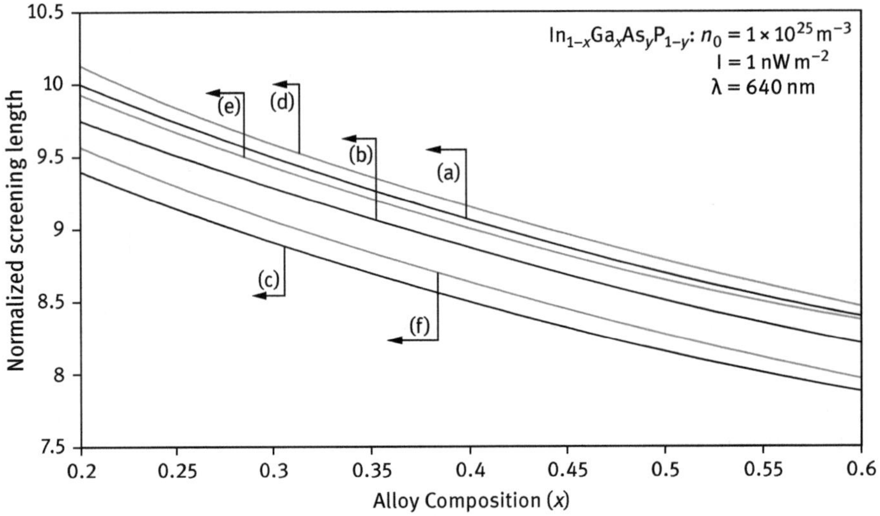

Figure 4.14: Plot of SL as a function of alloy composition for HD n-$In_{1-x}Ga_xAs_yP_{1-y}$ lattice matched to InP for all cases of 4.13.

three-and two-band models of Kane in the absence of photon field. Under certain limiting conditions, all the results of SLs for different materials having various band structures lead to the well-known classical SL equation. This indirect test not only exhibits the mathematical compatibility of our formulation but also shows the fact that our simple analysis is a more generalized one, since one can obtain the corresponding results for the relatively wide-gap non-degenerate materials having parabolic energy bands under certain limiting conditions from our present derivation. We can conclude that the influence of the presence of an external photoexcitation is to change radically the original band structure of the material. Because of this change the photon field causes to increase the band gap of semiconductor. The numerical results presented in this chapter would be different for other materials, but the nature of variation would be unaltered. The theoretical results as given here would be useful in analyzing various other experimental data related to this phenomenon.

4.2.4 2D SL systems of III-V, ternary, and quaternary semiconductors under external photoexcitation

4.2.4.1 Introduction

This section formulates the expression for the surface electron concentration and 2D SL for ultrathin films of the aforementioned materials in the presence of photoexcitation. In Section 4.2.4.2.2, SL in doping super lattices of HD Kane Type Semiconductors in the Presence of Light Waves has been studied. In Section 4.2.4.2.3, SL in accumulation and inversion layers of Kane Type Semiconductors in the Presence of Light Waves has been studied. In Section 4.2.4.2.4, 2D SL has been numerically investigated by taking HD ultrathin films of HD n-InAs and HD n-InSb as examples of III-V compounds, HD n-$Hg_{1-x}Cd_xTe$ as an example of ternary compounds and HD n-$In_{1-x}Ga_xAs_yP_{1-y}$ lattice matched to InP as examples of quaternary materials in accordance with the three- and the two-band models of Kane together with parabolic energy bands, respectively, for the purpose of relative comparison both in the presence and absence of photoexcitation.

4.2.4.2 Theoretical background

4.2.4.2.1 Formulation of 2D SL in the presence of light waves in ultrathin films of III-V, ternary, and quaternary semiconductors

SL in this case for HD materials can be written as

$$\frac{1}{L_D} = \frac{e^2}{2\varepsilon_{sc}} \text{Real Part of} \left[\frac{\partial n_0}{\partial(E_{F2DLHD} - E_{0H2D})}\right] \tag{4.9}$$

where E_{0H2D} is the sub-band energy in this case.

Using eqs. (1.136) and (4.9), we can study SL in this case.

2D SL in QWs of III-V, ternary, and quaternary materials in the absence of band tails, whose unperturbed band structure is defined by the three band model of Kane, in the presence of light waves, can be expressed as

$$\frac{1}{L_D} = \frac{e^2}{2\varepsilon_{sc}}\left[\frac{\partial n_0}{\partial E_{F2DL}}\right] \tag{4.10}$$

Using eqs. (1.140) and (4.10), we can write

$$\frac{1}{L_D} = \frac{e^2 m_c g_v}{2\varepsilon_{sc}\pi\hbar^2}\left[\sum_{n_z=1}^{n_{z\max}}[\beta'_0(E_{F2DL}, n_z, \lambda)]\right] \tag{4.11}$$

In the absence of band tails and light waves and for isotropic three-band model of Kane, 2D SL in this case can be written as

$$\frac{1}{L_D} = \frac{e^2 m_c g_v}{2\varepsilon_{sc}\pi\hbar^2}\left[\sum_{n_z=1}^{n_{z\max}}[T_{53}(E_{Fs}, n_z)]'\right] \tag{4.12}$$

2D SL in QWs of HD III-V, ternary, and quaternary materials, whose unperturbed band structure is defined by the two-band model of Kane, in the presence of light waves, can be investigated by using eqs. (4.9) and (1.152), respectively.

2D SL in QWs of III-V, ternary, and quaternary materials absence of band tails, whose unperturbed band structure is defined by the two-band model of Kane, in the presence of light waves, can be expressed as

$$\frac{1}{L_D} = \frac{e^2 m_c g_v}{2\varepsilon_{sc}\pi\hbar^2}\left[\sum_{n_z=1}^{n_{z\max}}[\tau_0(E_{F2DL}, n_z, \lambda)]'\right] \tag{4.13}$$

In the absence of light waves and heavy doping, 2D SL for isotropic two-band model of Kane can be written as

$$\frac{1}{L_D} = \frac{e^2 m_c g_v}{2\varepsilon_{sc}\pi\hbar^2}\left[\sum_{n_z=1}^{n_{z\max}}[(1+2\alpha E_{Fs})]\right] \tag{4.14}$$

SL at a finite temperature can be written as

$$\frac{1}{L_D} = \frac{e^2 m_c g_v}{2\varepsilon_c\pi\hbar^2}\left[\sum_{n_z=1}^{n_{z\max}}\left[(1+2\alpha E_{n_{z3}})F_{-1}(\eta_{n_1}) + 2\alpha k_B T F_0(\eta_{n_1})\right]\right] \tag{4.15}$$

Equation (4.15) is well known in the literature.

(iii) 2D *SL* in QWs of HD III-V, ternary, and quaternary materials, whose unperturbed band structure is defined by the parabolic energy bands in the presence of light waves, can be studied by using eqs. (1.164) and (4.9), respectively.

2D SL in QWs of III-V, ternary, and quaternary materials absence of band tails, whose unperturbed band structure is defined by the parabolic energy band in the presence of light waves, can be expressed as

$$\frac{1}{L_D} = \frac{e^2 m_c g_v}{2\varepsilon_{sc}\pi\hbar^2}\left[\sum_{n_z=1}^{n_{z\max}} [\rho_0(E_{F2DL}, n_z, \lambda)]'\right] \tag{4.16}$$

In the absence of light waves and heavy doping for isotropic parabolic energy band, 2D SL can be written as

$$\frac{1}{L_D} = \frac{e^2 m_c g_v}{2\varepsilon_{sc}\pi\hbar^2}\left[\sum_{n_z=1}^{n_{z\max}} [1]\right] \tag{4.17}$$

The DMR at a finite temperature can be written as

$$\frac{1}{L_D} = \frac{e^2 m_c g_v}{2\varepsilon_{sc}\pi\hbar^2 k_B T}\left[\sum_{n_z=1}^{n_{z\max}} [F_{-1}(\eta_{n_{11}})]\right] \tag{4.18}$$

4.2.4.2.2 SL in doping super lattices of HD Kane Type semiconductors in the presence of light waves

(i) The Sub-band energy (E_{3251}) in doping super lattices of HD III-V, ternary, and quaternary materials in the presence of external photoexcitation whose dispersion relation is given by eqs. (1.173) can be expressed as

$$T_1(E_{3251}, \eta_g, \lambda) = \left(n_i + \frac{1}{2}\right)\hbar\omega_{91HD}\left(E_{3251}, \eta_g, \lambda\right) \tag{4.19}$$

SL in this case can be written as

$$\frac{1}{L_D} = \frac{e^2}{2\varepsilon_{sc}}(\text{Real Part of })^x\left[\frac{\partial n_0}{\partial(E_{F2DLHDD} - E_{SE})}\right] \tag{4.20}$$

SL in this case can be studied by using eqs. (1.174), (4.19), (4.20), and $x=0$, respectively. In the absence of band tails sub-band energy in doping superlattices of III-V, ternary, and quaternary materials in the presence of external photoexcitation whose dispersion relation is given by eqs. (1.175) can be expressed as

$$\beta_0(E_{3251UP},\lambda) = \left(n_i + \frac{1}{2}\right)\hbar\omega_{911}(E_{3251UP},\lambda) \tag{4.21}$$

SL in this case for HD materials can be written as

$$\frac{1}{L_D} = \frac{e^2}{2\varepsilon_{sc}}\left[\frac{\partial n_0}{\partial(E_{F2DLD} - E_{3251UP})}\right] \tag{4.22}$$

where E_{3251UP} is the sub-band energy in this case.

Using eqs. (1.116), (4.21), and (4.22), we can study SL in this case.
SL in NIPI structures of III-V, ternary, and quaternary materials can be expressed in the absence of both band tails and light waves whose dispersion relation is given by eqs. (1.179) as

$$\frac{1}{L_D} = \frac{e^2}{2\varepsilon_{sc}}\left[\frac{\partial n_0}{\partial(\bar{E}_{Fn_i} - E_{2ni})}\right] \tag{4.23}$$

where $\bar{E}_{00}$ is the sub-band energy in this case.

The sub-band energy E_{2ni} in NIPI structures of III-V, ternary, and quaternary materials can be expressed in the absence of both band tails and light waves whose dispersion relation is given by eqs. (1.179) as

$$I_{11}(E_{2ni}) = \left(n_i + \frac{1}{2}\right)\hbar\omega_9(E_{2ni}) \tag{4.24}$$

Thus, using eqs. (4.23),(4.24), and (1.180), we can study SL in this case.

(ii) The sub-band energy (E_{3252}) in doping super lattices of HD III-V, ternary, and quaternary materials in the presence of external photoexcitation whose electrons are defined by eqs. (1.183) can be expressed as

$$T_2(E_{3252},\eta_g,\lambda) = \left(n_i + \frac{1}{2}\right)\hbar\omega_{92HD}(E_{3252},\eta_g,\lambda) \tag{4.25}$$

Using eqs. (4.20), (1.184), (4.25), and $x = 0$, we can investigate SL in this case.

In the absence of band tails, sub-band energy in doping super lattices of III-V, ternary, and quaternary materials in the presence of external photoexcitation whose unperturbed electrons are defined by eqs. (1.185) can be expressed asg

$$\tau_0(E_{3252UP},\lambda) = \left(n_i + \frac{1}{2}\right)\hbar\omega_{912}(E_{3252UP},\lambda) \tag{4.26}$$

SL in this case for HD materials can be written as

$$\frac{1}{L_D} = \frac{e^2}{2\varepsilon_{sc}}\left[\frac{\partial n_0}{\partial(E_{F2DLD} - E_{3252UP})}\right] \tag{4.27}$$

where E_{3252UP} is the sub-band energy in this case.
Using eqs. (4.26), (4.27), and (1.185), we can study SL in this case.

In the absence of band tails, sub-band energy(E_{3253UP}) in doping super lattices of III-V, ternary, and quaternary materials in the presence of external photoexcitation whose unperturbed electrons are defined by the two-band model of Kane can be expressed as

$$\tau_0(E_{3253UP}, \lambda) = \left(n_i + \frac{1}{2}\right)\hbar\omega_{912}(E_{3253UP}, \lambda) \tag{4.28}$$

SL in this case for HD materials can be written as

$$\frac{1}{L_D} = \frac{e^2}{2\varepsilon_{sc}}\left[\frac{\partial n_0}{\partial(E_{F2DLD} - E_{3253UP})}\right] \tag{4.29}$$

Using eqs. (1.187), (4.28), and (4.29), we can study SL in this case.

The sub-band energy(E_{188}) in NIPI structures of III-V, ternary, and quaternary materials can be expressed in the absence of both band tails and light waves and whose unperturbed dispersion relation is given by the two-band model of Kane as

$$E_{188}(1 + \alpha E_{188}) = \left(n_i + \frac{1}{2}\right)\hbar\omega_{10}(E_{188}) \tag{4.30}$$

SL in this case for HD materials can be written as

$$\frac{1}{L_D} = \frac{e^2}{2\varepsilon_{sc}}\left[\frac{\partial n_0}{\partial(\bar{E}_{Fn_i} - E_{188})}\right] \tag{4.31}$$

Using eqs. (1.190), (4.30), and (4.31), we can study SL in this case.

(iii) The sub-band energy (E_{3253}) in doping super lattices of HD III-V, ternary, and quaternary materials in the presence of external photoexcitation whose electrons are defined by eqs. (1.183) can be expressed as

$$T_3(E_{3255}, \eta_g, \lambda) = \left(n_i + \frac{1}{2}\right)\hbar\omega_{93HD}(E_{3255}, \eta_g, \lambda) \tag{4.32}$$

Using eqs. (4.20), (1.192), (4.32), and $x = 0$, we can investigate SL in this case.

In the absence of band tails, the sub-band energy in doping super lattices of III-V, ternary, and quaternary materials in the presence of external photoexcitation whose unperturbed electrons are defined by eq. (1.193) can be expressed as

$$\rho_0(E_{3255UP}, \lambda) = \left(n_i + \frac{1}{2}\right)\hbar\omega_{913}(E_{3255UP}, \lambda) \tag{4.33}$$

SL in this case for HD materials can be written as

$$\frac{1}{L_D} = \frac{e^2}{2\varepsilon_{sc}}\left[\frac{\partial n_0}{\partial(E_{F2DLD} - E_{3255UP})}\right] \tag{4.34}$$

Using eqs. (4.33), (4.34), and (1.195), we can study SL in this case.

In the absence of band tails, sub-band energy(E_{3256UP}) in doping super lattices of III-V, ternary, and quaternary materials in the presence of external photoexcitation whose unperturbed electrons are defined by the two-band model of Kane can be expressed as

$$\rho_0(E_{3256UP}, \lambda) = \left(n_i + \frac{1}{2}\right)\hbar\omega_{912}(E_{3256UP}, \lambda) \tag{4.35}$$

SL in this case for HD materials can be written as

$$\frac{1}{L_D} = \frac{e^2}{2\varepsilon_{sc}}\left[\frac{\partial n_0}{\partial(E_{F2DLD} - E_{3256UP})}\right] \tag{4.36}$$

Using eqs. (1.194), (4.35), and (4.36), we can study SL in this case.

The sub-band energy(E_{1881}) in NIPI structures of III-V, ternary, and quaternary materials can be expressed in the absence of both band tails and light waves and whose unperturbed dispersion relation is given by the two-band model of Kane as

$$E_{1881} = \left(n_i + \frac{1}{2}\right)\hbar\omega_{11} \tag{4.37}$$

SL in this case for HD materials can be written as

$$\frac{1}{L_D} = \frac{e^2}{2\varepsilon_{sc}}\left[\frac{\partial n_0}{\partial(\bar{E}_{Fn_i} - E_{1881})}\right] \tag{4.38}$$

Using eqs. (1.196), (4.37), and (4.38), we can study SL in this case.

4.2.4.2.3 SL in accumulation and inversion layers of Kane Type Semiconductors in the presence of light waves

(a) Under the weak electric field limit, SL in accumulation layers of HD III-V, ternary, and quaternary materials, whose unperturbed band structure is defined by the three-band model of Kane, in the presence of light waves, can be expressed as

$$\frac{1}{L_D} = \frac{e^2}{2\varepsilon_{sc}} \text{Real Part of} \left[\frac{\partial n_0}{\partial (E'_{fL} - E_{321})} \right] \tag{4.39}$$

Thus, using eqs. (1.256), (3.63), and (4.39), we can find SL in this case.

Under the weak electric field limit, SL in accumulation layers of HD III-V, ternary, and quaternary materials, whose unperturbed band structure is defined by the three-band model of Kane, in the absence of light waves, can be expressed as

$$\frac{1}{L_D} = \frac{e^2}{2\varepsilon_{sc}} \text{Real Part of} \left[\frac{\partial n_0}{\partial (E'_{f1} - E_{322})} \right] \tag{4.40}$$

Thus, using eqs. (1.261), (3.65a), and (4.40), we can find SL in this case.

In the absence of band tails and light waves, SL in III-V, ternary, and quaternary materials, whose bulk electrons obey the three-band model of Kane under the condition of weak electric field limit, assumes the form

$$\frac{1}{L_D} = \frac{e^2}{2\varepsilon_{sc}} \left[\frac{\partial n_0}{\partial (E_{Fiw} - E_{323})} \right] \tag{4.41}$$

Thus, using eqs. (1.266), (3.66), and (4.41), we can find SL in this case.

(b) Under the weak electric field limit, SL in accumulation layers of HD III-V, ternary, and quaternary materials, whose unperturbed band structure is defined by the two-band model of Kane, in the presence of light waves can be expressed as

$$\frac{1}{L_D} = \frac{e^2}{2\varepsilon_{sc}} \left[\frac{\partial n_0}{\partial (E'_{fL} - E_{3212})} \right] \tag{4.42}$$

Thus, using eqs. (1.270), (3.68), and (4.42), we can find SL in this case.

Under the weak electric field limit, SL in accumulation layers of HD III-V, ternary, and quaternary materials, whose unperturbed band structure is defined by the two-band model of Kane, in the absence of light waves, can be expressed as

$$\frac{1}{L_D} = \frac{e^2}{2\varepsilon_{sc}} \left[\frac{\partial n_0}{\partial (E'_{f1} - E_{3222})} \right] \tag{4.43}$$

Thus, using eqs. (1.275), (3.70), and (4.43), we can find SL in this case.

In the absence of band tails and light waves, SL in III-V, ternary, and quaternary materials, whose bulk electrons obey the two-band model of Kane under the condition of weak electric field limit, assumes the form

$$\frac{1}{L_D} = \frac{e^2}{2\varepsilon_{sc}}\left[\frac{\partial n_0}{\partial(E_{Fiw} - E_{3232})}\right] \tag{4.44}$$

Thus, using eqs. (1.280), (3.72), and (4.44), we can find SL in this case.

(c) Under the weak electric field limit, SL in accumulation layers of HD III-V, ternary, and quaternary materials, whose unperturbed band structure is defined by the parabolic band model, in the presence of light waves, can be expressed as

$$\frac{1}{L_D} = \frac{e^2}{2\varepsilon_{sc}}\left[\frac{\partial n_0}{\partial(E'_{fL} - E_{3213})}\right] \tag{4.45}$$

Thus, using eqs. (1.284), (3.74), and (4.45), we can find SL in this case.

Under the weak electric field limit, SL in accumulation layers of HD III-V, ternary, and quaternary materials, whose unperturbed band structure is defined by the parabolic band model in the absence of light waves, can be expressed as

$$\frac{1}{L_D} = \frac{e^2}{2\varepsilon_{sc}}\left[\frac{\partial n_0}{\partial(E'_{f1} - E_{3223})}\right] \tag{4.46}$$

Thus, using eqs. (1.289), (3.76), and (4.46), we can find SL in this case.

In the absence of band tails and light waves, SL in III-V, ternary, and quaternary materials, whose bulk electrons obey the parabolic band model under the condition of weak electric field limit, assumes the form

$$\frac{1}{L_D} = \frac{e^2}{2\varepsilon_{sc}}\left[\frac{\partial n_0}{\partial(E_{Fiw} - E_{3233})}\right] \tag{4.47}$$

Thus, using eqs. (1.294), (3.78), and (4.47), we can find SL in this case.

4.2.4.2.4 Results and discussion

Using the appropriate equations we have plotted 2D SL as a function of film thickness in the presence of photoexcitation for HD ultrathin films of n-InAs whose unperturbed electron dispersion laws are defined by the three- and two-band models of Kane together with the parabolic energy bands as shown by curves (a), (b), and (c) of Figure 4.15, respectively. The curves (d), (e), and (f) of the same figure exhibits the corresponding plots in the absence of photoexcitation. All the cases of Figure 4.15 have been drawn in

Figures 4.16, 4.17, and 4.18, for HD ultrathin films of HD n-InSb;n-$Hg_{1-x}Cd_xTe$ and HD n-$In_{1-x}Ga_xAs_yP_{1-y}$ lattice matched to InP, respectively. Figures 4.19, 4.20, 4.21, and 4.22 exhibit the plots for the aforementioned cases as a function of surface electron concentration for ultrathin films of the said materials. The dependence of 2D SL on the light intensity has been shown in Figures 4.23, 4.24, 4.25, and 4.26. Figures 4.27, 4.28, 4.29, and 4.30 exhibit the wavelength dependence of 2D SL of the said materials. From these figures we observe that 2D SL decreases from the light off case to the light on case, since the value of the Fermi energy in the presence of light waves becomes larger due to the increase in the carrier concentration as compared with the same in the absence of photoexcitation. Therefore, the numerical magnitude of 2D SL in the presence of light is smaller as compared with the same without light in the whole range of the appropriate variables as considered, although 2D SL decreases with increase in said variables.

The combined influence of the energy band constants on 2D SL for all the said compounds can easily be assessed from all figures. For the purpose of relative assessment, all the plots in the absence of light waves have further been drawn. In Figures 4.23–4.26, we observe that 2D SL decreases with increasing light intensity, whereas in the absence of external photoexcitation, the same is independent of intensity. Figures 4.27–4.30 exhibit the fact that 2D SL decreases as the wavelength shifts from red color to violet. For the ternary materials, we have taken the $x = 0.3$, since for $x < 0.17$, the band gap becomes negative in HD n-$Hg_{1-x}Cd_xTe$ leading to semimetallic state. The influence of quantum confinement on the aforementioned materials is immediately apparent from the all figures, since 2D SL depends strongly on the thickness of the size quantized materials, which is in direct contrast with their respective bulk specimens. Moreover, 2D SL for ultrathin films can become several orders of magnitude larger than of their bulk specimens, which is also a direct signature of quantum confinement. It appears from the said figures that 2D SL decreases with the increasing film thickness in a step-like manner both in the presence and absence of photoexcitation for all types of materials as considered here, although the numerical values vary widely, determined by the constants of the energy spectra. The oscillatory dependence is due to the crossing over of the Fermi level by the size quantized levels. For each coincidence of a size quantized level with the Fermi level, there would be a discontinuity in the density-of-states function resulting in a peak of oscillations. With large values of film thickness, the height of the steps decreases and 2D SL decreases with increasing film thickness in non-oscillatory manner and exhibits monotonic decreasing dependence. The height of step size and the rate of decrement are totally dependent on the band structure. The influence of light is immediately apparent from the plots in Figures 4.23–4.30, since 2D SL of ultrathin films of the aforementioned compounds depends strongly on I and λ, which is in direct contrast as compared with the corresponding cases for ultrathin films in the absence of external photoexcitation, respectively. The variations of 2D SL in all figures reflect the direct signature of the light waves on the electronic, optic, and the other band-structure-dependent properties of semiconducting materials in the presence of

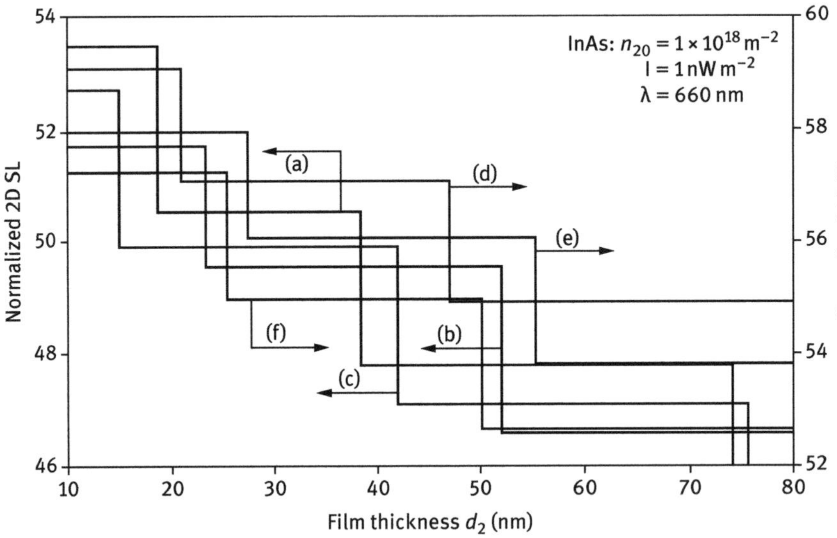

Figure 4.15: Plot of the normalized 2D SL as a function of film thickness for HD ultrathin films of n-InAs in the presence of light waves in which the curves (a), (b), and (c) represent the three- and two-band models of Kane and that of the parabolic energy bands, respectively. The plots (d), (e), and (f) represent the same in the absence of external photoexcitation.

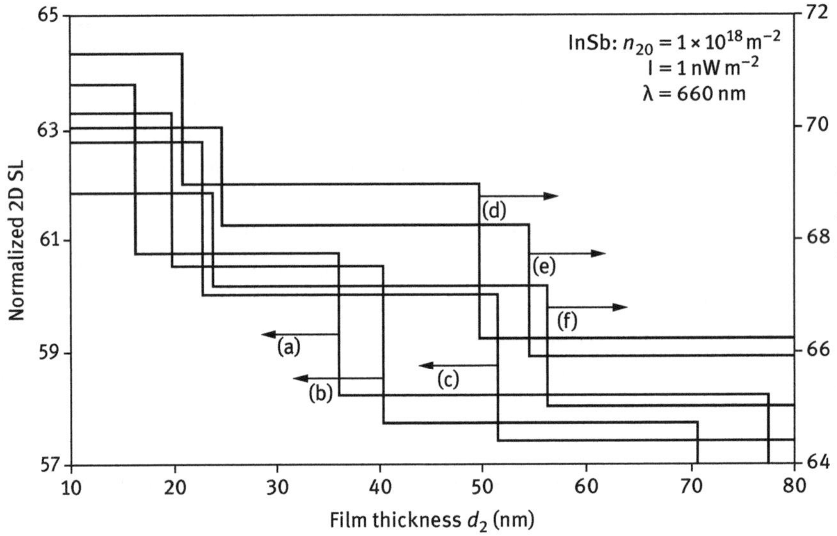

Figure 4.16: Plot of the normalized 2D SL as a function of film thickness for HD ultrathin films of n-InSb in the presence of light waves in which the curves (a), (b), and (c) represent the three- and two-band models of Kane and that of the parabolic energy bands, respectively. The plots (d), (e), and (f) represent the same in the absence of external photoexcitation.

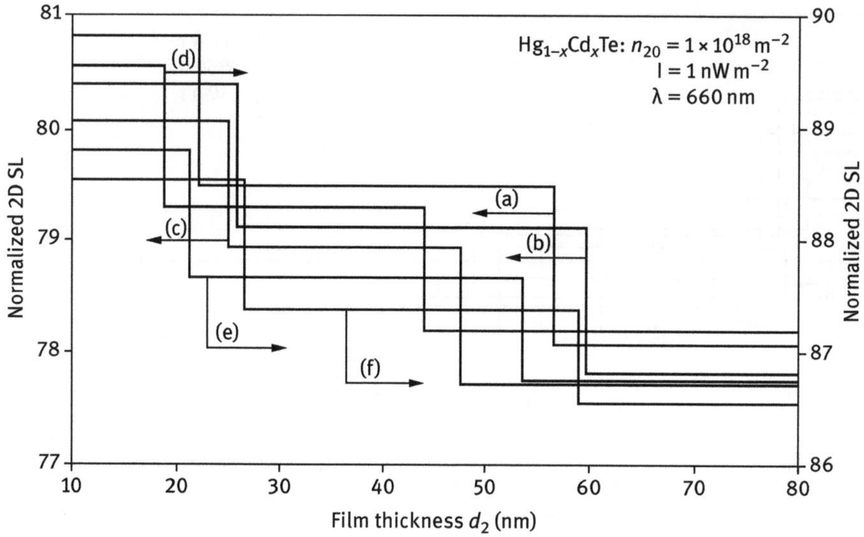

Figure 4.17: Plot of the normalized 2D SL as a function of film thickness for HD ultrathin films of n-$Hg_{1-x}Cd_xTe$ in the presence of light waves in which the curves (a), (b), and (c) represent the three- and two-band models of Kane and that of the parabolic energy bands, respectively. The plots (d), (e), and (f) represent the same in the absence of external photoexcitation.

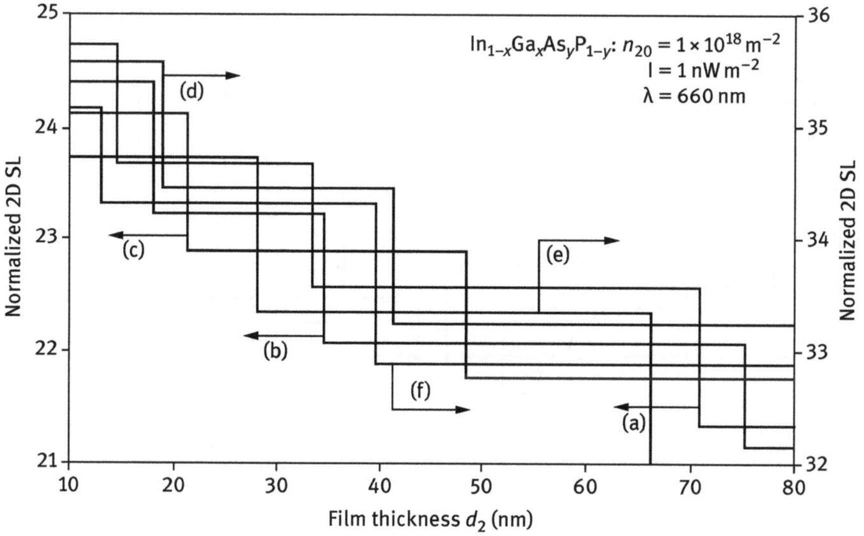

Figure 4.18: Plot of the normalized 2D SL as a function of film thickness for HD ultrathin films of n-$In_{1-x}Ga_xAs_yP_{1-y}$ lattice matched to InP in the presence of light waves in which the curves (a), (b), and (c) represent the three- and two-band models of Kane and that of the parabolic energy bands, respectively. The plots (d), (e), and (f) represent the same in the absence of external photoexcitation.

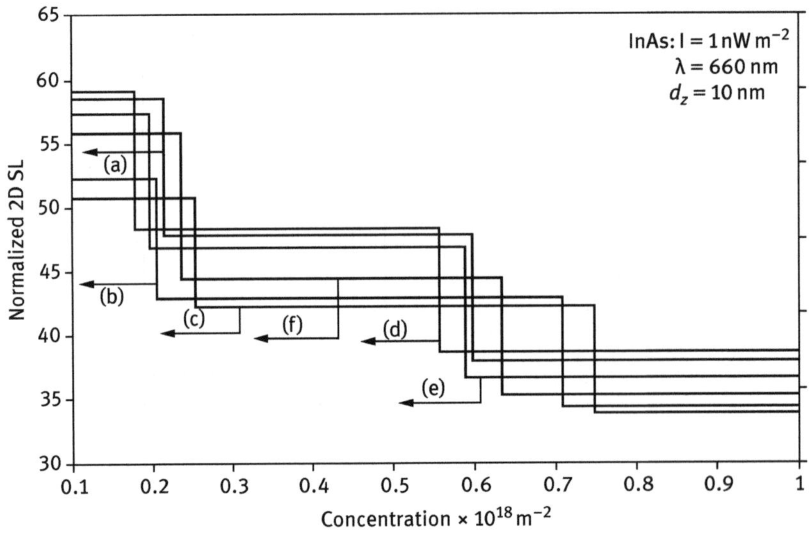

Figure 4.19: Plot of the normalized 2D SL as a function of surface electron concentration per unit area for HD ultrathin films of n-InAs in the presence of light waves in which the curves (a), (b), and (c) represent the three- and two-band models of Kane and that of the parabolic energy bands, respectively. The plots (d), (e), and (f) represent the same in the absence of external photoexcitation.

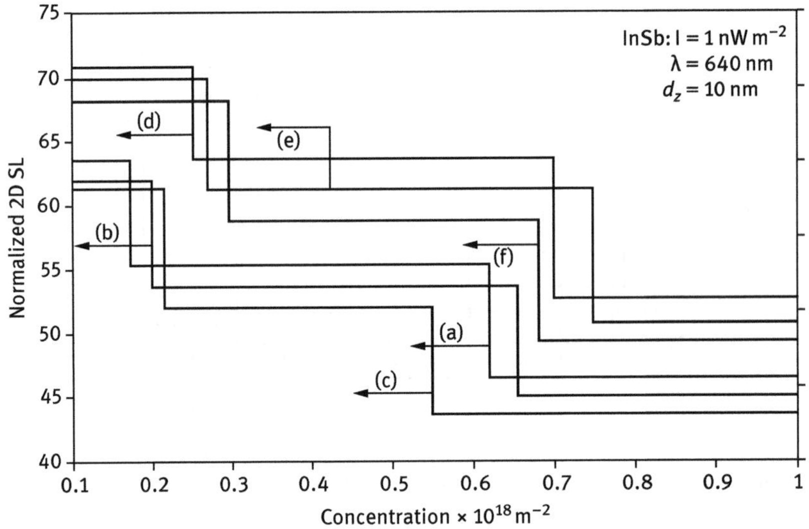

Figure 4.20: Plot of the normalized 2D SL as a function of surface electron concentration per unit area for HD ultrathin films of n-InSb in the presence of light waves in which the curves (a), (b), and (c) represent the three- and two-band models of Kane and that of the parabolic energy bands, respectively. The plots (d), (e), and (f) represent the same in the absence of external photoexcitation.

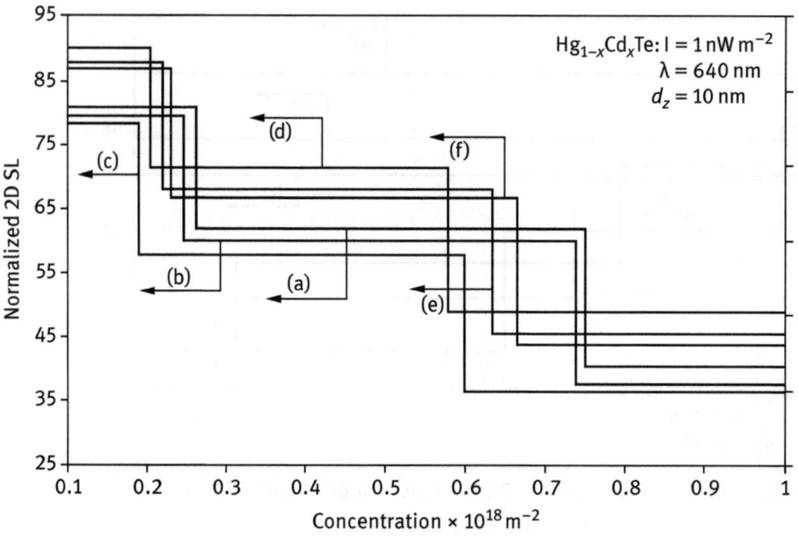

Figure 4.21: Plot of the normalized 2D SL as a function of surface electron concentration per unit area for HD ultrathin films of n-$Hg_{1-x}Cd_xTe$ in the presence of light waves in which the curves (a), (b), and (c) represent the three- and two-band models of Kane and that of the parabolic energy bands, respectively. The plots (d), (e), and (f) represent the same in the absence of external photoexcitation.

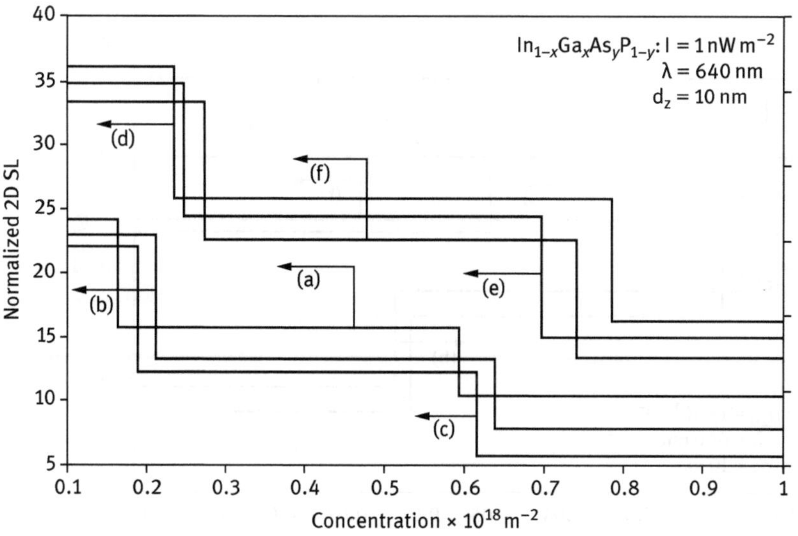

Figure 4.22: Plot of the normalized 2D SL as a function of surface electron concentration per unit area for HD ultrathin films of n-$In_{1-x}Ga_xAs_yP_{1-y}$ lattice matched to InP in the presence of light waves in which the curves (a), (b), and (c) represent the three- and two-band models of Kane and that of the parabolic energy bands, respectively. The plots (d), (e), and (f) represent the same in the absence of external photoexcitation.

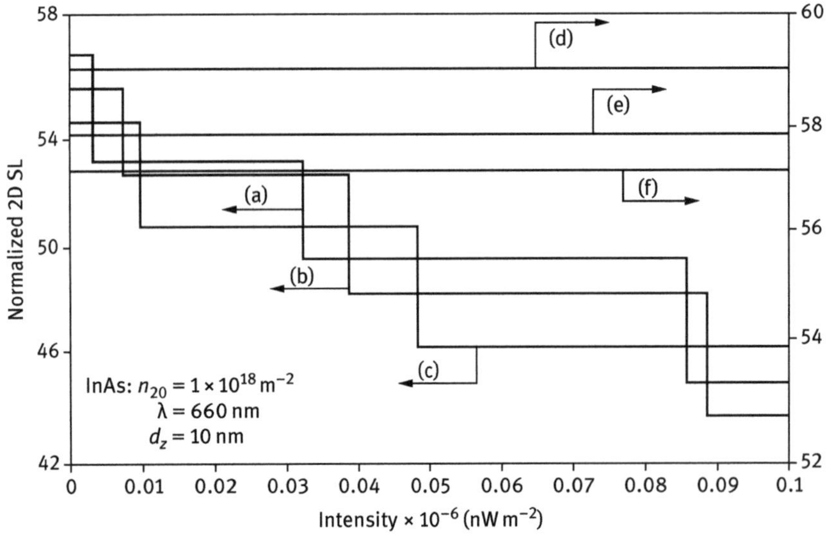

Figure 4.23: Plot of the normalized 2D SL as a function of light intensity for HD ultrathin films of n-InAs in which the curves (a), (b), and (c) represent the three- and two-band models of Kane and that of parabolic energy bands, respectively. The plots (d), (e), and (f) represent the same in the absence of external photoexcitation.

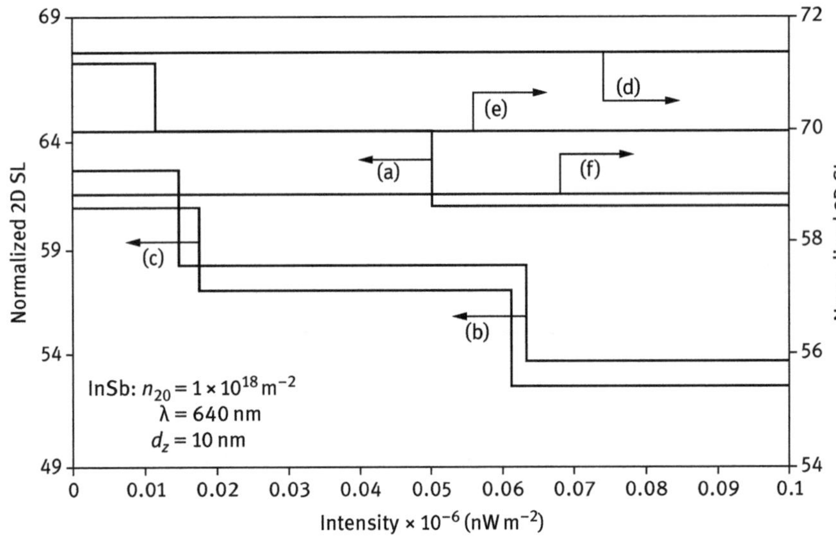

Figure 4.24: Plot of the normalized 2D SL as a function of light intensity for HD ultrathin films of n-InSb in which the curves (a), (b), and (c) represent the three- and two-band models of Kane and that of parabolic energy bands, respectively. The plots (d), (e), and (f) represent the same in the absence of external photoexcitation.

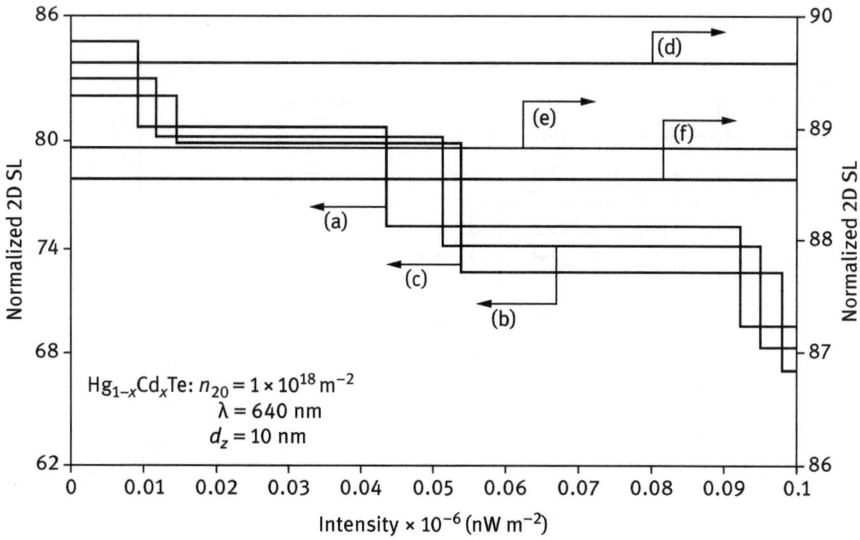

Figure 4.25: Plot of the normalized 2D SL as a function of light intensity for HD ultrathin films of n-$Hg_{1-x}Cd_xTe$ in which the curves (a), (b), and (c) represent the three- and two-band models of Kane and that of parabolic energy bands, respectively. The plots (d), (e), and (f) represent the same in the absence of external photoexcitation.

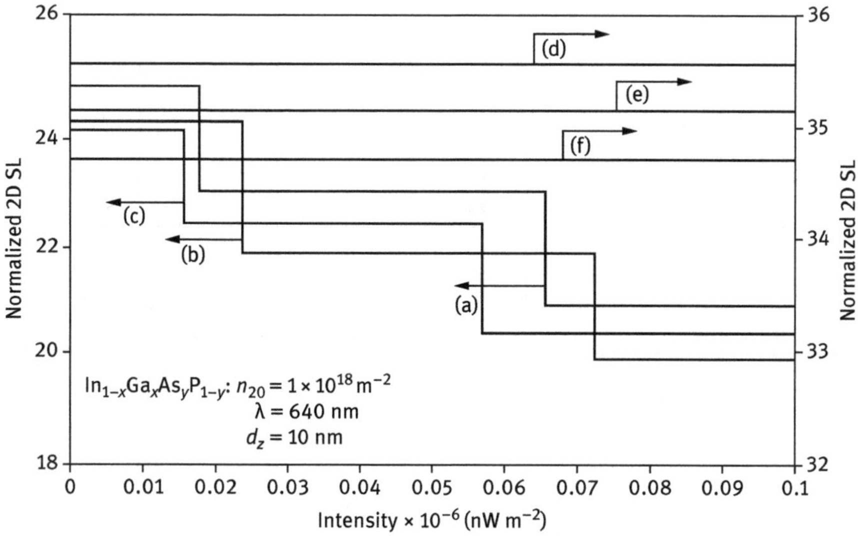

Figure. 4.26: Plot of the normalized 2D SL as a function of light intensity for HD ultrathin films of n-$In_{1-x}Ga_xAs_yP_{1-y}$ lattice matched to InP in which the curves (a), (b), and (c) represent the three- and two-band models of Kane and that of parabolic energy bands, respectively. The plots (d), (e), (f), and (f) represent the same in the absence of external photoexcitation.

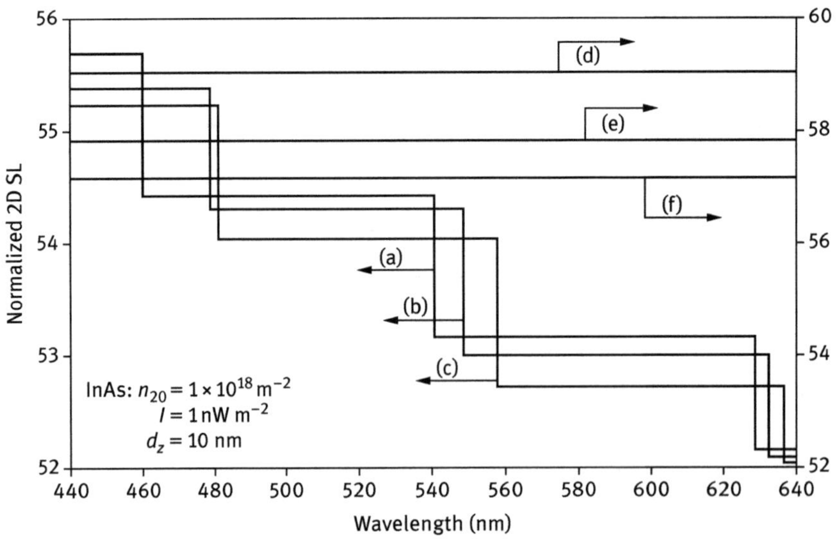

Figure 4.27: Plot of the normalized 2D SL as a function of wavelength for HD ultrathin films of n-InAs in which the curves (a), (b), and (c) represent the three- and two-band models of Kane and that of parabolic energy bands, respectively. The plots (d), (e), and (f) represent the same in the absence of external photoexcitation.

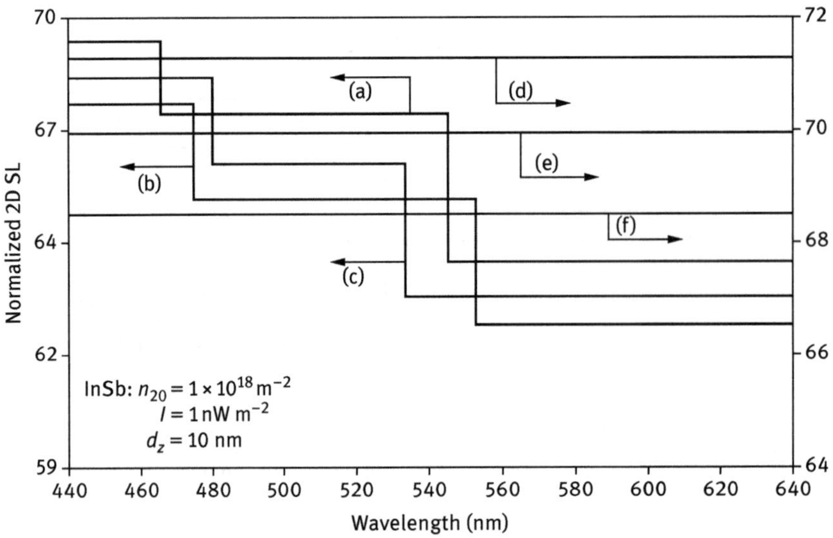

Figure 4.28: Plot of the normalized 2D SL as a function of wavelength for HD ultrathin films of n-InSb in which the curves (a), (b), and (c) represent the three- and two-band models of Kane and that of parabolic energy bands, respectively. The plots (d), (e), and (f) represent the same in the absence of external photoexcitation.

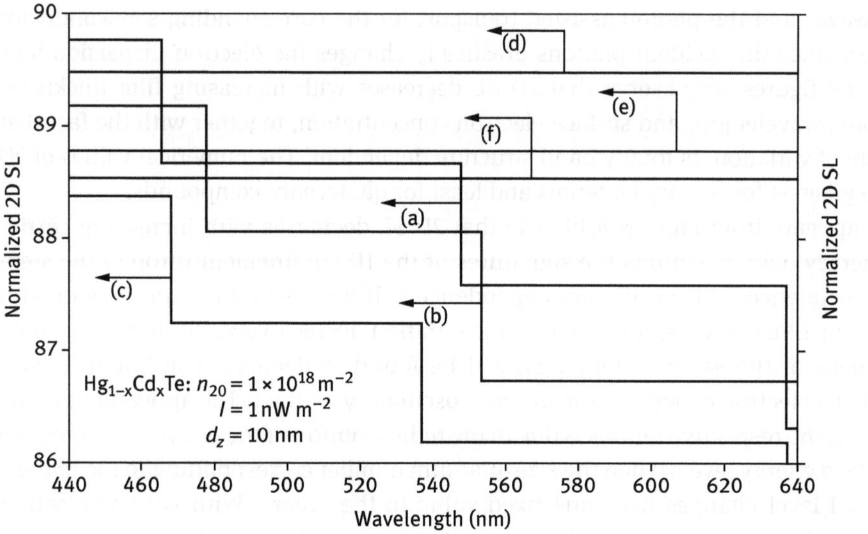

Figure 4.29: Plot of the normalized 2D SL as a function of wavelength for HD ultrathin films of n-$Hg_{1-x}Cd_xTe$ in which the curves (a), (b), and (c) represent the three- and two-band models of Kane and that of parabolic energy bands, respectively. The plots (d), (e), and (f) represent the same in the absence of external photoexcitation.

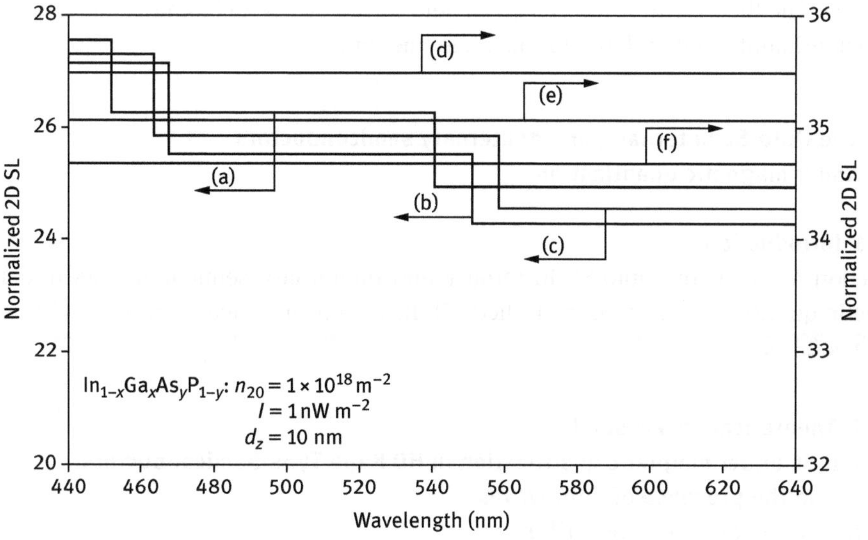

Figure 4.30: Plot of the normalized 2D SL as a function of wavelength for HD ultrathin films of n-$In_{1-x}Ga_xAs_yP_{1-y}$ lattice matched to InP in which the curves (a), (b), and (c) represent the three- and two-band models of Kane and that of parabolic energy bands, respectively. The plots (d), (e), and (f) represent the same in the absence of external photoexcitation.

light waves and the photon-assisted transport for the corresponding semiconductor devices, since the incident photons drastically changes the electron dispersion law. From the figures we observe that 2D SL decreases with increasing film thickness, intensity, wavelength, and surface electron concentration, together with the fact that the rate of variation, is totally band structure dependent. The numerical values of 2D SL are greatest for ternary materials and least for quaternary compounds.

It appears from Figures 4.19–4.22 that 2D SL decreases with increasing carrier degeneracy, which exhibits the signatures of the 1D confinement through the step-like dependence. This oscillatory dependence will be less and less prominent with increasing film thickness and carrier concentration, respectively. Ultimately, for bulk specimens of the same material, SL will be found to decrease continuously with increasing electron concentration in a non-oscillatory manner. The appearance of the humps of the respective figures is due to the redistribution of the electrons among the quantized energy levels when the size quantum number corresponding to the highest occupied level changes from one fixed value to the others. With varying electron concentration, a change is reflected in 2D SL through the redistribution of the electrons among the size-quantized levels.

We have not considered other types of optoelectronic materials and other external variables in order to keep the presentation brief. Besides, the influence of energy band models and the various band constants on 2D SL for different materials can also be studied from all figures of this chapter. The numerical results presented in this chapter would be different for other materials, but the nature of variation would be unaltered. The theoretical results as given here would be useful in analyzing various other experimental data related to this phenomenon.

4.2.5 The Opto-SL in ternary, and quaternary semiconductors under magnetic quantization

4.2.5.1 Introduction

In Section 4.2.5.2.1, the opto-SL in ternary and quaternary semiconductors under magnetic quantization has been studied. SL has been investigated numerically in Section 4.2.5.3.

4.2.5.2 Theoretical background

4.2.5.2.1 SL under magnetic quantization in HD Kane Type Semiconductors in the presence of light waves

(i) SL for HD materials can be written as

$$\frac{1}{L_D^2} = \frac{e^2}{\varepsilon_{sc}} \textit{Real Part of} \left[\left[\frac{\partial n_0}{\partial(E_{FHDLB} - E_{0HDB})}\right]\right] \tag{4.48}$$

Using eqs. (1.66a) and (4.48), we can study SL in this case.

In the absence of band-tails, we get

$$\frac{1}{L_D^2} = \frac{e^2 B g_v |e| \sqrt{2m_c}}{\varepsilon_{sc} \pi^2 \hbar^2} \left[\sum_{n=0}^{n_{\max}} \left[\left\{ \beta_0(E_{FLB}, \lambda) - (n + \frac{1}{2})\hbar\omega_0 \right\}^{1/2} \right]' \right] \tag{4.49}$$

In the absence of light waves and heavy doping, SL can be written as

$$\frac{1}{L_D^2} = \frac{e^2}{\varepsilon_{sc}} \left[\frac{\partial n_0}{\partial E_{FB}} \right] \tag{4.50}$$

Using eqs. (1.73) and (4.50), we get

$$\frac{1}{L_D^2} = \frac{e^2 B g_v |e| \sqrt{2m_c}}{\varepsilon_{sc} \pi^2 \hbar^2} \left[\sum_{n=0}^{n_{\max}} \left[\left\{ I_{11}(E_{FB}) - (n + \frac{1}{2})\hbar\omega_0 \right\}^{1/2} \right]' \right] \tag{4.51}$$

(ii) Using eqs. (1.79) and (4.48), the magneto-SL, in the absence of spin, for HD III-V, ternary, and quaternary semiconductors, in the presence of photoexcitation, whose unperturbed conduction electrons obey the two-band model of Kane, can be investigated in this case.

Using eqs. (1.81) and (4.50), the magneto-SL, in the absence of spin and band tails for III-V, ternary, and quaternary semiconductors, in the presence of photoexcitation, whose unperturbed conduction electrons obey the two-band model of Kane, is given by

$$\frac{1}{L_D^2} = \frac{e^2 B g_v |e| \sqrt{2m_c}}{\varepsilon_{sc} \pi^2 \hbar^2} \left[\sum_{n=0}^{n_{\max}} \left[\left\{ \tau_0(E_{FLB}, \lambda) - (n + \frac{1}{2})\hbar\omega_0 \right\}^{1/2} \right]' \right] \tag{4.52}$$

In the absence of light waves and band tails, SL for two-band model of Kane in the presence of magnetic quantization can be written using eqs. (1.84) and (4.50) as:

$$\frac{1}{L_D^2} = \frac{e^2 B g_v |e| \sqrt{2m_c}}{\varepsilon_{sc} \pi^2 \hbar^2} \left[\sum_{n=0}^{n_{\max}} \left[\left\{ E_{FB}(1 + \alpha E_{FB}) - (n + \frac{1}{2})\hbar\omega_0 \right\}^{1/2} \right]' \right] \tag{4.53}$$

(iii) Using eqs. (1.88) and (4.48) the magneto-SL, in the absence of spin, for HD III-V, ternary, and quaternary semiconductors, in the presence of photoexcitation whose unperturbed conduction electrons obey the parabolic energy bands can be investigated.

Using eqs. (4.50) and (1.90) the magneto-SL, in the absence of spin and band tails for III-V, ternary, and quaternary semiconductors, in the presence of photoexcitation whose unperturbed conduction electrons obey the parabolic energy bands is given by:

$$\frac{1}{L_D^2} = \frac{e^2 B g_v |e| \sqrt{2m_c}}{\varepsilon_{sc} \pi^2 \hbar^2} \left[\sum_{n=0}^{n_{\max}} \left[\left\{ \rho_0 (E_{FLB}, \lambda) - (n + \frac{1}{2}) \hbar \omega_0 \right\}^{1/2} \right]' \right] \tag{4.54}$$

In the absence of light waves and band tails, SL for isotropic parabolic energy bands can be written under magnetic quantization as:

$$\frac{1}{L_D^2} = \left(\frac{e^2 g_v e B \sqrt{2m_c}}{2\varepsilon_{sc} \pi^2 \hbar^2} \right) \sum_{n=0}^{n_{\max}} \left[E_{FB} - (n + \frac{1}{2}) \hbar \omega_0 \right]^{\frac{-1}{2}} \tag{4.55}$$

Equation (4.55) is well known in the literature

Under the condition $\alpha E_{FB} \ll 1$, SL at a finite temperature in this case can be expressed as:

$$\frac{1}{L_D^2} = \frac{e^2}{\varepsilon_{sc} k_B T} N_c \theta_{B1} \left[\sum_{n=0}^{n_{\max}} \frac{1}{\sqrt{a_{01}}} \left[\left(1 + \frac{3}{2} \alpha b_{01} \right) F_{\frac{-3}{2}} (\bar{\eta}_{B_1}) + \frac{3}{4} \alpha k_B T F_{\frac{-1}{2}} (\bar{\eta}_{B1}) \right] \right] \tag{4.56}$$

4.2.5.3 Results and discussion

Using the appropriate equations the plot of the normalized SL as a function of inverse magnetic field in the presence of light waves at $T = 4.2$ K is shown in Figures 4.31–4.34 by taking HD n-InAs, n-InSb, $Hg_{1-x}Cd_xTe$ and HD n-$In_{1-x}Ga_xAs_yP_{1-y}$ lattice matched to InP, respectively. Figures 4.35–4.38 exhibit the variation of the normalized SL as a function of electron concentration, under quantizing magnetic field in the presence of light waves for the aforementioned materials. The normalized inverse SL again shows the oscillatory dependence with different numerical magnitude emphasizing the influence of the energy band constants.

Figures 4.39–4.42 shows the variation of the normalized inverse SL as a function of light intensity in the presence of quantizing magnetic field, while Figures 4.43–4.46 exhibit the same as a function of wavelength, in which, the variations of the wavelengths are in the zone of visible region. One can observe that the normalized inverse SL decreases with increase of the light intensity and wavelengths in different ways, as found in Figures 4.39–4.46. The nature of variations in all cases depends strongly on the energy spectrum constants of the respective materials and the external physical conditions.

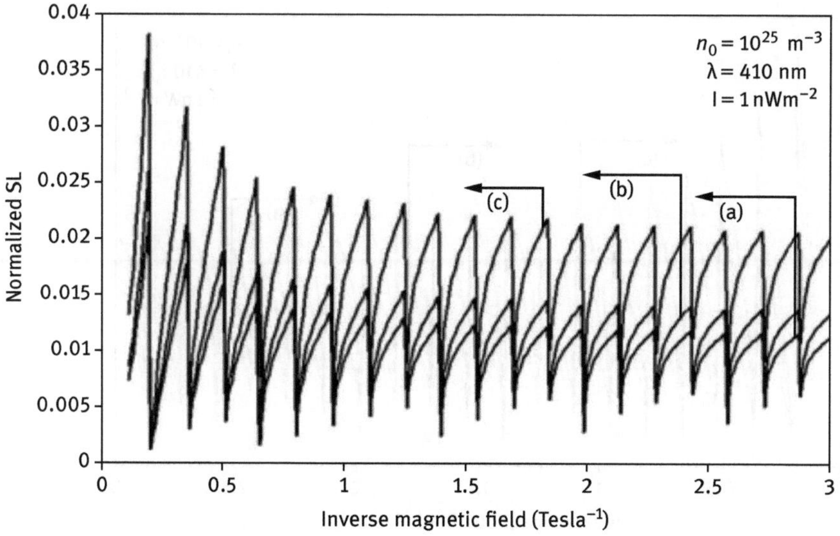

Figure 4.31: Plot of the normalized SL as a function of inverse quantizing magnetic field in the presence of light waves for HD n-InAs, in which the curves (a), (b), and (c) represent the perturbed three- and two-band models of Kane and that of parabolic energy bands, respectively.

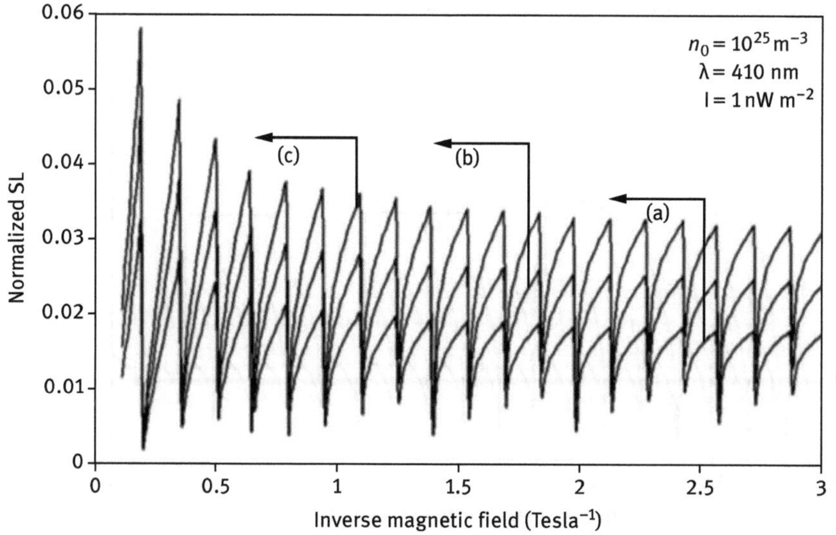

Figure 4.32: Plot of the normalized SL as a function of inverse quantizing magnetic field in the presence of light waves for HD n-InSb, in which the curves (a), (b), and (c) represent the perturbed three- and two-band models of Kane and that of parabolic energy bands, respectively.

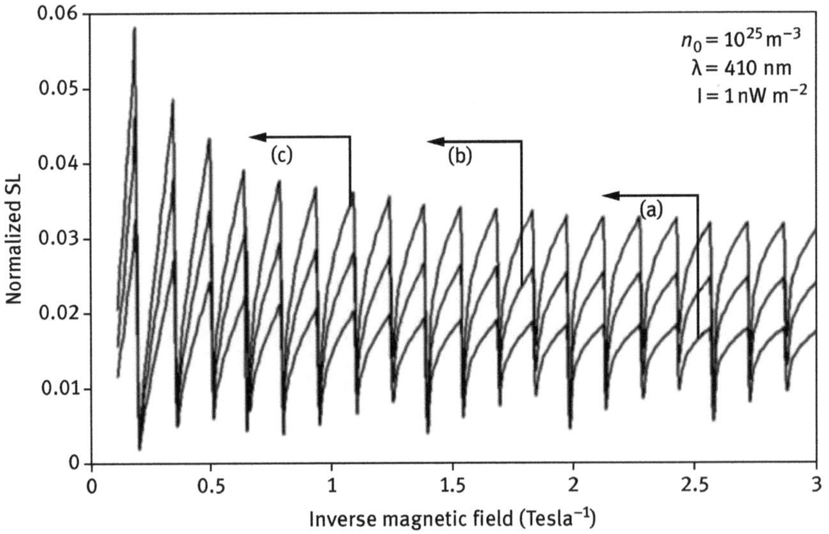

Figure 4.33: Plot of the normalized SL as a function of inverse quantizing magnetic field in the presence of light waves for HD n-$Hg_{1-x}Cd_xTe$, in which the curves (a), (b), and (c) represent the perturbed three- and two-band models of Kane and that of parabolic energy bands, respectively.

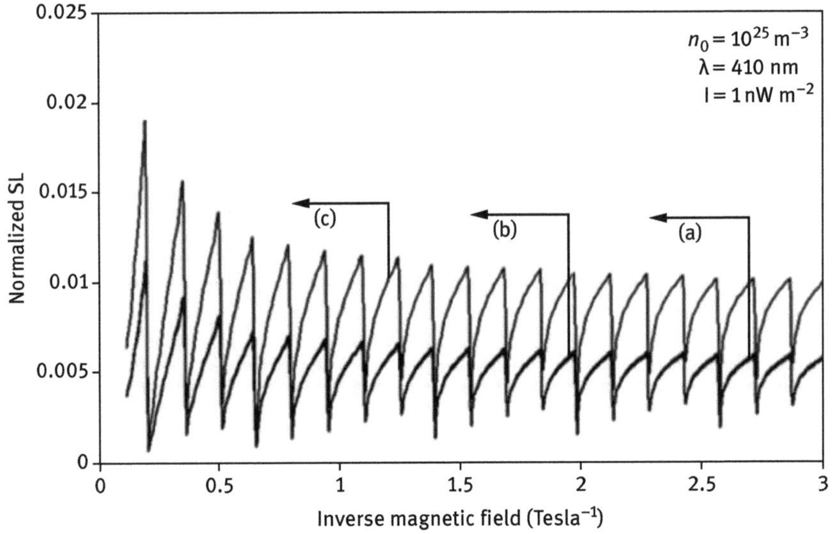

Figure 4.34: Plot of the normalized SL as a function of inverse quantizing magnetic field in the presence of light waves for HD n-$In_{1-x}Ga_xAs_yP_{1-y}$ lattice matched to InP, in which the curves (a), (b), and (c) represent the perturbed three- and two-band models of Kane and that of parabolic energy bands, respectively.

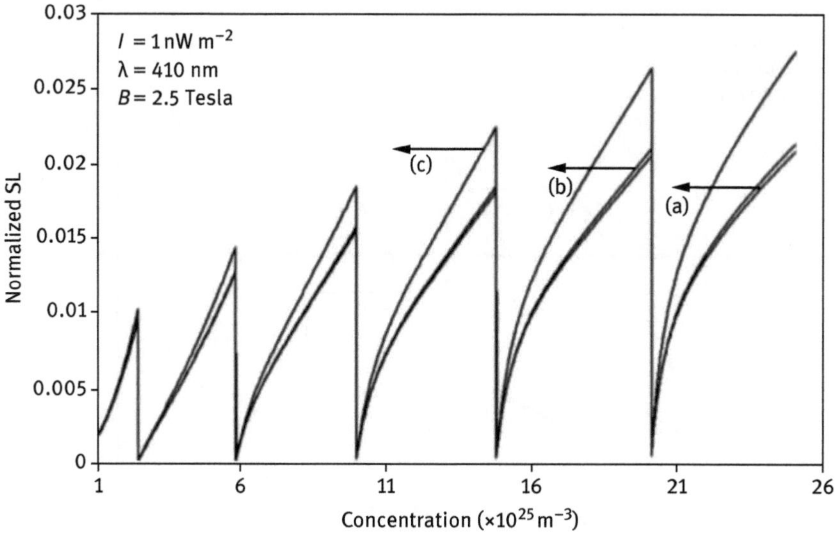

Figure 4.35: Plot of the normalized SL as a function of inverse quantizing magnetic field in the presence of light waves for HD n-$In_{1-x}Ga_xAs_yP_{1-y}$ lattice matched to InP, in which the curves (a), (b), and (c) represent the perturbed three- and two-band models of Kane and that of parabolic energy bands, respectively.

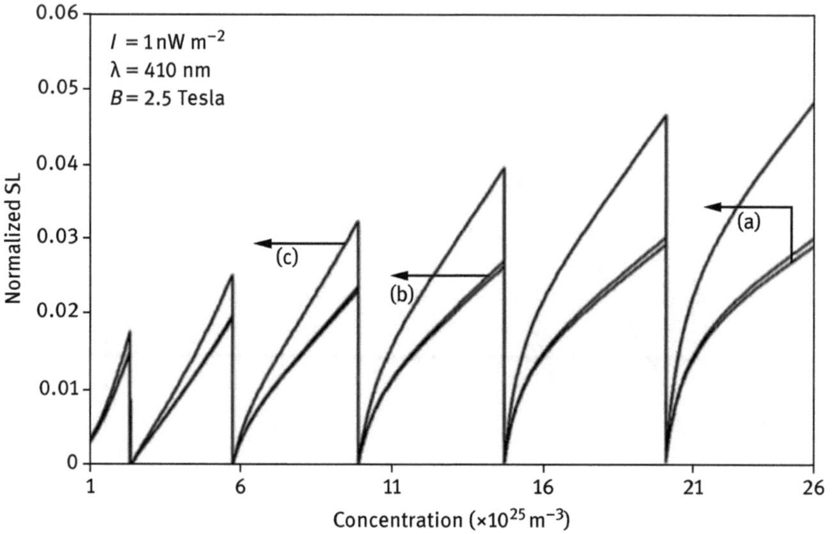

Figure 4.36: Plot of the normalized SL as a function of electron concentration under quantizing magnetic field in the presence of light waves for HD n-InSb, in which the curves (a), (b), and (c) represent the perturbed three- and two-band models of Kane and that of parabolic energy bands, respectively.

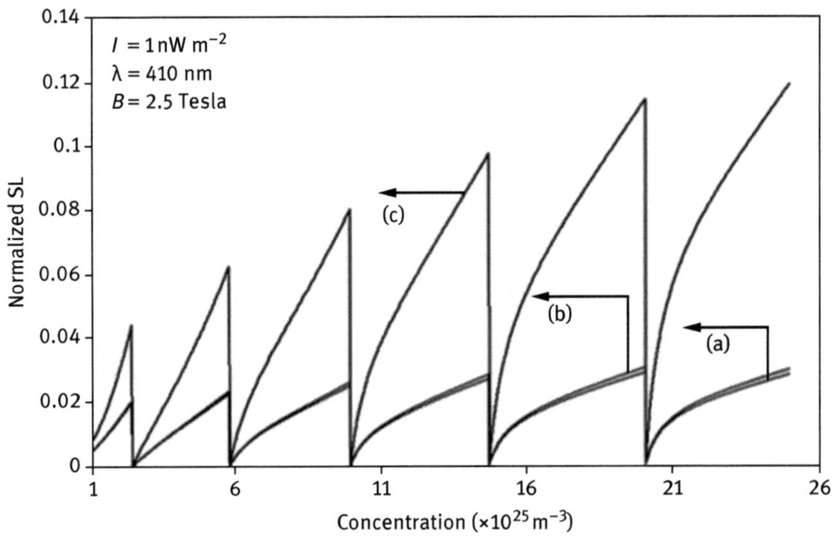

Figure 4.37: Plot of the normalized SL as a function of electron concentration under quantizing magnetic field in the presence of light waves for HD n-$Hg_{1-x}Cd_xTe$, in which the curves (a), (b), and (c) represent the perturbed three- and two-band models of Kane and that of parabolic energy bands, respectively.

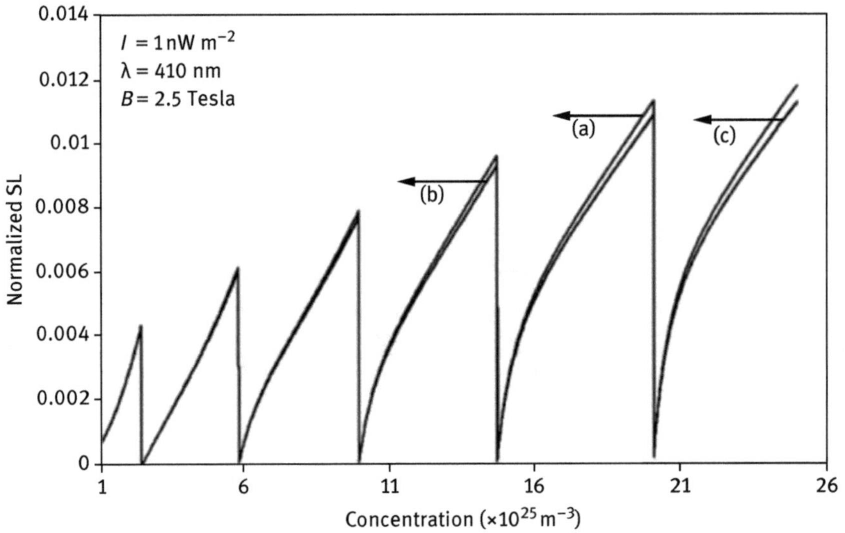

Figure 4.38: Plot of the normalized SL as a function of electron concentration under quantizing magnetic field in the presence of light waves for HD n-$In_{1-x}Ga_xAs_yP_{1-y}$ lattice matched to InP, in which the curves (a), (b), and (c) represent the perturbed three- and two-band models of Kane and that of parabolic energy bands, respectively.

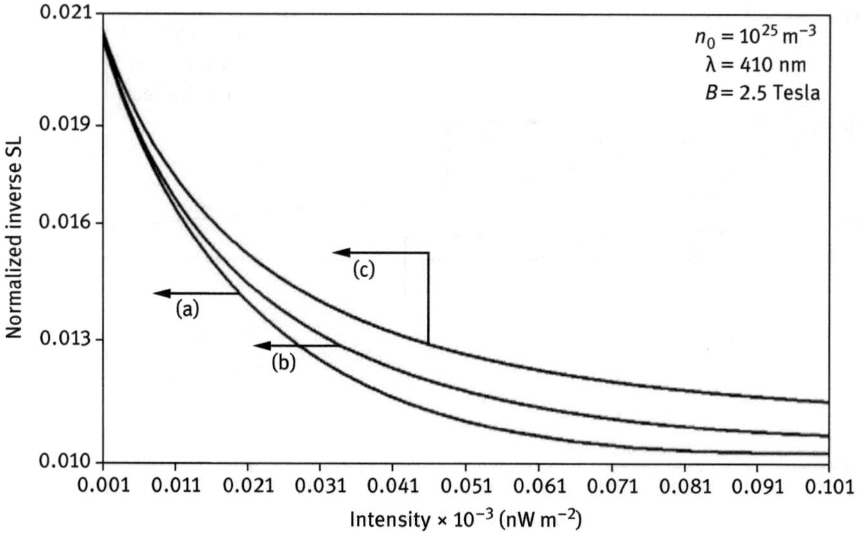

Figure 4.39: Plot of the normalized inverse SL as a function of light intensity under quantizing magnetic field for HD n-InAs, in which the curves (a), (b), and (c) represent the perturbed three- and two-band models of Kane and that of parabolic energy bands, respectively.

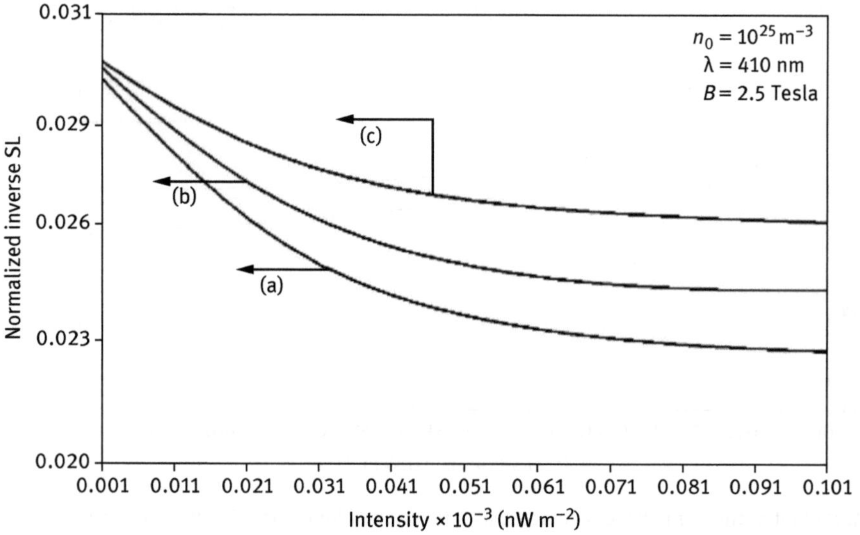

Figure 4.40: Plot of the normalized inverse SL as a function of light intensity under quantizing magnetic field for HD n-InSb, in which the curves (a), (b), and (c) represent the perturbed three- and two-band models of Kane and that of parabolic energy bands, respectively.

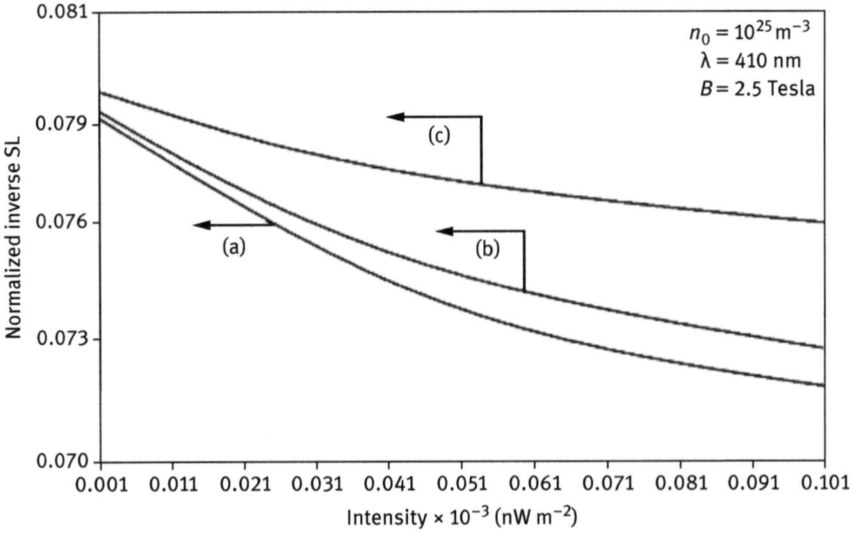

Figure 4.41: Plot of the normalized inverse SL as a function of light intensity under quantizing magnetic field for HD n-$Hg_{1-x}Cd_xTe$, in which the curves (a), (b), and (c) represent the perturbed three- and two-band models of Kane and that of parabolic energy bands, respectively.

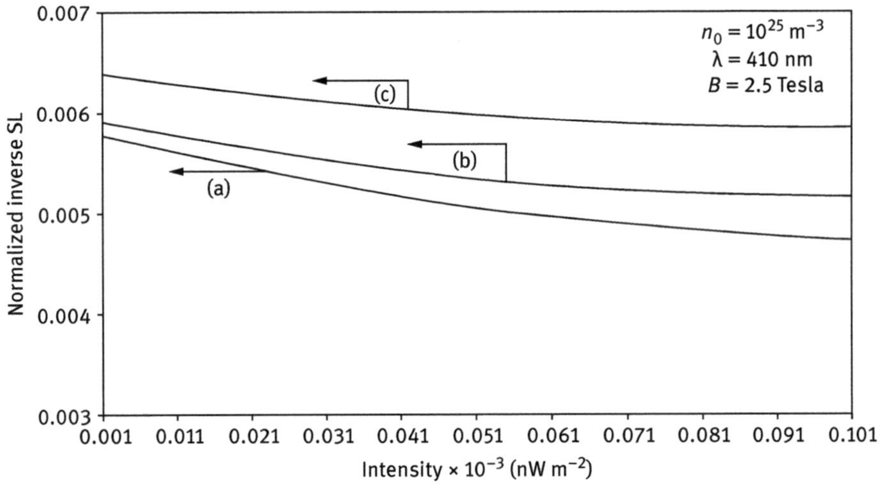

Figure 4.42: Plot of the normalized inverse SL as a function of light intensity under quantizing magnetic field for HD n-$In_{1-x}Ga_xAs_yP_{1-y}$ lattice matched to InP, in which the curves (a), (b), and (c) represent the perturbed three- and two-band models of Kane and that of parabolic energy bands, respectively.

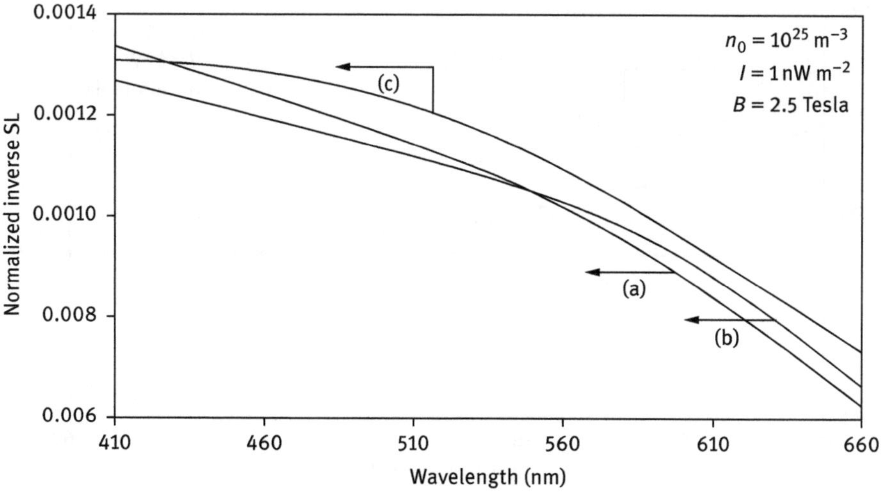

Figure 4.43: Plot of the normalized inverse SL as a function of wavelength under quantizing magnetic field for HD n-InAs, in which the curves (a), (b), and (c) represent the perturbed three- and two-band models of Kane and that of parabolic energy bands, respectively.

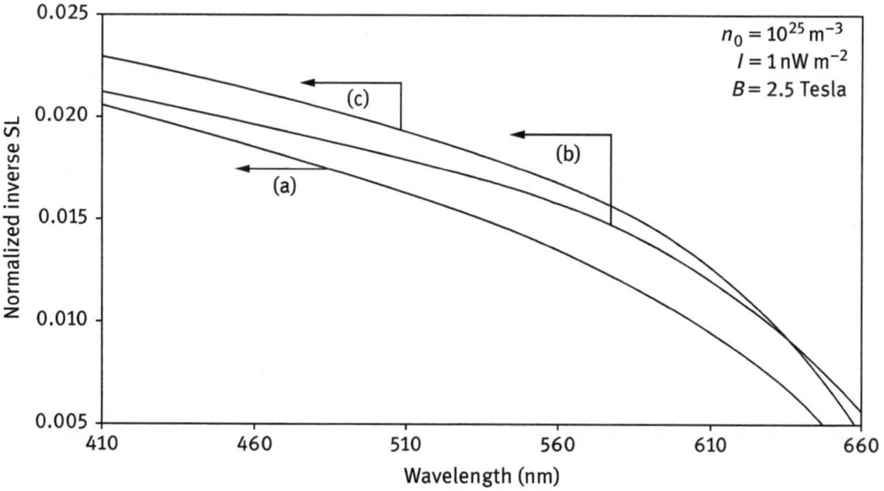

Figure 4.44: Plot of the normalized inverse SL as a function of wavelength under quantizing magnetic field for HD n-InSb, in which the curves (a), (b), and (c) represent the perturbed three- and two-band models of Kane and that of parabolic energy bands, respectively.

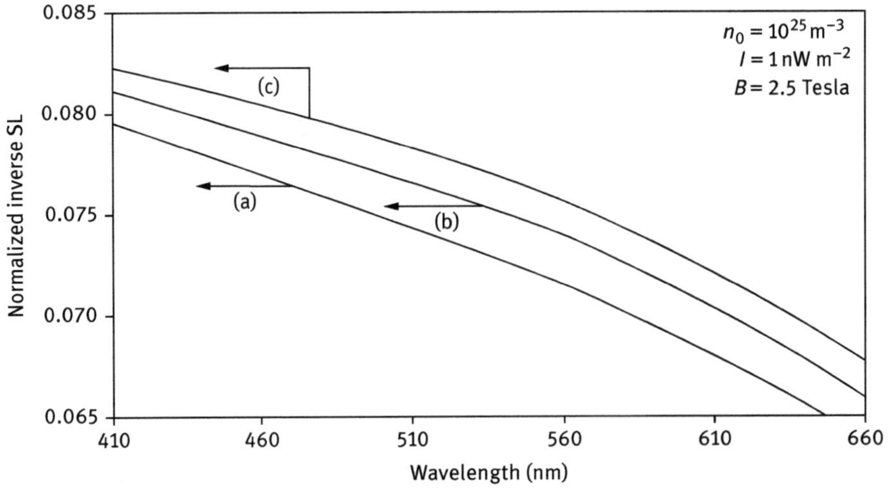

Figure 4.45: Plot of the normalized inverse SL as a function of wavelength under quantizing magnetic field for HD n-$Hg_{1-x}Cd_xTe$, in which the curves (a), (b), and (c) represent the perturbed three- and two-band models of Kane and that of parabolic energy bands, respectively.

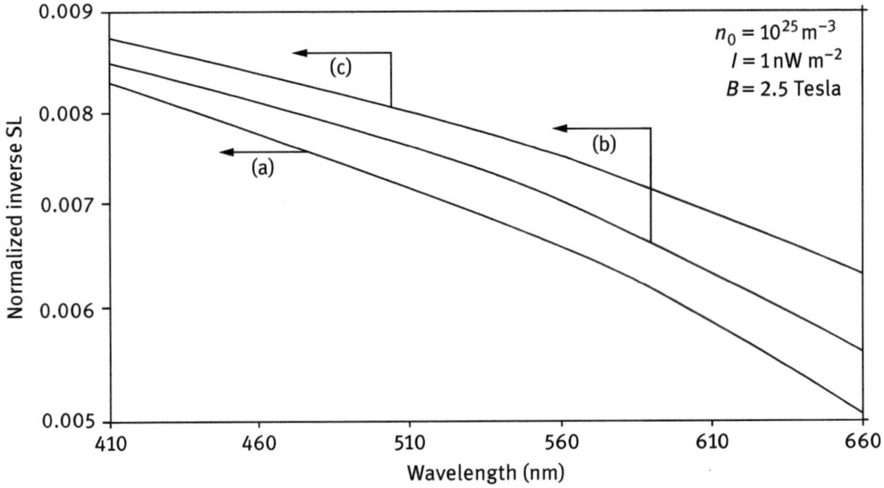

Figure 4.46: Plot of the normalized inverse SL as a function of wavelength under quantizing magnetic field for HD n-$In_{1-x}Ga_xAs_yP_{1-y}$ lattice matched to InP, in which the curves (a), (b), and (c) represent the perturbed three- and two-band models of Kane and that of parabolic energy bands, respectively.

4.2.6 The Opto-SL of III-V, ternary, and quaternary semiconductors under cross-field configuration

4.2.6.1 Introduction

The influence of a crossed electric and quantizing magnetic field on SL in HD III-V, ternary, and quaternary materials under external photoexcitation has been investigated in Section 4.2.6.2. SL has been investigated numerically in Section 4.2.6.3.

4.2.6.2 Theoretical background

(i) SL in this case can be written as:

$$\frac{1}{L_D^2} = \frac{e^2}{\varepsilon_{sc}} \text{Real Part of} \left[\frac{\partial n_0}{\partial (E_{F_{BLHDC}} - E_{0HDB1})}\right] \tag{4.57}$$

where E_{0HDB1} is obtained by putting $k_z(E) = 0$ and $k_y = 0$ in the corresponding dispersion relation under cross-field configuration.
Using eqs. (4.57) and (1.103), we can study SL in this case.

SL in the absence of band tails in the present case can be written as

$$\frac{1}{L_D^2} = \frac{e^2}{\varepsilon_{sc}} \left[\frac{\partial n_0}{\partial E_{F_{BLC}}}\right] \tag{4.58}$$

Using eqs. (4.58) and (1.106a) in this case, we get

$$\frac{1}{L_D^2} = \frac{2e^2 g_v B\sqrt{2m_c}}{3\varepsilon_{sc} L_x \pi^2 \hbar^2 E_0} \left[\sum_{n=0}^{n_{\max}} \left[M_{1612}(E_{F_{BLC}}, n, E_0, B, \lambda)\right]'\right] \tag{4.59}$$

(ii) SL in Kane type materials in the presence of light waves whose energy band structure in the present case is given by eqs. (1.109) can be studied by using eqs. (4.57) and (1.110), respectively.

SL in the present case in the absence of band tails is given by

$$\frac{1}{L_D^2} = \frac{2e^2 g_v B\sqrt{2m_c}}{3\varepsilon_{sc} L_x \pi^2 \hbar^2 E_0} \left[\sum_{n=0}^{n_{\max}} \left[M_{1614}(E_{F_{BLC}}, n, E_0, B, \lambda)\right]'\right] \tag{4.60}$$

(iii) SL in Kane-type materials in the presence of light waves whose energy band structure in the present case is given by eqs. (1.116) can be studied by using eqs. (4.57) and (1.117), respectively.

SL in the present case in the absence of band tails is given by

$$\frac{1}{L_D^2} = \frac{2e^2 g_v B\sqrt{2m_c}}{3\varepsilon_{sc} L_x \pi^2 \hbar^2 E_0} \left[\sum_{n=0}^{n_{\max}} \left[M_{1615}(E_{F_{BLC}}, n, E_0, B, \lambda)\right]'\right] \tag{4.61}$$

(iv) In the absence of light waves and heavy doping SL in III-V semi-conductors, whose energy band structures are defined by the three-band model of Kane, can be written using eqs. (1.124) and (4.57) in the presence of cross fields configuration as

$$\frac{1}{L_D^2} = \frac{2e^2 g_v B\sqrt{2m_c}}{3\varepsilon_{sc} L_x \pi^2 \hbar^2 E_0} \left[\sum_{n=0}^{n_{\max}} [T_{43}(n, \bar{E}_{FB})]'\right] \tag{4.62}$$

Under the condition $\Delta \gg E_g$, two-band Kane model SL in the present case can be expressed as

$$\frac{1}{L_D^2} = \frac{2e^2 g_v B\sqrt{2m_c}}{3\varepsilon_{sc} L_x \pi^2 \hbar^2 E_0} \left[\sum_{n=0}^{n_{\max}} [T_{45}(n, \bar{E}_{FB})]'\right] \tag{4.63}$$

For parabolic energy band $\alpha \to 0$, DMR in this case can be expressed at a finite temperature as

$$\frac{1}{L_D^2} = \frac{eN_c \phi g_v}{\varepsilon_{sc} E_0 L_x} \left[\sum_{n=0}^{n_{\max}} \left[F_{\frac{-1}{2}}(\eta_1) - F_{\frac{-1}{2}}(\eta_2)\right]\right] \tag{4.64}$$

4.2.6.3 Results and discussion

Using the appropriate equations, the plot of the normalized inverse SL as a function of inverse magnetic field under cross-field configurations in the presence of external photoexcitation at $T = 4.2$ K is shown in Figures 4.47–4.50 by taking HD n-InAs, n-InSb, $Hg_{1-x}Cd_xTe$, and HD n-$In_{1-x}Ga_xAs_yP_{1-y}$ lattice matched to InP, respectively. It appears that the normalized inverse SL oscillates with the inverse quantizing magnetic field with different numerical magnitudes for all the cases. Figures 4.51–4.61 exhibit the variation of the normalized inverse SL in this case as functions of electron concentration, light intensity, and wavelength, respectively. It appears from Figures 4.55–4.62 that the normalized inverse SL decreases with the increase in light intensity and the wavelength which is in the visible region. From Figures 4.63 to 4.66, it appears that the normalized inverse SL increases with the increase in the electric field. It should be noted that the rate of change of the normalized inverse SL in the respective cases are totally energy spectrum dependent.

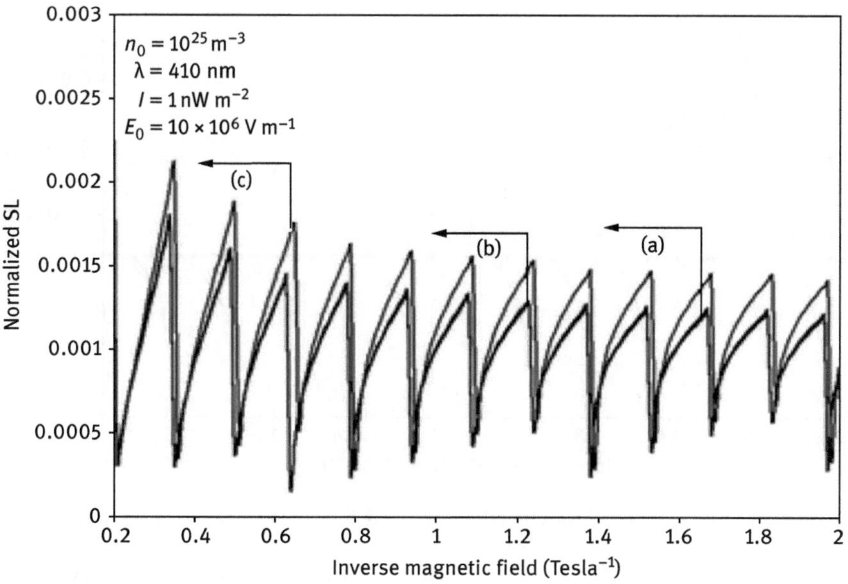

Figure 4.47: Plot of the normalized SL as a function of inverse quantizing magnetic field under cross-field configuration in external photoexcitation for HD n-InAs, in which the curves (a), (b), and (c) represent the perturbed three- and two-band models of Kane and that of parabolic energy bands, respectively.

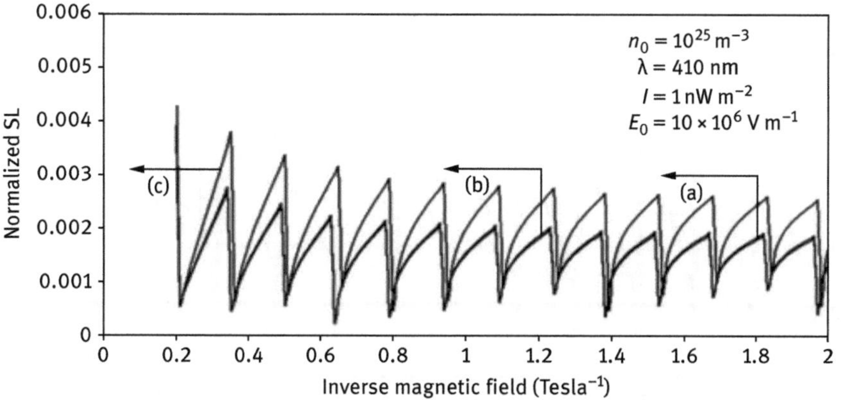

Figure 4.48: Plot of the normalized SL as a function of inverse quantizing magnetic field under cross-field configuration in external photoexcitation for HD n-InSb, in which the curves (a), (b), and (c) represent the perturbed three- and two-band models of Kane and that of parabolic energy bands, respectively.

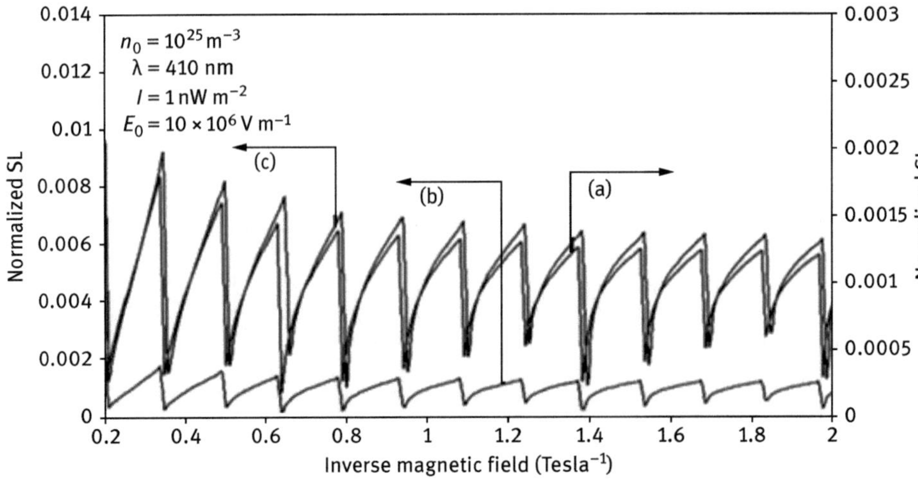

Figure 4.49: Plot of the normalized SL as a function of inverse quantizing magnetic field under cross-field configuration in external photoexcitation for HD n-$Hg_{1-x}Cd_xTe$, in which the curves (a), (b), and (c) represent the perturbed three- and two-band models of Kane and that of parabolic energy bands, respectively.

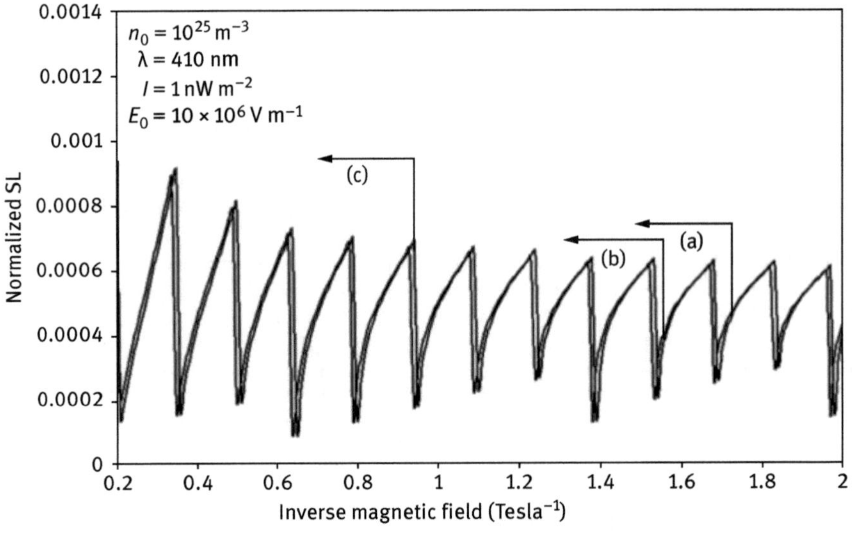

Figure 4.50: Plot of the normalized SL as a function of inverse quantizing magnetic field under cross-field configuration in external photoexcitation for HD n-$In_{1-x}Ga_xAs_yP_{1-y}$ lattice matched to InP, in which the curves (a), (b), and (c) represent the perturbed three- and two-band models of Kane and that of parabolic energy bands, respectively.

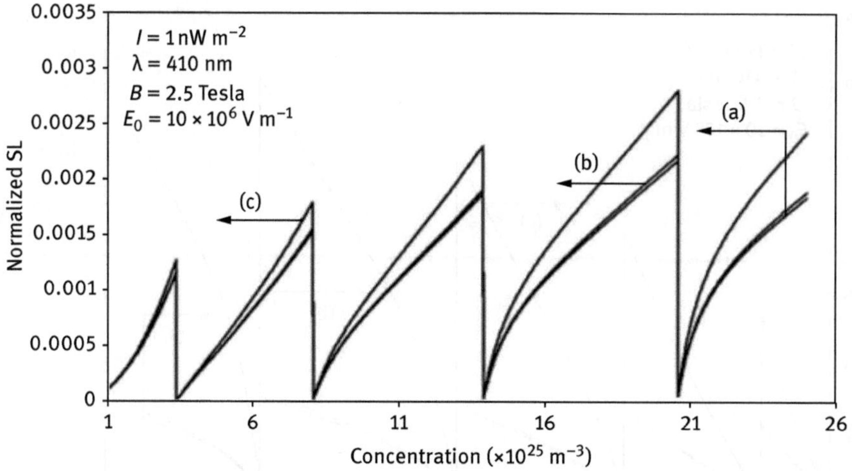

Figure 4.51: Plot of the normalized SL as a function of electron concentration field under cross-field configuration in external photoexcitation for HD n-InAs, in which the curves (a), (b), and (c) represent the perturbed three- and two-band models of Kane and that of parabolic energy bands, respectively.

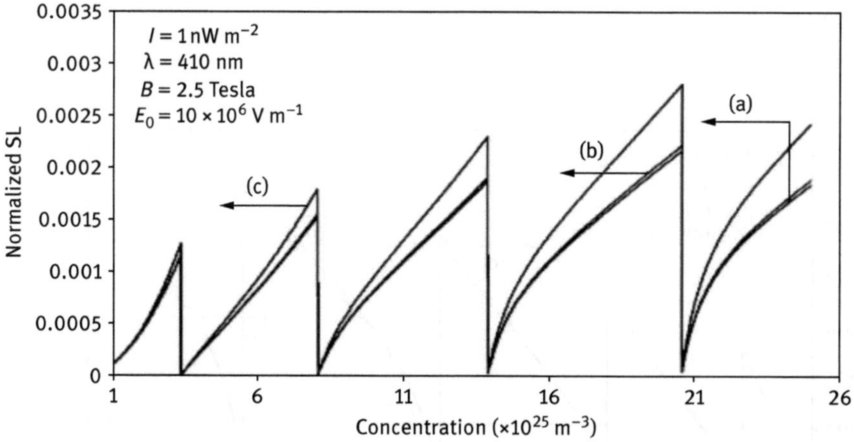

Figure 4.52: Plot of the normalized SL as a function of electron concentration field under cross-field configuration in external photoexcitation for HD n-InSb, in which the curves (a), (b), and (c) represent the perturbed three- and two-band models of Kane and that of parabolic energy bands, respectively.

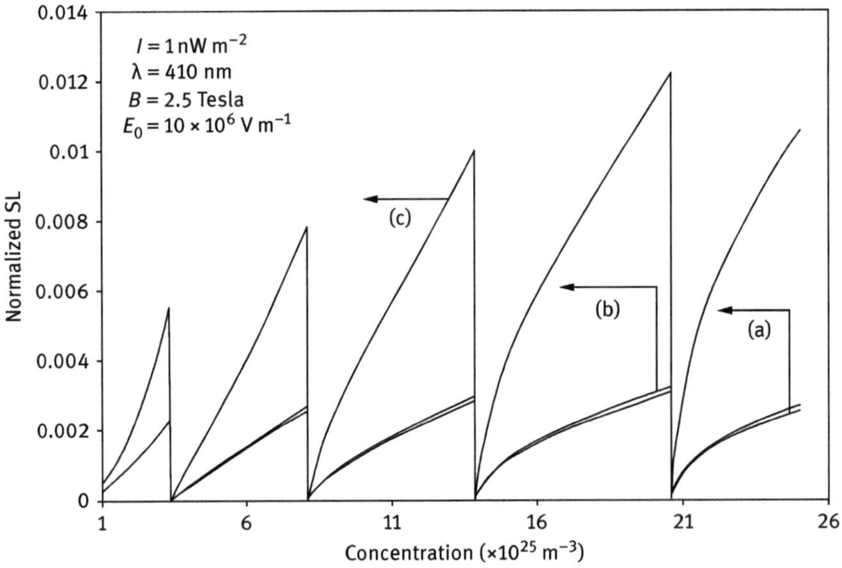

Figure 4.53: Plot of the normalized SL as a function of electron concentration field under cross-field configuration in external photoexcitation for HD n-$Hg_{1-x}Cd_xTe$ in which the curves (a), (b), and (c) represent the perturbed three- and two-band models of Kane and that of parabolic energy bands, respectively.

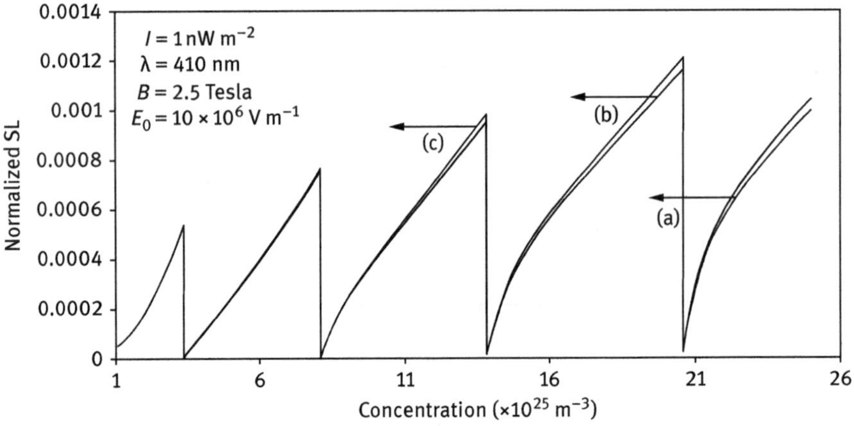

Figure 4.54: Plot of the normalized SL as a function of electron concentration field under cross-field configuration in external photoexcitation for HD n-$In_{1-x}Ga_xAs_yP_{1-y}$ lattice matched to InP in which the curves (a), (b), and (c) represent the perturbed three- and two-band models of Kane and that of parabolic energy bands, respectively.

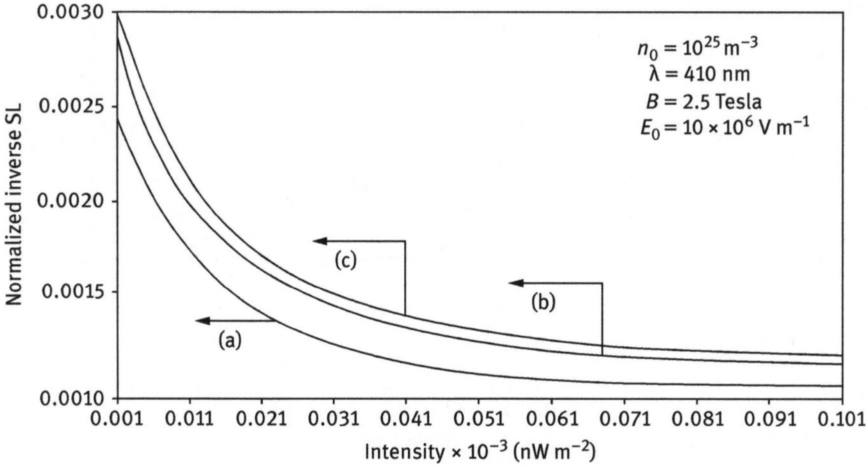

Figure 4.55: Plot of the normalized inverse SL as a function of light intensity under cross-field configuration in external photoexcitation for HD n-InAs, in which the curves (a), (b), and (c) represent the perturbed three- and two-band models of Kane and that of parabolic energy bands, respectively.

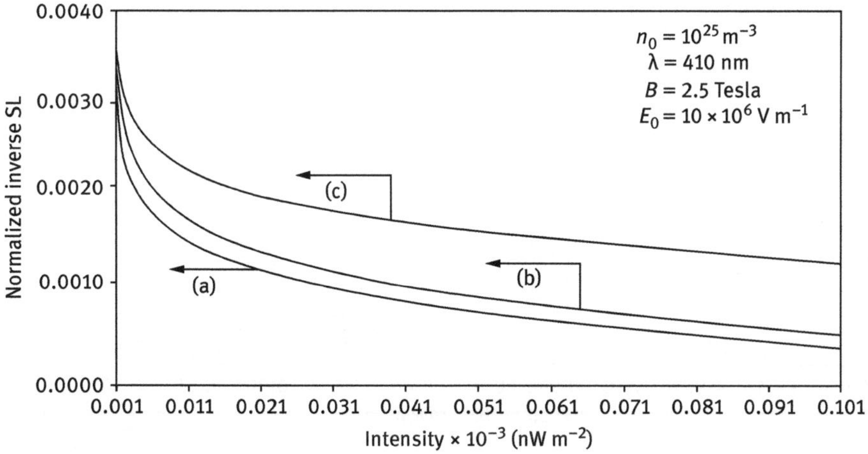

Figure 4.56: Plot of the normalized inverse SL as a function of light intensity under cross-field configuration in external photoexcitation for HD n-InSb, in which the curves (a), (b), and (c) represent the perturbed three- and two-band models of Kane and that of parabolic energy bands, respectively.

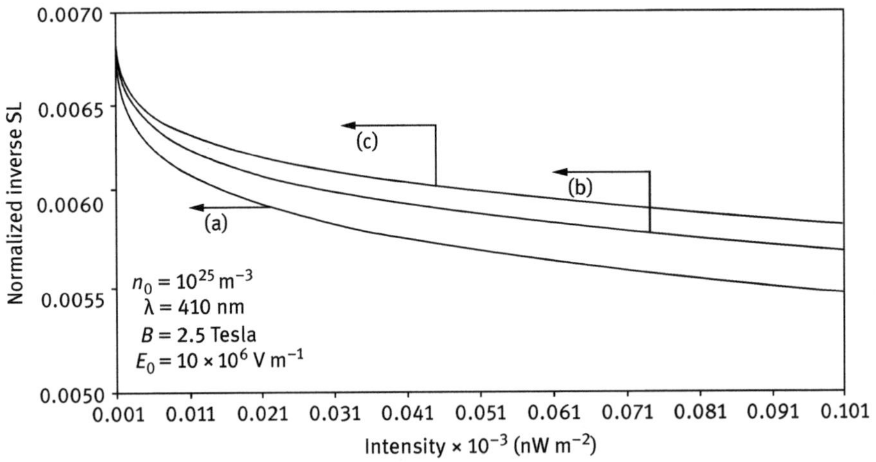

Figure 4.57: Plot of the normalized inverse SL as a function of light intensity under cross-field configuration in external photoexcitation for HD n-$Hg_{1-x}Cd_xTe$, in which the curves (a), (b), and (c) represent the perturbed three- and two-band models of Kane and that of parabolic energy bands, respectively.

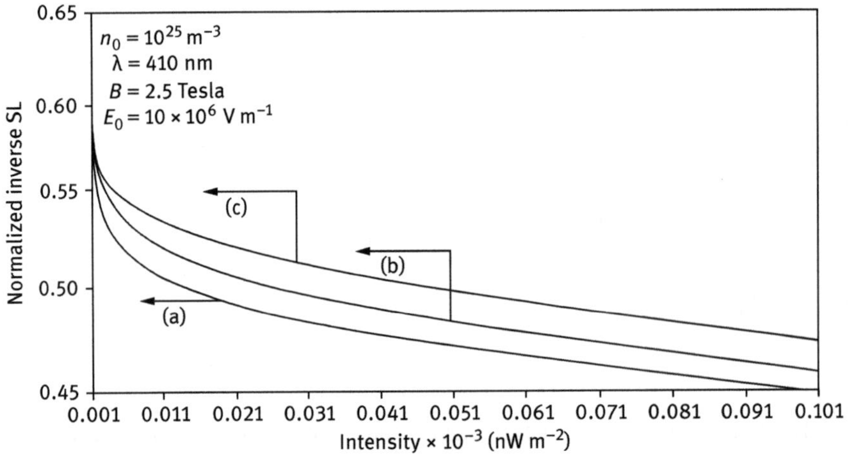

Figure 4.58: Plot of the normalized inverse SL as a function of light intensity under cross-field configuration in external photoexcitation for HD n-$In_{1-x}Ga_xAs_yP_{1-y}$ lattice matched to InP, in which the curves (a), (b), and (c) represent the perturbed three- and two-band models of Kane and that of parabolic energy bands, respectively.

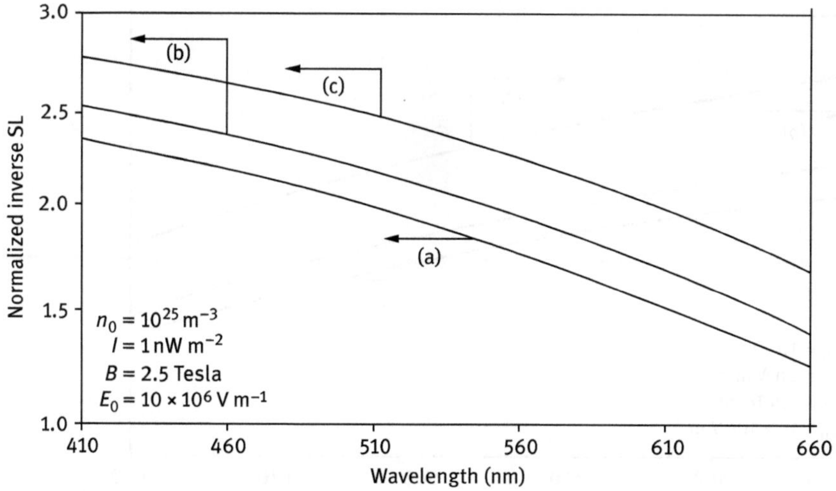

Figure 4.59: Plot of the normalized inverse SL as a function of wavelength under cross-field configuration in external photoexcitation for HD n-InAs, in which the curves (a), (b), and (c) represent the perturbed three- and two-band models of Kane and that of parabolic energy bands, respectively.

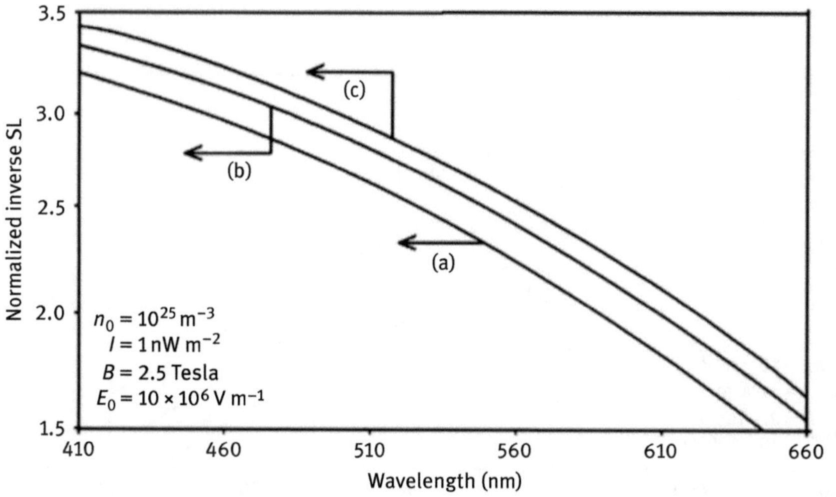

Figure 4.60: Plot of the normalized inverse SL as a function of wavelength under cross-field configuration in external photoexcitation for HD n-InSb, in which the curves (a), (b), and (c) represent the perturbed three- and two-band models of Kane and that of parabolic energy bands, respectively.

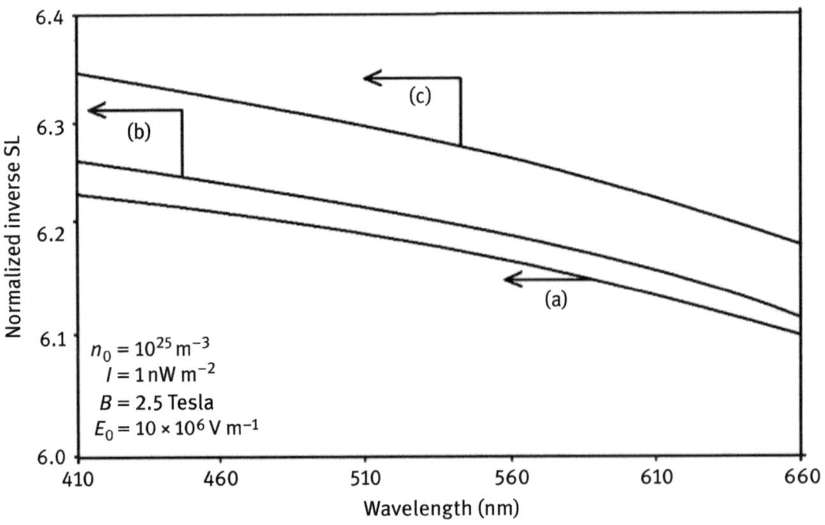

Figure 4.61: Plot of the normalized inverse SL as a function of wavelength under cross-field configuration in external photoexcitation for HD n-$Hg_{1-x}Cd_xTe$, in which the curves (a), (b), and (c) represent the perturbed three- and two-band models of Kane and that of parabolic energy bands, respectively.

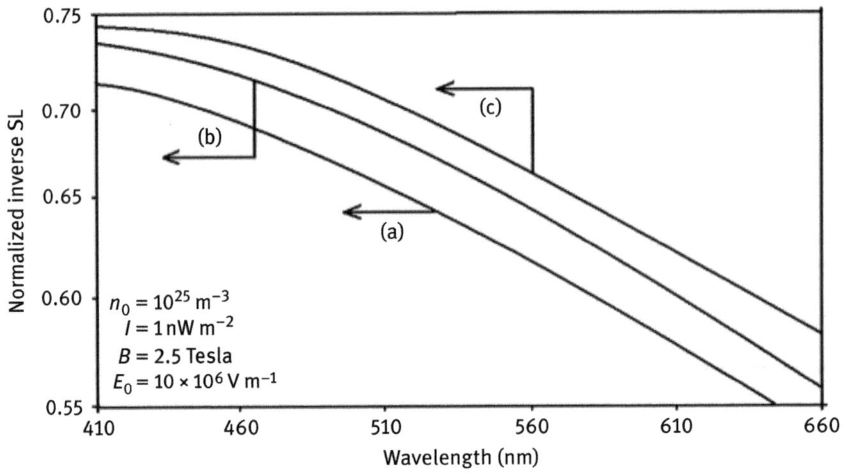

Figure 4.62: Plot of the normalized inverse SL as a function of wavelength under cross-field configuration in external photoexcitation for n-$In_{1-x}Ga_xAs_yP_{1-y}$ lattice matched to InP, in which the curves (a), (b), and (c) represent the perturbed three- and two-band models of Kane and that of parabolic energy bands, respectively.

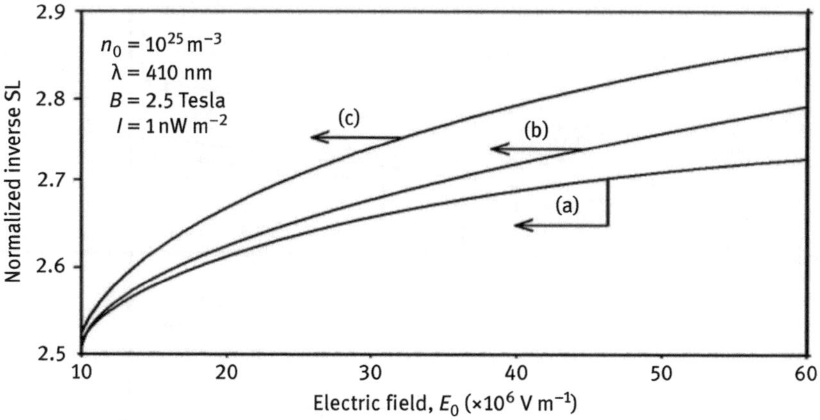

Figure 4.63: Plot of the normalized inverse SL as a function of electric field under cross-field configuration in external photoexcitation for HD n-InAs, in which the curves (a), (b), and (c) represent the perturbed three- and two-band models of Kane and that of parabolic energy bands, respectively.

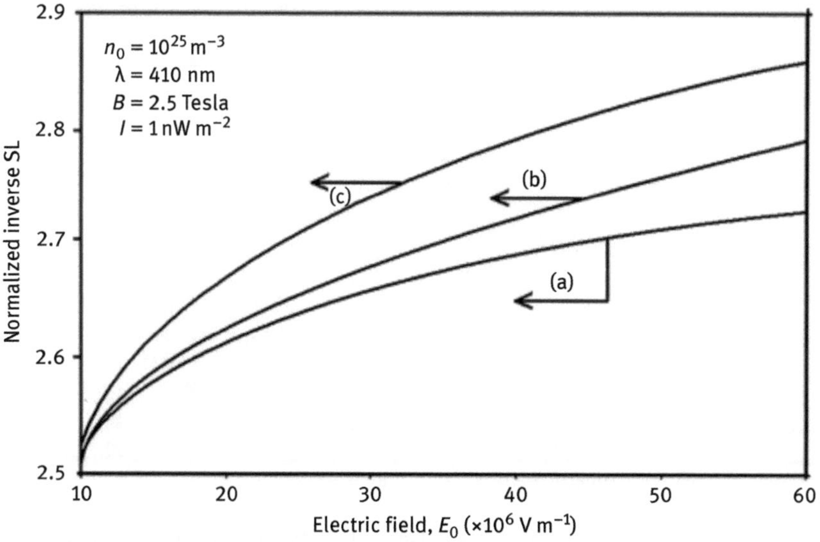

Figure 4.64: Plot of the normalized inverse SL as a function of electric field under cross-field configuration in external photoexcitation for HD n-InSb, in which the curves (a), (b), and (c) represent the perturbed three- and two-band models of Kane and that of parabolic energy bands, respectively.

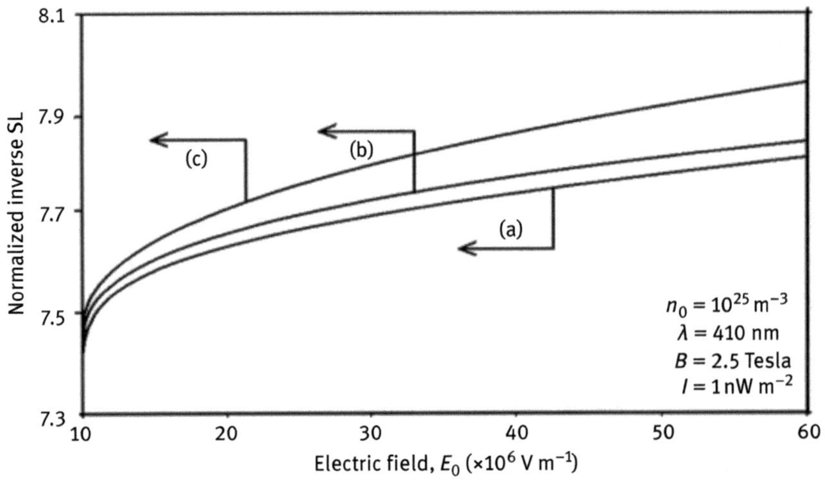

Figure 4.65: Plot of the normalized inverse SL as a function of electric field under cross-field configuration in external photoexcitation for HD n-$Hg_{1-x}Cd_xTe$, in which the curves (a), (b), and (c) represent the perturbed three- and two-band models of Kane and that of parabolic energy bands, respectively.

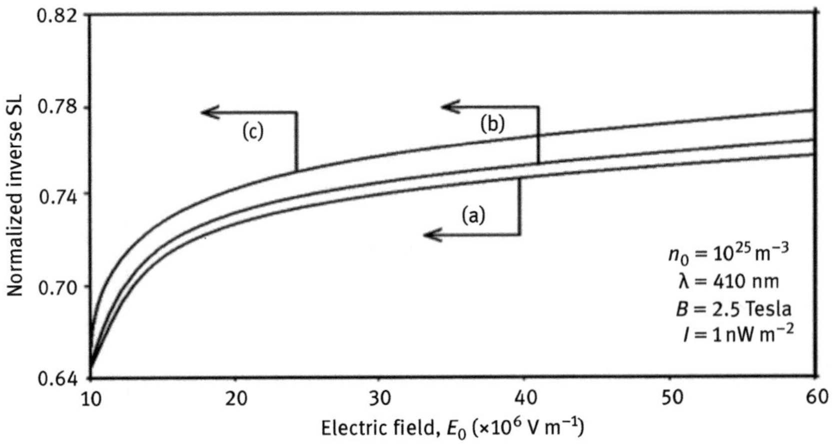

Figure 4.66: Plot of the normalized inverse SL as a function of electric field under cross-field configuration in external photoexcitation for HD n-$In_{1-x}Ga_xAs_yP_{1-y}$ lattice matched to InP, in which the curves (a), (b), and (c) represent the perturbed three- and two-band models of Kane and that of parabolic energy bands, respectively.

4.2.6.4 Open research problems

(R4.1) Investigate SL in the presence of intense external light waves for all the HD materials whose respective dispersion relations of the carriers in the absence of any field and heavy doping are given in (R1.1).

(R4.2) Investigate SL for the HD semiconductors in the presence of Gaussian, exponential, Kane, Halperian, Lax and Bonch-Burevich types of band tails [16] for all systems whose unperturbed carrier energy spectra are defined in (R1.1) in the presence of external light waves.

(R4.3) Investigate SL in the presence of external light waves for HD bulk specimens of the negative refractive index, organic, magnetic and other advanced optical materials in the presence of an arbitrarily oriented alternating electric field.

(R4.4) Investigate all the relevant problems of this chapter for a Dirac electron.

(R4.5) Investigate all the relevant problems of this chapter by including the many-body, broadening- and hot-carrier effects, respectively.

(R4.6) Investigate all the relevant problems of this chapter by removing all the mathematical approximations and establishing the respective appropriate uniqueness conditions.

(R4.7) Investigate SL for 2D HD systems for all semiconductors as considered in (R4.1) in the presence of arbitrarily oriented photo excitation.

(R4.8) Investigate SL for the 2D systems of HD semiconductors in the presences of Gaussian, exponential, Kane, Halperian, Lax and Bonch-Burevich types of band tails for all systems whose unperturbed carrier energy spectra are defined in (R1.1) in the presence of external light waves.

(R4.9) Investigate SL in the presence of external light waves for HD 2D systems of the negative refractive index, organic, magnetic and other advanced optical materials in the presence of an arbitrarily oriented alternating electric field.

(R4.10) Investigate SL in the presence of external light waves for the multiple HD 2D systems of semiconductors whose unperturbed carrier energy spectra are defined in (R1.1) and heavily–doped semiconductors in the presences of Gaussian, exponential, Kane, Halperian, Lax and Bonch-Burevich types of Band tails for all systems whose unperturbed carrier energy spectra are defined in the same problems, respectively.

(R4.11) Investigate SL in the presence of external light waves for all the appropriate HD 2D systems of this chapter in the presence of finite potential wells.

(R4.12) Investigate SL in the presence of external light waves for all the appropriate HD 2D systems of this chapter in the presence of parabolic potential wells.

(R4.13) Investigate SL in the presence of external light waves for all the appropriate HD 2D systems of this chapter forming quantum rings.

(R4.14) Investigate SL in the presence of external light waves for all the above appropriate problems in the presence of elliptical Hill and quantum square rings.

(R4.15) Investigate SL for all the appropriate systems from Chapter 1 up-to Chapter 5 in the presence of arbitrarily oriented light waves and strain.

References

[1] P.K. Basu, *Theory of Optical Process in Semiconductors, Bulk and Microstructures* (Oxford University Press, Oxford, 1997).
[2] K.P. Ghatak, S. Bhattacharya, S. Bhowmik, R. Benedictus, and S. Chowdhury, *J. Appl. Phys.* 103, 094314 (2008); K.P. Ghatak, S. Bhattacharya, *J. Appl. Phys.* 102, 073704 (2007); K.P. Ghatak, S. Bhattacharya, S.K. Biswas, A. De, and A.K.Dasgupta, *Phys. Scr.*, 75, 820 (2007).
[3] K. Seeger, Semiconductor Physics, 7th edn. (Springer Verlag, Germany, 2006).
[4] B.R. Nag, *Electron Transport in Compound Semiconductors* (Springer-Verlag, Germany, 1980).

5 Heisenberg's uncertainty principle and field emission in optoelectronic nanomaterials

Without hard work I achieve nothing. The prize will not be sent to me. I have to win it.

5.1 Introduction

The Fowler–Nordheim field emission (FE) is a well-known quantum-mechanical phenomenon that involves tunneling of electrons through a surface barrier because of the applications invovled in an intense external electric field. Normally, at field strengths of the order of 10^8 V/m (below the electrical breakdown), the potential barriers at the surface of metals and semiconductors usually become very thin and result in FE of electrons because of the tunnel effect [1]. This has been well-investigated with reference to three-dimensional electron gases in metals and semiconductors and the FE from quantum-confined structures has also been studied in this context [2–8]. Some of the significant features of the FE that have emerged from these investigations are as follows:

1. The FE increases with increasing electron concentration in bulk materials and are significantly influenced by the carrier energy spectra of different electronic materials
2. The FE increases with increasing electric field
3. The FE oscillates with film thickness for quantum-confined systems
4. The FE oscillates with inverse quantizing magnetic field in the presence of magnetic quantization because of the Shubnikov de Haas effect
5. For various types of superlattices of different materials, the FE shows composite oscillations with different system variables

In this chapter, Section 5.2.1 discusses the study of FE from heavily doped (HD) III–V, ternary and quaternary materials under magnetic quantization in the presence of strong photons and special cases in the absence of heavy doping and light have also been discussed. In Section 5.2.2, we have also investigated the FE in nanowires of HD III–V, ternary and quaternary materials in the presence of strong light waves where the special cases have also been discussed. In Section 5.2.3, the FE from HD effective mass superlattices whose constituent materials are III–V semiconductors has been investigated in the presence of light waves under magnetic quantization. The FE from HD quantum wire effective mass superlattices of the said materials in the presence of light waves has been studied in Section 5.2.4. Section 5.2.5 investigates FE from HD superlattices of III–V materials with graded interfaces in the presence of light waves under magnetic quantization. Section 5.2.6 discusses the FE from HD quantum wire

https://doi.org/10.1515/9783110610819-005

superlattices of the said materials with graded interfaces in the presence of light waves. Section 5.3 consists of results and discussion. Section 5.4 presents open research problems pertaining to this chapter.

5.2 Theoretical background

5.2.1 Field emission from HD III-V, ternary and quaternary materials under magnetic quantization in the presence of light waves

The field-emitted current density from HD III–V, ternary and quaternary materials under magnetic quantization in the presence of strong photons in accordance with three- and two-band models of Kane together with parabolic energy bands can, respectively, be written as follows:

$$J = \frac{e^2 B g_v}{2\pi^2\hbar^2} \text{ Real Part of} \sum_{n=0}^{n_{\max}} [E_{FBHDLB} - E_{51}] \exp(-\beta_{51}) \tag{5.1}$$

$$J = \frac{e^2 B g_v}{2\pi^2\hbar^2} \sum_{n=0}^{n_{\max}} [E_{FBHDLB} - E_{52}] \exp(-\beta_{52}) \tag{5.2}$$

$$J = \frac{e^2 B g_v}{2\pi^2\hbar^2} \sum_{n=0}^{n_{\max}} [E_{FBHDLB} - E_{53}] \exp(-\beta_{53}) \tag{5.3}$$

where E_{51} is the root of the equation

$$\begin{gathered} T_1(E_{51}, \eta_g, \lambda) = \left(n + \frac{1}{2}\right)\hbar\omega_0 \\ \beta_{51} = \frac{4\sqrt{2m_c}[T_1(V_0, \eta_g, \lambda) - (n + \frac{1}{2})\hbar\omega_0]^{3/2}}{3eF_{sx}\hbar T'_1(V_0, \eta_g, \lambda)} \end{gathered} \tag{5.4}$$

E_{52} is the root of the equation

$$\begin{gathered} T_2(E_{52}, \eta_g, \lambda) = \left(n + \frac{1}{2}\right)\hbar\omega_0 \\ \beta_{52} = \frac{4\sqrt{2m_c}[T_2(V_0, \eta_g, \lambda) - (n + \frac{1}{2})\hbar\omega_0]^{3/2}}{3eF_{sx}\hbar T'_2(V_0, \eta_g, \lambda)} \end{gathered} \tag{5.5}$$

E_{53} is the root of the equation

$$T_3(E_{53}, \eta_g, \lambda) = \left(n + \frac{1}{2}\right)\hbar\omega_0$$
$$\beta_{53} = \frac{4\sqrt{2m_c}[T_3(V_0, \eta_g, \lambda) - (n + \frac{1}{2})\hbar\omega_0]^{3/2}}{3eF_{sx}\hbar T'_3(V_0, \eta_g, \lambda)} \tag{5.6}$$

The field-emitted current density from III–V, ternary and quaternary materials under magnetic quantization in the presence of strong photons in accordance with three- and two-band model of Kane together with parabolic energy bands can, respectively, be written as follows:

$$J = \frac{e^2 B g_v}{2\pi^2\hbar^2}\sum_{n=0}^{n_{max}}[E_{FLB} - E_{54}]\exp(-\beta_{54}) \tag{5.7}$$

$$J = \frac{e^2 B g_v}{2\pi^2\hbar^2}\sum_{n=0}^{n_{max}}[E_{FLB} - E_{55}]\exp(-\beta_{55}) \tag{5.8}$$

$$J = \frac{e^2 B g_v}{2\pi^2\hbar^2}\sum_{n=0}^{n_{max}}[E_{FLB} - E_{56}]\exp(-\beta_{56}) \tag{5.9}$$

where E_{54} is the root of the equation

$$\beta_0(E_{54}, \eta_g, \lambda) = \left(n + \frac{1}{2}\right)\hbar\omega_0 \tag{5.10}$$

$$\beta_{54} = \frac{4\sqrt{2m_c}[\beta_0(V_0, \lambda) - (n + \frac{1}{2})\hbar\omega_0]^{3/2}}{3eF_{sx}\hbar\beta'_0(V_0, \lambda)}$$

E_{55} is the root of the equation

$$\tau_0(E_{55}, \lambda) = \left(n + \frac{1}{2}\right)\hbar\omega_0$$
$$\beta_{55} = \frac{4\sqrt{2m_c}[\tau_0(V_0, \lambda) - (n + \frac{1}{2})\hbar\omega_0]^{3/2}}{3eF_{sx}\hbar\tau'_0(V_0, \lambda)} \tag{5.11}$$

E_{56} is the root of the equation

$$\rho_0(E_{56}\lambda) = \left(n + \frac{1}{2}\right)\hbar\omega_0$$

$$\beta_{56} = \frac{4\sqrt{2m_c}[\rho_0(V_0, \lambda) - (n + \frac{1}{2})\hbar\omega_0]^{3/2}}{3eF_{sx}\hbar\rho'_0(V_0, \lambda)} \tag{5.12}$$

The field-emitted current density from III–V, ternary and quaternary materials under magnetic quantization in accordance with three- and two-band model of Kane together with parabolic energy bands can, respectively, be written as follows:

$$J = \frac{e^2 B g_v}{2\pi^2\hbar^2}\sum_{n=0}^{n_{\max}}[E_{FB} - E_{57}]\exp(-\beta_{57}) \tag{5.13}$$

$$J = \frac{e^2 B g_v}{2\pi^2\hbar^2}\sum_{n=0}^{n_{\max}}[E_{FB} - E_{58}]\exp(-\beta_{58}) \tag{5.14}$$

$$J = \frac{e^2 B g_v}{2\pi^2\hbar^2}\sum_{n=0}^{n_{\max}}[E_{FB} - E_{59}]\exp(-\beta_{59}) \tag{5.15}$$

where E_{57} is the root of the equation

$$\begin{gathered} I_{11}(E_{57}) = \left(n + \frac{1}{2}\right)\hbar\omega_0 \\ \beta_{57} = \frac{4\sqrt{2m_c}\left[I_{11}(V_0) - (n + \frac{1}{2})\hbar\omega_0\right]^{3/2}}{3eF_{sx}\hbar I'_{11}(V_0)} \end{gathered} \tag{5.16}$$

E_{58} is the root of the equation

$$\begin{gathered} E_{58}(1 + \alpha E_{58}) = \left(n + \frac{1}{2}\right)\hbar\omega_0 \\ \beta_{58} = \frac{4\sqrt{2m_c}\left[V_0(1 + \alpha V_0) - (n + \frac{1}{2})\hbar\omega_0\right]^{3/2}}{3eF_{sx}\hbar(1 + 2\alpha V_0)} \end{gathered} \tag{5.17}$$

E_{59} is given by

$$E_{59} = \left(n + \frac{1}{2}\right)\hbar\omega_0 \tag{5.18}$$

$$\beta_{59} = \frac{4\sqrt{2m_c}\left[V_0 - (n + \frac{1}{2})\hbar\omega_0\right]^{3/2}}{3eF_{sx}\hbar}$$

5.2.2 Field emission from HD nanowire (NW) III–V, ternary and quaternary materials in the presence of light waves

The field-emitted current from HD quantum wires of III–V, ternary and quaternary materials in the presence of strong photons in accordance with three- and two-band model of Kane together with parabolic energy bands can, respectively, be written as follows:

$$I = \frac{eg_v}{\pi\hbar} \text{Real Part of} \sum_{n_x=1}^{n_{x\max}} \sum_{n_y=1}^{n_{y\max}} [E_{F1HDNWL1} - E_{60}] \exp(-\beta_{60}) \tag{5.19}$$

$$I = \frac{eg_v}{\pi\hbar} \sum_{n_x=1}^{n_{x\max}} \sum_{n_y=1}^{n_{y\max}} [E_{F1HDNWL1} - E_{61}] \exp(-\beta_{61}) \tag{5.20}$$

$$I = \frac{eg_v}{\pi\hbar} \sum_{n_x=1}^{n_{x\max}} \sum_{n_y=1}^{n_{y\max}} [E_{F1HDNWL1} - E_{62}] \exp(-\beta_{62}) \tag{5.21}$$

where E_{60} is the root of the equation

$$\frac{\hbar^2(n_z\pi/d_z)^2}{2m_c} + \frac{\hbar^2(n_y\pi/d_y)^2}{2m_c} = T_1(E_{60}, \eta_g, \lambda) \tag{5.22}$$

and

$$\beta_{60} = \frac{4\sqrt{2m_c}\left[T_1(V_0, \eta_g, \lambda) - \left[\frac{\hbar^2(n_z\pi/d_z)^2}{2m_c} + \frac{\hbar^2(n_y\pi/d_y)^2}{2m_c}\right]\right]^{3/2}}{3eF_{sx}\hbar T'_1(V_0, \eta_g, \lambda)}$$

The E_{61} in eq. (5.20) is the root of the equation

$$\begin{gathered}\frac{\hbar^2(n_z\pi/d_z)^2}{2m_c} + \frac{\hbar^2(n_y\pi/d_y)^2}{2m_c} = T_2(E_{61}, \eta_g, \lambda)\\ \beta_{61} = \frac{4\sqrt{2m_c}\left[T_2(V_0, \eta_g, \lambda) - \left[\frac{\hbar^2(n_z\pi/d_z)^2}{2m_c} + \frac{\hbar^2(n_y\pi/d_y)^2}{2m_c}\right]\right]^{3/2}}{3eF_{sx}\hbar T'_2(V_0, \eta_g, \lambda)}\end{gathered} \tag{5.23}$$

The E_{62} in eq. (5.21) is the root of the equation

$$\frac{\hbar^2(n_z\pi/d_z)^2}{2m_c} + \frac{\hbar^2(n_y\pi/d_y)^2}{2m_c} = T_3(E_{62}, \eta_g, \lambda)$$

$$\beta_{62} = \frac{4\sqrt{2m_c}\left[T_3(V_0, \eta_g, \lambda) - \left[\frac{\hbar^2(n_z\pi/d_z)^2}{2m_c} + \frac{\hbar^2(n_y\pi/d_y)^2}{2m_c}\right]\right]^{3/2}}{3eF_{sx}\hbar T'_3(V_0, \eta_g, \lambda)} \tag{5.24}$$

The field-emitted current from quantum wires of III–V, ternary and quaternary materials in the presence of strong photons in accordance with three- and two-band model of Kane together with parabolic energy bands can, respectively, be written as follows:

$$I = \frac{eg_v}{\pi\hbar} \sum_{n_x=1}^{n_{x\max}} \sum_{n_y=1}^{n_{y\max}} [E_{F1NWL2} - E_{63}] \exp(-\beta_{63}) \tag{5.25}$$

$$I = \frac{eg_v}{\pi\hbar} \sum_{n_x=1}^{n_{x\max}} \sum_{n_y=1}^{n_{y\max}} [E_{F1NWL2} - E_{64}] \exp(-\beta_{64}) \tag{5.26}$$

$$I = \frac{eg_v}{\pi\hbar} \sum_{n_x=1}^{n_{x\max}} \sum_{n_y=1}^{n_{y\max}} [E_{F1NWL2} - E_{65}] \exp(-\beta_{65}) \tag{5.27}$$

where E_{63} for eq. (5.25) is the root of the equation

$$\frac{\hbar^2(n_z\pi/d_z)^2}{2m_c} + \frac{\hbar^2(n_y\pi/d_y)^2}{2m_c} = \beta_0(E_{63}, \lambda) \tag{5.28}$$

and

$$\beta_{63} = \frac{4\sqrt{2m_c}\left[\beta_0(V_0, \lambda) - \left[\frac{\hbar^2(n_z\pi/d_z)^2}{2m_c} + \frac{\hbar^2(n_y\pi/d_y)^2}{2m_c}\right]\right]^{3/2}}{3eF_{sx}\hbar\beta'_0(V_0, \lambda)}$$

The E_{64} in eq. (5.26) is the root of the equation

$$\frac{\hbar^2(n_z\pi/d_z)^2}{2m_c} + \frac{\hbar^2(n_y\pi/d_y)^2}{2m_c} = \tau_0(E_{64}, \lambda) \tag{5.29}$$

and

$$\beta_{64} = \frac{4\sqrt{2m_c}\left[\tau_0(V_0, \lambda) - \left[\frac{\hbar^2(n_z\pi/d_z)^2}{2m_c} + \frac{\hbar^2(n_y\pi/d_y)^2}{2m_c}\right]\right]^{3/2}}{3eF_{sx}\hbar\tau'_0(V_0, \lambda)}$$

The E_{65} in eq. (5.27) is the root of the equation

$$\frac{\hbar^2(n_z\pi/d_z)^2}{2m_c} + \frac{\hbar^2(n_y\pi/d_y)^2}{2m_c} = \rho_0(E_{65}, \lambda) \tag{5.30}$$

and

$$\beta_{65} = \frac{4\sqrt{2m_c}\left[\rho_0(V_0, \lambda) - \left[\frac{\hbar^2(n_z\pi/d_z)^2}{2m_c} + \frac{\hbar^2(n_y\pi/d_y)^2}{2m_c}\right]\right]^{3/2}}{3eF_{sx}\hbar\rho'_0(V_0, \lambda)}$$

The field-emitted current from quantum wires of III–V, ternary and quaternary materials in the absence of photons and heavy doping in accordance with three- and two-band model of Kane together with parabolic energy bands can be, respectively, written as follows:

$$I = \frac{eg_v}{\pi\hbar} \sum_{n_x=1}^{n_{x\max}} \sum_{n_y=1}^{n_{y\max}} [E_{F1NW2} - E_{66}] \exp(-\beta_{66}) \tag{5.31}$$

$$I = \frac{eg_v}{\pi\hbar} \sum_{n_x=1}^{n_{x\max}} \sum_{n_y=1}^{n_{y\max}} [E_{F1NW2} - E_{67}] \exp(-\beta_{67}) \tag{5.32}$$

$$I = \frac{eg_v}{\pi\hbar} \sum_{n_x=1}^{n_{x\max}} \sum_{n_y=1}^{n_{y\max}} [E_{F1NW2} - E_{68}] \exp(-\beta_{68}) \tag{5.33}$$

where the Fermi energy E_{F1NW2} can be determined from the following equation:

$$n_0 = \frac{2g_v}{\pi} \sum_{n_y=1}^{n_{y\max}} \sum_{n_z=1}^{n_{z\max}} [f_{100}(E_{F1NW2}, n_y, n_z)] \tag{5.34}$$

in which

$$f_{100}(E_{F1NW2}, n_y, n_z) = \left[\left[I_{11}(E_{F1NW2}) - \left[\frac{\hbar^2(n_z\pi/d_z)^2}{2m_c} + \frac{\hbar^2(n_y\pi/d_y)^2}{2m_c}\right]\right]\frac{2m_c}{\hbar^2}\right]^{\frac{1}{2}}$$

The E_{66} in eq. (5.31) is the root of the equation

$$\frac{\hbar^2(n_z\pi/d_z)^2}{2m_c} + \frac{\hbar^2(n_y\pi/d_y)^2}{2m_c} = I_{11}(E_{66}) \tag{5.35}$$

and

$$\beta_{66} = \frac{4\sqrt{2m_c}\left[I_{11}(V_0) - \left[\frac{\hbar^2(n_z\pi/d_z)^2}{2m_c} + \frac{\hbar^2(n_y\pi/d_y)^2}{2m_c}\right]\right]^{3/2}}{3eF_{sx}\hbar I'_{11}(V_0)}$$

Again E_{67} for eq. (5.32) can be determined from the following equation:

$$\frac{\hbar^2(n_z\pi/d_z)^2}{2m_c} + \frac{\hbar^2(n_y\pi/d_y)^2}{2m_c} = [E_{67}(1 + \alpha E_{67})] \tag{5.36}$$

For eq. (5.32), the Fermi energy E_{F1NW2} can be determined from the following equation:

$$n_0 = \frac{2g_v}{\pi} \sum_{n_y=1}^{n_{y\max}} \sum_{n_z=1}^{n_{z\max}} [f_{101}(E_{F1NW2}, n_y, n_z)] \tag{5.37}$$

in which

$$f_{101}(E_{F1NW2}, n_y, n_z) = \left[\left[E_{F1NW2}(1+\alpha E_{F1NW2}) - \left[\frac{\hbar^2(n_z\pi/d_z)^2}{2m_c} + \frac{\hbar^2(n_y\pi/d_y)^2}{2m_c}\right]\right]\frac{2m_c}{\hbar^2}\right]^{\frac{1}{2}}$$

and

$$\beta_{67} = \frac{4\sqrt{2m_c}\left[V_0(1+\alpha V_0) - \left[\frac{\hbar^2(n_z\pi/d_z)^2}{2m_c} + \frac{\hbar^2(n_y\pi/d_y)^2}{2m_c}\right]\right]^{3/2}}{3eF_{sx}\hbar(1+2\alpha V_0)}$$

Besides E_{68} for eq. (5.33) can be determined from the following equation:

$$\frac{\hbar^2(n_z\pi/d_z)^2}{2m_c} + \frac{\hbar^2(n_y\pi/d_y)^2}{2m_c} = [E_{68}] \tag{5.38}$$

For eq. (5.33), the Fermi energy E_{F1NW2}can be determined from the following equation:

$$n_0 = \frac{2g_v}{\pi} \sum_{n_y=1}^{n_{y\max}} \sum_{n_z=1}^{n_{z\max}} [f_{102}(E_{F1NW2}, n_y, n_z)] \tag{5.39}$$

in which

$$f_{102}(E_{F1NW3}, n_y, n_z) = \left[\left[E_{F1NW2} - \left[\frac{\hbar^2(n_z\pi/d_z)^2}{2m_c} + \frac{\hbar^2(n_y\pi/d_y)^2}{2m_c}\right]\right]\frac{2m_c}{\hbar^2}\right]^{\frac{1}{2}}$$

and

$$\beta_{67} = \frac{4\sqrt{2m_c}\left[V_0 - \left[\frac{\hbar^2(n_z\pi/d_z)^2}{2m_c} + \frac{\hbar^2(n_y\pi/d_y)^2}{2m_c}\right]\right]^{3/2}}{3eF_{sx}\hbar}$$

5.2.3 Field emission from HD effective mass superlattices of III–V semiconductors in the presence of light waves under magnetic quantization

(a) The field-emitted current density under magnetic quantization and in the presence of light waves from HD effective mass superlattices whose constituent materials are defined by the three-band model of Kane can be expressed as follows:

$$J = \frac{e^2 B g_v}{2\pi^2 \hbar^2} \text{Real Part of} \sum_{n=0}^{n_{\max}} [E_{fSLHDB} - E_{70}] \exp(-\beta_{70}) \tag{5.40}$$

where E_{70} is the root of the equation

$$[\rho_{4HD1}(n, E_{70}, \lambda)] = 0 \tag{5.41}$$

and

$$\beta_{70} = \frac{4[\rho_{4HD1}(n, V_0, \lambda)]^{3/2}}{3eF_{sx}[\rho_{4HD1}(n, V_0, \lambda)]'}$$

(b) The field-emitted current density under magnetic quantization and in the presence of light waves from HD effective mass superlattices whose constituent materials are defined by the two-band model of Kane can be expressed as follows:

$$J = \frac{e^2 B g_v}{2\pi^2 \hbar^2} \sum_{n=0}^{n_{\max}} [E_{fSLHDB} - E_{71}] \exp(-\beta_{71}) \tag{5.42}$$

where E_{71} is the root of the equation

$$[\rho_{4HD2}(n, E_{71}, \lambda)] = 0 \tag{5.43}$$

and

$$\beta_{71} = \frac{4[\rho_{4HD2}(n, V_0, \lambda)]^{3/2}}{3eF_{sx}[\rho_{4HD2}(n, V_0, \lambda)]'}$$

(c) The field-emitted current density under magnetic quantization and in the presence of light waves from HD effective mass superlattices whose constituent materials are defined by the parabolic energy bands can be expressed as follows:

$$J = \frac{e^2 B g_v}{2\pi^2 \hbar^2} \sum_{n=0}^{n_{\max}} [E_{fSLHDB} - E_{72}] \exp(-\beta_{72}) \tag{5.44}$$

where E_{72} is the root of the equation

$$[\rho_{4HD3}(n, E_{72}, \lambda)] = 0 \tag{5.45}$$

and

$$\beta_{72} = \frac{4[\rho_{4HD3}(n, V_0, \lambda)]^{3/2}}{3eF_{sx}[\rho_{4HD3}(n, V_0, \lambda)]'}$$

5.2.4 The field-emitted current from nanowire heavily doped (NWHD) effective mass superlattices of Kane type semiconductors in the presence of light waves

The field-emitted current from NWHD effective mass superlattices of Kane type semiconductors in the presence of light waves, the dispersion relations of whose constituent materials in the absence of any perturbation are defined by the three-band model of Kane can be written as follows:

$$I = \frac{e g_v}{\pi \hbar} \textit{Real Part} \text{ of} \sum_{n_y=1}^{n_{y\max}} \sum_{n_z=1}^{n_{z\max}} [E_{F1.2.13} - E_{73}] \exp(-\beta_{73}) \tag{5.46}$$

where E_{73} is the root of the following equation:

$$\left[\frac{1}{L_0^2}\left\{\cos^{-1}\left(f_{HD1}\left(E_{73}, \frac{n_y\pi}{d_y}, \frac{n_z\pi}{d_z}, \lambda\right)\right)\right\}^2 - \left[\left(\frac{n_y\pi}{d_y}\right)^2 + \left(\frac{n_z\pi}{d_z}\right)^2\right]\right] = 0 \tag{5.47}$$

and

$$\beta_{73} = \frac{-4\left[\frac{1}{L_0^2}\left\{\cos^{-1}\left(f_{HD1}\left(V_0, \frac{n_y\pi}{d_y}, \frac{n_z\pi}{d_z}, \lambda\right)\right)\right\}^2 - \left[\left(\frac{n_y\pi}{d_y}\right)^2 + \left(\frac{n_z\pi}{d_z}\right)^2\right]\right]^{3/2}}{3eF_{sx}\left[\frac{1}{L_0^2}\left\{\cos^{-1}\left(f_{HD1}\left(V_0, \frac{n_y\pi}{d_y}, \frac{n_z\pi}{d_z}, \lambda\right)\right)\right\}^2 - \left[\left(\frac{n_y\pi}{d_y}\right)^2 + \left(\frac{n_z\pi}{d_z}\right)^2\right]\right]'}$$

(a) The field-emitted current from NWHD effective mass superlattices of Kane type semiconductors in the presence of light waves, the dispersion relations of whose constituent materials in the absence of any perturbation are defined by the two-band model of Kane can be written as follows:

$$I = \frac{eg_v}{\pi\hbar} \sum_{n_y=1}^{n_{y\max}} \sum_{n_z=1}^{n_{z\max}} [E_{F1.2.13} - E_{74}] \exp(-\beta_{74}) \tag{5.48}$$

$$\left[\frac{1}{L_0^2}\left\{\cos^{-1}\left(f_{HD2}\left(E_{74}, \frac{n_y\pi}{d_y}, \frac{n_z\pi}{d_z}, \lambda\right)\right)\right\}^2 - \left[\left(\frac{n_y\pi}{d_y}\right)^2 + \left(\frac{n_z\pi}{d_z}\right)^2\right]\right] = 0 \tag{5.49}$$

and

$$\beta_{74} = \frac{-4\left[\frac{1}{L_0^2}\left\{\cos^{-1}\left(f_{HD2}\left(V_0, \frac{n_y\pi}{d_y}, \frac{n_z\pi}{d_z}, \lambda\right)\right)\right\}^2 - \left[\left(\frac{n_y\pi}{d_y}\right)^2 + \left(\frac{n_z\pi}{d_z}\right)^2\right]\right]^{3/2}}{3eF_{sx}\left[\frac{1}{L_0^2}\left\{\cos^{-1}\left(f_{HD2}\left(V_0, \frac{n_y\pi}{d_y}, \frac{n_z\pi}{d_z}, \lambda\right)\right)\right\}^2 - \left[\left(\frac{n_y\pi}{d_y}\right)^2 + \left(\frac{n_z\pi}{d_z}\right)^2\right]\right]'}$$

where E_{74} is the root of the equation

(b) The field-emitted current from NWHD effective mass superlattices of Kane type semiconductors in the presence of light waves, the dispersion relations of whose constituent materials in the absence of any perturbation are defined by the parabolic energy bands can be written as follows:

$$I = \frac{eg_v}{\pi\hbar} \sum_{n_y=1}^{n_{y\max}} \sum_{n_z=1}^{n_{z\max}} [E_{F1.2.13} - E_{75}] \exp(-\beta_{75}) \tag{5.50}$$

where E_{75} is the root of the equation

$$\left[\frac{1}{L_0^2}\left\{\cos^{-1}\left(f_{HD3}\left(E_{75}, \frac{n_y\pi}{d_y}, \frac{n_z\pi}{d_z}, \lambda\right)\right)\right\}^2 - \left[\left(\frac{n_y\pi}{d_y}\right)^2 + \left(\frac{n_z\pi}{d_z}\right)^2\right]\right] = 0 \tag{5.51}$$

and

$$\beta_{75} = \frac{-4\left[\frac{1}{L_0^2}\left\{\cos^{-1}\left(f_{HD3}\left(V_0, \frac{n_y\pi}{d_y}, \frac{n_z\pi}{d_z}, \lambda\right)\right)\right\}^2 - \left[\left(\frac{n_y\pi}{d_y}\right)^2 + \left(\frac{n_z\pi}{d_z}\right)^2\right]\right]^{3/2}}{3eF_{sx}\left[\frac{1}{L_0^2}\left\{\cos^{-1}\left(f_{HD3}\left(V_0, \frac{n_y\pi}{d_y}, \frac{n_z\pi}{d_z}, \lambda\right)\right)\right\}^2 - \left[\left(\frac{n_y\pi}{d_y}\right)^2 + \left(\frac{n_z\pi}{d_z}\right)^2\right]\right]'}$$

5.2.5 Field emission in the presence of strong light waves from HD superlattices of III–V, ternary and quaternary constituent materials with graded interfaces under magnetic quantization

The field-emitted current density is given by the following equation:

$$J = \frac{e^2 B g_v}{2\pi^2 \hbar^2} \text{Real Part of} \sum_{n=0}^{n_{max}} [E_{F8.17,54} - E_{76}] \exp(-\beta_{76}) \tag{5.52}$$

where E_{76} is the root of the equation

$$[[G_{8,17,54} + iH_{8,17,54}]|_{E=E_{76}}] = 0 \tag{5.53}$$

and

$$\beta_{76} = \frac{-4[[G_{8,17,54} + iH_{8,17,54}]|_{E=V_0}]^{3/2}}{3eF_{sx}[[G_{8,17,54} + iH_{8,17,54}]|_{E=V_0}]'}$$

5.2.6 Field emission from HD quantum wire superlattices of III–V semiconductors with graded interfaces

The field-emitted current in this case is given by the following equation:

$$I = \frac{e g_v}{\pi \hbar} \text{Real Part of} \sum_{n_x=1}^{n_{x_{max}}} \sum_{n_y=1}^{n_{y_{max}}} [E_{F8.17,51} - E_{8.17.52}] \exp(-\beta_{77}) \tag{5.54}$$

Where

$$\beta_{77} = \frac{-4[[G_{8,17,50} + iH_{8,17,50}]|_{E=V_0}]^{3/2}}{3eF_{sx}[[G_{8,17,50} + iH_{8,17,50}]|_{E=V_0}]'}$$

5.3 Results and discussion

Using appropriate equations, we have plotted the field-emitted current density from n-InSb under magnetic quantization as functions of 1/B, concentration, wavelength, intensity, and electric field as shown in Figures 5.1–5.5 in accordance with both three- and two-band models of Kane. Figures 5.6–5.8 represent the field-emitted current

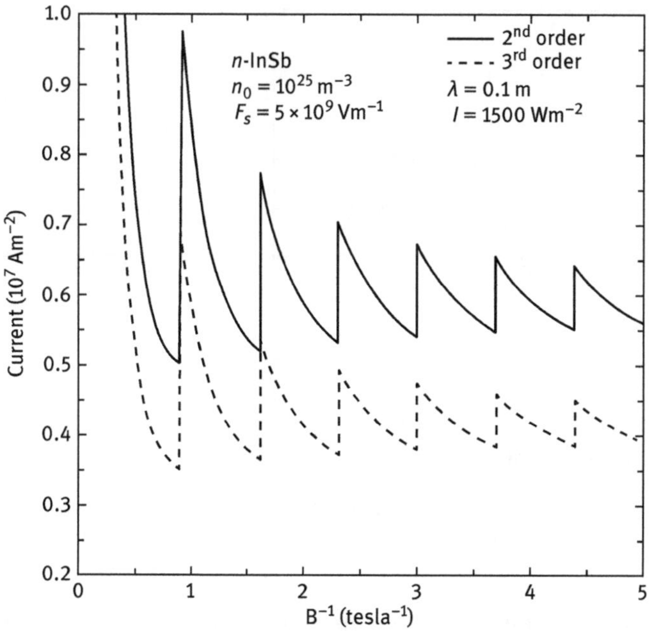

Figure 5.1: Plot of the field-emitted current density as a function of inverse magnetic field for HD *n*-InSb in the presence of light waves for both three- and two-band models of Kane.

from quantum wires of *n*-InSb in accordance with the three- and two-band models of Kane as functions of film thickness, concentration, and electric field respectively. Figures 5.9 and 5.10 exhibit the field-emitted current density from GaAs/AlGaAs superlattices with graded interfaces and also its effective mass counterpart under magnetic quantization as functions of 1/B and carrier concentration, respectively. Figures 5.11 and 5.12 exhibit the field-emitted current from GaAs/AlGaAs quantum wire superlattices with graded interfaces and also its effective mass counterpart as functions of film thickness and concentration respectively.

From Figure 5.1, we observe that the field-emitted current density from *n*-InSb under magnetic quantization exhibits oscillations with 1/B, the background physics of which has already been explained. We note that although the generation Landau sub-bands remains same within the given bandwidth, nature of the orientation of the curves is radically different. We note that when the wavelength of the incident light waves stays in the regime of radiowave zone and whose intensity lies within that of the solar intensity at the earth surface, the degeneracy increases. This implies that in the presence of radiowaves, the Fermi energy increases leading to a decrease in the variation of the field-emitted current density per sub-band. This was not in the case with the corresponding

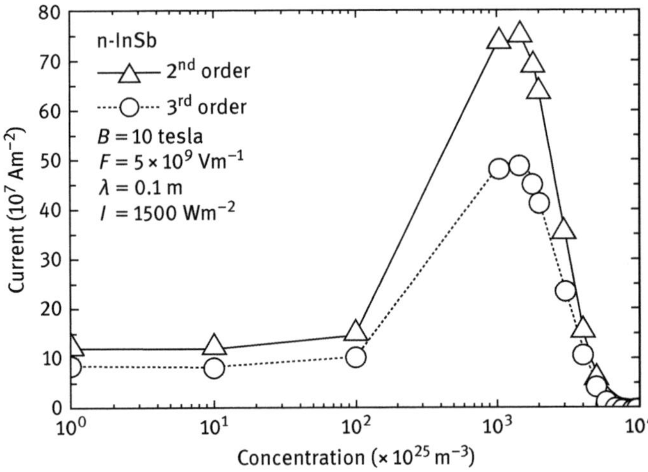

Figure 5.2: Plot of the field-emitted current density as a function of carrier concentration for HD *n*-InSb in the presence of light waves and external magnetic field for both three- and two-band models of Kane.

figure in Chapter 3. Figure 5.2 exhibits the variation of the field-emitted current density at magnetic quantum limit as a function of carrier concentration for the said case. We observe a peak in the current density near the value of the electron concentration 10^{25} m^{-3} at low temperatures for both the models. The field-emitted current density remains almost constant below the degeneracy of about 10^{25} m^{-3}. It should be noted that the said peak may alter its position with the variation of both wavelength and intensity respectively. Similar nature of the dependence of the field-emitted current density on the wavelength has been shown in Figure 5.3. We note that the peak happens in the radiowave zone for an intensity of 1,500Wm^{-2}. In Figure 5.4, we observe an almost constant field-emitted current density with respect to the light intensity upto $10^4 Wm^{-2}$. As the intensity level increases, initially the current density starts increasing slowly, whereas beyond $10^4 Wm^{-2}$ exhibits very large rise. The effect of the electric field on the field-emitted current density has been plotted in Figure 5.5 and it appears that an application of radiowaves increases the cut-in field in *n*-InSb under magnetic quantization as exhibited in the same figure. Composite oscillations in the field-emitted current as a function of film thickness in the presence of light waves has been exhibited for quantum wires of HD *n*-InSb in Figure 5.6. Few tenths of microamperes of current has been observed in the same figure for 0.1m wavelength and 1,500Wm^{-2} for a carrier concentration of 10^{10} m^{-1}. In this case, we note that there exist both increment and decrement in the field-emitted current, the reasons of which have already been stated in Chapter 1 and the curves can be compared to the corresponding figures therein. The influence of electron concentration on the field-emitted current in this case has been shown

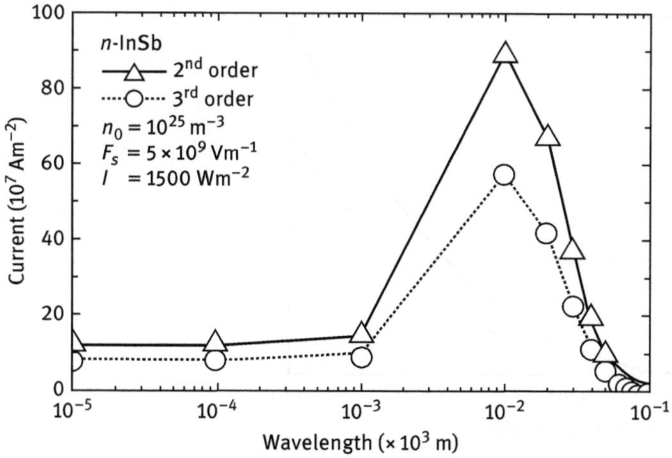

Figure 5.3: Plot of the field-emitted current density as a function of wavelength for HD *n*-InSb in the presence of light waves and external magnetic field for both three- and two-band models of Kane.

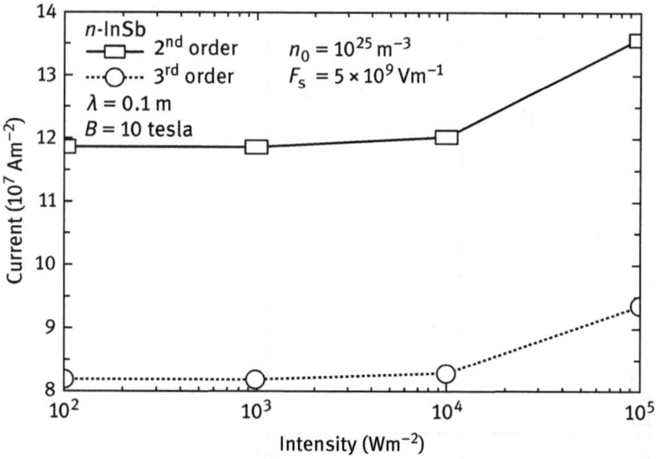

Figure 5.4: Plot of the field-emitted current density as a function of intensity for HD *n*-InSb in the presence of light waves and external magnetic field for both three- and two-band models of Kane.

in Figure 5.7 for the quantum limit for the two- and three-band models of Kane in the presence of light waves. We observe that the current rises to a peak near the value of about 10^{11}m^{-1} after which the field-emitted current falls sharply. The cut-in fields in the present perturbed case are near 10^7Vm^{-1}, which can be compared with the unperturbed one.

The influence of light waves increases the magnitude of the field-emitted current density in both effective mass superlattices and superlattices with graded interface

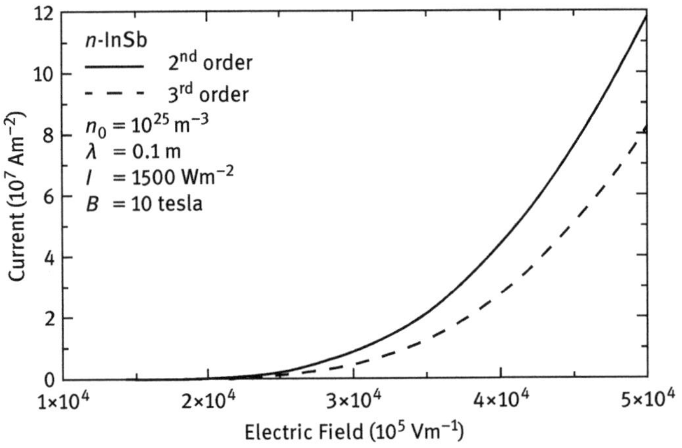

Figure 5.5: Plot of the field-emitted current density as a function of electric field for HD n-InSb in the presence of light waves and external magnetic field for both three- and two-band models of Kane.

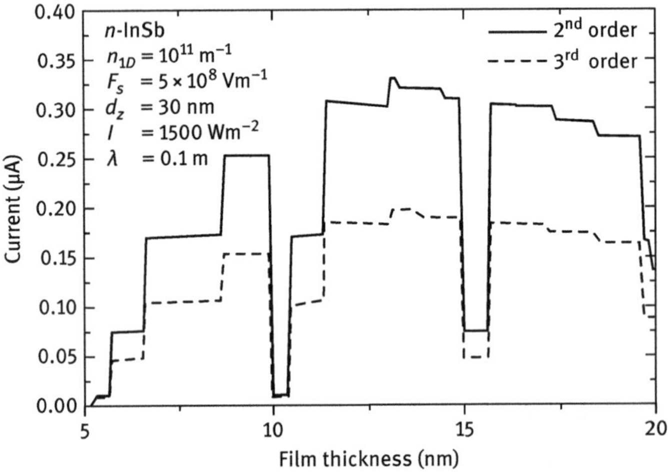

Figure 5.6: Plot of the field-emitted current as a function of film thickness from quantum wires of HD n-InSb in the presence of light waves for both three- and two-band models of Kane.

respectively, which appears from Figures 5.8, 5.9 and 5.10 in the presence of quantizing magnetic field. With the increase in the electron concentration, the field-emitted current density increases in a nonperiodic manner. In the case of quantum wires, the drastic reduction of field-emitted current to an order of nano amperes because of the high increase in the Fermi energy in Figure 5.11. Incidentally, for low concentration level, we note from Figure 5.12 that the field-emitted current increases slowly and

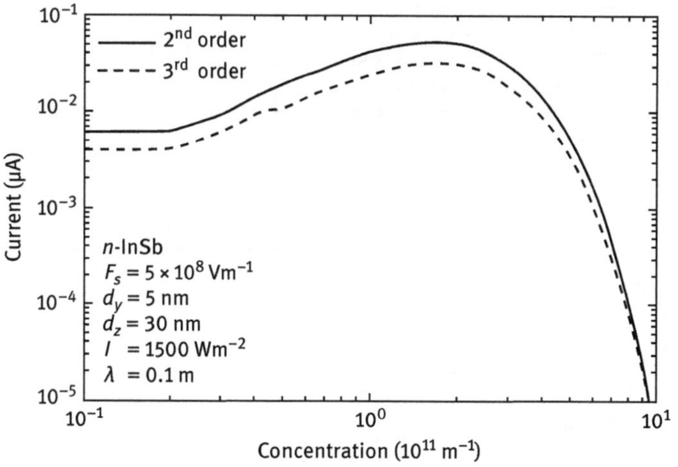

Figure 5.7: Plot of the field-emitted current as a function of electron concentration from quantum wires of HD *n*-InSb in the presence of light waves for both three- and two-band models of Kane.

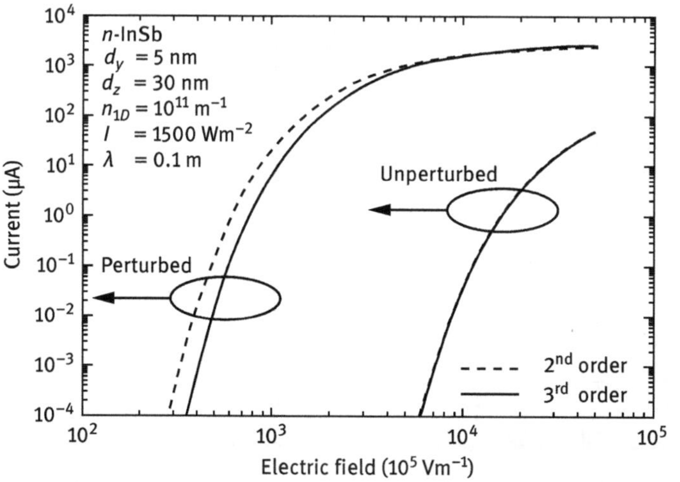

Figure 5.8: Plot of the field-emitted current as a function of electric field from quantum wires of HD *n*-InSb in the presence of light waves for both three- and two-band models of Kane.

sharply falls off above a carrier degeneracy of 10^{11}m^{-1} for both types of quantum wire superlattices as discussed in this chapter. It may be noted that although ternary and quaternary materials are primarily known as optoelectronic materials, their conduction electron energy band models in the absence of any field is the same as

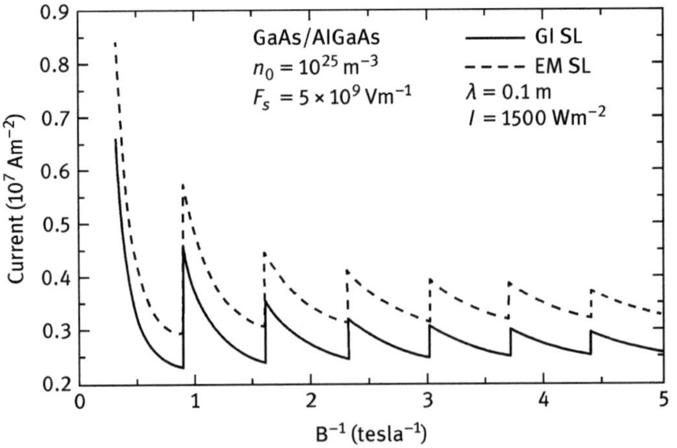

Figure 5.9: Plot of the field-emitted current as a function of inverse magnetic field from HD GaAs/AlGaAs effective mass superlattices (EM SL) and superlattices with graded interfaces (GI SL) in the presence of light waves, the constituent materials of which obey the unperturbed two-band model of Kane.

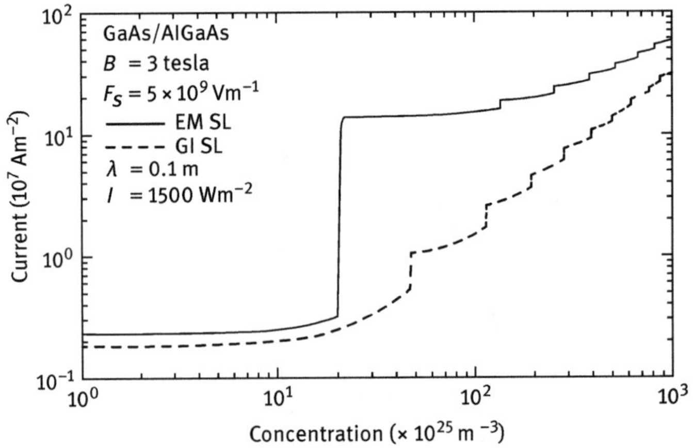

Figure 5.10: Plot of the field-emitted current as a function of carrier concentration in the presence of a magnetic field from HD GaAs/AlGaAs effective mass superlattices (EM SL) and superlattices with graded interfaces (GI SL) in the presence of light waves, the constituent materials of which obey the unperturbed two-band model of Kane.

that of III–V semiconductors whose dispersion relations obey the three- and two-band models of Kane. All the results in this chapter are also equally valid for ternary and quaternary materials and only the numerical values will be different.

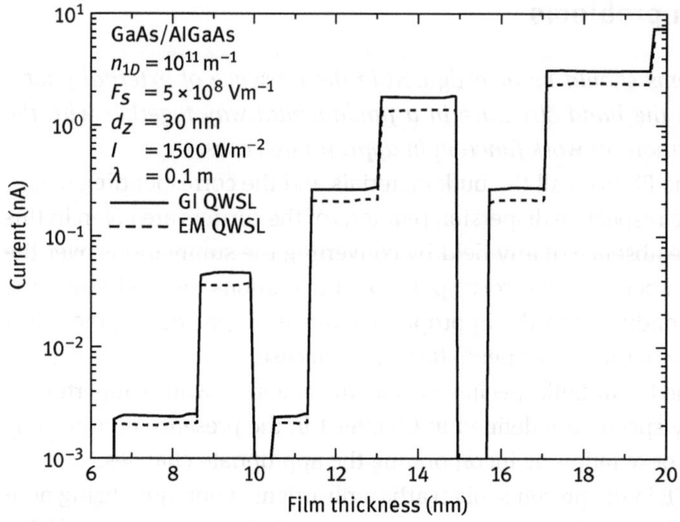

Figure 5.11: Plot of the field-emitted current as a function of film thickness from HD GaAs/AlGaAs quantum wire effective mass superlattices (EM QWSL) and quantum wire superlattices with graded interfaces (GI QWSL) in the presence of light waves, the constituent materials of which obey the unperturbed two-band model of Kane.

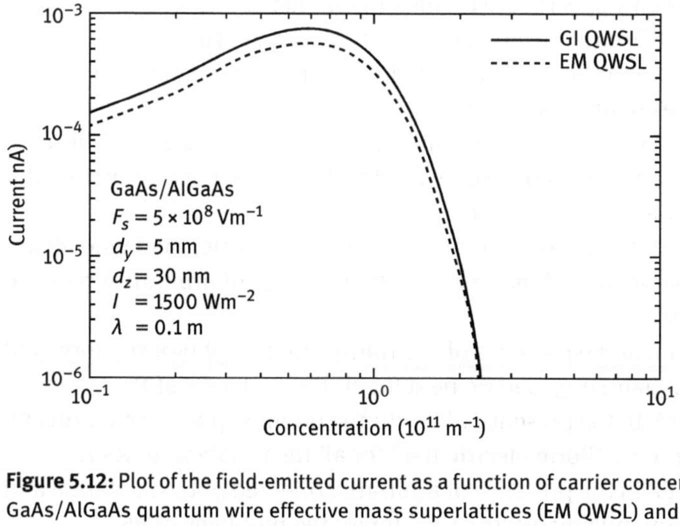

Figure 5.12: Plot of the field-emitted current as a function of carrier concentration per unit length from GaAs/AlGaAs quantum wire effective mass superlattices (EM QWSL) and quantum wire superlattices with graded interfaces (GI QWSL) in the presence of light waves, the constituent materials of which obey the unperturbed two-band model of Kane.

5.4 Open research problems

All the following problems should be investigated in the presence of external photoexcitation that changes the band structure in a fundamental way together with the proper inclusion of variations of work function in appropriate cases.

(R.5.1) (a) Investigate the FE from all the bulk materials and the corresponding superlattices whose respective dispersion relations of the carriers are given in this chapter in the absence of any field by converting the summations over the quantum numbers to the corresponding integrations by including the uniqueness conditions in the appropriate cases and considering the effect of image force in the subsequent study in each case.

(b) Investigate the FE for bulk specimens of all the materials whose unperturbed carrier energy spectra are defined in Chapter 1 in the presence of arbitrarily oriented photoexcitation by incorporating the appropriate changes.

(R5.2) Investigate the FE in the presence of an arbitrarily oriented non-quantizing nonuniform electric field and photoexcitation, respectively for all the cases of R5.1.

(R5.3) Investigate the FE in the presence of arbitrarily oriented non-quantizing alternating electric field and photoexcitation, respectively for all the cases of R5.1.

(R5.4) Investigate the FE for arbitrarily oriented photoexcitation from the heavily doped materials in the presence of Gaussian, exponential, Kane, Halperin, Lax, and Bonch–Bruevich types of band tails for all materials whose unperturbed carrier energy spectra are defined in Chapter 1.

(R5.5) Investigate the FE from all the materials in the presence of arbitrarily oriented non-quantizing nonuniform electric field and photoexcitation for all the appropriate cases of problem R5.4.

(R5.6) Investigate the FE from all the materials in the presence of arbitrarily oriented non-quantizing alternating electric field and photoexcitation for all the appropriate cases of problem R5.4.

(R5.7) Investigate the FE from negative refractive index, organic, magnetic, disordered, and other advanced materials in the presence of arbitrarily oriented photoexcitation.

(R5.8) Investigate the FE in the presence of arbitrarily oriented photoexcitation and alternating nonquantizing electric field for all the problems of R5.7.

(R5.9) Investigate the FE in the presence of arbitrarily oriented photoexcitation and non-quantizing nonuniform electric field for all the problems of R5.7.

(R5.10) Investigate the FE in the presence of arbitrarily oriented photoexcitation and alternating nonquantizing electric field for all the problems of R5.7.

(R5.11) Investigate the FE from quantum dots of all the materials whose bulk dispersion relations are given in Chapter 1 in the presence of arbitrarily oriented photoexcitation and quantizing magnetic field, respectively.

(R5.12) Investigate the FE from quantum dots of all the materials whose bulk dispersion relations are given in Chapter 1 in the presence of an arbitrarily oriented non-quantizing nonuniform electric field, photoexcitation, and quantizing magnetic field, respectively.

(R5.13) Investigate the FE from quantum dots of all the materials whose bulk dispersion relations are given in Chapter 1 in the presence of an arbitrarily oriented non-quantizing alternating electric field, photoexcitation, and quantizing magnetic field, respectively.

(R5.14) Investigate the FE from quantum dots of all the materials whose bulk dispersion relations are given in Chapter 1in the presence of an arbitrarily oriented non-quantizing alternating electric field, photoexcitation, and quantizing alternating magnetic field, respectively.

(R5.15) Investigate the FE from quantum dots of all the materials whose bulk dispersion relations are given in Chapter 1 in the presence of an arbitrarily oriented photoexcitation and crossed electric and quantizing magnetic fields, respectively.

(R5.16) Investigate the FE for arbitrarily oriented photoexcitation and quantizing magnetic field from the heavily doped materials in the presence of Gaussian, exponential, Kane, Halperin, Lax, and Bonch–Bruevich types of band for all materials whose unperturbed carrier energy spectra are defined in Chapter 1.

(R5.17) Investigate the FE for arbitrarily oriented photoexcitation and quantizing alternating magnetic field for all the cases of R5.16.

(R5.18) Investigate the FE for arbitrarily oriented photoexcitation and non-quantizing alternating electric field and quantizing magnetic field for all the cases of R5.16.

(R5.19) Investigate the FE for arbitrarily oriented photoexcitation and nonuniform alternating electric field and quantizing magnetic field for all the cases of R5.16.

(R5.20) Investigate the FE for arbitrarily oriented photoexcitation and crossed electric and quantizing magnetic fields for all the cases of R5.16.

(R5.21) Investigate the FE from negative refractive index, organic, magnetic, heavily doped, disordered, and other advanced optical materials in the presence of arbitrary oriented photoexcitation and quantizing magnetic field.

(R5.22) Investigate the FE in the presence of arbitrarily oriented photoexcitation, quantizing magnetic field, and alternating non-quantizing electric field for all the problems of R5.21.

(R5.23) Investigate the FE in the presence of arbitrarily oriented photoexcitation, quantizing magnetic field, and non-quantizing nonuniform electric field for all the problems of R5.21.

(R5.24) Investigate the FE in the presence of arbitrary oriented photoexcitation, alternating quantizing magnetic field, and crossed alternating non-quantizing electric field for all the problems of R5.21.

(R5.25) Investigate the FE from all the quantum confined materials (i.e, multiple quantum wells, wires, and dots) whose unperturbed carrier energy spectra are defined in Chapter 1 in the presence of arbitrarily oriented photoexcitation and quantizing magnetic field, respectively.

(R5.26) Investigate the FE in the presence of arbitrarily oriented photoexcitation and alternating quantizing magnetic field respectively for all the problems of R5.25.

(R5.27) Investigate the FE in the presence of arbitrarily oriented photoexcitation, alternating quantizing magnetic field, and an additional arbitrary oriented non-quantizing nonuniform electric field respectively for all the problems of R5.25.

(R5.28) Investigate the FE in the presence of arbitrarily oriented photoexcitation, alternating quantizing magnetic field, and additional arbitrary oriented non-quantizing alternating electric field respectively for all the problems of R5.25.

(R5.29) Investigate the FE in the presence of arbitrarily oriented photoexcitation, crossed quantizing magnetic, and electric fields respectively for all the problems of R5.25.

(R5.30) Investigate the FE for arbitrarily oriented photoexcitation and quantizing magnetic field from the entire quantum confined heavily doped materials in the presence of exponential, Kane, Halperin, Lax, and Bonch–Bruevich types of band tails for all materials whose unperturbed carrier energy spectra are defined in Chapter 1.

(R5.31) Investigate the FE for arbitrarily oriented photoexcitation and alternating quantizing magnetic field for all the cases of R5.30.

(R5.32) Investigate the FE in the presence of arbitrarily oriented photoexcitation, alternating quantizing magnetic field, and an additional arbitrarily oriented non-quantizing nonuniform electric field for all the cases of R5.30.

(R5.33) Investigate the FE in the presence of arbitrary oriented photoexcitation, alternating quantizing magnetic field, and additional arbitrary oriented non-quantizing alternating electric field respectively for all the cases of R5.30.

(R5.34) Investigate the FE in the presence of arbitrary oriented photoexcitation, crossed quantizing magnetic, and electric fields respectively for all the cases of R5.30.

(R5.35) Investigate the FE for all the appropriate problems from R5.25 to R5.34 in the presence of finite potential wells.

(R5.36) Investigate the FE for all the appropriate problems from R5.25 to R5.34 in the presence of parabolic potential wells.

(R5.37) Investigate the FE for all the above-mentioned appropriate problems for quantum rings.

(R5.38) Investigate the FE for all the above-mentioned appropriate problems in the presence of elliptical Hill and quantum square rings respectively.

(R5.39) Investigate the FE from carbon nanotubes in the presence of arbitrary photoexcitation.

(R5.40) Investigate the FE from carbon nanotubes in the presence of arbitrary photoexcitation and non-quantizing alternating electric field.

(R5.41) Investigate the FE from carbon nanotubes in the presence of arbitrary photoexcitation and non-quantizing alternating magnetic field.

(R5.42) Investigate the FE from carbon nanotubes in the presence of arbitrary photoexcitation and crossed electric and quantizing magnetic fields.

(R5.43) Investigate the FE from heavily doped semiconductor nanotubes in the presence of arbitrary photoexcitation for all the materials whose unperturbed carrier dispersion laws are defined in Chapter 1.

(R5.44) Investigate the FE from heavily doped semiconductor nanotubes in the presence of non-quantizing alternating electric field and arbitrary photoexcitation for all the materials whose unperturbed carrier dispersion laws are defined in Chapter 1.

(R5.45) Investigate the FE from heavily doped semiconductor nanotubes in the presence of non-quantizing alternating magnetic field and arbitrary photoexcitation for all the materials whose unperturbed carrier dispersion laws are defined in Chapter 1.

(R5.46) Investigate the FE from heavily doped semiconductor nanotubes in the presence of arbitrary photoexcitation and nonuniform electric field for all the materials whose unperturbed carrier dispersion laws are defined in Chapter 1.

(R5.47) Investigate the FE from heavily doped semiconductor nanotubes in the presence of arbitrary photoexcitation and alternating quantizing magnetic fields for all the materials whose unperturbed carrier dispersion laws are defined in Chapter 1.

(R5.48) Investigate the FE from heavily doped semiconductor nanotubes in the presence of arbitrary photoexcitation and crossed electric and quantizing magnetic fields for all the materials whose unperturbed carrier dispersion laws are defined in Chapter 1.

(R5.49) Investigate the FE in the presence of arbitrary photoexcitation for all the appropriate nipi structures of the materials whose unperturbed carrier energy spectra are defined in Chapter 1.

(R5.50) Investigate the FE in the presence of arbitrary photoexcitation for all the appropriate nipi structures of the materials whose unperturbed carrier energy spectra are defined in Chapter 1 in the presence of an arbitrarily oriented non-quantizing nonuniform additional electric field.

(R5.51) Investigate the FE for all the appropriate nipi structures of the materials whose unperturbed carrier energy spectra are defined in Chapter 1 in the presence of an arbitrarily oriented photoexcitation and non-quantizing alternating additional magnetic field.

(R5.52) Investigate the FE for all the appropriate nipi structures of the materials whose unperturbed carrier energy spectra are defined in Chapter 1 in the presence of an arbitrarily oriented photoexcitation and quantizing alternating additional magnetic field.

(R5.53) Investigate the FE for all the appropriate nipi structures of the materials whose unperturbed carrier energy spectra are defined in Chapter 1 in the presence of an arbitrarily oriented photoexcitation and crossed electric and quantizing magnetic fields.

(R5.54) Investigate the FE from heavily doped nipi structures for all the appropriate cases of all the above-mentioned problems.

(R5.55) Investigate the FE in the presence of arbitrary photoexcitation for the appropriate inversionlayers of all the materials whose unperturbed carrier energy spectra are defined in Chapter 1.

(R5.56) Investigate the FE in the presence of arbitrary photoexcitation for the appropriate inversion layers of all the materials whose unperturbed carrier energy spectra are defined in Chapter 1 in the presence of an arbitrarily oriented nonquantizing nonuniform additional electric field.

(R5.57) Investigate the FE for the appropriate inversion layers of all the materials whose unperturbed carrier energy spectra are defined in Chapter 1 in the presence of an arbitrarily oriented photoexcitation and non-quantizing alternating additional magnetic field.

(R5.58) Investigate the FE for the appropriate inversion layers of all the materials whose unperturbed carrier energy spectra are defined in Chapter 1 in the presence of an arbitrarily oriented photoexcitation and quantizing alternating additional magnetic field.

(R5.59) Investigate the FE for the appropriate inversion layers of all the materials whose unperturbed carrier energy spectra are defined in Chapter 1 in the presence of an arbitrarily oriented photoexcitation and crossed electric and quantizing magnetic fields by considering electron spin and broadening of Landau levels.

(R5.60) Investigate the FE in the presence of arbitrary photoexcitation for the appropriate accumulation layers of all the materials whose unperturbed carrier energy spectra are defined in Chapter 1 by modifying the above-mentioned appropriate problems.

(R5.61) Investigate the FE in the presence of arbitrary photoexcitation from wedge shaped and cylindrical QDs of all the materials whose unperturbed carrier energy spectra are defined in Chapter 1.

(R5.62) Investigate the FE in the presence of arbitrary photoexcitation from wedge shaped and cylindrical QDs of all the materials whose unperturbed carrier energy spectra are defined in Chapter 1 in the presence of an arbitrarily oriented non-quantizing nonuniform additional electric field.

(R5.63) Investigate the FE from wedge shaped and cylindrical QDs of all the materials whose unperturbed carrier energy spectra are defined in Chapter 1 in the presence of an arbitrarily oriented photoexcitation and non-quantizing alternating additional magnetic field.

(R5.64) Investigate the FE from wedge shaped and cylindrical QDs of all the materials whose unperturbed carrier energy spectra are defined in Chapter 1 in the presence of an arbitrarily oriented photoexcitation and quantizing alternating additional magnetic field.

(R5.65) Investigate the FE from wedge shaped and cylindrical QDs of all the materials whose unperturbed carrier energy spectra are defined in Chapter 1 in the presence of an arbitrarily oriented photoexcitation and crossed electric and quantizing magnetic fields.

(R5.66) Investigate the FE from wedge shaped and cylindrical QDs for all the appropriate cases of the above-mentioned problems.

(R5.67) Investigate all the problems from R5.25 to R5.66 by removing all the mathematical approximations and establishing the respective appropriate uniqueness conditions.

(R5.68) Investigate the FE from quantum confined III–V, II–VI, IV–VI, HgTe/CdTe effective mass superlattices together with short period, strained layer, random, Fibonacci, poly type, and sawtooth superlattices in the presence of arbitrarily oriented photoexcitation.

(R5.69) Investigate the FE in the presence of arbitrarily oriented photoexcitation and quantizing magnetic field respectively for all the cases of R5.68.

(R5.70) Investigate the FE in the presence of arbitrarily oriented photoexcitation and non-quantizing nonuniform electric field respectively for all the cases of R5.68.

(R5.71) Investigate the FE in the presence of arbitrarily oriented photoexcitation and non-quantizing alternating electric field respectively for all the cases of R5.68.

(R5.72) Investigate the FE in the presence of arbitrarily oriented photoexcitation and crossed electric and quantizing magnetic fields respectively for all the cases of R5.68.

(R5.73) Investigate the FE from heavily doped quantum confined superlattices for all the problems of R5.68.

(R5.74) Investigate the FE in the presence of arbitrarily oriented photoexcitation and quantizing magnetic field respectively for all the cases of R5.73.

(R5.75) Investigate the FE in the presence of arbitrarily oriented photoexcitation and non-quantizing nonuniform electric field respectively for all the cases of R5.73.

(R5.76) Investigate the FE in the presence of arbitrarily oriented photoexcitation and non-quantizing alternating electric field respectively for all the cases of R5.73.

(R5.77) Investigate the FE in the presence of arbitrarily oriented photoexcitation and crossed electric and quantizing magnetic fields respectively for all the cases of R5.73.

(R5.78) Investigate all the problems from R5.68 to R5.77 by removing all the mathematical approximations and establishing the respective appropriate uniqueness conditions.

(R5.79) Investigate the FE from quantum confined III–V, II–VI, IV–VI, HgTe/CdTe superlattices with graded interfaces together with short period, strained layer, random, Fibonacci, polytype, and sawtooth superlattices in this context in the presence of arbitrarily oriented photoexcitation.

(R5.80) Investigate the FE in the presence of arbitrarily oriented photoexcitation and quantizing magnetic field respectively for all the cases of R5.79.

(R5.81) Investigate the FE in the presence of arbitrarily oriented photoexcitation and non-quantizing nonuniform electric field respectively for all the cases of R5.79.

(R5.82) Investigate the FE in the presence of arbitrarily oriented photoexcitation and non-quantizing alternating electric field respectively for all the cases of R5.79.

(R5.83) Investigate the FE in the presence of arbitrarily oriented photoexcitation and crossed electric and quantizing magnetic fields respectively for all the cases of R5.79.

(R5.84) Investigate the FE from heavily doped quantum confined superlattices for all the problems of R5.79.

(R5.85) Investigate the FE in the presence of arbitrarily oriented photoexcitation and quantizing magnetic field respectively for all the cases of R5.84.

(R5.86) Investigate the FE in the presence of arbitrarily oriented photoexcitation and non-quantizing nonuniform electric field respectively for all the cases of R5.84.

(R5.87) Investigate the FE in the presence of arbitrarily oriented photoexcitation and non-quantizing alternating electric field respectively for all the cases of R5.84.

(R5.88)

(a) Investigate the FE in the presence of arbitrarily oriented photoexcitation and crossed electric and quantizing magnetic fields respectively for all the cases of R5.84.

(b) Investigate the FE from multiple wall carbon nanotubes in presence of an arbitrarily oriented alternating electric field.

(c) Investigate the FE from heavily doped semiconductor nanotubes in the presence of an arbitrarily oriented alternating electic field for all the materials whose unperturbed carrier energy spectra are defined in R1.1 and R1.2 respectively.

(R5.89) Investigate all the problems of this chapter by removing all the mathematical approximations and establishing the respective appropriate uniqueness conditions.

References

[1] R.H. Fowler, L. Nordheim, *Proc. Roy. Soc. London A*, 119, 173 (1928).

[2] A. Van Der Ziel, *Solid State Physical Electronics* (Prentice-Hall, Englewood Cliffs, 1957), p. 176.

[3] B. Mitra, K.P. Ghatak, *Phys. Lett. A*, 357, 146 (1990).

[4] B. Mitra, K.P. Ghatak, *Phys. Lett. A*, 142A, 401 (1989).

[5] K.P. Ghatak, M. Mondal, *J. Mag. Mag. Mat.*, 74, 203 (1988).

[6] K.P. Ghatak, B. Mitra, *Phys. Lett.*, 156A, 233 (1991).

[7] K.P. Ghatak, A. Ghosal, S.N. Biswas, M. Mondal, *Proc. SPIE.*, 1308, 356 (1990).

[8] V.T. Binh, Ch. Adessi, *Phys. Rev. Lett.* 85, 864 (2000).

6 Conclusion and scope for future research

The best contribution that I can make to me is to improve myself.

In Volume 1 of this book, we have investigated the carrier contribution to the elastic constants, Einstein's Photoemission, the diffusivity–mobility ratio, the screening length, and the field emission from heavily doped optoelectronic nanomaterials by using the Heisenberg's uncertainty principle. *We have suggested the experimental methods of determining the carrier contribution to the elastic constants, the diffusivity–mobility ratio, and the screening length since these physical properties affect the carrier transport and the analysis of nanodevices in general.*

Our analyses are valid under single electron approximation. The quantitative comparison between the theoretical formulations of the said quantities for various materials under different physical conditions and the suggestion for the experimental determinations for them is not possible in many cases, since the experimental data of G are not available in the literature for all the materials considered here. Thus, the detailed experimental works are needed not only to uncover the phenomena, but also for in-depth probing of the band structures of the different quantized materials which, in turn, control the key, that is, the Boltzmann transport equation. In spite of such constraints, the new concepts, which have emerged from the present investigation are really amazing in general and are discussed throughout the book.

We are presenting our readers with the following last set of open challenging research problems:

(6.1) Investigate all the physical properties of all the heavily doped (HD) systems as discussed from Chapter 1 up to Chapter 5 by removing all the mathematical approximations and establishing appropriate uniqueness conditions.

(6.2) Investigate all the problems of (6.1) in the presence of many body effects.

(6.3) Investigate all the problems of (6.1) for 3D quantizations of the wave vector space of the charge carriers.

(6.4) Investigate all the problems of (6.1) for magneto-accumulation layers in the presence of spin and broadening and considering the effects of surface states.

(6.5) Investigate all the problems of (6.1) in the presence of an arbitrarily oriented quantizing magnetic field and considering the influences of electron spin and broadening.

(6.6) Investigate all the problems of (6.1) for quantum dot superlattices.

(6.7) Investigate all the problems of (6.1) for all types of quantum wire superlattices in the presence of an arbitrarily oriented quantizing magnetic field by including spin and broadening.

(6.8) Investigate all the problems of (6.1) for all types of quantum well superlattices in the presence of an arbitrarily oriented quantizing magnetic field.

https://doi.org/10.1515/9783110610819-006

(6.9) Investigate the higher order diffusivity–mobility ratio for all the systems of Chapter 3.
(6.10) Investigate all the problems of (6.1) for nonlinear charge transport for all the systems from Chapters 1 to 5.
(6.11) Investigate the higher order diffusivity–mobility ratio after proper modifications for all the systems of Chapter 1.
(6.12) Investigate the higher order diffusivity–mobility ratio after proper modifications for all the systems in Chapter 3 in the presence of arbitrarily oriented high electric field.
(6.13) Investigate the higher order diffusivity–mobility ratio after proper modifications for all the systems in Chapter 3 in the presence of hot electron effects.
(6.14) Investigate all the problems of (6.1) for p-type materials for all the systems from Chapters 1 to 5.
(6.15) Investigate the nonlocal diffusivity–mobility ratio after proper modifications for all the systems in Chapter 3.
(6.16) Investigate all the problems of (6.1) from Chapters 1 to 5 introducing new theoretical formalisms for amorphous systems.
(6.17) Investigate the nonequilibrium diffusivity–mobility ratio for all the cases from Chapter 3 introducing new theoretical formalisms.
(6.18) Investigate all the problems of (6.1) in the presence of many body effects.
(6.19) Introducing new theoretical formalisms, investigate all appropriate problems of this book for quantum systems, where the Boltzmann transport equation is invalid.

The formulation of* the Electronic Properties *for all types of HD materials and their quantum confined counterparts considering the influence of all the bands created because of all types of quantizations after removing all the assumptions and establishing the respective appropriate uniqueness conditions is, in general, an extremely difficult problem. 200 open research problems *have been presented in this monograph and we hope that the readers will not only solve them but also will generate new concepts, both theoretical and experimental. Incidentally, we can easily infer how little is presented and how much more is yet to be investigated in this exciting topic, which is the signature of coexistence of new physics and advanced mathematics combined with the inner fire for performing creative researches in this context from the young scientist such as Kikoin* [1]: *we firmly believe that* ***"A young scientist is no good if his teacher learns nothing from him and gives his teacher nothing to be proud of".*** *In the mean time our research interest has been shifted and we are leaving this particular beautiful topic with the hope that (6.19) alone is sufficient to draw the attention of the researchers from diverse fields and our readers are surely in tune with the fact that* ***"Exposition, criticism, appreciation is the work for second-rate minds"*** [2].

References

[1] I.K. Kikoin, *Science for Everyone: Encounters with Physicists and Physics* (Mir Publishers, Russia, 1989), p. 154.

[2] G.H. Hardy, *A Mathematician's Apology* (Cambridge University Press, Oxford, 1990), p. 61.

Appendix
The numerical values of the energy band constants of few materials

	Materials	Numerical values of the energy band constants
1	The conduction electrons of *n*-cadmium germanium arsenide can be described by three types of band models	1. The values of the energy band constants in accordance with the generalized electron dispersion relation of nonlinear optical materials are as follows: $E_{g_0}=0.57eV, \Delta_P=0.30eV, \Delta_\perp=0.36eV, m^*_P=0.034m_0, m^*_\perp=0.039m_0, T=4K, \delta=-0.12\,eV, g_v=1$ [1, 2], $\varepsilon_{sc}=18.4\varepsilon_0$ [3] (ε_{sc} and ε_0 are permittivities of the semiconductor material and free space, respectively), and $W=4\,eV$ [4]. 2. In accordance with the three-band model of Kane the spectrum constants are given in $\Delta=(\Delta_\parallel+\Delta_\perp)/2=0.33eV, E_{g_0}=0.57eV, m^*=\left(m^*_\parallel+m^*_\perp\right)/2=0.0365m_0$, and $\delta=0\,eV$. 3. In accordance with the two-band model of Kane, $E_{g_0}=0.57eV$ and $m^*=0.0365m_0$.
2	*n*-Indium arsenide	The values $E_{g_0}=0.36eV, \Delta=0.43eV, m^*=0.026m_0, g_v=1, \varepsilon_{sc}=12.25\varepsilon_0$ [5], and $W=5.06eV$ [6] are valid for the three-band model of Kane.
3	*n*-Gallium arsenide	The values $E_{g_0}=1.55eV, \Delta=0.35eV, m^*=0.07m_0, g_v=1, \varepsilon_{sc}=12.9\varepsilon_0$ [5], and $W=4.07eV$ [7] are valid for the three-band model of Kane; values $\alpha_{13}=-1.97\times10^{-37}eVm^4$ and $\alpha_{15}=-2.3\times10^{-34}eVm^4$ [8] are valid; and values $\alpha_{11}=-2,132\times10^{-40}\,eVm^4, \alpha_{12}=9,030\times10^{-50}eVm^5, \beta_{11}=-2,493\times10^{-40}eVm^4, \beta_{12}=12,594\times10^{-50}eVm^5, \gamma_{11}=30\times10^{-30}eVm^3, \gamma_{12}=-154\times10^{-42}eVm^4$ [9] are valid.
4	*n*-Gallium aluminum arsenide	$E_{g_0}=(1.424+1.266x+0.26x^2)eV, \Delta=(0.34-0.5x)eV, m^*=[0.066+0.088x]m_0, g_v=1, \varepsilon_{sc}=[13.18-3.12x]\varepsilon_0$ [10], and $W=(3.64-0.14x)eV$ [11].
5	*n*-Mercury cadmium telluride	$E_{g_0}=(-0.302+1.93x+5.35\times10^{-4}(1-2x)T-0.810x^2+0.832x^3)eV, \Delta=(0.63+0.24x-0.27x^2)eV, m^*=0.1m_0E_{g_0}(eV)^{-1}, g_v=1, \varepsilon_{sc}=[20.262-14.812x+5.22795x^2]\varepsilon_0$ [12], and $W=(4.23-0.813(E_{g_0}-0.083))\ eV$ [13].
6	*n*-Indium gallium arsenide phosphide lattice matched to indium phosphide	$E_{g_0}=(1.337-0.73y+0.13y^2)eV, \Delta=(0.114+0.26y-0.22y^2)eV, m^*=(0.08-0.039y)m_0, y=(0.1896-0.4052x)/(0.1896-0.0123x), g_v=1, \varepsilon_{sc}=[10.65+0.1320y]\varepsilon_0$, and $W(x,y)=[5.06(1-x)y+4.38(1-x)(1-y)+3.64xy+3.75\{x(1-y)\}]eV$ [14].

(continued)

https://doi.org/10.1515/9783110610819-007

(continued)

	Materials	Numerical values of the energy band constants
7	*n*-Indium antimonide	$E_{g_0} = 0.235eV$, $\Delta = 0.81eV$, $m^* = 0.01359m_0$, $g_v = 1$, $\varepsilon_{sc} = 15.56\varepsilon_0$ [5], and $W = 4.72eV$ [6].
8	*n*-Gallium antimonide	The values of $E_{g_0} = 0.81eV$, $\Delta = 0.80eV$, $P = 9.48 \times 10^{-10} eVm$, $\bar{\varsigma}_0 = -2.1$, $\bar{\nu}_0 = -1.49$, $\bar{\omega}_0 = 0.42$, $g_v = 1$, [15] and $\varepsilon_{sc} = 15.85\varepsilon_0$ [15, 16] are valid for the model of Seiler et al. The values $\bar{E}_1 = 1.024eV$, $\bar{E}_2 = 0eV$, $\bar{E}_3 = -1.132eV$, $\bar{E}_4 = 0.05eV$, $\bar{E}_5 = 1.107eV$, $\bar{E}_6 = -0.113eV$, and $\bar{E}_7 = -0.0072eV$ [16] are valid for the model of Zhang.
9	*n*-Cadmium sulfide	$m_p^* = 0.7m_0$, $m_\perp^* = 1.5m_0$, $\bar{\lambda}_0 = 1.4 \times 10^{-8} eVm$(value changed), $g_v = 1$ [5], $\varepsilon_{sc} = 15.5\varepsilon_0$ [17], and $W = 4.5eV$ [6].
10	*n*-Lead telluride	The values $m_t^- = 0.070m_0$, $m_l^- = 0.54m_0$, $m_t^+ = 0.010m_0$, $m_l^+ = 1.4m_0$, $P_\parallel = 141\,\text{meV nm}$, $P_\perp = 486\ \text{meV nm}$, $E_{g_0} = 190\ \text{meV}$, $g_v = 4$ [5], $\varepsilon_{sc} = 33\varepsilon_0$ [5, 18], and $W = 4.6eV$ [19] are valid for the Dimmock model. The values $(\bar{R})^2 = 2.3 \times 10^{-19} (eV\text{m})^2$, $E_{g_0} = 0.16eV$, $(\bar{s})^2 = 4.6(\bar{R})^2$, $\Delta'_c = 3.07eV$, $(\bar{Q})^2 = 1.3(\bar{R})^2$, $\Delta''_c = 3.28eV$, $(\bar{A})^2 = 0.83 \times 10^{-19}\ (eVm)^2$ [21], and $W = 4.21eV$ [6] are valid for the model of Bangert and Kastner. The values $m_{tv} = 0.0965m_0$, $m_{lv} = 1.33m_0$, $m_{tc} = 0.088m_0$, and $m_{lc} = 0.83m_0$ [19] are valid for the model of Foley et al. The values $m_1 = 0.0239m_0$, $m_2 = 0.024m_0$, $m'_2 = 0.31m_0$, and $m_3 = 0.24m_0$ [22] are valid for the Cohen model.
11	Stressed *n*-indium antimonide	The values $m^* = 0.048m_o$, $E_{g_0} = 0.081eV$, $B_2 = 9 \times 10^{-10}\ eVm$, $C_1^c = 3eV$, $C_2^c = 2eV$, $a_0 = -10eV$, $b_0 = -1.7eV$, $d = -4.4eV$, $S_{xx} = 0.6 \times 10^{-3}\ (kbar)^{-1}$, $S_{yy} = 0.42 \times 10^{-3}\ (kbar)^{-1}$, $S_{zz} = 0.39 \times 10^{-3}\ (kbar)^{-1}$, $S_{xy} = 0.5 \times 10^{-3}\ (kbar)^{-1}$, $\varepsilon_{xx} = \sigma S_{xx}$, $\varepsilon_{yy} = \sigma S_{yy}$, $\varepsilon_{zz} = \sigma S_{zz}$, $\varepsilon_{xy} = \sigma S_{xy}$, σ is the stress in kilobar, and $g_v = 1$ [24] are valid for the model of Seiler et al.
12	Bismuth	$E_{g_0} = 0.0153eV$, $m_1 = 0.00194m_0$, $m_2 = 0.313m_0$, $m_3 = 0.00246m_0$, $m'_2 = 0.36m_0$, $g_v = 3$, $g_s = 2$ [25], $M_2 = 0.128m_0$, $M'_2 = 0.80m_0$ [26], and $W = 4.34eV$.
13	Mercury telluride	$m_v^* = 0.028m_0$, $g_v = 1$, $\varepsilon_\infty = 15.2\varepsilon_0$, [27] and $W = 5.5eV$ [28].
14	Platinum antimonide	For valence bands, along <111> direction, $\lambda_1 = 0.33eV$, $l_1 = 1.09eV$, $\nu_1 = 0.17eV$, $\bar{\bar{n}} = 0.22eV$, $\bar{\bar{a}} = 0.643\text{nm}$, $I_0 = 0.30\ (eV)^2$, $\delta'_0 = 0.33eV$, $g_v = 8$ [29], $\varepsilon_{sc} = 30\varepsilon_0$ [30], and $\phi \approx 3.0eV$ [31].

Induced effects in Semiconductors Nauka, Russia (1972). (in Russian); M. Mondal and K. P. Ghatak, Phys. Stat. Sol. (b) 135, K21 (1986).
[25] C.C. Wu and C.J. Lin, J. Low. Temp. Phys. 57, 469 (1984).
[26] S. Takaoka, H. Kawamura, K. Murasa, S. Takano, Phys. Rev. B, 13, 1428 (1976).
[27] V.I. Ivanov-Omskii, A.Sh. Mekhtisev, S.A. Rustambekova and E.N. Ukraintsev, Phys. Stat. Sol. (b), 119, 159 (1983).
[28] H. Kim, K. Cho, H. Song, B. Min, J. Lee, G. Kim, S. Kim, S.H. Kim and T. Noh, Appl. Phys. Lett., 83, 4619 (2003).
[29] P.R. Emtage, Phys. Rev. A, 246, 138 (1965).
[30] R.A. Reynolds, M.J. Brau, R. A. Chapman, J. Phys. Chem. Solids, 29, 755, (1968).
[31] J. O'Shaughnessy and C. Smith, Solid State Commun., 8, 481, (1970).
[32] G.J. Rees, *Physics of Compounds*, Proceedings of the 13th International National Conference. F. G. Fumied., North Holland Company, (1976), p. 1166.
[33] J. Bouat and J.C. Thuillier, Surface Sci., 73, 528 (1978).
[34] M.V. Ortenberg and K.J. Button, Phys. Rev. B, 16, 2618 (1977).
[35] G. Haeffler, A.E. Klinkmüller, J. Rangell, U. Berzinsh and D. Hanstorp, Z. Phys. D, 38, 211, (1996).
[36] N.B. Brandt, V.N. Davydov, V.A. Kulbachinskii and O.M. Nikitina, Sov. Phys. Sol. Stat., 29, 1014, (1987).
[37] L.M. Viculis, J.J. Mack, O.M. Mayer, H.T. Hahn and R.B. Kaner, J. Mater. Chem., 15, 974 (2005).
[38] L.A. Vassilev, Phys. Stat. Sol. (b), 121, 203 (1984); S. Takaoka and K. Murase, Phys. Rev. B, Phys. Rev. B, 20, 2823 (1979).
[39] D.L. Partin, Superlattices and Microstructures, 1, 131 (1985).
[40] Y. Yamada, J. Phys. Japan, 35, 1600 (1973); M. Singh, P. R. Wallace, S. D. Jog and E. Arushanov, J. Phys. Chem. Solids, 45, 409 (1984).
[41] W. J. Turner, A. S. Fischler, and W. E. Reese, Phys. Rev., 121, 759 (1961).
[42] G. P. Chuiko, Sov. Phys. Semi., 19 (12), 1381 (1985).
[43] W. E. Swank, P. G. Le Comber, Phys. Rev. 153, 844 (1967).
[44] M. Stordeur and W. Kuhnberger, Phys. Stat. Sol. (b), 69, 377 (1975); D. R. Lovett, *Semimetals and Narrow-Bandgap Semiconductors*, Pion Limited, 185 (1977); H. Köhler, Phys. Stat. Sol. (b), 74, 591 (1976).
[45] D. Haneman, J. Phys. Chem. Solids, 11, 205, (1959).
[46] N. Wei, G. Wu and J. Dong, Phys. Lett. A 325, 403 (2004).
[47] S. Reich, J. Maultzsch, C. Thomsen and P. Ordejo´n, Phys. Rev. B, 66, 035412, (2006).
[48] M. S. Lundstrom and J. Guo, *Nanoscale Transistors: Device Physics, Modeling and Simulation* (Springer, USA, 2006); J. W. Mintmire and C. T. White, Phys. Rev. Lett., 81, 2506 (1998)
[49] F. Buonocore, F. Trani, D. Ninno, A. Di Matteo, G. Cantele and G. Iadonisi, Nanotechnology, 19, 025711, (2008).
[50] J.B. Ketterson, Phys. Rev. 129, 18 (1963).
[51] R.C. Vilão, J. M. Gil, A. Weidinger, H. V. Alberto, J. Piroto Duarte, N. A. de Campos, R. L. Lichti, K. H. Chow, S. P. Cottrell and S. F. J. Cox, Phys. Rev. B, 77, 235212, (2008).
[52] I. Kang and F. W. Wise, Phys. Rev. B, J. Opt. Soc. Am. B, 14, 1632, (1997).
[53] D. Cui, J. Xu, S.-Y. Xu, G. Paradee, B.A. Lewis and M. D. Gerhold, IEEE, Trans. Elect. Dev. 5, 362 (2006).

Materials Index

https://doi.org/10.1515/9783110610819-008

Subject Index

Bei Fragen zur Produktsicherheit wenden Sie sich bitte an:
If you have any questions regarding product safety,
please contact:

Walter de Gruyter GmbH
Genthiner Straße 13
10785 Berlin
productsafety@degruyterbrill.com